Behaviour and Ecology of Spiders

Carmen Viera • Marcelo O. Gonzaga
Editors

Behaviour and Ecology of Spiders

Contributions from the Neotropical Region

Springer

Editors
Carmen Viera
Entomología, Biología Animal, Facultad de Ciencias
Universidad de la República
Montevideo, Uruguay

Laboratorio Ecología del Comportamiento
Instituto de Investigaciones Biológicas Clemente Estable
Montevideo, Uruguay

Marcelo O. Gonzaga
Universidade Federal de Uberlândia
Biology Institute
Uberlândia, Minas Gerais, Brazil

ISBN 978-3-319-88089-1 ISBN 978-3-319-65717-2 (eBook)
https://doi.org/10.1007/978-3-319-65717-2

Printed on acid-free paper

This Springer imprint is published by Springer Nature
The registered company is Springer International Publishing AG
The registered company address is: Gewerbestrasse 11, 6330 Cham, Switzerland

To Latin-American arachnologists
who developed their studies in difficult times

Preface

Arachnology in the Neotropical region has been greatly benefited by the relatively recent development and establishment of research groups in Central and South America and, especially, by the collaborative network involving these groups during the last few decades. For centuries, all the research developed in the Neotropics was almost exclusively restricted to studies conducted by Europeans and North Americans during short sampling expeditions, or based on material deposited in museums. Only a few institutions located in countries within the Neotropics hosted laboratories dedicated to the study of systematics and ecology of spiders and other arachnids. However, even in these locations, most studies were conducted by a few very productive researchers, and the prospects for stable and collaborative groups remained inexistent for a long time.

We have had the privilege to witness the first changes in this scenario and to take part in the process. One keystone was the implementation of regular international meetings of arachnologists from South and Central America. In 1997, exactly 20 years ago, in Montevideo (Uruguay), the first "Encontro de Aracnólogos del Cono Sur" attracted 40 researchers (including postgraduate and graduate students) from Argentina, Brazil, Peru, and Uruguay. In the subsequent editions (in 1999, 2002, and 2003) there was a significant increase in the number of participants and studies presented (respectively 149 and 201 in the fourth edition). From these initial meetings, named based on the relatively limited geographical distribution of their participants (the "southern cone" or "Cono Sur" is an artificial geographic region which includes Brazil, Paraguay, Argentina, Chile, and Uruguay), the events turned into the Latin-American Congress of Arachnologists (the first also taking place in Uruguay), and Brazil will host the fifth edition in December 2017. At each edition, the congress becomes bigger and more productive.

This brief historical retrospective of arachnological meetings in the Neotropics reflects the growth in number and quality of the studies performed by researchers from institutions located within the countries of this region. Research in several countries from South and Central America is still periodically threatened by budget cuts and political crisis, but even though sometimes encountering instable situations, contributions in this area seems to resist and are very important to overall knowledge on the behavior and ecology of spiders. This is especially true considering the great diversity of this taxonomic group within the region.

Considering the efforts of many researchers to continue their studies in the Neotropics and the importance of these studies, we decided to organize this book, inviting many colleagues to present general reviews on the topics they are working now or have interest in developing in their countries. The guiding principle was to highlight the contributions of studies from the Neotropical region to the knowledge of each specific topic. In some chapters, particularly well studied cases have been used to illustrate one theme, while in others the authors decided to present general pictures of the current status and future prospects of research in the area. Information on relevant studies developed in other biogeographical regions is often mentioned, especially in cases in which there is a gap of information regarding Neotropical species, or when these studies are particularly important to illustrate one theme. We considered that this approach was required in order to better present understandable general reviews without the omission of relevant information.

We hope that the efforts dedicated by all the authors involved in this project will act as motivation to students and other researchers to develop the themes presented here in their future studies. We could say that we have experienced the beginning of a new phase in arachnology within the Neotropics. Now is the right time to tell a little about what has been discovered during these last few decades of work and what remains to be discovered, and to show to students from Central and South America interested in starting their careers studying spiders and other arachnids how relevant their contributions can be to overall knowledge on the subject.

Montevideo, Uruguay Carmen Viera
Uberlândia, Minas Gerais, Brazil Marcelo O. Gonzaga

Acknowledgments

We are grateful to many researchers who revised initial versions of the manuscripts and offered their expertise in order to improve the quality of this book: Anita Aisenberg, Andrea Armendano, Alexandre Bonaldo, Pedro Cardoso, Kleber Del Claro, William Eberhard, Nelson Ferretti, Luis Fernando García, Felipe M. Gawryszewski, José Paulo Guadanucci, Nuria Macías, Fernando Pérez-Miles, Dinesh Rao, Gustavo Romero, Adalberto Santos, Miguel Simó, João Vasconcellos-Neto, Anne Wignall, and Rodrigo Willemart. We also thank Joao Pildervasser and Raagai Priya Chandrasekaran for their patience and assistance in several editorial tasks and stages. Finally, we acknowledge our institutions and supporting agencies for allowing our dedication to this project. Marcelo Gonzaga was supported by Fundação de Amparo à Pesquisa do Estado de Minas Gerais (Procs. APQ-02104-14, CRA-30058/12), Instituto Nacional de Ciência e Tecnologia dos Hymenoptera Parasitoides da Região Sudeste (HYMPAR/Sudeste e CNPq/CAPES/Fapesp) and Conselho Nacional de Desenvolvimento Científico e Tecnológico (Procs. 306,157/2014–4, 403,733/2012–0, 445,832/2014–2, 441,225/2016–0). Carmen Viera was supported by Comisión Sectorial de Investigación (CSIC) de la Universidad de la República (UdelaR); Fondos para el Progreso de la Tecnología Agraria (FPTA) Instituto Nacional de Investigación Agraria (INIA); Agencia Nacional de Investigación e Innovación (ANII); Programa de Postgrado em Ciencias Básicas (PEDECIBAS) by projects, grants, fellowships, and scholarships.

Contents

Contributors

Ingi Agnarsson Department of Biology, University of Vermont, Burlington, VT, USA

Anita Aisenberg Laboratorio de Etología, Ecología y Evolución, Instituto de Investigaciones Biológicas Clemente Estable, Montevideo, Uruguay

María José Albo Laboratório de Etología, Ecología y Evolución, Instituto de Investigaciones Biológicas Clemente Estable, Montevideo, Uruguay

Miquel Arnedo Biodiversity Research Institute UB, Departament de Biologia Evolutiva, Ecologia i Ciències Ambientals, Universitat de Barcelona, Barcelona, Spain

Marco Antonio Benamú Centro Universitario de Rivera, Universidad de la República, Rivera, Uruguay

Laboratorio Ecología del Comportamiento, Instituto de Investigaciones Biológicas Clemente Estable, Montevideo, Uruguay

Leticia Bidegaray-Batista Laboratorio de Etología, Ecología y Evolución, Instituto de Investigaciones Biológicas Clemente Estable, Montevideo, Uruguay

Departamento de Biodiversidad y Genética, Instituto de Investigaciones Biológicas Clemente Estable, Montevideo, Uruguay

Sean J. Blamires Evolution & Ecology Research Centre, School of Biological, Earth & Environmental Sciences, The University of New South Wales, Sydney, Australia

Department of Life Science, Tunghai University, Taichung, Taiwan

Fedra Bollatti Instituto de Diversidad y Ecología Animal, CONICET – Universidad Nacional de Córdoba, Córdoba, Argentina

Antonio D. Brescovit Laboratório Especial de Coleções Zoológicas, Instituto Butantan, São Paulo, SP, Brazil

Lucía Calbacho-Rosa Instituto de Diversidad y Ecología Animal, CONICET – Universidad Nacional de Córdoba, Córdoba, Argentina

Franco Cargnelutti Instituto de Diversidad y Ecología Animal, CONICET – Universidad Nacional de Córdoba, Córdoba, Argentina

Ana Carlozzi Laboratorio de Etología, Ecología y Evolución, Instituto de Investigaciones Biológicas Clemente Estable, Montevideo, Uruguay

Consejo de Educación Secundaria, Montevideo, Uruguay

Sofía Copperi Instituto de Ciencias Biológicas y Biomédicas del Sur (INBIOSUR), Universidad Nacional del Sur-CONICET, San Juan, Argentina

Luiz Ernesto Costa-Schmidt Programa de Pós-Graduação em Biologia, Universidade do Vale do Rio dos Sinos (UNISINOS), São Leopoldo, RS, Brazil

Kleber Del-Claro Universidade Federal de Uberlândia, Instituto de Biologia, Laboratório de Ecologia Comportamental e de Interações, Uberlândia, MG, Brazil

Nelson Ferretti Centro de Estudios Parasitológicos y de Vectores (CEPAVE) UNLP, CONICET, Boulevard, Argentina

Oscar Francke Departamento de Zoología, Instituto de Biología, Universidad Nacional Autónoma de México, Coyoacán, Mexico

Luis Fernando García Centro Universitario de la Región Este, sede Treinta y Tres, Universidad de la República, Treinta y Tres, Uruguay

Laboratorio Ecología del Comportamiento, Instituto de Investigaciones Biológicas Clemente Estable, Montevideo, Uruguay

Felipe M. Gawryszewski Universidade Federal de Goiás, Goiás, Brazil

Marcelo O. Gonzaga Instituto de Biologia, Universidade Federal de Uberlândia, Uberlândia, Minas Gerais, Brazil

Matías Izquierdo Instituto de Diversidad y Ecología Animal, CONICET – Universidad Nacional de Córdoba, Córdoba, Argentina

Hilton F. Japyassú Instituto de Biologia, Universidade Federal da Bahia, Salvador, BA, Brazil

Carolina Jorge Instituto Superior de Estudios Forestales, CENUR NORESTE, Sede Tacuarembó, Universidad de la República, Tacuarembó, Uruguay

Thiago G. Kloss Universidade Federal do Espírito Santo, Centro de Ciências Agrárias, Departamento de Biologia, Alegre, Espírito Santo, Brazil

Álvaro Laborda Sección Entomología, Universidad de la República, Montevideo, Uruguay

Mariángeles Lacava Centro Universitario de Rivera, UdelaR, Rivera, Uruguay

Laboratorio Ecología del Comportamiento, Instituto de Investigaciones Biológicas Clemente Estable, Montevideo, Uruguay

Yuri Fanchini Messas Programa de Pós-Graduação em Biologia Animal, IB, UNICAMP, Campinas, Brazil

Carlos Víquez Núñez Independent Researcher, Heredia, Costa Rica

Larissa Nahas Universidade Federal de Uberlândia, Instituto de Biologia, Laboratório de Ecologia Comportamental e de Interações, Uberlândia, MG, Brazil

Universidade Federal de Uberlândia, Escola de Educação Básica, Uberlândia, MG, Brazil

Ubirajara Oliveira Departamento de Zoologia, Instituto de Ciências Biológicas, Universidade Federal de Minas Gerais, Belo Horizonte, MG, Brazil

Centro de Sensoriamento Remoto, Instituto de Geociências, Universidade Federal de Minas Gerais, Belo Horizonte, MG, Brazil

Márcia de Oliveira-Tomasi Departamento de Zoologia, Instituto de Ciências Biológicas, Universidade Federal de Minas Gerais, Belo Horizonte, MG, Brazil

Fernando Pérez-Miles Sección Entomología, Facultad de Ciencias, Universidad de la República, Montevideo, Uruguay

Griselda Montiel Parra Departamento de Zoología, Instituto de Biología, Universidad Nacional Autónoma de México, Coyoacán, Mexico

Stano Pekár Department of Botany and Zoology, Faculty of Science, Masaryk University, Brno, Czech Republic

Carlos Perafán Sección Entomología, Facultad de Ciencias, Universidad de la República, Montevideo, Uruguay

Alfredo V. Peretti Instituto de Diversidad y Ecología Animal, CONICET – Universidad Nacional de Córdoba, Córdoba, Argentina

Patricio Pliscoff Instituto de Geografía, Facultad de Historia, Geografía y Ciencia Política, Pontificia Universidad Católica de Chile y Departamento de Ecología, Facultad de Ciencias Biológicas, Pontificia Universidad Católica de Chile, Santiago, Chile

Rodrigo Postiglioni Laboratorio de Etología, Ecología y Evolución, Instituto de Investigaciones Biológicas Clemente Estable, Montevideo, Uruguay

Dinesh Rao Inbioteca, Universidad Veracruzana, Xalapa, Veracruz, Mexico

Gustavo Quevedo Romero Depto de Biologia Animal, IB, UNICAMP, Campinas, Brazil

Philip Russo Departamento de Zoologia, Instituto de Ciências Biológicas, Universidade Federal de Minas Gerais, Belo Horizonte, MG, Brazil

Pós-graduação em Zoologia, Universidade Federal de Minas Gerais, Belo Horizonte, MG, Brazil

Martín Santana Entomología, Biología Animal, Facultad de Ciencias, UdelaR, Montevideo, Uruguay

Laboratorio Ecología del Comportamiento, Instituto de Investigaciones Biológicas Clemente Estable, Montevideo, Uruguay

Adalberto J. Santos Departamento de Zoologia, Instituto de Ciências Biológicas, Universidade Federal de Minas Gerais, Belo Horizonte, MG, Brazil

Cristina Scioscia División Aracnología, Museo Argentino de Ciencias Naturales "Bernardino Rivadavia" (CONICET), Ciudad Autónoma de Buenos Aires, Argentina

Hebert da Silva Souza Programa de Pós-Graduação em Biologia Animal, IB, UNICAMP, Campinas, Brazil

Miguel Simó Sección Entomología, Facultad de Ciencias, Universidad de la República, Montevideo, Uruguay

Jober F. Sobczak Universidade da Integração Internacional da Lusofonia Afro-Brasileira, Instituto de Ciências Exatas e da Natureza, Acarape, Ceará, Brazil

Vanessa Stefani Universidade Federal de Uberlândia, Instituto de Biologia, Laboratório de Ecologia Comportamental e de Interações, Uberlândia, MG, Brazil

Helena Maura Torezan-Silingardi Universidade Federal de Uberlândia, Instituto de Biologia, Laboratório de Ecologia Comportamental e de Interações, Uberlândia, MG, Brazil

I-Min Tso Department of Life Science, Tunghai University, Taichung, Taiwan

Center for Tropical Ecology & Biodiversity, Tunghai University, Taichung, Taiwan

João Vasconcellos-Neto Depto de Biologia Animal, IB, UNICAMP, Campinas, Brazil

Carmen Viera Entomología, Biología Animal, Facultad de Ciencias, Universidad de la República, Montevideo, Uruguay

Laboratorio Ecología del Comportamiento, Instituto de Investigaciones Biológicas Clemente Estable, Montevideo, Uruguay

German Antonio Villanueva-Bonila Depto de Biologia Animal, IB, UNICAMP, Campinas, Brazil

Programa de Pós-Graduação em Biologia Animal, IB, UNICAMP, Campinas, Brazil

Rodrigo H. Willemart Universidade de São Paulo, USP Leste, Rua Arlindo Béttio, Sao Paulo, SP, Brazil

Shichang Zhang Department of Life Science, Tunghai University, Taichung, Taiwan

About the Editors

Carmen Viera is a professor and director of the Department of Animal Biology of Faculty of Sciences, Universidad de la República, and director of the Laboratory of Behavioral Ecology of the Instituto de Investigaciones Biológicas Clemente Estable, Uruguay. She earned her PhD in 1995 and is a researcher for the Sistema Nacional de Investigadores (SNI) Level II, researcher and coordinator of Zoology in the Uruguayan post-graduate development program in basic science (PEDECIBA). She is working on a procedure to improve the teaching of behavioral sciences, and collaborates in PhD courses for her area of expertise in the region. She has experience in behavioral ecology and sociobiology and has supervised numerous theses in this field.

Marcelo O. Gonzaga has been a professor at the Universidade Federal de Uberlândia, Minas Gerais, Brazil, since 2008. He earned his bachelor's degree at Universidade Federal de Minas Gerais (1996), and his master's and doctorate at Universidade Estadual de Campinas (1999, 2004). He had a post-doctoral fellowship at Universidade Estadual de Campinas (2006), and developed a project in the Program "Young Researchers Awards" of FAPESP in the Department of Ecology and Evolutionary Biology of Universidade Federal de São Carlos (2007–2008). He was the coordinator of the postgraduate program in Ecology and Conservation of Natural Resources of Universidade Federal de Uberlândia (2012–2014) and has experience in behavioral ecology, working mainly in the following subjects: construction and variation of spider webs, sociobiology, host behavioral manipulation by parasitoids, and parental care in spiders.

Chapter 1
Curves, Maps and Hotspots: The Diversity and Distribution of Araneomorph Spiders in the Neotropics

Adalberto J. Santos, Antonio D. Brescovit, Márcia de Oliveira-Tomasi, Philip Russo, and Ubirajara Oliveira

Abstract The infraorder Araneomorphae comprises more than nine-tenths of spider diversity, including most of the better known web-weaving spiders. As observed for other taxa, the group is particularly diverse in the Neotropics, where it can be find in any terrestrial ecosystem. In this chapter we synthesize the current taxonomic and biogeographic knowledge on the Neotropical araneomorphs, based on a large database of species described since 1758. We describe the pattern of described species accumulation through time in the Neotropics, and explore factors responsible for variations in species discovery in space and time. We also use statistical methods to predict the total number of species in the Neotropics and discuss the challenges involved in the description of the remaining Neotropical species. Finally, we describe biogeographical patterns throughout the Neotropics, based on detailed species distribution data for well-known spider families.

A.J. Santos (✉) • M.de. Oliveira-Tomasi
Departamento de Zoologia, Instituto de Ciências Biológicas, Universidade Federal de Minas Gerais, Av. Antonio Carlos 6627, CEP 31270-901 Belo Horizonte, MG, Brazil
e-mail: ajsantos@ufmg.br

A.D. Brescovit
Laboratório Especial de Coleções Zoológicas, Instituto Butantan,
Av. Vital Brasil 1500, CEP 05503-900 São Paulo, SP, Brazil

P. Russo
Departamento de Zoologia, Instituto de Ciências Biológicas, Universidade Federal de Minas Gerais, Av. Antonio Carlos 6627, CEP 31270-901 Belo Horizonte, MG, Brazil

Pós-graduação em Zoologia, Universidade Federal de Minas Gerais,
Belo Horizonte, MG, Brazil

U. Oliveira
Departamento de Zoologia, Instituto de Ciências Biológicas, Universidade Federal de Minas Gerais, Av. Antonio Carlos 6627, CEP 31270-901 Belo Horizonte, MG, Brazil

Centro de Sensoriamento Remoto, Instituto de Geociências, Universidade Federal de Minas Gerais, Av. Antonio Carlos 6627, CEP 31270-901 Belo Horizonte, MG, Brazil

C. Viera, M.O. Gonzaga (eds.), *Behaviour and Ecology of Spiders*,
https://doi.org/10.1007/978-3-319-65717-2_1

Spiders are among the most diverse animal groups on Earth. The 46,618 currently recognized species (World Spider Catalog 2017) does not rival the megadiverse insect orders, such as beetles (392,415 species); butterflies and moths (158,570 species); wasps, bees, and ants (155,517 species); or, among arachnids, mites and ticks (over 55,000 species; all numbers from Zhang 2013). However, spiders are diverse, widespread, and ubiquitous enough to be considered a particularly successful lineage. The high diversity of the group, along with its abundance and its importance as a group of exclusively terrestrial predators, make them an interesting model for ecology and biogeography (see chapters in Penney 2013).

Spider classification is currently a matter of intense debate, with traditional family-level groupings undergoing profound review (Dimitrov et al. 2017; Wheeler et al. 2017). However, the deep lineages within the order, which were discovered early with the advent of phylogenetic inference methods (Platnick and Gertsch 1976), are still solidly recognized as monophyletic taxa (Garrison et al. 2016; Wheeler et al. 2017). Spiders are classified into three very easily identified infraorders. The Liphistiomorphae, the sole lineage within the suborder Mesothelae, consists of 96 species of primitive-looking spiders with an externally segmented abdomen, and all are restricted to eastern and southeastern Asia (Xu et al. 2015). The two remaining infraorders are grouped in the suborder Opisthothelae. The Mygalomorphae includes 2,860 species of tarantulas and trapdoor spiders, which are found worldwide (Foelix 2011; Bond et al. 2012; World Spider Catalog 2017). The last group is also the most species-rich. The Araneomorphae (historically but inappropriately known as true spiders) unites the remaining 43,662 species, constituting 94% of all known spider species.

The figures presented above indicate that the taxonomic diversity of spiders is not distributed uniformly among its main lineages. Araneomorphae contains more than 15 times the number of species of its sister group, the mygalomorphs. Because the two groups have, by definition, the same geologic age (Penney and Selden 2011), it is obvious that araneomorphs experienced a much higher diversification rate (see some possible but incomplete explanations in Craig et al. 1994; Bond and Opell 1998; Blackledge et al. 2009). This resulted in an impressive diversity of size, shape, colour, and habits (Fig. 1.1). Araneomorphs occupy all terrestrial habitats on the planet (besides freshwater and seashore microhabitats) in all continents except Antarctica. They also show an amazing, and still insufficiently described, diversity of prey capture, mating, and parental care behaviours. In this chapter, we explore the diversity and distribution of these remarkable spiders in the Neotropics.

It is a well-established fact that the terrestrial biological diversity in the tropics far exceeds that of temperate and boreal environments. This biogeographical pattern, called the latitudinal diversity gradient, has been repeatedly reported for several terrestrial animal lineages (Hillebrand 2004). Spiders are no exception, although worldwide data on spider species richness variation are still lacking. The extent of the geographic variation in spider species richness is well illustrated by the Afrotropical region, which contains over 7,000 spider species, well above the 4,000 species in Europe (Jocqué et al. 2013). It is a bit more difficult to make a similar comparison for the Neotropics, because no similar cataloguing effort has been

Fig. 1.1 A very superficial glimpse of the diversity of araneomorph spiders. (**a**) *Sicarius* (Sicariidae), a genus of sand-dwelling spiders, endemic from semiarid to desert places in the Neotropics. (**b**) Oonopidae, a family of small-sized spiders, found mainly in the leaf litter and on the foliage of humid forests. (**c**) *Phoneutria* (Ctenidae), a large-sized, active hunting spider genus known mostly for its aggressiveness and the effect of its venom on people. The photo shows a female caring for her egg sac over the leaf litter. (**d**) *Lyssomanes* (Salticidae), a foliage-dwelling

attempted. However, country-level checklists provide a hint regarding the magnitude of differences in diversity within the Americas. According to these scarce sources, Colombia contains 914 spider species (Barriga and Moreno 2013); Costa Rica has 513 (Zúñiga-Vega 1980), and Panama is home to an impressive 1,223 spider species (Nentwig 1993). These figures might seem unremarkable compared with the near 3,700 spider species reported from USA and Canada (Cushing 2005), but this impression changes considerably when the area of each country is considered: North America has approximately 0.0002 spider species/km^2. This is well below the figure for Colombia (0.0008 species/km^2), Costa Rica (0.01), and Panama (0.02). As we discuss below, the New World's temperate–tropical disparity in spider diversity is even more pronounced because the North American spider fauna is much better known than its third world, Latin American counterpart.

Spider Databasing: How Do We Know What We Know?

Spider taxonomists are privileged among most of their peers. By tradition, spider taxonomic knowledge has been periodically compiled in species catalogues, starting with Petrunkevitch (1911) and culminating in a thorough on-line database updated almost in real time (World Spider Catalog 2017). The World Spider Catalog (WSC) not only lists all the currently known spider species with all their bibliographic and nomenclatural information, but also contains a depository where all taxonomic literature on the group is freely available for download. It is impossible to overstate the importance of this initiative for the advancement of spider taxonomy, and how great it would be to have a similar catalogue for every taxon on Earth. However, for biogeographic purposes, the WSC is still not enough, because it does not compile detailed species distribution data.

The geographic distribution of species is the raw material of biogeography, macroecology, and conservation science. The place of each species on Earth can be delimited only through point species records, which represent sites where the species has been found and (ideally) specimens have been collected and permanently stored in scientific collections. Unfortunately, the geographic distribution of most animal species is still insufficiently known, generating the so-called Wallacean

Fig. 1.1 (continued) jumping spider known mostly from humid forests. Jumping spiders comprise the most species-rich spider family, and are renowned for their acute eyesight. (**e**) *Misumenops* (Thomisidae), a crab spider, here feeding on a bee over the petals of a sunflower. Although thomisids live associated with a wide variety of microhabitats, the family is better known from species that ambush insects over flowers. (**f**) *Micrathena sexspinosa* (Araneidae), a spiny orb-weaving spider widely distributed in the Neotropical forests. Araneidae is the third most diverse spider family, and includes species with impressive colours and shapes, such as the members of *Micrathena*. (**g**) An orb-web fixed to delicate tree brushes. Although spiders build a remarkable diversity of web structures, the orb-web is surely the most famous, and is built by thousands of species from six families, including Araneidae. Photo credits: (**a**) — A.J. Santos; (**b, f**) — Pedro H. Martins; (**c–e, g**) — U. Oliveira knowledge shortfall (Hortal et al. 2015). This situation is particularly severe for invertebrate species, which are usually considered less well-known (at least in proportional terms) than charismatic, furry and feathery vertebrates (Cardoso et al. 2011, but see also Oliveira et al. 2016). Consequently, any compilation effort for invertebrate species distributions should be considered of the utmost importance.

Spiders are affected by the Wallacean knowledge shortfall as much as any other arthropod taxa. It is fair to say that delimiting the exact distribution range of any spider species in the world is virtually impossible, except for most European species (Nentwig et al. 2017). This situation is, in part, a result of insufficient collection effort and poor taxonomic knowledge. However, for the most diverse tropical regions of the World, the situation is even worse because the scarce knowledge available is scattered in the taxonomic literature. Simply put, we do not know how much we know about spider distribution. In an attempt to fix this impediment, we (the authors of this chapter) have been working for a few years on a large database containing the distributions of Neotropical spider species. Our intent is to assemble all (reliable) spider species distribution records from the Neotropics, available in the taxonomic and biodiversity survey literature. Such a database, which is still a work in progress, will complement the World Spider Catalog by providing much-needed information on species distribution. In plain English, we hope to be able to accurately answer questions such as "where does this spider species occur in the Neotropics?" or even "how many spider species are known to occur in this specific part (country, state, region, etc.) of the Neotropical Region?" Although we are still far from the completion of this goal, we hope to show in this chapter the potential of our initiative.

The Neotropical Spider Database is currently incomplete, but growing rapidly. We already have country records for all spider species, for all Neotropical countries (including portions of Mexico considered part of the Nearctic region — see Cox 2001). We also have detailed species distribution records for Brazil (Oliveira et al. 2017), and our efforts are now directed toward the compilation of detailed records for all the Neotropical countries. In the following topics, we discuss the state of knowledge of Neotropical araneomorph spiders in general, and focus more attention on the geographic distribution of two araneomorph families, Oonopidae and Araneidae (Fig. 1.1b, f, respectively).

Neotropical Araneomorph Diversity Through Time

The modern taxonomic study of Neotropical araneomorphs began together with the animal taxonomy itself, when Carolus N. Linnaeus described the araneids *Aranea cancriformis* (today *Gasteracantha cancriformis*) and *Aranea spinosa* (*Micrathena spinosa*), both reported to be from "America" (with no further data). These species were described in the same book that marks the beginning of the currently accepted zoological classification system (Linnaeus 1758). If, on one hand, it is fair to say that Neotropical spider taxonomy is as old as can be, on the other hand, its initial development was a bit slow. World biodiversity exploration was much more challenging in the eighteenth century than it is nowadays, as the taxonomic enterprise was completely concentrated in the most influential European countries. The description of species from the hyperdiverse tropical regions of the world was dependent upon the collection and shipment of specimens by expensive and frequently life-threatening expeditions to the far places on the planet (Levi 1964; Conniff 2011). It is thus unsurprising that knowledge of the Neotropical araneomorph spiders grew very slowly until the early nineteenth century (Fig. 1.2a). The uneven, eurocentric geographic distribution of taxonomists remained almost unchanged until the early twentieth century. However, the development of navigation technology, which facilitated worldwide travelling and trading, made it progressively easier to acquire specimens from around the planet. Although this factor certainly contributed to the increase in Neotropical araneomorph species descriptions (as seen in Fig. 1.2a), the mid-1800s saw a more important change in spider taxonomy: the emergence of the spider taxonomist. It was approximately at that time when European naturalists started to dedicate most, if not all, of their efforts to the study of spiders (Savory 1961). In this epoch, important scholars such as the German Eugen von Keyserling (who described Neotropical araneomorphs over a period of 28 years, between 1865 and 1893), the French Eugène Simon (37 years, 1864–1901), the British Octavius Pickard-Cambridge (32 years, 1870–1902), and his nephew, Frederick Octavius Pickard-Cambridge (7 years, 1897–1904) effectively started the seemingly never-ending, ascending curve of Neotropical araneomorph species accumulation that resulted in the 10,755 species known by 2009 (Fig. 1.2a).

The taxonomic study of Neotropical araneomorphs achieved another important step forward from the late 1800s to the mid-twentieth century, when the participation of local, New World arachnologists became truly relevant. At the beginning, North American taxonomists such as Elisabeth B. Bryant (who described Neotropical araneomorphs over a period of 25 years, from 1923 to 1948) and Ralph Chamberlin (39 years, 1916–1955), followed by Arthur M. Chickering (35 years, 1937–1972) and Willis G. Gertsch (53 years, 1929–1982), described hundreds of araneomorph species from Central America, the Caribbean, and South America. The scientific legacy of these authors served as the basis for the descriptive efforts of more recent arachnologists, such as Herbert W. Levi and Norman I. Platnick (more on them below), and marked a continuum in the accumulation of Neotropical araneomorph species until today.

The establishment of local, Latin American araneomorph taxonomists took much more time, a delay easily explained by economic development discrepancies throughout the continent. Although the first Latin America-born arachnologist (the Argentinean Eduardo L. Holmberg) appeared in the literature in the mid-nineteenth century, it was not until the taxonomic activities of the Brazilian Cândido F. de Mello-Leitão (active during 32 years, from 1915 to 1947) that Latin American countries really started to contribute to the description of araneomorph spiders. Despite the timid start, today's Latin America-based spider taxonomists are as productive in terms of species description as their European and North American colleagues are.

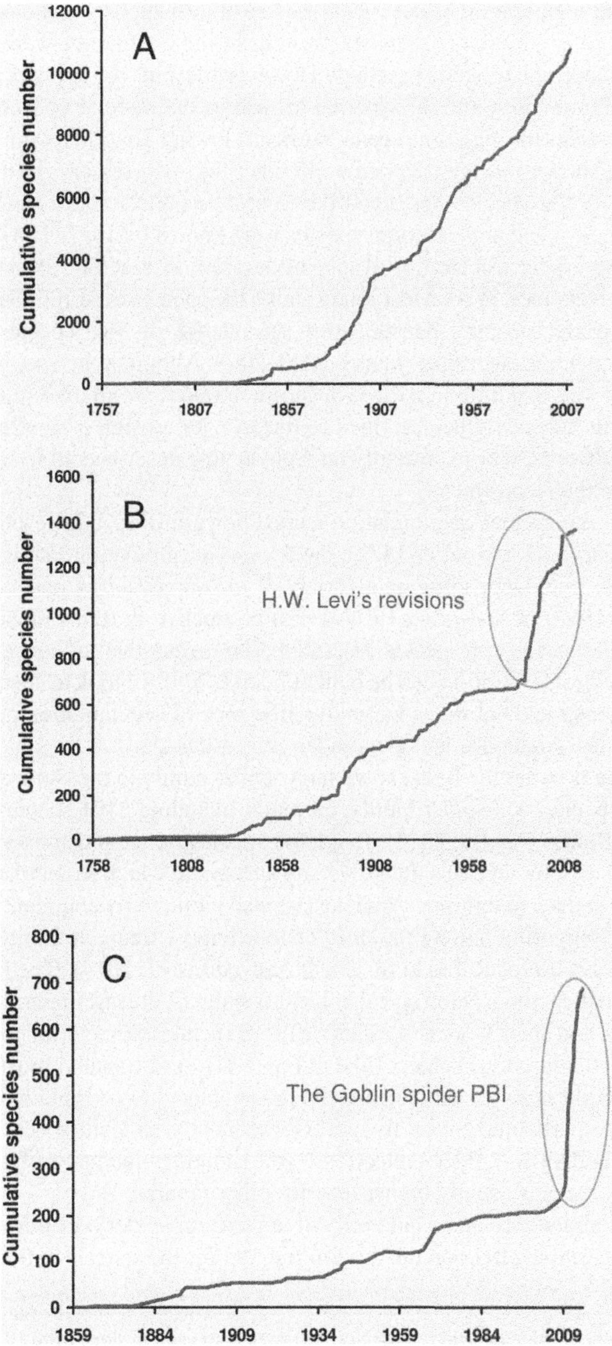

Fig. 1.2 The progress on the description of Neotropical araneomorph spiders, expressed as species accumulation curves. The curves show only species currently considered valid, thus already excluding synonymous species names. (**a**) Species accumulation curve for Araneomorphae species

The preceding paragraphs describe a picture of ever-increasing taxonomic effort, in which more and more taxonomists, from differing continents, were dedicated to describing Neotropical spider species. However, despite the growth in the spider taxonomist community, the incremental growth in the number of Neotropical araneomorph species through time seems far from slowing down. The species accumulation curve shown in Fig. 1.2a can be divided into two sectors. In the first, from 1757 to 1832, the species described per year was almost stable, and less than a thousand Neotropical araneomorph species were known by 1832. The second sector shows an impressive acceleration in species description, reaching an average of 73.8 species per year since 1950. Most importantly, the species description rate seems to be increasing decade after decade, from an average of 56.9 species per year in 1950–1959 to 93.2 species per year in 2000–2009. Although the species accumulation curve shows variation in species description rate through time, the sector from 1832 onwards can easily be described as linear, with no sign of nearing an asymptote. In other words, we are currently far from having described all the araneomorph species from the Neotropics.

Although the species accumulation curve shown in Fig. 1.2a resembles more a line than a curve (at least after 1832), the species accumulation rate is visibly variable through time. Depending on a variety of factors, differing periods were more (e.g., 1873–1903) or less (e.g., 1903–1929) productive in terms of the number of Neotropical araneomorph species described. The factors that influence the progress in spider species description can be better understood if we look at specific families. This is demonstrated below, as we analyse the species accumulation curves for two model families, Araneidae and Oonopidae (Fig. 1.2b, c).

Araneidae is by far the best-known large spider family in the Neotropics. It is the third most species-rich spider family, currently including 3101 species in 169 genera (World Spider Catalog 2017). Its global species richness can also be observed locally, as shown by several spider diversity surveys in which Araneidae was among the three most diverse families. Araneids are also particularly abundant in field sampling, usually figuring among the third or fourth most frequent families in species lists (e.g., Silva 1996; Rubio et al. 2008; Azevedo et al. 2014). The high share of Araneidae in the known Neotropical spider diversity is, at least in part, related to the morphology and habits of its species. Most araneids are medium to large sized, colourful, and ubiquitous spiders (like in Fig. 1.1f). Additionally, most members of the family build highly visible orb-webs in the understory vegetation (Fig. 1.1g), making them particularly easy to spot. Consequently, araneid specimens are particularly abundant in spider collections, and sampling intensity of this group in structured surveys is usually higher than for other families.

The high abundance and availability of araneids in spider collections had apparently no discernible effect on the description rate for the species until recently. The

Fig 1.2 (continued) from the Neotropics. Note the steep increment in slope from 1832. (**b**) Species accumulation curve for Neotropical Araneidae. The ellipsis marks the period of taxonomic activity of Herbert W. Levi, the most productive araneid spider taxonomist. (**c**) Species accumulation curve for Neotropical species of Oonopidae. Note the meager species accumulation rate until the beginning of the Goblin Spider PBI project (ellipsis)

Neotropical araneid species accumulation curve strongly resembles the total Araneomorphae curve until the middle of the twentieth century (Fig. 1.2a, b). From 1958 to 1983, the known araneid species richness from the Neotropics nearly stabilized, suggesting an exhaustion of the undescribed species stock. However, a different trend emerged from 1983, when the number of Neotropical araneid species described per year increased in an unprecedented way. This radical change of course can be explained by reference to a single name: Herbert Walter Levi.

Our current knowledge of the Neotropical Araneidae is intimately tied to the career of the German-American arachnologist H.W. Levi, from the Museum of Comparative Zoology at Harvard University (see more on his life and career in Leibensperger 2016). Over a period of 41 years, he produced a series of detailed taxonomic monographs on American Araneidae (most of them listed in Levi 2002). These monographs are characteristic not only because of the careful revision of the taxonomy of each araneid genus (usually one per publication), but also because of the meticulous sampling of biological collections, including those located in Neotropical countries. As a result, he described (between 1973 and 2009) 649 of the 1110 currently known Neotropical araneid species. His absolute record was the description of 127 new species in a single paper (Levi 2007). Levi's effect on Neotropical araneid knowledge was so accentuated that a drop in the species accumulation rate appears on Fig. 1.2b from 2008, just after the publication of his last monograph (Levi 2007). Of course, taxonomists are still discovering new araneid species from the Neotropics (e.g., Saturnino et al. 2015; Lise et al. 2015; Magalhães et al. 2017), but the current species accumulation rate for the family is much lower.

Our second model family, the Oonopidae, is in certain aspects the exact opposite of the Araneidae. Oonopids are tiny spiders, usually with no more than 2 mm in body length. They also have secretive habits, living in the leaf litter, and more rarely on the understory vegetation or the tree canopy (Fannes et al. 2008, and references therein). Additionally, they are active hunters (Korenko et al. 2014; Alligand and Henrard 2016) that do not use silk to build prey-capture webs. These traits make them much harder to find and, consequently, scarcer in collections than araneids and other aerial web-weaving families. As could be expected, the rate of Neotropical oonopid species discovery has been much lower than that seen for Araneidae or even araneomorph spiders in general (Fig. 1.2). The oonopid species accumulation curve from the Neotropics follows a slow, steady rhythm from the description of *Oonops globosus* (today *Paradysderina globosa*), the first truly Neotropical oonopid (Keyserling 1877), until 2004 (Fig. 1.2c). The only remarkable exception was an increase in the species description rate between 1968 and 1972, when 54 (currently valid) species were described by the American arachnologist A.M. Chickering. During this short interval, he published eight monographs on the Central American (mostly Panama) and the Caribbean oonopids (Chickering 1968a, b, c, d, 1969, 1970, 1971, 1972). Although his contribution is no match for H.W. Levi's effect on Araneidae, it is another example of the positive effect a single taxonomist can have on the knowledge on a certain group. However, the most impressive change in oonopid taxonomy was still to come; it was the result of collective effort, and is the main reason we choose to focus on this family.

Starting in 2004 (when only 212 Neotropical species were known), the rate of Neotropical oonopid species description increased in a way that could be described as unprecedented for any spider family: in just 11 years, over twice the number of species described in the preceding 127 years were described (Fig. 1.2c). This impressive step forward in oonopid systematics was the result of the *Goblin Spider PBI* (http://research.amnh.org/oonopidae), an international initiative funded by the American NSF's *Planetary Biodiversity Inventory* (*PBI*) program, and coordinated by Norman I. Platnick, from the American Museum of Natural History (himself one of the main describers of oonopid species). The project congregated 54 arachnologists from 19 institutions throughout the world, all dedicated to the advancement of the systematics of Oonopidae worldwide. For the first time in history, several taxonomists from different parts of the planet joined efforts to study a single spider group, producing (up to now) 103 papers and 5344 pages of (mostly) taxonomic information (N.I. Platnick, personal communication). Incidentally, the project also introduced the common name "goblin spiders" for the oonopids. On examining the results of this initiative, it becomes obvious that the 1993–2003 slow rate of Neotropical oonopid species description was not indicative that the family was sufficiently known, but a result of scant taxonomic effort. Most importantly, considering the results of the project (which are still under publication), one cannot avoid asking "what have they done differently?" The secret of the success of the *Goblin Spider PBI* was most likely the careful global planning and division of tasks among its members, which facilitated specimen sharing and access to type material, along with the use of cybertaxonomy tools. These procedures resulted in intensive sampling of spider collections, which was complemented by field collecting. Some of the descriptions by the project's collaborators of new species were based on unsorted and unidentified specimens already available in spider collections, particularly specimens collected in spider diversity inventories. Nevertheless, several species were also first discovered through field sampling in poorly explored and/or particularly diverse places. For instance, a single expedition conducted in Ecuador in 2009 resulted (again, up to now) in the discovery of three new genera and 42 new species (Platnick and Dupérré 2010, 2011a, b; Platnick et al. 2013a, b, Grismado and Ramírez 2013; Moss et al. 2016).

The comparison between the species accumulation curves shown in Fig. 1.2 demonstrated that the description of araneomorph spider fauna of the Neotropics is a collective work in progress, with no apparent slowdown in the species description rate. Despite the apparent monotonous increment in the number of species described since 1832, we have shown that the rhythm of species description can be extremely variable among families. Most importantly, the progress in species descriptions for each family is in part dictated by variation in study effort, in such a way that a single taxonomist can make a meaningful difference in the knowledge of a particular group. In the next topic, we intend to deepen this discussion by putting the current knowledge of Neotropical araneomorphs in a geographical context. By doing this, we add an additional layer of complexity to the theme, by looking not only at the number of species known from the Neotropics, but also at where to find them.

Araneomorph Spider Diversity Through the Neotropical Region

In the preceding topic, we discussed the rate of araneomorph species description for the Neotropical region as whole. The species accumulation curves in Fig. 1.2 represent how the world taxonomic community came to the 10,755 Neotropical araneomorph species known by 2009, and specifically how the currently known 1110 Neotropical species of Araneidae and 595 species of Oonopidae came to be described. However, it is obvious that differing parts of the Neotropical region have their own history of species description, and that araneomorph diversity will vary geographically for a variety of reasons. In fact, araneomorph species richness can vary by an order of magnitude among the Neotropical countries (Fig. 1.3). In this topic, we are interested in understanding the factors that can explain such a large discrepancy.

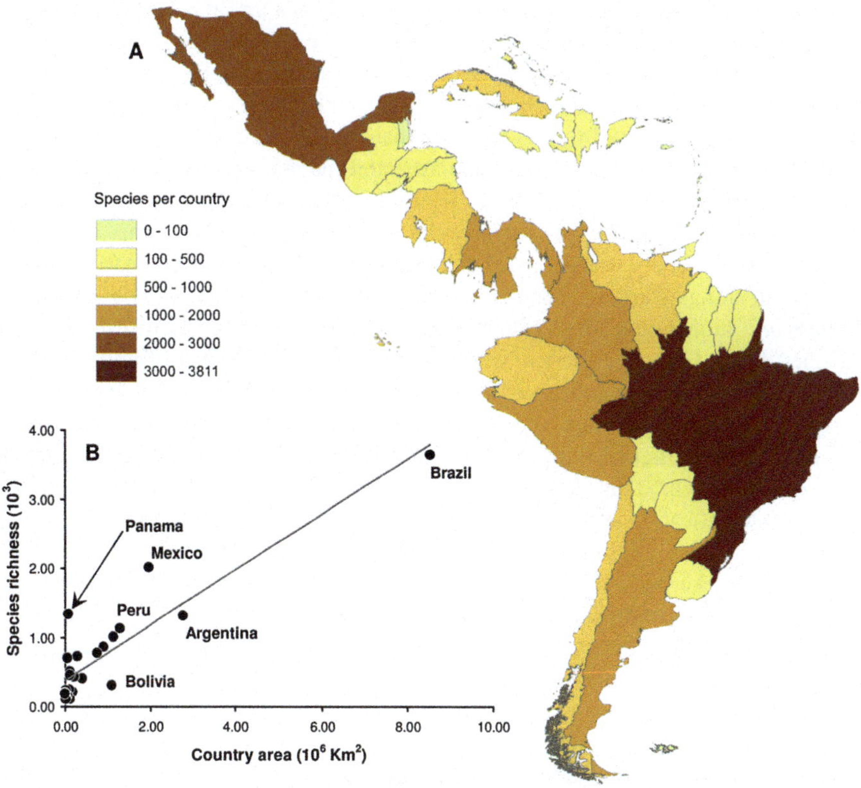

Fig. 1.3 Araneomorph spider diversity in the Neotropical region. (**a**) Cartogram representing Neotropical country size distorted proportionally to its araneomorph species richness. (**b**) Relationship between Neotropical country area and spider species richness. The dots represent only countries larger than 5000 km^2

To explain species richness variation among countries, we must first consider total country size. It is no surprise that Brazil, Argentina, and Mexico are the Neotropical countries with more araneomorph species, because they are the three largest countries in the region (Fig. 1.3b). Larger countries should have more species simply because they encompass more space to contain species (as predicted by the passive sampling explanation for species–area relationship, see Connor and McCoy 2001). They could also contain more environmental heterogeneity, which can affect species endemism and diversity (Finch et al. 2008; Meng et al. 2008; Melo et al. 2009; Foord et al. 2011). These factors certainly explain part of the variation in araneomorph species richness among Neotropical countries, but we must also consider the effects of the past taxonomic effort applied in each country.

Figure 1.3b shows a high correlation between the number of araneomorph spider species and the size of each Neotropical country. There are solid theoretical reasons to expect such a positive relationship, but a few countries are particularly distant from the expected species richness, given their size. The most impressive of them is Panama (Fig. 1.3a, b), which is 0.9% the size of Brazil, but contains (according to our database) 1342 araneomorph species. This accounts for 36.8% of the total species number of Brazil (3645 species). If we were to express that in terms of species per unit area, Panama would have 0.018 species per km^2, compared with 0.0004 in Brazil. The explanation for such disparity is not in Panama itself, which is covered mostly by tropical humid forests (nothing that could not be found in Brazil and its South American neighbors), but in the attention the country received from spider taxonomists.

Panama has been in a privileged position regarding the study of its biota. The country's biodiversity was not particularly attractive for naturalists in early colonial times, but things started to change with the publication of the *Biologia Centrali Americana* monograph series (Selander and Vaurie 1962), in which several spider species from Central America were described by O. Pickard-Cambridge and F.O. Pickard-Cambridge between 1889 and 1904 (Godman 1915). In the early twentieth century, the construction of the Panama Canal turned the country, the Canal Zone in particular, into an American laboratory for tropical biology (Hagen 1990). This benefited the taxonomic study of Panamanian spiders through an unparalleled intensification of field sampling by biologists in general, including some prominent arachnologists based in the USA (such as Alexander A. Petrunkevitch and A.M. Chickering). The highly concentrated study effort on the Panamanian spiders resulted in the publication of several taxonomic monographs (e.g., Banks 1929; Chickering 1946) and generated the impressive number of spider species described (or just recorded) from there (Nentwig 1993).

The history told above requires three important clarifications. First, Panama is not the only small, Neotropical country with a comparatively well-known spider fauna. Costa Rica, Cuba, Haiti, the Dominican Republic, Puerto Rico, and Jamaica also hold more araneomorph species than would be expected by their size (Fig. 1.3b), though not as many species as Panama. These countries have also been the subject of intense field sampling and taxonomic study, mostly by USA-based arachnologists (e.g., Petrunkevitch 1930; Bryant 1940, 1948). Thus, they are species-rich because, among other factors, they have been more intensively sampled. The second clarification is a more important one: we do not intend to imply that the high araneomorph species richness of Panama and its neighbors is just a consequence of sampling. Local species inventories show that the spider diversity in Panama is comparable to other Neotropical forest places. For instance, intensive sampling in a 0.48 hectare plot of humid forest in Panama resulted in 394 spider species (Basset et al. 2012), which is comparable to surveys with similar collecting effort in South America (Azevedo et al. 2014: Table 2). This means that, although we currently know more species from Panama, other parts of the Neotropics could be equally diverse, but only insufficiently studied. This leads us to the third clarification: the spider species diversity in Panama is an interesting demonstration of how high Neotropical species diversity can be, but that does not mean the Panamanian araneofauna is completely known. Instead, new araneomorph species are still being described from Panama (e.g., Zhang and Maddison 2012; Platnick and Berniker 2014; Platnick et al. 2014).

Our data on species diversity by Neotropical country, together with historical information, shows that explaining geographic variation in biodiversity requires particular attention to sampling bias. That the geographic variation in collection intensity can strongly affect attempts to compare or estimate species richness is already a well-known caveat of biodiversity studies (Boakes et al. 2010). However, it can also affect our perception of species distribution, and so exert an undesired effect on biogeographic studies and conservation policies (Oliveira et al. 2016). In the next topic, we explore the spatial variation in araneomorph species richness in more detail, with focus on our two model families.

Spider Diversity (or Sampling) Hotspots in the Neotropics

Throughout this chapter, we have been examining the variation in Neotropical araneomorph species richness through time, since the beginning of the formal taxonomic study of spiders, and at a coarse geographic scale, by comparing species richness among countries. This approach ignores the variation in species richness inside each country, and precludes any discussion on the effects of environmental heterogeneity on the diversity and distribution of araneomorphs. We will now examine the araneomorph diversity and the distribution of species records at a smaller scale, focusing on Araneidae and Oonopidae.

To compare the spatial variation of diversity between the two families, their species richness is expressed as species counts inside hexagonal, 0.5-degree grid cells overlapping a map (Figs. 1.4a and 1.5a). This representation illustrates that the number of species recorded within each country can vary from a single species to more than a hundred in each grid cell. With regard to the Araneidae, species richness is particularly high in a few places in the Brazilian Atlantic Forest, parts of the Amazon, southwestern Colombia, Peru, and Panama (Fig. 1.4a). For the Oonopidae,

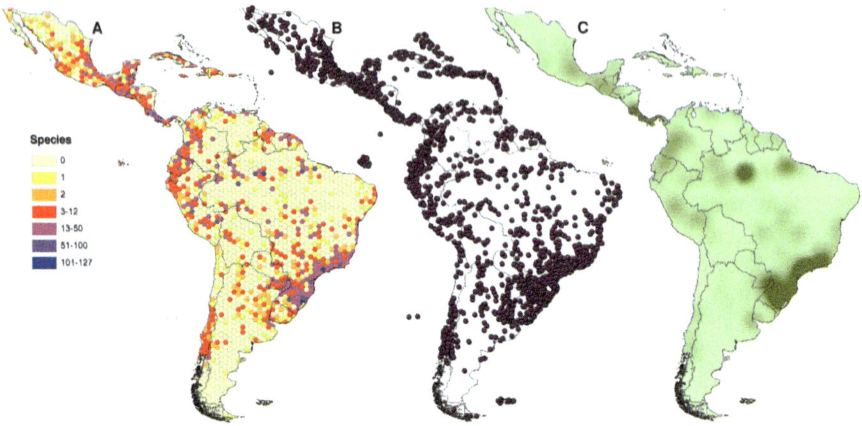

Fig. 1.4 Species richness and distribution records of the spider family Araneidae in the Neotropics. (**a**) Araneid species richness within hexagonal, 0.5-degree grid cells. (**b**) Distribution records of araneid species. Each dot represents at least one record of a single species. (**c**) Kernel-estimated density of distribution records from (**b**)

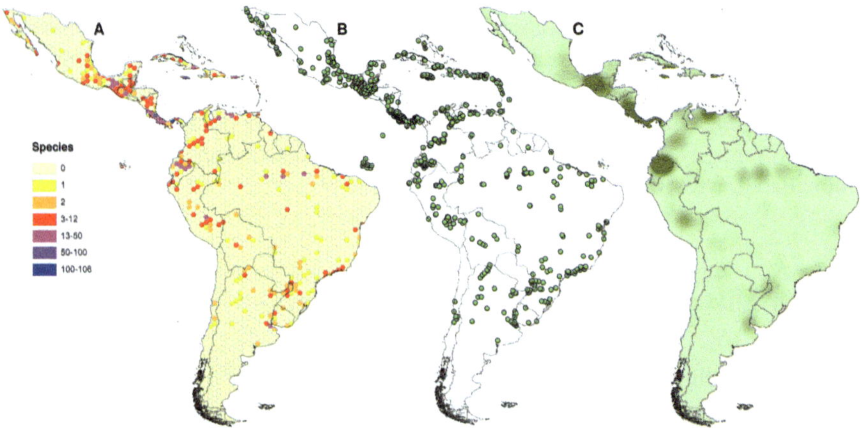

Fig. 1.5 Species richness and distribution records of the spider family Oonopidae in the Neotropics. (**a**) Oonopid species richness within hexagonal, 0.5-degree grid cells. (**b**) Distribution records of oonopid species. Each dot represents at least one record of a single species. (**c**) Kernel-estimated density of distribution records from (**b**)

high-diversity locations are far more restricted, with the most species-rich grid cells located in Panama, Costa Rica, Venezuela, Ecuador, and the Brazilian Amazonia (Fig. 1.5a). Nevertheless, the most evident difference between the two families is the larger portion of the oonopid map without any species records. As mentioned above, araneids are particularly well represented in biological collections, so the knowledge on the family has a much denser spatial coverage than that of the cryptic, hardly sampled Oonopidae.

Despite the large gaps in the diversity data shown in Figs. 1.4a and 1.5a, these results can provide insights on the factors that influence spatial variation in species richness. However, as advised above, we first must consider the effects of sampling bias, which can be as influential at the 0.5-degree spatial scale, as seen on our country-based comparison. A better way to assess sampling bias is to analyze the distribution of species records and its correlation with spatial variation in species richness. Each dot on Figs. 1.4b and 1.5b represents a distribution record of at least one species of Araneidae and Oonopidae respectively. As expected, these maps mirror the species richness maps in the larger record-free portion for Oonopidae compared to Araneidae. Most of these empty zones are covered by habitats suitable for these spiders to live, meaning that they are a visual representation of sampling insufficiency. In plain words, araneids and oonopids most probably occur in these areas, but nobody has sampled them there.

The point-distribution maps shown here are illustrative of species distribution; thus, it is evident why this kind of representation is commonly used to represent species ranges in taxonomic studies. However, this kind of map has a limitation as an instrument to represent the distribution of species records (instead of representing species distribution per se). Each dot on these maps can represent a single record of a species, or a pile of records of several species captured at the same locality. Thus, the real density of records cannot be represented this way. To better represent this important parameter, we applied a kernel density estimator; a mathematical procedure that expresses the variation in the density of dots on a map through colour shades (ESRI 2017). This analysis shows that most areas loaded with species records in the Araneidae map (Fig. 1.4b) actually have low record density, but it also reveals areas where araneids were most intensively sampled. Specifically, araneid records are particularly concentrated in the Atlantic Forest of southern and southeastern Brazil, central and eastern Brazilian Amazonia, and (again) in Panama and Costa Rica (Fig. 1.4c). The equivalent map for Oonopidae shows sites with the highest record density in Ecuador, northern Venezuela, Jamaica, and Puerto Rico, a long stretch from Panama to southern Nicaragua, in southern Mexico, and Guatemala (Fig. 1.5c).

The interpretation of geographic variation in species richness must necessarily consider the effect of variation in sampling effort, as it makes sense that more densely sampled locations will have more species records (as discussed above, and reinforced in Figs. 1.4 and 1.5, for Panama). In the absence of direct measures of sampling effort, the number and density of species records are customarily used as a proxy (Oliveira et al. 2017). Although helpful, this resource must be used with caution, because the number of distribution records is necessarily correlated with species richness. After all, it is exactly the presence of different species in a locality that results in the report of different species records there. In our particular case, we are more interested in analyzing how frequently the most species-rich grid cells coincide with larger areas with high record density.

The Brazilian Atlantic Forest seems to be particularly rich in araneid species and distribution records (Fig. 1.4). The humid forests that cover southeastern and southern Brazil, and the Argentinean province of Misiones, have an extremely high concentration of distribution records (Fig. 1.4c). Thus, it is not a surprise that this portion is also covered by highly species-rich grid cells (Fig. 1.4a). This pattern is, at least in part, caused by what has been called the "Museum Effect". Because taxonomists tend to collect more intensely in the vicinity of their work place, areas near important research institutions (particularly natural history museums) tend to show a higher concentration of species distribution records (Moerman and Estabrook 2006; Sobral and Stehmann 2009). In fact, the Atlantic Forest area in Fig. 1.4 is home to five Brazilian institutions with long traditions of arachnological research (Brescovit 1999). In addition, the area coincides with the most economically developed Brazilian states, and financial wealth is a factor known to be correlated with biodiversity knowledge (Amano and Sutherland 2013). However, we have good reasons to discard sampling bias as the sole explanation for the high araneid diversity in the Atlantic Forest. This biome is known for its high diversity and endemism level for animal and plant taxa, including groups with better sampling histories (Fonseca et al. 2004; Oliveira et al. 2017 — and references therein). The high biodiversity of this biome is possibly related to its topographic and hydrologic heterogeneity (Dantas et al. 2011), which resulted in species with narrow distribution ranges and in the differentiation of several small areas of endemism (Silva et al. 2012; Oliveira et al. 2015; DaSilva et al. 2015; Hoffmeister and Ferrari 2016; Saiter et al. 2016). Additionally, the fact that part of this forest was environmentally stable during recent glaciations also created conditions for species accumulation (Carnaval et al. 2009). Thus, we believe the high araneomorph diversity in the Atlantic Forest, compared to other Neotropical biomes, is not merely a sampling artifact, but a result of the history and geoclimatic characteristics of this biome.

The Amazonian biota is probably as diverse as that of the Atlantic Forest (Mittermeier et al. 2003), though a comparison between these two large forest areas is hindered by a difference in sampling intensity. Biodiversity sampling in the Amazon is strongly biased towards the vicinity of its most important cities and access routes, primarily major rivers (Heyer et al. 1999; Hopkins 2007; Oliveira et al. 2017). This is particularly evident in our araneid and (though to a lesser extent) oonopid data, which show species records strongly concentrated along the Amazon River (Figs. 1.4b and 1.5b). Both groups also have their Amazonian diversity and sampling hotspots around the Brazilian cities of Manaus and Belém, the largest Amazonian cities. Once again, we have a "museum effect" strongly biasing our knowledge of the geographic variation in spider diversity. Manaus and Belém are home to two important biodiversity research institutions (Brescovit 1999), and forest reserves near these cities have been intensively sampled by arachnologists (Adis et al. 2002; Bonaldo et al. 2009a, b; Cafofo et al. 2013). Our results (and those of Oliveira et al. 2017) leave no doubt with regard to the contribution of Amazonia to spider diversity in the Neotropics, but also show that we are still far from understanding the true spatial variation in spider diversity within this biome.

Among the remaining high-diversity areas depicted in Figs. 1.4 and 1.5, two deserve particular attention. Our data show that oonopids are particularly diverse (and densely sampled) in northern Ecuador (Fig. 1.5a, c). This part of the Neotropics is particularly diverse, probably because of the high environmental variation in such a small area, which includes a portion of the western Amazon, the tropical Andes, and part of the Tumbes–Chocó–Magdalena biodiversity hotspot (Rodríguez-Mahecha et al. 2004a, b). It is interesting to note that high oonopid diversity in northern Ecuador is apparently not correlated with particularly high araneid species richness (Fig. 1.4a, c). Even more intriguingly, although the Atlantic Forest is so rich in araneid species, it does not seem to be as important in terms of oonopid diversity. Considering that these are relatively well-sampled areas, these results probably reveal particularities of the diversification history of each group.

A second high-diversity area worth mentioning is Central America, which has relatively high species richness and sampling coverage for both families. Once more, a considerable number of species are known from Panama and Costa Rica, primarily because of their sampling history. It is not a coincidence that the Panamanian Canal Zone marks a high species-rich grid cell for Araneidae, and the most species-rich cell for Oonopidae (Figs. 1.4a and 1.5a). However, most of Central America is currently recognized for its high species richness and endemism (Mittermeier et al. 2004), so its elevated spider diversity is probably part of a biogeographic pattern.

Up to this point, we have been emphasizing the spatial variation in species richness throughout the Neotropics. However, biodiversity exploration is not simply a matter of species numbers. Knowing species distributions accurately is essential to understanding biodiversity patterns and guiding conservation decisions. This parameter can also be of central importance to better understanding the geographic variation in species richness. Thus, we now turn our focus from "how many spiders are here?" to "where can we find them?".

Spider Distribution Knowledge and Sampling Bias

Most of the discussion above refers to the so-called Linnaean knowledge shortfall, which is the degree of ignorance of the scientific community with regard to the number of species in the world (or, said another way, how many species are yet to be described — Hortal et al. 2015). In this topic, we return to the Wallacean knowledge shortfall, which was mentioned in the introduction. It is beyond any disagreement that we should aim to describe all the species on the planet (whether or not it is possible to do that is something we will talk about below). However, naming species is just part of the challenge, because knowing their geographic distribution is equally important if we intend to fully understand and preserve the world's biota (Whittaker et al. 2005).

A species distribution can be documented only through occurrence records, each one being a historical, material indication of the presence of a species in a locality. However, it is well known that distribution records are an incomplete representation of a species distribution range. Thus, biogeographers and macroecologists have developed tools specifically aimed at estimating the total distribution of a species, based on occurrence reports (Fortin et al. 2005). Herein, we choose to represent araneid and

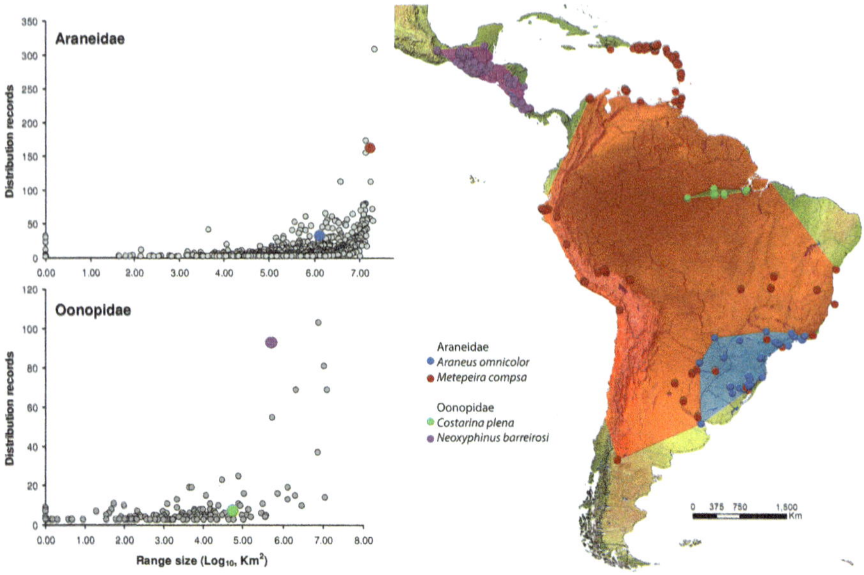

Fig. 1.6 Distribution range and occurrence records of Neotropical species of Araneidae and Oonopidae. The map show examples of distribution records, and minimum convex polygon estimation of distribution range, for two araneid and two oonopid species. The graphs show the relationship between estimated range size and occurrence record number for all species of both families. The species represented in the map are highlighted using the same colours in the graphs. Only species with at least three records are included, as range size estimation is impossible for species with fewer records

oonopid distribution ranges using the most intuitive of these methods, the minimum convex polygon (Burgman and Fox 2003). This method consists, in simple terms, of connecting the most external dots representing a species distribution, thereby drawing a polygon that contains all the known records of the species (Fig. 1.6). Despite the limitations of this approach (Gaston and Fuller 2009), we think it is good enough for assessing how well we know Neotropical spider distribution.

Perhaps the most impressive aspect of our knowledge of spider species distribution in the tropics is the high proportion of species known from a single occurrence record. This is the situation for 392 of the 1110 species of Araneidae in the Neotropics, which represent 35.3% of the family's species known from the region. One can easily presume these species are narrow endemics, with extremely restrictive habitat requirements and/or low dispersal capacity. This hypothesis might be true for part of them, but it is not difficult to find evidence that most of these apparently endemic species are merely insufficiently known. An example is *Hypognatha belem*, a small araneid originally known (for sure) only from its type locality in the eastern Brazilian Amazonia (Levi 1996). The apparent extremely restricted distribution of this species was subsequently refuted, when Santos (2002) reported this species from two Atlantic Forest localities near the Brazilian eastern coast, over 2000 km away. Later reports increased the distribution records of the species to seven, which delimited a 3.2 million km^2 distribution polygon. Thus, it is reasonable to see the high number of single-record species more as an index of our ignorance than as an indication of the level of endemism. In fact, the proportion of species known from a single record is even larger for Oonopidae (330, 55.5% of the 595 Neotropical species). Considering their lower availability in collections, and the large oonopid record-free zone in the Neotropics (Fig. 1.5b), it seems only logical to presume that oonopid species distributions are, on average, larger than we know today.

By estimating species ranges, we intended to analyse how accurate are species distributions based on the relationship between species range size and their number of occurrence records. We also think that, despite the influence of sampling bias, this procedure could permit inferences of biogeographic patterns for each family. The estimated distribution range and number of records for each species show an apparently exponential relationship, both for Araneidae and Oonopidae (Fig. 1.6). The absolute champion in distribution range size among both families is the garden spider *Argiope argentata* (Araneidae), which occurs over an estimated 20.7 million km^2 within the Neotropical countries (which together span 45 million km^2). The distribution range of the species is actually a bit larger than this, as it also occurs in southern USA (Levi 2004). Its high dispersal capacity by ballooning (Bell et al. 2005) and the generalist ecological requirements of this species explain its wide distribution, as it occurs in natural habitats as well as disturbed ones, such as agroecosystems and even domestic gardens. In fact, its ubiquity, high local abundance, and ease of observation and manipulation have made *Argiope argentata* one of the most studied behavioural models within spiders. Unfortunately, this species represents more an exception than the rule even among the better-known Araneidae in the Neotropics.

The data compiled from the taxonomic literature show that most Neotropical Araneidae and Oonopidae species have few occurrence records, but with wide variation in their estimated range size (Fig. 1.6). For Araneidae, species occur on average over 2.45 million km^2 (±3.86 km^2 standard deviation), and have 16.27 ± 23.04 occurrence records (calculated only for species with three or more records, for which the range size could be estimated). These figures are markedly above those observed for Oonopidae, which have 0.41 ± 1.8 million km^2 of range size based on 8.8 ± 15.25 distribution records. Despite the fact that Oonopidae are evidently lesser known than Araneidae, we cannot avoid speculating whether these numbers indicate a real biogeographic pattern. It makes sense to imagine that a group of tiny, predominantly litter-dwelling spiders with lower dispersal capacity (Bell et al. 2005) would show on average smaller distribution ranges than orb-weaving and ballooning spiders. In fact, oonopids seem to have much fewer species with wide distributions represented by a few records when compared with Araneidae (Fig. 1.6). This pattern would mean that Araneidae could be a good model for macroecological studies because of the wide distribution and high sampling coverage on a geographic scale for its species. On the other hand, perhaps groups such as Oonopidae could be better models for historical biogeography, because its species are on average more restricted geographically, better preserving the historical signal on their distributions. We think this question deserves attention in future biogeographic studies on spiders.

In conclusion, our data show that even more intensively studied araneomorph families such as Araneidae and Oonopidae are victims of the Wallacean knowledge shortfall. However, because the situation could be even worse for families particularly deficient of taxonomic study, Araneidae and Oonopidae are certainly among the better candidates for biogeographic studies involving spiders in the Neotropics. The uncertainty in species range estimation is probably higher for species with wide distribution ranges and very few distribution records, which are particularly common among Araneidae. A tight correlation between range size and distribution records should not be expected, because species can also vary regarding their spatial range occupancy (Gaston and Fuller 2009). However, it seems evident that species such as araneid *Pronous tuberculifer* (with only three records in a 10.9 million km^2 range) are undersampled. The estimated geographic distribution of species with such knowledge shortage must be seen with extreme caution, as the reliability of species range estimation depends on the number of distribution records.

Closing Remarks

In the preceding paragraphs, we intended to make it clear that Araneomorphae is a particularly diverse arthropod lineage in the Neotropics. We have shown through a species accumulation curve (Fig. 1.2a) that we are far from having described all the species in the group. Additionally, by focusing on two families, we demonstrated that even the described species are in general poorly known, as their geographic distribution is probably insufficiently represented by the currently published distribution records (Fig. 1.6). In this last topic, we hope to offer possible solutions to these limitations, as well as suggest that the data currently available could still be used for biogeographic, macroecological, and conservation studies.

With regard to the progress in the taxonomic description of Araneomorphae, it is obvious that sampling bias strongly affects our knowledge of this group. As shown above, a single researcher (H.W. Levi) had an impressive effect on the Araneidae species description rate through time (Fig. 1.2b). Something similar happened with the Oonopidae: an apparently species-poor family of minute, cryptic spiders revealed itself as particularly diverse after a few years of collective taxonomic effort by the *Goblin Spider PBI* team (Fig. 1.2c). A similar effect can be seen at a geographic scale, as we show that Panama is the better-known Neotropical country with regard to its araneomorph fauna (Fig. 1.3) because it received concentrated field sampling and taxonomic effort. We believe that the main message that can be taken from these examples is that taxonomy pays off. Even though a lot of Neotropical species are still waiting to be described, the focused effort of a few researchers can make a considerable difference in our knowledge within a moderate time period (see additional examples in Platnick and Raven 2013).

The recent progress on the description of Neotropical araneomorph spiders depicted above inevitably leads to the question: would it be possible to describe all the araneomorph spider species from the Neotropics in a reasonable time? By reasonable time we mean before human-caused environmental degradation leads most species to extinction before they are even described (see Costello et al. 2013). This apparently utopic goal could be attainable for moderately diverse groups such as spiders, especially considering that this is one of the few fully-catalogued animal lineages (World Spider Catalog 2017). The estimated total number of spider species worldwide varies from 90,000 to 170,000 (Coddington and Levi 1991; Planick 1999). A recent quantitative approach, based on a set of simple assumptions and on recent progress in systematics, reached an intermediate estimation of 120,000 species (Agnarsson et al. 2013). By naively applying the same equation to our data, we estimate that the Neotropics probably contain approximately 30,780 araneomorph spider species. Although there is much to debate concerning the parameters used to reach this estimation, it seems reasonable, considering the high proportion of newly described species in recent taxonomic monographs, to presume that we currently know only a third of the total araneomorph spider fauna. Accepting this estimate as true, if we maintain the description rate of 93 species per year, the last unknown Neotropical araneomorph species should be described by the year 2339. This figure does not really inspire much optimism, but it also leads us to an important question. Describing all the spider species is desirable, but it is really necessary in the short- to mid-term?

The incredible effort that will be necessary to completely describe the world's biota has stimulated consideration on the feasibility of reaching a level of knowledge that, although incomplete, could be sufficient to support other biodiversity studies and guide conservation decisions (Huber 2014). Once again, considering the dimensions of the known and estimated spider diversity, the group is a good candidate to reach this compromise goal, if taxonomic and sampling efforts are directed correctly. As predicted for other taxonomic groups (e.g., Fontaine et al. 2012), a considerable number of undescribed spider species have already been collected and await discovery in scientific collections. The results of H.W. Levi's monographs are an excellent example of this. Thus, much of the taxonomic effort to be applied in the future will certainly involve studying the unsorted material available in arachnid collections, particularly those located in Neotropical countries.

Despite the availability of unsorted spider specimens in collections, there is no doubt that additional field sampling will be necessary to find most undescribed species. As has been shown for several arthropods, species with large body size and wide distributions tend to be described first (Scheffers et al. 2012). In fact, a recent survey on the taxonomy of the araneomorph spider family Pholcidae indicated that most short-range species still wait collection in the field (Huber 2014). The central question is, where to collect? It is possible to imagine that sampling could be more fruitful if directed to poorly known places such as the record-empty zones shown herein for Araneidae and Oonopidae (Figs. 1.4 and 1.5). However, recent evidence shows that even better sampled and extremely species-rich areas, such as the Brazilian Atlantic forest, still hold a lot of undescribed and narrowly distributed species (Huber 2014, 2015, 2016). Thus, it seems that defining sampling priorities would need a more sophisticated approach.

As stressed above, knowing the biodiversity is more than simply cataloguing species, as their geographic distribution is also a parameter of high interest. As we see it, in the short- to mid-term it would be preferable to have only a fraction of the world's spider species described, if in exchange their geographic distribution is better known. Of course, there is no real conflict between describing species and characterizing their distribution range. In fact, taxonomic monographs (like Levi 2002) are the main source of both new species and additional distribution data. This kind of study should be more often complemented by the important, but undervalued, species lists and new distribution reports (such as Carvalho et al. 2017). Additionally, the maintenance and improvement of collection databases, such as GBIF (http://www.gbif.org/) and SpeciesLink (http://splink.cria.org.br/), could contribute to greater spider distribution knowledge. Of course, all of these require the work of taxonomists, which will continue to be the most essential assets in the mission to document and understand the world's biodiversity.

To attain a better representation of Neotropical spider distribution, a change in habits will probably be necessary. It is well known that biodiversity data are frequently spatially biased because taxonomists tend to sample near home (e.g., Moerman and Estabrook 2006) or in easily accessible places (Oliveira et al. 2016). Additionally, spatial bias can be generated and reinforced by researchers that collect repeatedly in places previously expected to be species-rich (Sastre and Lobo 2009) — a kind of biodiversity confirmation bias. Although field sampling is limited by financial factors, and even the better-known areas can still have taxonomic novelties, we think araneomorph spider sampling should be the subject of more planning, based on the knowledge already available. More importantly, we are aware that intensive sampling throughout the Neotropics would not be realistic, but perhaps creating more sampling hotspots (such as those in Figs. 1.5 and 1.6), evenly spread through the region, would be enough to allow reasonable estimates of species distributions.

We would like to finish with a positive message: the incompleteness of our current knowledge should not be seen as a barrier for macroecological, biogeographic, and (mainly) conservation decisions involving Neotropical araneomorph spiders. Even though data on spiders (as well as other invertebrates) are incomplete, they could be included in large-scale studies (Diniz-Filho et al. 2010). In addition, data on vertebrates and angiosperms, the preferred models at geographic scales, are not that much better (Oliveira et al. 2016). We think that assembling the data already available in the literature and scientific collections would be of paramount importance to assess our current knowledge of spiders, measure the limitations and deficiencies of this knowledge, and filter data partitions that could be reliably used to infer biological patterns or test biogeographic hypotheses, discounting the effects of sampling bias (Oliveira et al. 2017). Additionally, databases can also be useful to guide future sampling efforts (Funk et al. 2005). In short, considering all that has been done already regarding the study of Neotropical araneomorph spiders, it seems particularly important to know how much we in fact know at this moment.

Acknowledgements We are grateful to Marcelo O. Gonzaga and Carmen Viera for the opportunity of preparing this revision, for the confidence they placed in us and, most importantly, their patience waiting to receive the first draft. Thanks are also extended to Pedro H. Martins, for providing oonopid and araneid photos, and to Alexandre B. Bonaldo for reviewing the first draft of the text. The data discussed here is part of a long-term project from the authors, which has been continuously funded by CNPq (Procs. 407288/2013-9; 306222/2015-9 to AJS, and 301776/2004-0 to ADB), FAPEMIG (PPM-00651-15, AJS), FAPESP (2011/50689-0, ADB), and Instituto Nacional de Ciência e Tecnologia dos Hymenoptera Parasitóides da Região Sudeste Brasileira (http://www.hympar.ufscar.br/). AJS and ADB also received financial support from the *Goblin Spider PBI* (http://research.amnh.org/oonopidae). MOT was financially supported by an undergraduate research fellowship from PIBIC/CNPq, PRS from PRONOTURNO/FUMP/UFMG and a CAPES Master's Dissertation Fellowship, and UO from a CAPES doctoral grant.

References

Adis J, Bonaldo AB, Brescovit AD et al (2002) Arachnida at 'Reserva Ducke', central Amazonia/ Brazil. Amazoniana 17:1–14

Agnarsson I, Coddington JA, Kuntner M (2013) Systematics — progress in the study of spider diversity and evolution. In: Penney D (ed) Spider research in the 21st century. Siri Scientific press, Manchester, pp 58–111

Alligand B, Henrard A (2016) Sur l'observation de prédation par *Tapinesthis inermis* (Simon, 1882) (Araneae, Oonopidae) sur un psoque et mise à jour de sa distribution en France. Rev Arachnol 2:21–26

Amano T, Sutherland WJ (2013) Four barriers to the global understanding of biodiversity conservation: wealth, language, geographical location and security. Proc R Soc Lond B Biol Sci 280:20122649

Azevedo GHF, Faleiro BT, Magalhães ILF et al (2014) Effectiveness of sampling methods and further sampling for accessing spider diversity: a case study in a Brazilian Atlantic rainforest fragment. Insect Conserv Divers 7:381–391

Banks N (1929) Spiders from Panama. Bull Mus Comp Zool 69:53–96

Barriga JC, Moreno AG (2013) Listado de las arañas de Colombia (Arachnida: Araneae). Biota Colombiana 14:21–33

Basset Y, Cizek L, Cuénoud P et al (2012) Arthropod diversity in a tropical forest. Science 338:1481–1484

Bell JR, Bohan DA, Shaw EM, Weyman GS (2005) Ballooning dispersal using silk: world fauna, phylogenies, genetics and models. Bull Entomol Res 95:69–114

Blackledge T, Coddington JA, Agnarsson I (2009) Fecundity increase supports adaptive radiation hypothesis in spider web evolution. Commun Integr Biol 2:1–5

Boakes EH, McGowan PJ, Fuller RA et al (2010) Distorted views of biodiversity: spatial and temporal bias in species occurrence data. PLoS Biol 8:e1000385. https://doi.org/10.1371/journal.pbio.1000385

Bonaldo AB, Brescovit AD, Höfer H et al (2009a) A Araneofauna (Arachnida, Araneae) da Reserva Florestal Adolfo Ducke, Manaus, Amazonas, Brasil. In: Chilson EF, Magalhães C (eds) A fauna de Artrópodos da Reserva Florestal Adolpho Ducke. Estado atual do conhecimento taxonômico e biológico. Instituto Nacional de Pesquisa da Amazonia, Manaus, pp 201–217

Bonaldo AB, Carvalho LS, Pinto-da-Rocha R et al (2009b) Inventário e história natural dos aracnídeos da Floresta Nacional de Caxiuanã. In: Lisboa PLB (ed) Caxiuanã: desafios para a conservação de uma Floresta Nacional na Amazônia. Museu Paraense Emílio Goeldi, Belém, pp 577–621

Bond JA, Opell BD (1998) Testing adaptive radiation and key innovation hypotheses in spiders. Evolution 52:403–414

Bond JE, Hendrixson BE, Hamilton CA et al (2012) A reconsideration of the classification of the spider infraorder Mygalomorphae (Arachnida: Araneae) based on three nuclear genes and morphology. PLoS One 7:e38753. https://doi.org/10.1371/journal.pone.0038753

Brescovit AD (1999) Araneae. In: Joly CA, Bicudo CEM (eds) Biodiversidade do Estado de São Paulo, Brasil: síntese do conhecimento ao final do século XX, Invertebrados terrestres, vol 5. FAPESP, São Paulo, pp 47–56

Bryant EB (1940) Cuban spiders in the museum of comparative zoology. Bull Mus Comp Zool 86:247–532

Bryant EB (1948) The spiders of Hispaniola. Bull Mus Comp Zool 100:329–447

Burgman MA, Fox JC (2003) Bias in species range estimates from minimum convex polygons: implications for conservation and options for improved planning. Anim Conserv 6:19–28

Cafofo EG, Saturnino R, Santos AJ, Bonaldo AB (2013) Riqueza e composição em espécies de aranhas da Floresta Nacional de Caxiuanã. In: Lisboa PLB (ed) Caxiuanã: paraíso ainda preservado. Museu Paraense Emílio Goeldi, Belém, pp 539–561

Cardoso P, Erwin TL, Borges PA, New TR (2011) The seven impediments in invertebrate conservation and how to overcome them. Biol Conserv 144:2647–2655

Carnaval AC, Moritz C, Hickerson M et al (2009) Stability predicts diversity in the Brazilian Atlantic forest hotspot. Science 323:785–789

Carvalho LS, Martins PH, Schneider MC, Cabra-García JJ (2017) New records of spiders (Arachnida, Araneae) from the state of Roraima, northern Brazil. Check List 13. 10.15560/13.1.2040

Chickering AM (1946) The Salticidae of Panama. Bull Mus Comp Zool 97:1–474

Chickering AM (1968a) The genus *Dysderina* (Araneae, Oonopidae) in Central America and the West Indies. Breviora 296:1–37

Chickering AM (1968b) The genus *Ischnothyreus* (Araneae, Oonopidae) in Central America and the West Indies. Psyche 75:77–86

Chickering AM (1968c) The genus *Scaphiella* (Araneae, Oonopidae) in Central America and the West Indies. Psyche 75:135–156

Chickering AM (1968d) The genus *Triaeris* Simon (Araneae, Oonopidae) in Central America and the West Indies. Psyche 75:351–359

Chickering AM (1969) The genus *Stenoonops* (Araneae, Oonopidae) in Panama and the West Indies. Breviora 339:1–35

Chickering AM (1970) The genus *Oonops* (Araneae, Oonopidae) in Panama and the West Indies. Part 1. Psyche 77:487–512

Chickering AM (1971) The genus *Oonops* (Araneae, Oonopidae) in Panama and the West Indies. Part 2. Psyche 78:203–214

Chickering AM (1972) The genus *Oonops* (Araneae, Oonopidae) in Panama and the West Indies. Part 3. Psyche 79:104–115

Coddington JA, Levi HW (1991) Systematics and evolution of spiders (Araneae). Annu Rev Ecol Syst 22:565–592

Conniff R (2011) The species seekers: heroes, fools, and the mad pursuit of life on earth. WW Norton and Company, London

Connor EF, McCoy ED (2001) Species–area relationships. In: Levin S (ed) Encyclopedia of biodiversity, vol 5. Elsevier, London, pp 397–411

Costello MJ, May RM, Stork NE (2013) Can we name Earth's species before they go extinct? Science 339:413–416

Cox CB (2001) The biogeographic regions reconsidered. J Biogeogr 28:511–523

Craig CL, Bernard GD, Coddington JA (1994) Evolutionary shifts in the spectral properties of spider silks. Evolution 48:287–296

Cushing PE (2005) Introduction. In: Ubick D, Paquin P, Cushing PE et al (eds) Spiders of North America: an identification manual. American Arachnological Society, Poughkeepsie, pp 1–17

Dantas GPM, Cabanne GS, Santos FR (2011) How past vicariant events can explain the Atlantic Forest biodiversity? In: Grillo O, Venora G (eds) Ecosystems biodiversity. InTech, Rijeka, pp 429–442

DaSilva MB, Pinto-da-Rocha R, DeSouza AM (2015) A protocol for the delimitation of areas of endemism and the historical regionalization of the Brazilian Atlantic rain forest using harvestmen distribution data. Cladistics 31:692–705

Dimitrov D, Benavides Silva LR, Arnedo MA et al (2017) Rounding up the usual suspects: a standard target-gene approach for resolving the interfamilial phylogenetic relationships of ecribellate orb-weaving spiders with a new family–rank classification (Araneae, Araneoidea). Cladistics 33:221–250

Diniz-filho JAF, De Marco P Jr, Hawkins BA (2010) Defying the curse of ignorance: perspectives in insect macroecology and conservation biogeography. Insect Conserv Divers 3:172–179

ESRI (2017) How Kernel Density works. http://pro.arcgis.com/en/pro-app/tool-reference/spatial-analyst/how-kernel-density-works.htm. Accessed 15 Apr 2017

Fannes W, De Bakker D, Loosveldt K et al (2008) Estimating the diversity of arboreal oonopid spider assemblages (Araneae, Oonopidae) at Afrotropical sites. J Arachnol 36:322–330

Finch OD, Blick T, Schuldt A (2008) Macroecological patterns of spider species richness across Europe. Biodivers Conserv 17:2849–2868

Foelix R (2011) Biology of spiders, 3rd edn. Oxford University Press, Oxford

Fonseca GAB, Rylands A, Paglia A et al (2004) Atlantic forest. In: Mittermeier RA, Gil PR, Hoffman M et al (eds) Hotspots revisited. Earth's biologically richest and most endangered terrestrial ecoregions. CEMEX, Mexico City, pp 84–91

Fontaine B, Perrard A, Bouchet P (2012) 21 years of shelf life between discovery and description of new species. Curr Biol 22:R943–R944

Foord SH, Dippenaar-Schoeman AS, Haddad CR (2011) South African spider diversity: African perspectives on the conservation of a mega-diverse group. In: Grillo O, Venora G (eds) Changing diversity in changing environment. Intech, Rijeka, pp 163–182

Fortin MJ, Keitt TH, Maurer BA et al (2005) Species' geographic ranges and distributional limits: pattern analysis and statistical issues. Oikos 108:7–17

Funk VA, Richardson KS, Ferrier S (2005) Survey-gap analysis in expeditionary research: where do we go from here? Biol J Linn Soc Lond 85:549–567

Garrison NL, Rodriguez J, Agnarsson I et al (2016) Spider phylogenomics: untangling the spider tree of life. PeerJ 4:e1719. https://doi.org/10.7717/peerj.1719

Gaston KJ, Fuller RA (2009) The sizes of species' geographic ranges. J Appl Ecol 46:1–9

Godman FD (1915) Introductory volume. In: Godman FD, Salvim O (eds) Biologia Centrali-Americana. R.H. Porter, London

Grismado CJ, Ramírez MJ (2013) The New World goblin spiders of the new genus *Neotrops* (Araneae: Oonopidae), Part 1. Bull Am Mus Nat Hist 383:1–150

Hagen JB (1990) Problems in the institutionalization of tropical biology: the case of the Barro Colorado Island biological laboratory. Hist Philos Life Sci 12:225–247

Heyer WR, Coddington JA, Kress WJ et al (1999) Amazonian biotic data and conservation decisions. Cienc Cult 51:372–385

Hillebrand H (2004) On the generality of the latitudinal diversity gradient. Am Nat 163:192–211

Hoffmeister CH, Ferrari A (2016) Areas of endemism of arthropods in the Atlantic Forest (Brazil): an approach based on a metaconsensus criterion using endemicity analysis. Biol J Linn Soc Lond 119:126–144

Hopkins MJ (2007) Modelling the known and unknown plant biodiversity of the Amazon Basin. J Biogeogr 34:1400–1411

Hortal J, de Bello F, Diniz-Filho JAF et al (2015) Seven shortfalls that beset large-scale knowledge of biodiversity. Annu Rev Ecol Evol Syst 46:523–549

Huber BA (2014) Progress and prospects in taxonomy: what is our goal and are we ever going to reach it? J Arachnol 42:142–147

Huber BA (2015) Small scale endemism in Brazil's Atlantic Forest: 14 new species of *Mesabolivar* (Araneae, Pholcidae), each known from a single locality. Zootaxa 3942:1–60

Huber BA (2016) Spider diversity and endemism in a south American hotspot: 20 new species of *Carapoia* (Araneae: Pholcidae) from Brazil's Atlantic Forest. Zootaxa 4177:1–69

Jocqué R, Alderweireldt M, Dippenaar-Schoeman A (2013) Biodiversity: an African perspective. In: Penney D (ed) Spider research in the 21st century: trends and perspectives. Siri Scientific Press, Manchester, pp 18–57

Keyserling E (1877) Amerikanische Spinnenarten aus den Familien der Pholcoidae, Scytodoidae und Dysderoidae. Verh Zool Bot Ges Wien 27:205–234
Korenko S, Hamouzová K, Pekár S (2014) Trophic niche and predatory behavior of the goblin spider *Triaeris stenaspis* (Oonopidae): a springtail specialist? J Arachnol 42:74–78
Leibensperger LB (2016) Herbert Walter Levi (1921–2014) and Lorna Levi (1928–2014). Breviora 551:1–37
Levi HW (1964) Nineteenth century South American araneology. Pap Avulsos Zool 16:9–19
Levi HW (1996) The American orb weavers *Hypognatha*, *Encyosaccus*, *Xylethrus*, *Gasteracantha*, and *Enacrosoma* (Araneae, Araneidae). Bull Mus Comp Zool 155:89–157
Levi HW (2002) Keys to the genera of araneid orbweavers (Araneae, Araneidae) of the Americas. J Arachnol 30:527–562
Levi HW (2004) Comments and new records for the American genera *Gea* and *Argiope* with the description of new species (Araneae: Araneidae). Bull Mus Comp Zool 158:47–65
Levi HW (2007) The orb weaver genus *Mangora* in South America (Araneae, Araneidae). Bull Mus Comp Zool 159:1–144
Linnaeus C (1758) Systema naturae per regna tria naturae, secundum classes, ordines, genera, species cum characteribus differentiis, synonymis, locis. Editio decima, reformata. Holmiae
Lise AA, Kesster CC, Silva ELC (2015) Revision of the orb-weaving spider genus *Verrucosa* McCook, 1888 (Araneae, Araneidae). Zootaxa 3921:1–105
Magalhães ILF, Martins PH, Nogueira AA et al (2017) Finding hot singles: matching males to females in dimorphic spiders (Araneidae: *Micrathena*) using phylogenetic placement and DNA barcoding. Invertebr Syst 31:8–36
Melo AS, Rangel TFL, Diniz-Filho JAF (2009) Environmental drivers of beta-diversity patterns in New-World birds and mammals. Ecography 32:226–236
Meng K, Li S, Murphy RW (2008) Biogeographical patterns of Chinese spiders (Arachnida: Araneae) based on a parsimony analysis of endemicity. J Biogeogr 35:1241–1249
Mittermeier R, Mittermeier CG, Gil PR et al (2003) Wilderness: Earth's last wild places. Conservation International, Washington, DC
Mittermeier R, Schipper J, Davidse G et al (2004) Mesoamerica. In: Mittermeier RA, Gil PR, Hoffman M et al (eds) Hotspots revisited. Earth's biologically richest and most endangered terrestrial ecoregions. CEMEX, Mexico City, pp 103–112
Moerman DE, Estabrook GF (2006) The botanist effect: counties with maximal species richness tend to be home to universities and botanists. J Biogeogr 33:1969–1974. https://doi.org/10.1111/j.1365-2699.2006.01549.x
Moss DF, Feitosa NM, Bonaldo AB et al (2016) Description of eleven new species of the goblin spider genus *Neoxyphinus* Birabén, 1953 (Araneae, Oonopidae). Zootaxa 4098:95–133
Nentwig W (1993) Spiders of Panama: biogeography, investigation, phenology, check list, key and bibliography of a tropical spider fauna. Sandhill Crane Press, Gainesville
Nentwig W, Blick T, Gloor D et al (2017) Spiders of Europe. www.araneae.unibe.ch. Accessed 22 Feb 2017
Oliveira U, Brescovit AD, Santos AJ (2015) Delimiting areas of endemism through kernel interpolation. PLoS One 10:e0116673
Oliveira U, Paglia AP, Brescovit AD et al (2016) The strong influence of collection bias on biodiversity knowledge shortfalls of Brazilian terrestrial biodiversity. Glob Ecol Biogeogr 22:1232–1244
Oliveira U, Brescovit AD, Santos AJ (2017) Sampling effort and species richness assessment: a case study on Brazilian spiders. Biodivers Conserv (in press). https://doi.org/10.1007/s10531-017-1312-1
Penney D (2013) Spider research in the 21st century: trends and perspectives. Siri Scientific Press, Manchester
Penney D, Selden PA (2011) Fossil spiders: the evolutionary history of a mega-diverse order. Siri Scientific Press, Manchester
Petrunkevitch A (1911) A synonymic index-catalogue of spiders of North, Central and South America with all adjacent islands, Greenland, Bermuda, West Indies, Terra del Fuego, Galapagos, etc. Bull Am Mus Nat Hist 29:1–791

Petrunkevitch A (1930) The spiders of Porto Rico, vol 31. Connecticut Academy of Arts and Sciences, New Haven, pp 1–191

Platnick NI (1999) Dimensions of biodiversity: targeting megadiverse groups. In: Cracraft J, Grifo T (eds) The living planet in crisis: biodiversity science and policy. Columbia University Press, New York, pp 33–52

Platnick NI, Berniker L (2014) The goblin spider genus *Costarina* (Araneae, Oonopidae), Part 3. Am Mus Novit 3819:1–67

Platnick NI, Dupérré N (2010) The Andean goblin spiders of the new genera *Niarchos* and *Scaphios* (Araneae, Oonopidae). Bull Am Mus Nat Hist 345:1–120. https://doi.org/10.1206/727.1

Platnick NI, Dupérré N (2011a) The Andean goblin spiders of the new genera *Paradysderina* and *Semidysderina* (Araneae, Oonopidae). Bull Am Mus Nat Hist 364:1–121

Platnick NI, Dupérré N (2011b) The Andean goblin spiders of the new genus *Scaphidysderina* (Araneae, Oonopidae), with notes on *Dysderina*. Am Mus Novit 3712:1–51

Platnick NI, Gertsch WJ (1976) The suborders of spiders: a cladistic analysis (Arachnida, Araneae). Am Mus Novit 2607:1–18

Platnick NI, Raven RJ (2013) Spider systematics: past and future. Zootaxa 3683:595–600

Platnick NI, Berniker L, Bonaldo AB (2013a) The South American goblin spider genera *Dysderina* and *Tridysderina* (Araneae, Oonopidae). Am Mus Novit 3772:1–52

Platnick NI, Dupérré N, Berniker L et al (2013b) The goblin spider genera *Prodysderina*, *Aschnaoonops*, and *Bidysderina* (Araneae, Oonopidae). Bull Am Mus Nat Hist 373:1–102

Platnick NI, Berniker L, Víquez C (2014) The goblin spider genus *Costarina* (Araneae, Oonopidae), Part 2: the Costa Rican fauna. Am Mus Novit 3794:1–75

Rodríguez-Mahecha JV, Salaman P, Jørgensen O et al (2004a) Tropical Andes. In: Mittermeier RA, Gil PR, Hoffman M et al (eds) Hotspots revisited. Earth's biologically richest and most endangered terrestrial ecoregions. CEMEX, Mexico City, pp 73–79

Rodríguez-Mahecha JV, Salaman P, Jørgensen O et al (2004b) Tumbes-Chocó-Magdalena. In: Mittermeier RA, Gil PR, Hoffman M et al (eds) Hotspots revisited. Earth's biologically richest and most endangered terrestrial ecoregions. CEMEX, Mexico City, pp 80–84

Rubio GD, Corronca JA, Damborsky MP (2008) Do spider diversity and assemblages change in different contiguous habitats? A case study in the protected habitats of the humid chaco ecoregion, northeast Argentina. Environ Entomol 37:419–430

Saiter FZ, Brown JL, Thomas WW et al (2016) Environmental correlates of floristic regions and plant turnover in the Atlantic Forest hotspot. J Biogeogr 43:2322–2331. https://doi.org/10.1111/jbi.12774

Santos AJ (2002) Description of the male of *Hypognatha belem* (Araneae, Araneidae). Iheringia Ser Zool 92:91–92

Sastre P, Lobo JM (2009) Taxonomist survey biases and the unveiling of biodiversity patterns. Biol Conserv 142:462–467

Saturnino R, Rodrigues BVB, Bonaldo AB (2015) *Alpaida* (Araneae: Araneidae) from the Amazon Basin and Ecuador: new species, new records and complementary descriptions. Fortschr Zool 32:241–256

Savory TH (1961) Spiders, men, and scorpions: being the history of arachnology. University of London Press, London

Scheffers BR, Joppa LN, Pimm SL et al (2012) What we know and don't know about Earth's missing biodiversity. Trends Ecol Evol 27:501–510

Selander RB, Vaurie P (1962) A gazetteer to accompany the Insecta volumes of the biologia Centrali-Americana. Am Mus Novit 2099:1–70

Silva D (1996) Species composition and community structure of Peruvian rainforest spiders: a case study from a seasonally inundated forest along the Samiria river. Rev Suisse Zool, vol hors serie:597–610

Silva S, Moraes-Barros N, Ribas CC et al (2012) Divide to conquer: a complex pattern of biodiversity depicted by vertebrate components in the Brazilian Atlantic Forest. Biol J Linn Soc Lond 107:39–55

Sobral M, Stehmann JR (2009) An analysis of new angiosperm species discoveries in Brazil (1990–2006). Taxon 58:227–232

Wheeler WH, Coddington JA, Crowley LM et al (2017) The spider tree of life: phylogeny of Araneae based on target-gene analyses from an extensive taxon sampling. Cladistics 2016:1–43. https://doi.org/10.1111/cla.12182

Whittaker RJ, Araújo MB, Jepson P et al (2005) Conservation biogeography: assessment and prospect. Divers Distrib 11:3–23

World Spider Catalog (2017) World Spider Catalog. Natural History Museum Bern, version 18.0. http://wsc.nmbe.ch. Accessed 8 Apr 2017

Xu X, Liu FX, Chen J et al (2015) A genus-level taxonomic review of primitively segmented spiders (Mesothelae, Liphistiidae). ZooKeys 488:121–151

Zhang Z-Q (2013) Phylum Arthropoda. In: Zhang, Z-Q (ed) animal biodiversity: an outline of higher-level classification and survey of taxonomic richness (addenda 2013). Zootaxa 3703:1–82

Zhang JX, Maddison WP (2012) New euophryine jumping spiders from Central and South America (Araneae: Salticidae: Euophryinae). Zootaxa 3578:1–35

Zúñiga-Vega CM (1980) Lista anotada de especies de arañas de Costa Rica. Brenesia 18:301–352

Chapter 2
Behavior and Biology of Mygalomorphae

Fernando Pérez-Miles and Carlos Perafán

Abstract The infraorder Mygalomorphae comprises almost 3000 species. It is widely distributed all over the world, and about one third of the species included in this infraorder is represented in the Neotropics. The knowledge on several aspects of the biology of Mygalomorphae is relatively scarce in comparison with the Araneomorphae. However, recent studies developed all over the world, but especially in the Neotropical region, are continuously contributing to add information on foraging strategies, communication, reproductive biology, habitat selection, and defenses against natural enemies, minimizing the differences in the status of knowledge between these groups. In this chapter, we review some the main topics on Mygalomorphae biology, including behavior and ecology.

The infraorder Mygalomorphae comprises about 3000 species of spiders included in 16 families (Bond et al. 2012; World Spider Catalog 2017) and distributed all over the world. This diverse group of relative large and long-lived spiders includes tarantulas, trap-door spiders, sheet-web spiders and others that build diverse silk constructions. Mygalomorphae is an ancient lineage, considered the sister group of a more highly diverse and better known group of spiders, the Araneomorphae. Mygalomorphae retain several features considered primitive for spiders, such as two pairs of book lungs, paraxial chelicerae, simple silk spigots, and biomechanically weak silk (Bond et al. 2012; Garrison et al. 2016). However, several characteristics support the monophyly of the infraorder: the presence of labial and maxillary cuspules, fewer palpal bulb sclerites, the presence of sternal sigillae, segmentation of posterior lateral spinnerets, reduction of the anterior lateral spinnerets, and the loss of anterior median spinnerets (Raven 1985; Goloboff 1993).

The fossil record of the Mygalomorphae lineage extends back to the lower Triassic Period, but molecular studies suggest inter-familiar divergences may be as old as 300 million years (My) (Ayoub and Hayashi 2009). The infraorder is widely distributed and particularly diverse in the Neotropics, southern Africa, and Australasia; however, mygalomorphs are sedentary and have very limited dispersal mechanisms, being excellent subjects for historical biogeographical studies.

F. Pérez-Miles (✉) • C. Perafán
Sección Entomología, Facultad de Ciencias, Universidad de la República,
Iguá 4225, 11400 Montevideo, Uruguay
e-mail: myga@fcien.edu.uy

C. Viera, M.O. Gonzaga (eds.), *Behaviour and Ecology of Spiders*,
https://doi.org/10.1007/978-3-319-65717-2_2

Since early taxonomic papers, Mygalomorphae has been considered a monophyletic group (Bond et al. 2012); however, the infraorder shows several taxonomic difficulties because of its relative conservative morphology and the usual homoplastic characters (Bond and Opell 2002; Hedin and Bond 2006). Raven (1985) made the first cladistic analysis of the group, Eskov and Zonshtein (1990) evaluated Raven's hypothesis using some morphological characters, and later Goloboff (1993) reanalyzed the phylogeny of the infraorder with computational tools. More recently, the molecular studies by Hedin and Bond (2006) and Bond et al. (2012) were important contributions. However, the monophyly of some families and the inter-familiar relationships remain poorly supported, probably as a consequence of taxonomic difficulties for identification of species; knowledge on the biology of Mygalomorphae is relatively scarce in comparison with that of the Araneomorphae.

The growing fashion of maintaining spiders in captivity as pets involves negative factors such as illegal commerce. Conversely, awareness of the importance of their conservation has stimulated several studies of mygalomorph biology in different parts of the world. In this chapter, we review the main aspects of Mygalomorphae biology, including behavior and ecology.

Communication, Sex and Reproductive Biology

Almost all mygalomorph spiders have poor vision (Land 1985, Dippenaar-Schoeman 2002), which is probably related mainly to their nocturnal habits. Contact pheromones are widespread in sexual communication by spiders (Schultz 2004, 2013; Gaskett 2007). These chemical signals are usually associated with silk threads. Tactile-chemical and vibration signals are certainly the main channels of communication among mygalomorphs, and several sensory organs such as setae and trichobothria have been described for the group (Barth 1985; Guadanucci 2012; Foelix et al. 2013). Tactile-chemical communication was suggested in Mygalomorphae by Baerg (1958), but remained controversial until the 1970s, when Platnick (1971) discredited its existence, probably following the ideas of Petrunkevitch (1911) and Baerg (1928).

In recent decades, several authors have found that under laboratory conditions, male courtship is usually elicited by the silk threads of a female's web, so tactile-chemical communication seems to be widespread throughout the infraorder (Minch 1979; Costa and Pérez-Miles 1992; Pérez-Miles and Costa 1992; Coyle and Icenogle 1994; Prentice 1997; Schillington and Verrell 1997; Costa and Pérez-Miles 1998; Yañez et al. 1999; Costa and Pérez-Miles 2002; Postiglioni and Costa 2006; Pérez-Miles et al. 2007; Ferretti and Ferrero 2008; Almeida-Silva et al. 2008; Copperi et al. 2012; Ferretti et al. 2013). However, the persistence of tactile-chemical signals in natural conditions is poorly known; consequently, it has been reported only for Theraphosidae in a recent study by Costa et al. (2015). These authors found long-term persistence and water resistance of such sexual cues in *Eupalaestrus weijenberghi*, lasting about 2 months in outdoor conditions.

Information about the persistence of chemical signals from female spiders under natural conditions provides fundamental data for understanding how males respond to such cues over time in the field, and for deducing the sexual tactics involved in mate searching by both sexes. Climatic factors such as sun, wind, dew, and rain are expected to wash or degrade the semiochemicals, as was suggested by Dondale and Hegdeckar (1973), Wilder et al. (2005), Gaskett (2007), and Baruffaldi et al. (2010). Unexpectedly, Costa et al. (2015) found a long-term persistence (46–55 days) and water resistance of female sex cues in the tarantula *Eupalaestrus weijenberghi* (Thorell 1894). This persistence is unexpectedly high, considering the persistence of pheromone in the wolf spider *Schizocosa malitiosa* (Tullgren 1905) outdoors is about a week (Baruffaldi et al. 2010). The water resistance of tarantula cues seems to be related with the fact that males of some species intently search for females during stormy and rainy days (Pérez-Miles et al. 2005). Conversely, the only pheromone reported for a theraphosid is a polar substance soluble in acetonitrile (Fukushima et al. 2003).

Although tactile-chemical communication plays an important role during searching and during early stages of courtship, the main communication channel among tarantulas seems to be vibratory signals, including seismic communication in burrowing species (Quirici and Costa 2005, 2007) and vibrations conducted through the silk in the sheet-tube web of mygalomorphs (Costa and Pérez-Miles 1998; Ferretti et al. 2013).

Sperm induction is the maneuver through which the male transfers sperm from the genital opening to the palpal copulatory organs. It is usually performed by males soon after maturation, and they can recharge their palpal organs several times (e.g., Baerg 1958; Pérez-Miles et al. 2005). In some species, with only one sperm induction the male is capable of mating with more than one female (Pérez-Miles et al. 2007). Usually, males which perform more than one sperm induction during their lives copulate more frequently than males which perform sperm induction only once (Perez-Miles et al. 2007). Sperm induction is achieved by depositing a sperm drop onto a sperm web (Baerg 1958; Costa and Pérez-Miles 2002). The sperm is then absorbed by the palpal organ, in which it is stored until mating. After this maneuver, adult males change their sedentary life-style to become wanderers in search of females.

Courtship and mating usually take place during warm months, in spring, summer, or fall, with some exceptions (Coyle and Icenogle 1994; Costa and Pérez-Miles 1998, 2002; Pérez-Miles et al. 2005). Male courtship mainly includes body vibrations, palpal drumming, and leg tapping; depending on the family it can also include leg beating, scratching, stretching, spasmodic beating, and brushing. A detailed description of these behaviors as well as the comparison with their nomination by several previous authors was given by Ferretti et al. (2013, see Table 2.1). Females play an important role in courtship by indicating their disposition to mating or by rejecting males. Receptive responses can include body vibrations and leg tapping, while rejection responses include "piston behavior" (forward and backward movements of the female in the burrow) that has been described for Dipluridae and Theraphosidae (Coyle and O'Shields 1990; Prentice 1992; Stradling 1994; Yañez

Table 2.1 Common sexual behaviors of Mygalomorphae: brief description, equivalencies of terms across authors, and families

Behavior	Brief description	Equivalency among authors	Families
Male Courtship Behaviors			
Palpal drumming	Up and down alternating movements of pedipalps	*tapping* (Coyle and O'Shields 1990) *Pawing* (Coyle and Icenogle 1994) *pedipalp drumming* (Stradling 1994; Yañez et al. 1999) *palpating* (Ferretti and Ferrero 2008)	Antrodiaetidae, Mecicobothriidae, Dipluridae, Theraphosidae
Leg tapping	Vigorous hits with legs I and sometimes with legs II or both against the substrate, the female's silk mat, or the female's body	*tapping* (Coyle and O'Shields 1990; Shillington and Verrel 1997; Punzo and Henderson 1999) *pawing front legs* (Coyle and Icenogle 1994) *leg drumming* (Yañez et al. 1999) *beating* (Ferretti et al. 2011)	Antrodiaetidae, Mecicobothriidae, Dipluridae, Nemesiidae, Theraphosidae, Microstigmatidae
Leg beating	Quick upward and downward movements of the legs, with tibia, metatarsi, with extended tarsi and beating and scraping female coxae	Ferretti et al. (2012)	Microstigmatidae
Scratching	Extending legs I or pedipalps, touching the substrate, and then moving the leg backward over the substrate, removing the soil and piling it	Bertani et al. (2008b), Ferretti et al. (2011)	Nemesiidae, Theraphosidae
Stretching	Legs I upward and downward to contact the web, flexing reaching an angle of 45° between femur-patellae	*quivering* (Ferretti et al. 2012)	Microstigmatidae, Nemesiidae
Body vibrations	High-frequency movement of the body or legs, usually caused by the inward contractions of legs III with femora positioned almost vertically	*twitching/quivering/body jerking* (Coyle and O'Shields 1990) *vibratory movements* (Stradling 1994) *quivering* (Shillington and Verrel 1997) *stridulating behavior* (Prentice 1992) *shaking body* (Punzo and Henderson 1999) *shaking* (Yañez et al. 1999)	Mecicobothriidae, Dipluridae, Theraphosidae
Spasmodic beats	Extending legs II or III and making vigorous backward and forward movements, reaching the legs of female	Pérez-Miles and Costa (1992), Costa and Pérez-Miles (2002), Quirici and Costa (2005), Pérez-Miles et al. (2007), Ferretti and Ferrero (2008)	Theraphosidae Nemesiidae

(continued)

Table 2.1 (continued)

Behavior	Brief description	Equivalency among authors	Families
Brushing	Raising, flexing and extending legs I, II and pedipalps and scratching against those of the female	*leg fencing* (Coyle and O'Shields 1990, Shillington and Verrel 1997) *gentle movements* (Ferretti et al. 2011)	Dipluridae, Nemesiidae, Theraphosidae
Female Courtship Behaviors			
Body vibrations	Quick (high-frequency) vibratory movements of the entire body or legs	*twitching/quivering/body jerking* (Coyle and O'Shields 1990) *vibratory movements* (Stradling 1994) *shaking* (Yañez et al. 1999) *body jerks* (Ferretti et al. 2011)	Dipluridae, Theraphosidae
Leg tapping	Leg flexing, lifting and lowering, contacting the ground	*drumming forelegs* (Prentice 1992) *active female display* (Costa and Pérez-Miles 2002) *faint tapping* (Bertani et al. 2008a)	Theraphosidae
Male Mating Behaviors			
Clasping	The male use claspers on his legs I to clasp the female's fangs, pedipalps, or legs when she raise her body up	Coyle and Icenogle (1994), Costa and Pérez-Miles (1998, 2002), Coyle and O'Shields (1990), Ferretti et al. (2011), Ferretti et al. (2012)	Antrodiaetidae, Mecicobothriidae, Dipluridae, Nemesiidae, Microstigmatidae, Theraphosidae
Palpal boxing	Alternating up and down movement of pedipalps, contacting the female's sternum	*pedipalp boxing* (Yañez et al. 1999) *palpal movements* (Costa and Pérez-Miles 2002)	Theraphosidae, Nemesiidae, Microstigmatidae
Biting fangs	Male and female interweave forelegs and open chelicerae and fangs; then the male presses the female fangs with his own	Postiglioni and Costa (2006)	Theraphosidae
Female Mating Behaviors			
Catalepsis	An immobile state during and after mating, in which females turn quiescent and let a male push and pull her around	*cataleptic state* (Coyle and O'Shields 1990) *quiescence* (Pérez-Miles et al. 2007) *passive state* (Costa and Pérez-Miles 1998) *unmoving* (Shillington and Verrel 1997)	Mecicobothriidae, Dipluridae, Theraphosidae, Microstigmatidae
Body jerks	High-amplitude twitching of all legs and pedipalps	*quivering* (Coyle and O'Shields 1990)	Nemesiidae, Dipluridae

Reproduced from Ferretti et al. 2013, with authorization of the author

Fig. 2.1 Couple of the theraphosid *Grammostola quirogai* in late stages of courtship, adopting mating position (female at left, male at right) (Photo: A. Laborda)

et al. 1999; Costa and Pérez-Miles 2002; Bertani et al. 2008b; Ferretti 2011; Ferretti et al. 2013).

Just prior to mating, the male performs body movements, downward pushing the female body, and the female performs threat-like behavior (raising the carapace and opening her fangs). In species with male tibial apophyses, the previous behaviors are followed by male clasp of female open fangs with their tibial apophyses. Then the male usually pushes the female and raises her body, and extends his palps (Fig. 2.1). Frequently at this time the male performs palpal drumming on female body and legs. After that, the female arches backwards with a dorsal flexion of her body that can reach a 90° angle between her carapace and abdomen. Copulation is brief, with few, alternate palpal insertions.

Prey Capture and Locomotion

Mygalomorphs have limited dispersal mechanisms and low vagility (Janowski-Bell and Horner 1990; Raven 2010; Ferretti et al. 2014). Juveniles and females are usually sedentary, while adult males increase their motility during mate search in the reproductive season (Coyle and Icenogle 1994; Prentice 1997; Punzo and Henderson 1999; Costa and Pérez-Miles 2002). Females and juveniles only leave their burrows for short distances and during brief time periods, usually for prey capture and for discarding prey rests (Dippenaar-Schoeman 2002; Alvarez et al. 2016). Females and juveniles usually remain at the entrance of the burrow and catch prey that wander near the retreat (Fig. 2.2). Some mygalomorphs extend the prey detection area by using silk threads with twigs, grass, or debris around the burrow entrances (Main 1978). Trap-door mygalomorphs do not abandon the burrow to hunt. The claws of the fourth legs keep a firm grip on the rim, allowing them to capture prey from the entrance by partially opening the door (Dippenaar-Schoeman 2002). A few families of mygalomorph spiders use the funnel or sheet webs for prey capture, such as diplurids and some Nemesiids and Mecicobothriids.

Fig. 2.2 Prey capture of tarantulas. (**a**) *Eupalaestrus weijenberghi* capturing *Diloboderus abderus* (Coleoptera) at the entrance of its burrow. (**b**) *Grammostola quirogai* capturing male *Blaptica dubia* (Blattaria) (Photo **b**: A. Laborda)

The mechanics of walking among mygalomorphs is similar to the diagonal rhythm described for araneomorphs (Wilson 1967). They move two sets of legs alternately; for example, right legs I and III together with left legs II and IV. Kinematics and energetics of the locomotion among mygalomorphs have been mainly studied on theraphosids (Biancardi et al. 2011; Booster et al. 2015; Grossi et al. 2016).

Several Mygalomorphae families have specialized adhesive setae called scopula, and claw tufts on the ventral surfaces of distal leg segments (tarsi, metatarsi). These setae are related to adhesion for prey capture and/or with climbing ability on vertical smooth surfaces (Fig. 2.3) (Homann 1957; Rovner 1978, 1980; Dunlop 1995;

Fig. 2.3 *Grammostola quirogai* climbing on a vertical glass plate

Foelix 2011; Pekar et al. 2011; Foelix et al. 2012; Niederegger 2013; Wolff and Gorb 2012a; Wolff et al. 2013; Wolff and Gorb 2015; Lapinski et al. 2015; Eggs et al. 2015). The mechanics of setae adhesion has been intensively studied in spiders (Homann 1957; Foelix and Chu-Wang 1975; Hill 1977; Rovner 1978; Kesel et al. 2003, 2004; Foelix 2011; Niederegger and Gorb 2006; Varenberg et al. 2010; Wolff and Gorb 2012a, b; Wolff et al. 2013; Niederegger 2013; Wohlfart et al. 2014; Wolff and Gorb 2015; Lapinski et al. 2015; Pérez-Miles et al. 2015). Scopula and claw tuft setae are covered on the proximal face by microtrichia, which provide thousands of contact points for adhesion. It has been proposed that when the leg pulls, setae are curved and the adhesive face of each one contacts with the surface, producing adhesion. When the leg pushes, setae are curved in the opposite direction and adhesion does not occur. However, occasional observations and also some recent studies (Pérez-Miles et al. 2015) show that adhesion is produced both when legs pull or push. Consequently, adhesive setae could play important roles both for prey capture and also for climbing. These results enhanced further studies on the detailed morphology of adhesive setae. We have observed that the adhesive setae of the distal tarsus have microtrichia on the proximal face, while the rest of the tarsus have microtrichia on the distal face, in agreement with the pulling–pushing adhesion mechanism (Pérez-Miles et al., *in prep.*).

Hairy attachment devices of spiders have traditionally been interpreted as dry adhesion devices relying on the Van Der Waals interactions between setae spatulae and substrate (Autumn et al. 2000; Kesel et al. 2003). Scopula adhesion is not supplemented by any secretory substance but facilitated by the humidity of the environment (Homann 1957; Huber et al. 2005). Such dry adhesion was seriously questioned when Gorb et al. (2006) proposed the secretion of silk from theraphosid legs as an additional adhesive mechanism for climbing on vertical smooth surfaces. They supported their proposal by indicating the presence of silk lines associated to spider legs and the occurrence of specialized setae on spider tarsi.

Pérez-Miles et al. (2009) experimentally tested this possibility in theraphosid spiders with the spinnerets sealed with wax. These authors did not find silk line in footprints, and they interpreted the specialized tarsal setae as sensory traits. The controversy continued when Rind et al. (2011) tested climbing on smooth vertical surfaces in the theraphosid *Grammostola rosea* (Walckenaer 1837). After the spider had been gently shaken the spider, silk threads were supposedly extruded from tarsal 'spigots'. These authors also proposed that those fine silk threads would prevent falls when the spider slips when walking on smooth surfaces. In contrast with Pérez-Miles et al. (2009) and Pérez-Miles and Ortíz-Villatoro (2012), in all those experiments the tarantulas had normally functioning spinnerets.

Peattie et al. (2011) also studied the adhesion of spiders, solpugids, and mites during locomotion. They found fluid footprints when arachnids climbed on vertical surfaces. In experiments with the tarantula *G. rosea* they also found threads apparently released from tarsal hairs, but they could not confirm their silken nature. Most recently, Foelix et al. (2011, 2012) provided good morphological evidence that the alleged tarsal spigots (Gorb et al. 2006; Rind et al. 2011) are sensory hairs, probably contact chemoreceptors. Finally, Pérez-Miles and Ortiz (2012) tested tarsal silk secretion in four species of tarantulas (one arboreal and three terrestrial species) under extreme conditions of climbing on smooth surfaces, with and without sealed spinnerets. Their study with confocal microscopy revealed the presence of fluid footprints but no silk traces when spider spinnerets were sealed. Whether fluid footprints are a consequence of humidity from the environment or a special secretion of the spider remains to be tested.

Habitat

Mygalomorphae are widely distributed throughout the world, but they are especially diverse and abundant in tropical regions and temperate austral regions of South America, Southern Africa and Australasia (Raven 1985; World Spider Catalog 2017). Several groups have been isolated on the continents by continental drift, or have reduced their geographical distribution due to events of glaciation, orogenic activity, and habitat fragmentation. Some groups present a Gondwanic distribution or origin, disjointed or endemic distributions (Raven 1984, 1985, 2010; Griswold 1985).

Most mygalomorph spiders have low vagility, limited dispersion mechanisms, and sedentary habits, and live in strictly defined environments, so they show high values of local endemicity (Raven 2010). They are mainly terrestrial, although some groups present arboreal habits. Most mygalomorphs live in silk-lined burrows or modified natural retreats used for prey capture and shelter (Coyle 1986; Jocqué and

Dippenaar-Schoeman 2006). Microstigmatids and some paratropids are free-living spiders that hide under rocks or logs and leaf litter in the top layers of the soil (Raven 1999; Jocqué and Dippenaar-Schoeman 2006). Mygalomorph retreats can be vertical burrows of various shapes dug in the soil or sac-like chambers constructed under rocks, vegetation, or tree bark, or inside moss; and on occasion mygalomorphs can use epiphytes as shelters (Dippenaar-Schoeman 2002; Bertani 2012).

The retreats can stay open or closed, usually by a trapdoor, and spiders often build extensions to the entrance that include lids, signal threads, collars, turrets, or catch webs. These structures extend the range at which the substrate vibration receptors work and allow the spider to detect the prey, which is captured close to the entrance of the retreat (Coyle 1986; Álvarez et al. 2016). The construction of trapdoors and other structures around the retreats have evolved independently many times (Coyle et al. 1992).

The walls of the burrow are usually covered by a thick layer of silk. The silk layer prevents the soil from collapsing, and provides a well-balanced microclimate. The eggs are usually deposited at the bottom of the burrow and are covered with a layer of silk to protect them from predators, parasites, and microbial infections. Normally, tarantulas dig only one hole during their lifetime, and enlarge them as they grow older (Dippenaar-Schoeman 2002). Life in retreats provides protection against predators and parasites. Spider burrows are generally used as breeding chambers, where egg-sacs are cared for until spiderling emergence. The protection of the burrow is also essential during the molting process, when the animal is still exposed, and during courtship. The burrow also protects the spiders during inactive periods, especially in winter, during flooding due to the waterproof condition of the silk, or against thermal stress. Furthermore, during fires the spiders can withdraw deep into the burrow to protect themselves (Ghione et al. 2006). The shelter also reduces fungal and bacterial attack due to the antibiotic and antifungal properties of the silk (the macromolecular structure of the silk is inert, and most enzymes cannot decompose it) (Decae 1996; Dippenaar-Schoeman 2002).

The mygalomorphs use a variety of excavating methods. The fangs and rastellum of the chelicerae are used by most trapdoor spiders to loosen soil (Coyle et al. 1992), while spiders without rastellum (e.g., Theraphosidae) use their chelicerae and fangs. Spiders with rastellum are able to initiate and excavate new burrows, while spiders without rastellum are usually opportunistic burrowers and adapt existing holes in the ground to be used as burrows (Dippenaar-Schoeman 2002).

The shape of the burrow of the ground-dwelling spiders and the microhabitat in which it is constructed differ between genera and families. The majority of the species prefer relatively stable ground surfaces (Fig. 2.4) (some cyrtaucheniids, barychelids, nemesiids, paratropids, and theraphosids); however, some species (e.g., nemesiids *Chaco* Tullgren 1905, *Acanthogonathus* Karsch 1880, *Diplothelopsis* Tullgren 1905) can live in sand dunes (Fig. 2.5) (Goloboff 1995; Montes de Oca and Pérez-Miles 2013). The depth and shape of the burrows varies between taxa and is affected by obstacles occurring in the ground, and depends on the size of the spider, hardness or softness of the substrate, soil type, and slope of the ground.

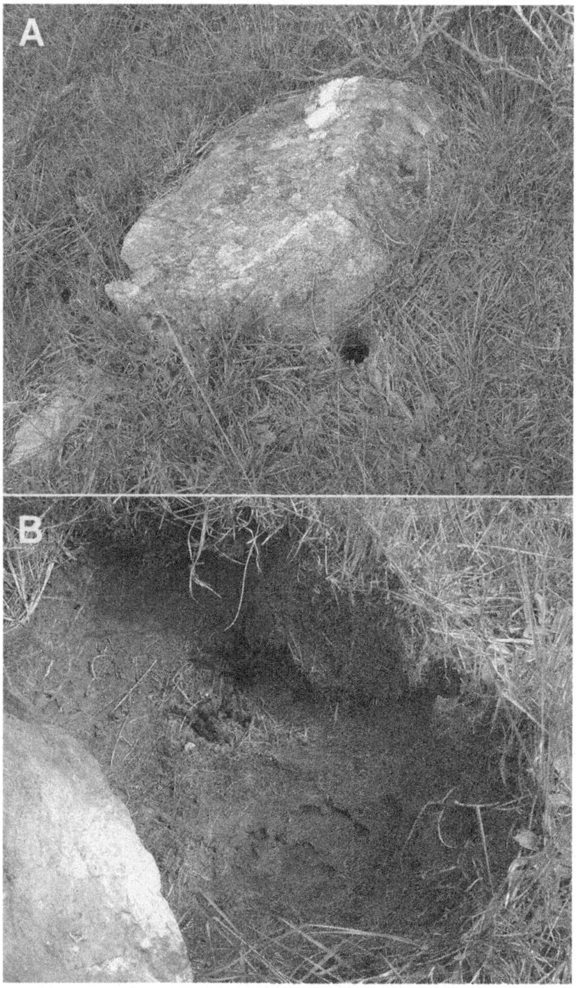

Fig. 2.4 Burrow of *Grammostola anthracina* at Pan de Azúcar, Maldonado, Uruguay. (**a**) Burrow entrance. (**b**) Burrow shape after removing the rock

Likewise, the shape, size, and thickness of the trapdoor in spiders that close their shelters with a door varies between taxa. Ctenizids, idiopids and some actinopodids and migids build a cork-lid trapdoor (Fig. 2.6); some cyrtaucheniids, other actinopodids, barychelids, and nemesiids build a soft flap or wafer-lid trapdoor. Some migids are referred to as tree/bank trapdoor spiders, which build well-camouflaged trap-door nests on the trunks and buttresses of trees (Dippenaar-Schoeman 2002).

Nemesiids and most diplurids build tubular silk retreats between vegetation or under rocks, with extensions to catch prey (Fig. 2.7). Nemesiids such as *Acanthogonatus* and several species of the Dipluridae subfamilies Ischnothelinae and Euagrinae build arrays of short tunnels that connect single or numerous small sheets that capture mostly ambulatory prey (Coyle 1986, 1988, 1995; Coyle and Ketner 1990). However, some diplurids build some of the most elaborate prey

Fig. 2.5 Burrow entrance of nemesiid *Chaco costai* in the sand dunes of La Perla, Rocha, Uruguay

Fig. 2.6 Burrow entrance of trapdoor spider *Actinopus* sp. at Sogamoso, Boyacá, Colombia. (Photo: Y. Cifuentes)

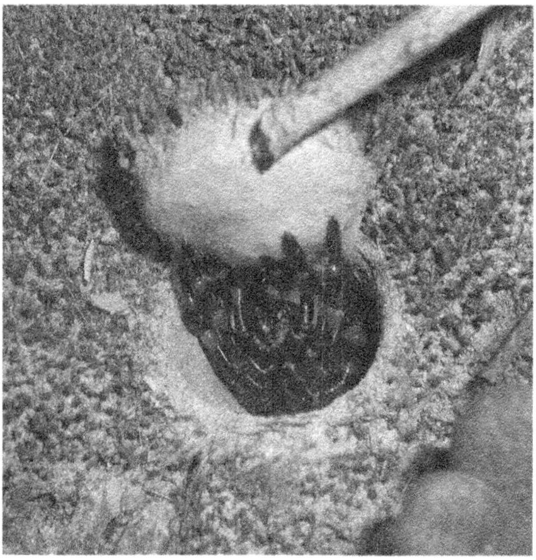

Fig. 2.7 Sheet–funnel web of the diplurid *Ischnothele* sp. at Costa Rica (Photo: A. Aisenberg)

capture webs among mygalomorph spiders (Coyle 1986). Some species in the subfamily Diplurinae, including species from the genera *Linothele* Karsch 1879, *Trechona* C.L. Koch 1850 and *Diplura* C.L. Koch 1850, construct a single large horizontal sheet with a tubular retreat. Some of these sheets are suspended in the air from many centimeters up to 2 m above the ground, and have tangles that extend up to a meter or more above the sheet, while others are built on the surface of the leaf litter or on top of some other substrate (Coyle 1986; Paz 1988; Viera et al. 2007).

Most atypids present peculiar habits. They live permanently in purse-like webs above the ground, which they also use to capture their prey. In some genera, the burrows extend into a tough, tubular, prolonged segment, while in others the top of the burrow is covered with a tough layer of silk, covered with soil particles; some can also build tubular webs attached to the bases of trees (Beatty 1986; Dippenaar-Schoeman 2002).

Compared with Araneomorphae and other arachnid groups, few mygalomorph species with troglobite habits have been reported. They present the typical troglobite characteristics such as reduction or complete loss of eyes and pigmentation; some theraphosids from caves have also been shown to lose urticating setae (Pérez-Miles and Locht 2003). In the Neotropics, cave mygalomorphs such as *Actinopus* Perty 1833 (Actinopodidae), *Psalistops* Simon 1889 and *Troglothele* Fage 1929 (Barychelidae), *Linothele*, *Masteria* L. Koch 1873, *Euagrus* Ausserer 1875, *Harmonicon* F. O. Pickard-Cambridge 1896, *Trechona*, and *Troglodiplura* Main 1969 (Dipluridae), *Speloctenliza* Gertsch 1982 (Microstigmatidae), *Melloina* Brignoli 1985 (Paratropididae), and *Hemirrhagus* Simon 1903, *Tmesiphantes* Simon 1892, *Trichopelma* Simon 1888, *Phormictopus* Pocock 1901, *Holothele* Karsch 1879, and *Lasiodora* C.L. Koch 1850 (Theraphosidae) have been described (Gertsch 1982; Goloboff 1994; Bernardi et al. 2007; de Armas et al. 2009; Bertani 2013; Bertani et al. 2013; Mendoza 2014; Pedroso and Baptista 2014; Guadanucci et al. 2016; Guadanucci et al. *in press*).

Enemies and Defenses (See Also Chap. 15)

Enemies

Spiders at all stages are attacked by a wide variety of predators, parasitoids, and parasites, some of which have taken advantage of the sedentary habits of mygalomorphs. At present, the theraphosid spiders face another threat because of them being hunted illegally because of their increasing commercial value, due to their attractiveness as ornamental pets or souvenirs.

Mygalomorph spiders are a potential food source for a number of animals such as birds, centipedes, reptiles (lizards, chameleons), insectivorous mammals (honey badgers, shrews, bats, mice, and baboons), and other arachnids such as scorpions, solpugids, and even other spiders. Members of the Araneomorph family

Palpimanidae are araneophagic spiders frequently found associated with trapdoor spiders (Dippenaar-Schoeman 2002).

Also, mygalomorphs are frequently attacked by various fungi (Rong and Grobbelaar 1998). Killer fungi of the genus *Cordiceps* are parasites of mygalomorph spiders that construct burrows with lids (e.g., Barychelidae, Idiopidae, *pers. obs.*). A number of insects and mites are specialized predators or parasites of spiders in general: Hymenoptera (Sphecidae, Pompilidae, Ichneumonidae), dipterous parasitoids (Drosophilidae, Phoridae, Chloropidae), egg predators (Sarcophagidae), and endoparasites (Acroceridae) (Dippenaar-Schoeman 2002). Spiders also have endoparasites such as parasitic nematodes of the family Mermithidae.

Pompilid wasps are a distinct group of wasps that have an interesting natural history where single spiders are used as hosts. Pompilids show energetic search behaviors, performing short flights involving rapid movements of their wings and antennae. These wasps are characterized by notably robust bodies and long spiny legs (Fernández 2000). Most spider wasps hunt and paralyze their prey with the venom of their sting, although a few species lay an egg on the living spider without paralyzing it. Some species simply deposit their eggs in the spider after paralyzing it, while others drag the spider to a nest where they store it to feed their larvae (Evans 1953).

Prey-carrying mechanisms in hunter wasps vary considerably. The female wasp transports the spider to the nest by dragging it backwards or forwards, or even flying with it. *Auplopus* Spinola 1841 and related genera have apparently streamlined transport by amputating the legs of their hosts (Evans 1953; Evans and West-Eberhard 1970). Rego et al. (2004) reported an *Avicularia* sp. immobilized and transported by a Pompilid wasp along 30 m in less than 5 min to a burrow beside tree roots. Pompilids also vary in where they choose to locate a nest, and the nest can be prepared before or after hunting. Many spider wasps nest in the ground, some in preexisting cavities; others dig their own nests, and a few spider wasps use the spider's nest as their own. Most pompilids prepare only a single cell per nest; however, some species make multiple cells per nest (Evans 1953; Restrepo-Giraldo et al. 2012; Costa et al. 2004).

Defenses

Mygalomorphs use different mechanisms to defend themselves against their enemies, adopting an active or passive defense. When threatened, most mygalomorph spiders react by adopting an aggressive posture in which they raise their front legs and expose their large fangs by curving the front part of their bodies backwards, while releasing a drop of venom in some cases (Fig. 2.8). Some theraphosids show a different defensive threat display by raising their abdomen and sometimes legs IV vertically and extending the spinnerets (Fig. 2.9) (Pérez-Miles et al. 2005; Bertani

Fig. 2.8 Defensive display of the theraphosid *Acanthoscurria suina*, lateral view

Fig. 2.9 Defensive display of the theraphosid *Eupalaestrus weijenberghi*, lateral view

et al. 2008a). In both displays, the spiders can maintain the posture (freeze) for some minutes.

Other families such as Microstigmatidae and Paratropididae (Paratropidinae) have camouflage mechanisms of defense. These spiders have a cuticle that allows them to encrust particles of soil, which enables them to camouflage themselves with the substrate. This peculiarity, together with the fact that they often remain immobile against a threat, allows them to camouflage themselves from potential predators (Raven and Platnick 1981).

Stridulation

Stridulation can be defined as the process of sound production by friction of one rigid body part (the scraper) across a second part (the file) (Uetz and Stratton 1982). The stridulatory organs are widely distributed in spider species (Uetz and Stratton 1982), but in mygalomorph spiders only Diplurid: *Diplura*, *Trechona*, and the Theraphosidae family exhibits those structures (Marshall et al. 1995 and *pers. obs.*). The evidence thus far suggests that in theraphosid spiders the stridulatory setae perform solely a defensive role within a behavioral deployment (Marshall et al. 1995; Pérez-Miles et al. 2005; Bertani et al. 2008b). When alarmed, some members of the Theraphosidae stridulate, showing a great diversity in volume and sound production. The sound is produced by the specialized setae principally localized on proximal segments of front appendages; e.g., rubbing the chelicerae against the palp, or palp against front legs.

The stereotyped defensive behaviors described in the previous section are sometimes accompanied by stridulation. It has been suggested that the defensive movements of the tarantulas additionally would be able to produce stridulation since they involve appendages containing stridulatory setae that could rub together, as has been reported for *Theraphosa blondi* (Latreille, 1804) (Goliath bird-eater tarantula), *Acanthoscurria suina* Pocock, 1903, and *Aguapanela arvi* Perafán, Cifuentes, and Estrada 2015 (Marshall et al. 1995; Pérez-Miles et al. 2005; Perafán et al. 2015). In *T. blondi*, the anterior legs and palps do not remain extended while performing the defensive behavior but in fact are drawn backwards, putting the plumose hairs in contact with bearing surfaces. Marshall et al. (1995) suggested that in *T. blondi* the stridulation was an acoustic aposematism, since the spider produces sound and sheds urticating hairs during defensive behavior.

Stridulatory features have been widely used in systematics; many theraphosid taxa are characterized by the type and/or position of their stridulatory apparatus (Pocock 1897, 1899; Simon 1903; Raven 1985; Pérez-Miles et al. 1996; Schmidt 1999, 2000; Bertani et al. 2008a). In the Neotropics, there are species with bacilliform setae located at the outer chelicerae, such as *Psalmopoeus* Pocock 1895 (Aviculariinae) and some Theraphosinae species with plumose setae located principally on coxa, trochanter, and/or femur of palp and front legs. *Acanthoscurria* Ausserer 1871, *Aguapanela* Perafán and Cifuentes 2015, *Brachypelma* Simon 1891, *Citharacantus* Pocock 1901, *Cyrtopholis* Simon 1892, *Grammostola* Simon 1892, *Lasiodora* Koch 1850, *Longilyra* Gabriel 2014, *Phormictopus* Pocock 1901, *Pterinopelma* Thorell 1869, and *Theraphosa* Thorell 1870, are Theraphosinae reported to have stridulatory apparatus.

Another mechanism for stridulating has been reported in *Pamphobeteus crassifemur* Bertani, Fukushima and Silva 2008, and observed in other *Pamphobeteus* Pocock 1901 species (Bertani et al. 2008a; CP pers. obs., and Weinmman pers. comm.). Bertani et al. (2008a) described a characteristic stridulating sound similar to a whistle at the same time that the tarantula elevates its hind legs in the typical posture of shedding urticating setae. They explained that this sound is produced by contact between the many spiniform setae on retrolateral coxae and trochanter III and the prolateral spiniform setae of the same segments of leg IV. However, this sound could also be produced by the friction of the spines of the hind legs against the long hairs of the abdomen (CP pers. obs., and Weinmman pers. comm.). This behavioral display seems to respond to the same aposematic behavior described above for *T. blondi*, because the spider also sheds urticating setae while stridulating. Some species of *Brachypelma* also produce a sound similar to rubbing hairs when the spider is disturbed (Mendoza, pers. comm.).

Urticating Setae

The release of urticating setae as a mechanism of defense is a unique characteristic of most New World theraphosids (Cooke et al. 1972). This defense mechanism is found in roughly 540 of the 600 Neotropical theraphosid species (Bertani and Guadanucci 2013) and is absent in theraphosids from other regions.

Representatives of all known species of the subfamily Theraphosinae, as well as species of the Aviculariinae genera *Avicularia* Lamarck 1818, *Ephebopus* Simon 1892, *Iridopelma* Pocock 1901, *Pachistopelma* Pocock 1901, and *Typhochlaena* C.L. Koch 1850 have urticating setae. The arboreal tarantulas *Tapinauchenius* Ausserer 1871 and *Psalmopoeus* as well as Ischnocolinae and Schismathotelinae genera are the New World theraphosids that lack urticating setae (Bertani and Guadanucci 2013).

Seven different types of urticating setae have been described based on their morphology, ornamentation, length, and releasing mechanism; two types are known in Aviculariinae (II and V), and another five in Theraphosinae (I, III, IV, VI, and VII) (Cooke et al. 1972; Marshall and Uetz 1990; Pérez-Miles 1998; Perafán et al. 2016). Except for type V which occurs on the distal prolateral surface of the palpal femora (Marshall and Uetz 1990; Foelix et al. 2009), all other types are found on the dorsum of the abdomen, and they can be located in one (the majority) or two specific areas (some Theraphosinae). Type V urticating setae, present in the Aviculariinae genus *Ephebopus*, are released by scratching the palps against the basal segments of the chelicerae (Marshall and Uetz 1990; Foelix et al. 2009).

Urticating setae (Fig. 2.10) differ from other setae by the insertion feature through a stalk (types I, II, III, IV, and VII) or attachment in a specialized socket (types V and VI) that facilitates detachment, plus the presence of a penetrating acute tip, and barbs or scales that aid embedding them into their attackers (Cooke et al. 1972; Bertani and Guadanucci 2013; Perafán et al. 2016).

In Theraphosinae, two types of setae can occur simultaneously in the same specimen, and each type is segregated into distinct regions of the spider's abdomen. There may be co-occurrence of the types I with III and III with IV, and usually type III is located in a medial region of the abdomen whereas the types I or IV have been found in regions surrounding the patch of type III setae (Bertani and Guadanucci 2013).

Sexual dimorphism in urticating setae has also been described. One of the sexes can present the two types of urticating setae, while the other presents only one, or

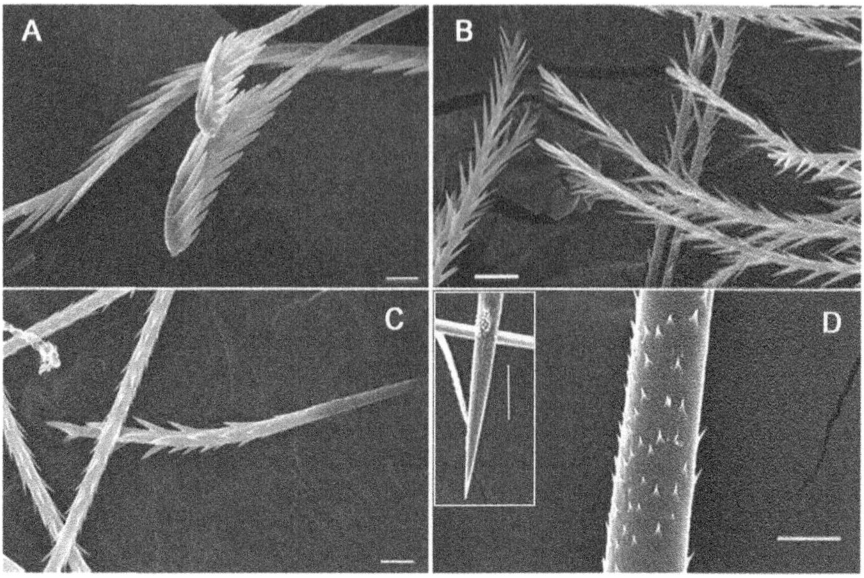

Fig. 2.10 Urticating setae of theraphosid spiders. (**a**) Type I. (**b**) Type III. (**c**) Types III and IV (*arrow*). (**d**) Type VII (*close-up*, penetration tip)

each sex presents a different type of urticating setae. Sexual dimorphism in the size of the urticating setae has also been described (Bertani 2002; Pérez-Miles 2002; Bertani and Guadanucci 2013).

The urticating setae can be thrown into the air by friction of the hind legs against the dorsum of the abdomen (Fig. 2.11), or can be inserted by direct contact with potential predators, causing irritation when it enters the skin or mucous membranes. Urticating setae dislodged by friction and air-transported are restricted to burrowing spiders of the subfamily Aviculariinae (lighter type II) and most of the Theraphosinae (type I, III, IV), whereas contact urticating setae are used by arboreal spiders of the Aviculariinae (type II) and *Kankuamo* Perafán et al. 2016 genus (type VII) (Bertani and Marques 1996; Perafán et al. 2016); those tarantulas direct the abdomen towards the stimulus and transfer the urticating setae when the abdomen of the spider touches against the target. Bertani and Marques (1996) proposed that the differences of the shape, size, and thickness between urticating setae could explain the two releasing mechanisms. The morphological characteristics of types I, III, and IV (the urticating setae types in Theraphosinae), particularly their light weight and aerodynamics, would allow them to float through the air and meet their targets. In contrast, most of the known urticating setae type II, and the type VII, have a greater length/width ratio.

The urticating setae can also be used as passive defense when they are transferred to the silk of the burrows, molting web, or eggsacs, functioning as a defense principally against other arthropods such as flying larvae or ants (Marschall and Uetz 1990; Bertani and Guadanucci 2013).

Fig. 2.11 *Grammostola quirogai* releasing urticating setae by friction with hindlegs. (**a**) release with left leg, front view. (**b**) Release with right leg, lateral view (Photos: A. Laborda)

The morphological characteristics of urticating setae have been used in the taxonomy and systematics of Theraphosidae, being useful as a set of characters for differentiation of subfamilies and genera, as shown in phylogenetic analysis (Cooke et al. 1972; Raven 1985; Pérez-Miles et al. 1996; Pérez-Miles 2002; Perafán 2010; Bertani and Guadanucci 2013).

Final Considerations

Knowledge concerning the biology of Mygalomorphae was given less focus in comparison with Araneomorphae, probably due to the history of taxonomic confusion of the group. The geographic distribution of Mygalomorphae, biased toward developing (poor) countries, also explains the slow production of biological and mainly field studies. The recent increase in taxonomic studies in the group and the increasing research teams in developing countries are good signals of the promising future for the study of biology of Mygalomorphae.

Acknowledgements The authors thank Anita Aisenberg (IIBCE) for her critical reading and help with the English. We also thank two anonymous reviewers for their valuable comments and suggestions on the manuscript.

References

Almeida-Silva LM, Camacho A, Brescovit AD, Lucas SM, Brazil TK (2008) Redescription and notes on the natural history of the arboreal tarantula *Iridopelma seladonium* (Araneae: Theraphosidae: Aviculariinae). Rev Bras Zoo 25:728–736

Álvarez L, Perafán C, Pérez-Miles F (2016) At what time, for what distance, and for how long does the tarantula *Eupalaestrus weijenberghi* (Araneae, Theraphosidae) leave its burrow during the mating season? Arachnology 17:152–154

Autumn K, Liang YA, Hsieh ST, Zesch W, Chan W-P, Kenny WT, Fearing R, Full RJ (2000) Adhesive force of a single gecko foot-hair. Nature 405:681–685

Ayoub NA, Hayashi CY (2009) Spiders (Araneae). In: Hedges SB, Kumar S (eds) The time tree of life. Oxford University Press, Oxford, pp 255–259

Baerg WJ (1928) The life cycle and mating habits of the male tarantula. Q Rev Biol 3:109–116

Baerg WJ (1958) The tarantula. University of Kansas Press, Lawrence, p 88

Barth FG (1985) Neurobiology of arachnids. Springer, Berlin, p 385

Baruffaldi L, Costa FG, Rodríguez A, González A (2010) Chemical communication in *Schizocosa malitiosa*: evidence of a female contact sex pheromone and persistence in the field. J Chem Ecol 36:759–767

Beatty JA (1986) Web structure and burrow location of *Sphodros niger* (Hentz) (Araneae, Atypidae). J Arachnol 14:130–132

Bernardi LPO, Ferreira RL, Silva MS (2007) Aspectos da ecologia de uma população de *Lasiodora* sp. (Araneae:Theraphosidae) em caverna granítica. Espeleo-Tema 19:65–80

Bertani R (2002) Morfologia e evolução das cerdas urticantes em Theraphosidae (Areaneae). Thesis (unpublished), Instituto de Biociências, Universidade de São Paulo, Brasil

Bertani R (2012) Revision, cladistic analysis and biogeography of *Typhochlaena* C. L. Koch, 1859, *Pachistopelma* Pocock, 1901 and *Iridopelma* Pocock, 1901 (Araneae, Theraphosidae, Aviculariinae). Zoo Keys 230:1–94

Bertani R, Bichuette ME, Pedroso DR (2013) *Tmesiphantes hypogeus* sp. nov. (Araneae, Theraphosidae), the first troglobitic tarantula from Brazil. An Acad Bras Ciênc 85:235–243

Bertani R, Fukushima CS, da Silva PI Jr (2008a) Two new species of *Pamphobeteus* Pocock 1901 (Araneae: Mygalomorphae: Theraphosidae) from Brazil, with a new type of stridulatory organ. Zootaxa 1826:45–28

Bertani R, Fukushima CS, Silva Júnior PI (2008b) Mating behaviour of *Sickius longibulbi* (Araneae, Theraphosidae, Ischnocolinae), a spider that lacks spermathecae. J Arachnol 36:331–335

Bertani R, Guadanucci JPL (2013) Morphology, evolution and usage of urticating setae by tarantulas (Araneae: Theraphosidae). Fortschr Zool 30:403–418

Bertani R, Marques OAV (1996) Defensive behaviors in mygalomorph spiders: release of urticating hairs by some Aviculariinae (Araneae, Theraphosidae). Zool Anz 234:161–165

Bertani R (2013) A new species of *Melloina* (Araneae: Paratropididae) from Venezuela. Fortschr Zool 30:101–106

Biancardi CM, Fabrica CG, Polero P, Loss JF, Minetti AE (2011) Biomechanics of octopedal locomotion: kinematic and kinetic analysis of the spider *Grammostola mollicoma*. J Exp Biol 214:3433–3442

Bond JE, Hendrixson BE, Hamilton CA, Hedin M (2012) A reconsideration of the classification of the spider infraorder Mygalomorphae (Arachnida: Araneae) based on three nuclear genes and morphology. PLoS One 7(6):e38753

Bond JE, Opell BD (2002) Phylogeny and taxonomy of the genera of south-western North American Euctenizinae trapdoor spiders and their relatives (Araneae: Mygalomorphae: Cyrtaucheniidae). Zool J Linn Soc–Lond 136:487–534

Booster NA, Su FY, Adolph SC, Ahn AN (2015) Effect of temperature on leg kinematics in sprinting tarantulas (*Aphonopelma hentzi*): high speed may limit hydraulic joint actuation. J Exp Biol 218:977–982

Cooke JA, Roth VD, Miller F (1972) The urticating hairs of Theraphosidae. Am Mus Novit 2498:1–43

Copperi S, Ferretti N, Pompozzi G, Pérez-Miles F (2012) Can't you find me? Female sexual response in an Argentinean tarantula (Araneae, Theraphosidae). Rev Colomb Entomol 38:164–166

Costa FG, Pérez-Miles F (1992) Notes on mating and reproductive success of *Ceropelma longisternalis* (Araneae, Theraphosidae) in captivity. J Arachnol 20:129–133

Costa FG, Pérez-Miles F (1998) Behavior, life cycle and webs of *Mecicobothrium thorelli* (Araneae, Mygalomorphae, Mecicobothriidae). J Arachnol 26:317–329

Costa FG, Pérez-Miles F (2002) Reproductive biology of Uruguayan theraphosids (Araneae, Theraphosidae). J Arachnol 30:571–587

Costa FG, Pérez-Miles F, Mignone A (2004) Pompilid wasp interactions with burrowing tarantulas: *Pepsis cupripennis* versus *Eupalaestrus weijenberghi* and *Acanthoscurria suina* (Araneae, Theraphosidae). Stud Neotrop Fauna Environ 39:37–43

Costa GF, Curbelo B, Pérez-Miles F (2015) Long-term persistence and water resistance of female sex cues in the tarantula *Eupalaestrus weijenberghi* (Araneae: Theraphosidae). Arachnology 16:311–313

Coyle FA (1986) The role of silk in prey capture by nonaraneomorph spiders, pp 269–305. In: Shear WA (ed) Spiders: webs, behaviour, and evolution. Stanford University Press, California, p 492

Coyle FA (1988) A revision of the American funnel-web mygalomorph spider genus *Euagrus* (Araneae, Dipluridae). Bull Am Mus Nat Hist 187:203–292

Coyle FA (1995) A revision of the funnelweb Mygalomorph spider subfamily Ischnothelinae (Araneae, Dipluridae). Bull Am Mus Nat Hist 226:1–113

Coyle FA, Dellinger RE, Bennett RG (1992) Retreat architecture and construction behaviour of an East African idiopine trapdoor spider (Araneae, Idiopidae). Bull Br Arachnol Soc 9:99–104

Coyle FA, Icenogle WR (1994) Natural history of the Californian trapdoor spider genus *Aliatypus* (Araneae, Antrodiaetidae). J Arachnol 22:225–255

Coyle FA, Ketner ND (1990) Observations on the prey and prey structure behaviour of the funnel web mygalomorph spider genus *Ischnothele* (Araneae, Dipluridae). Bull Br Arachnol Soc 8:97–104

Coyle FA, O'Shields TC (1990) Courtship and mating behaviour of *Thelechoris karschi* (Araneae, Dipluridae), an African funnel web spider. J Arachnol 18:281–296

de Armas LF, Alayón García G, Ramos Hernández JM (2009) Aracnofauna (excepto Acari) del macizo Guamuhaya, Cuba Central. Primera aproximación. Boln SEA 45:135–146

Decae AE (1996) Variation in burrow morphology of Mediterranean trapdoor spiders (Ctenizidae, Cyrtaucheniidae, Nemesiidae). Rev Suisse Zool (hors série) 1:135–140

Dippenaar-Schoeman AS (2002) Baboon and trapdoor spiders of Southern Africa: an identification manual. Plant protection research institute handbook no. 13. Agricultural Research Council, Pretoria, p 129

Dondale CD, Hegdeckar BM (1973) The contact sex pheromone of *Pardosa lapidicina* Emerton (Araneida: Lycosidae). Can J Zool 51:400–401

Dunlop JA (1995) Movements of scopulate claw tufts at the tarsus tip of a tarantula spider. Neth J Zool 45:513–520

Eggs W, Wolff JW, Khun-Nentwig L, Gorb SN, Nentwig W (2015) Hunting without a web: how lycosoid spiders subdue their prey. Ethology 121:1166–1177

Eskov KY, Zonhstein SL (1990) A new classification for the order Araneida (Arachnida: chelicerata). Acta Zool Fenn 190:129–137

Evans HE (1953) Comparative ethology and the systematics of spider wasps. Syst Zool 2:155–172
Evans HE, West-Eberhard MJ (1970) The wasps. University of Michigan Press, Ann Arbor, p 265
Fernández F (2000) Avispas cazadoras de arañas (Hymenoptera: Pompilidae) de la region Neotropical. Biota Col 1:3–24
Ferretti N, Ferrero A (2008) Courtship and mating behavior of *Grammostola schulzei* (Schmidt 1994) a burrowing tarantula from Argentina. J Arachnol 36:480–483
Ferretti N, Pérez-Miles F, González A (2014) Historical relationships among Argentinean biogeographic provinces based on mygalomorph spider distribution data (Araneae: Mygalomorphae). Stud Neotrop Fauna Envirom 49:1–10
Ferretti N, Pompozzi G, Copperi S, González A, Pérez-Miles F (2013) Sexual behaviour of mygalomorph spiders: when simplicity becomes complex, an update of the last 21 years. Arachnology 16:85–93
Ferretti N, Pompozzi G, Copperi S, Pérez-Miles F, González A (2012) Copulatory behavior of Microstigmatidae (Araneae: Mygalomorphae): a study with *Xenonemesia platensis* from Argentina. J Arachnol 40:252–255
Ferretti N, Pompozzi G, Pérez-Miles F (2011) Sexual behavior of Acanthogonatus centralis (Araneae: Mygalomorphae: Nemesiidae) from Argentina, with some notes on their burrows. J Arachnol 39:533–536
Foelix R, Erb B, Rast B (2013) Alleged silk spigots on tarantula feet: electron microscopy reveals sensory innervation, no silk. Arthropod Struct Dev 42(3):209–217
Foelix R, Rast B, Erb B, Wullschleger B (2011) Spinnspulen auf den Tarsen von Vogelspinnen? Eine Gegendarstellung. Arachne 16:4–9
Foelix R, Rast B, Peattie AM (2012) Silk secretion from tarantula feet revisited: alleged spigots are probably chemoreceptors. J Exp Biol 215:1084–1089
Foelix RF (2011) Biology of spiders. Oxford University Press, New York, p 419
Foelix RF, Bastian R, Erb B (2009) Palpal urticating hairs in the tarantula *Ephebopus*: fine structure and mechanism of release. J Arachnol 37:292–298
Foelix RF, Chu-Wang IW (1975). The structure of scopula hairs in spiders. Proc 6th Int. Congr. Arachnol., Amsterdam, pp 56–58
Fukushima CS, Bertani R, Da Silva Jr PI (2003) Feromônio sexual em aranhas migalomorfas? Actas IV Encontro de Aracnólogos do Cone Sul, São Pedro, Brazil: 201
Garrison et al (2016) Spider phylogenomics: untangling the spider tree of life. Peer J 4:e1719
Gaskett AC (2007) Spiders sex pheromones: emission, reception, structures, and functions. Biol Rev 82:27–48
Gertsch WJ (1982) The troglobitic mygalomorphs of the Americas (Arachnida, Araneae). AMCS Bulletin 8:79–94
Ghione S, Aisenberg A, Costa F, Montes de Oca L, Pérez-Miles F, Postiglioni R, Quirici V, Useta G (2006) Efecto de los incendios sobre la aracnofauna criptozoica (Araneae, Scorpiones y Opiliones) en una zona de serranías de Sierra De Las Ánimas, Maldonado, Uruguay. Bol Soc Zool Uruguay 16:26–31
Goloboff PA (1993) A reanalysis of mygalomorph spider families (Araneae). Am Mus Novit 3056:1–32
Goloboff PA (1994) *Linothele cavicola*, a new diplurine spider (Araneae, Dipluridae) from caves in Ecuador. J Arachnol 22:70–72
Goloboff PA (1995) A revision of the South American spiders of the family Nemesiidae (Araneae, Mygalomorphae). Part I: species from Peru, Chile, Argentina, and Uruguay. Bull Am Mus Nat Hist 224:189
Gorb SN, Niederegger S, Hayashi CY, Summers AP, Vötsch W, Walther P (2006) Biomaterials: silk-like secretion from tarantula feet. Nature 443:407
Griswold CE (1985) A revision of African spiders of the family Microstigmatidae (Araneae: Mygalomorphae). Ann Natal Mus 27:1–37
Grossi B, Veloso C, Tacuaré-Ríos A, Canals M (2016) Allometry of locomotor organs and sexual size dimorphism in the mygalomorph spider *Grammostola rosea* (Walckenaer, 1837) (Araneae, Theraphosidae). J Arachnol 44:99–102

Guadanucci JPL (2012) Trichobothrial morphology of Theraphosidae and Barychelidae spiders (Araneae, Mygalomorphae). Zootaxa 3439:1–42

Guadanucci JPL, Fonseca-Ferreira R, Cerqueira Baptista RL, Pedroso DR (2016) An unusual new species of *Trechona* (Araneae: Mygalomorphae: Dipluridae), from quartzitic caves of the Diamantina Plateau, Minas Gerais, Brazil, with a key to the known species. J Nat Hist 50:2487–2497

Hedin M, Bond J (2006) Molecular phylogenetics of the spider infraorder Mygalomorphae using nuclear rRNA genes (18S and 28S): conflict and agreement with the current system of classification. Mol Phylogenet Evol 41:454–471

Hill DE (1977) The pretarsus of salticid spiders. Zool J Linn Soc-Lond 60:319–338

Homann H (1957) Haften Spinnen an einer Wasserhaut? Naturwissenschaften 44:318–319

Huber G, Mantz H, Spolenak R, Mecke K, Jacobs K, Gorb SN, Arzt E (2005) Evidence for capillarity contributions to gecko adhesion from single spatula: nanomechanical measurements. Proc Natl Acad Sci U S A 102:16293–16296

Janowski-Bell ME, Horner NV (1990) Movement of the male brown tarantula, *Aphonopelma hentzi* (Araneae, Theraphosidae), using radio telemetry. J Arachnol 27:503–512

Jocqué R, Dippennar-Schoeman (2006) Spiders families of the world. Royal Museum for Central Africa, p 336

Kesel AB, Martin A, Seidl T (2003) Adhesion measurements on the attachment devices of the jumping spider *Evarcha arcuata*. J Exp Biol 206:2733–2738

Kesel AB, Martin A, Seidl T (2004) Getting a grip on spider attachment: an AFM approach to microstructure adhesion in arthropods. Smart Mater Struct 13:512–518

Land MF (1985) The morphology and optics of spider eyes. In: Barth FG (ed) Neurobiology of arachnids. Springer, Berlin, pp 53–78

Lapinski W, Walther P, Tschapka M (2015) Morphology reflects microhabitat preferences in an assemblage of Neotropical wandering spiders. Zoomorphology. https://doi.org/10.1007/s00435-015-0247-8

Main BY (1978) Biology of the arid-adapted Australian trapdoor spider *Anidiops villosus* (Rainbow). Bull Brit Arachnol Soc 4:161–175

Marshall SD, Thoms EM, Uetz GW (1995) Setal entanglement: an undescribed method of stridulation by a Neotropical tarantula (Araneae: Theraphosidae). J Zool 235:587–595

Marshall SD, Uetz GW (1990) The pedipalpal brush of *Ephebopus* sp. (Araneae, Theraphosidae): evidence of a new site for urticating hairs. Bull Br Arachnol Soc 8:122–124

Mendoza Marroquin JI (2014) Taxonomic revision of *Hemirrhagus* Simon, 1903 (Araneae: Theraphosidae, Theraphosinae), with description of five new species from Mexico. Zool J Linn Soc-Lond 170:634–689

Minch EW (1979) Reproductive behaviour of the tarantula *Aphonopelma chalcodes* (Araneae: Theraphosidae). Bull Br Arachnol Soc 4:416–420

Montes de Oca L, Pérez-Miles F (2013) Two new species of *Chaco* Tullgren from the Atlantic coast of Uruguay (Araneae, Mygalomorphae, Nemesiidae). Zookeys 337:73–87

Niederegger S (2013) Functional aspects of spider scopulae. In: Nentwig W (ed) Spider ecophysiology. Springer, Berlin

Niederegger S, Gorb S (2006) Friction and adhesion in the tarsal and metatarsal scopulae of spiders. J Comp Physiol A 192:1223–1332

Paz N (1988) Ecología y aspectos del comportamiento en *Linothele* sp. (Araneae, Dipluridae). J Arachnol 16:5–22

Peattie AM, Dirks JH, Henriques S, Federle W (2011) Arachnids secrete a fluid over their adhesive pads. PLoS One 6:e20485

Pedroso DR, Baptista RLC (2014) A new troglomorphic species of *Harmonicon* (Araneae, Mygalomorphae, Dipluridae) from Pará, Brazil, with notes on the genus. Zoo Keys 389:77–88

Pekar S, Sobotnik J, Lubin J (2011) Armoured spiderman: morphological and behavioural adaptations of a specialised araneophagous predator (Araneae: Palpimanidae). Naturwissenschaften 98:593–603

Perafán C (2010) Revisión taxonómica y análisis filogenético de un grupo basal de Theraphosinae (Araneae, Mygalomorphae, Theraphosidae). MSc Thesis, Universidad de la República, Montevideo

Perafán C, Cifuentes-Gil Y, Estrada-Gómez S (2015) *Aguapanela*, a new tarantula genus from the Colombian Andes (Araneae, Theraphosidae). Zootaxa 4033:529–542

Perafán C, Galvis W, Gutiérrez M, Pérez-Miles F (2016) Kankuamo, a new Theraphosid genus from Colombia (Araneae, Mygalomorphae), with a new type of urticating setae and divergent male genitalia. ZooKeys 601:89–109

Pérez-Miles F (1998) Notes on the systematics of the little known theraphosid spider *Hemirrhagus cervinus*, with a description of a new type of urticating hair. J Arachnol 26:120–123

Pérez-Miles F (2002) The occurrence of abdominal urticating hairs during development in Theraphosinae (Araneae, Thraphosidae): phylogenetic implications. J Arachnol 30:316–320

Pérez-Miles F, Costa FG (1992) Interacciones intra e intersexuales en *Grammostola mollicoma* (Araneae, Theraphosidae) en condiciones experimentales. Bol Soc Zool Uruguay 7:71–72

Pérez-Miles F, Costa FG, Toscano-Gadea C, Mignone A (2005) Ecology and behaviour of the "road tarantulas" *Acanthoscurria suina* and *Eupalaestrus weijenberghi* (Araneae, Theraphosidae) from Uruguay. J Nat Hist 39:483–498

Pérez-Miles F, Locht A (2003) Revision and cladistic analysis of the genus *Hemirrhagus* Simon, 1903 (Araneae, Theraphosidae, Theraphosinae). Bull Br Arachnol Soc 12:365–375

Pérez-Miles F, Lucas SM, da Silva PI Jr, Bertani R (1996) Systematic revision and cladistic analysis of Theraphosinae (Araneae: Theraphosidae). Mygalomorph 1:33–68

Pérez-Miles F, Montes de Oca L, Postiglioni R, Baruffaldi L, Costa F (2007) Mating system in the tarantula spider Eupalaestrus weijenberghi (Thorell, 1894): evidences of monandry and polygyny. Zoology 110:253–260

Pérez-Miles F, Ortíz-Villatoro D (2012) Tarantulas do not shoot silk from their legs: experimental evidence in four species of New World tarantulas. J Exp Biol 215:1749–1752

Pérez-Miles F, Panzera A, Ortíz-Villatoro D, Perdomo C (2009) Silk production from tarantula feet questioned. Nature 461:E9–E10

Pérez-Miles F, Perafán C, Santamaría L (2015) Tarantulas (Araneae: Theraphosidae) use different adhesive pads complementary during climbing on smooth surfaces: experimental approach in eight arboreal and burrower species. Biol Open 4:1643–1648

Pérez-Miles F, Postiglioni R, Montes De Oca L, Baruffaldi L, Costa FG (2007) Mating system in the tarantula spider *Eupalaestrus weijenberghi* (Thorell, 1894): evidences of monandry and polygyny. Fortschr Zool 110:253–260

Petrunkevitch A (1911) Sense of sight, courtship and mating in *Dugesiella hentzi* (Girard), a theraphosid spider from Texas. Zool Jahrb (Systematik) 31:355–376

Platnick N (1971) The evolution of courtship behaviour in spiders. Bull Br Arachnol Soc 2:40–70

Pocock RI (1897) On the spiders of the suborder Mygalomorphae from the Ethiopian Region contained in the collection of the British Museum. Proc Zool Soc Lond:724–774

Pocock RI (1899) A new stridulating theraphosid spider from South America. Ann Mag Nat Hist 3:347–349

Postiglioni R, Costa FG (2006) Reproductive isolation among three populations of the genus *Grammostola* from Uruguay (Araneae, Theraphosidae). Iheringia 96:71–74

Prentice TR (1992) A new species of North American tarantula, *Aphonopelma paloma* (Araneae, Mygalomorphae, Theraphosidae). J Arachnol 20:189–199

Prentice TR (1997) Theraphosidae of the Mojave desert west and north of the Colorado River (Araneae, Mygalomorphae, Theraphosidae). J Arachnol 25:137–176

Punzo F, Henderson L (1999) Aspects of the natural history and behavioural ecology of the tarantula spider *Aphonopelma hentzi* (Girard, 1854) (Orthognatha, Theraphosidae). Bull Br Arachnol Soc 11:121–128

Quirici V, Costa FG (2005) Seismic communication during courtship in two burrowing tarantula spiders: an experimental study on *Eupalaestrus weijenberghi* and *Acanthoscurria suina*. J Arachnol 33:159–166

Quirici V, Costa FG (2007) Seismic sexual signals design of two sympatric burrowing tarantula spiders from meadows of Uruguay: *Eupalaestrus weijenberghi* and *Acanthoscurria suina* (Araneae, Theraphosidae). J Arachnol 35:38–45

Raven RJ (1984) Systematics and biogeography of the mygalomorph spider family Migidae (Araneae) in Australia. Aust J Zool 32:379–390

Raven RJ (1985) The spider infraorder Mygalomorphae (Araneae): cladistics and systematics. Bull Am Mus Nat Hist 182:1–180

Raven RJ (1999) Review of the mygalomorph genus *Melloina* Brignoli (Paratropididae: Araneae). Mem Queensl Mus 43:819–825

Raven RJ (2010) A review of the Mygalomorphae: biology, morphology and systematics. Book of Abstracts of the 18th International Congress of Arachnology, 2010 Jul 11–17. Siedle, Poland

Raven RJ, Platnick NI (1981) A revision of the American spiders of the family Microstigmatidae (Araneae, Mygalomorphae). Am Mus Novit 2707:1–20

Rego FN, Rheims CA, Venticinque EM (2004) Notes on the predation of an Aviculariine Spider (Araneae: Theraphosidae, Avicularia sp.) by Pepsis frivaldszkyi (Hymenoptera: Pompilidae) in Brazilian Amazonia. J Br Tarantula Soc 20:18–25

Restrepo-Giraldo C, Rodriguez J, Pitts JP (2012) Temporal activity patterns of the spider wasp *Pepsis montezuma* Smith (Hymenoptera: Pompilidae) in a disturbed lower montane rainforest (Manizales, Colombia). Psyche, Volume 2012, Article ID 516419

Rind FC, Birkett CL, Duncan BJ, Ranken AJ (2011) Tarantulas cling to smooth vertical surfaces by secreting silk from their feet. J Exp Biol 214:1874–1879

Rong IH, Grobbelaar E (1998) South African records of associations between fungi and arthropods. Afr Plant Prot 4:43–60

Rovner JS (1978) Adhesive hairs in spiders: behavioral functions and hydraulically mediated movement. Symp Zool Soc Lond 42:99–108

Rovner JS (1980) Morphological and ethological adaptations for prey capture in wolf spiders (Araneae, Lycosidae). J Arachnol 8:201–215

Schmidt G (1999) Eine Klassifizierung der Stridulationsorgane. Mitt Dtsch Arachnol Ges 4:3–5

Schmidt G (2000) Zur Klassifizierung der Stridulationsorgane bei Vogelspinnen (Araneae: Theraphosidae). Entomol Z 110:58–61

Schultz S (2013) Spider pheromones — a structural perspective. J Chem Ecol 39:1–14

Schulz S (2004) Semiochemistry of spiders. In: Carde RT, Miller JG (eds) Advances in insect chemical ecology. Cambridge University Press, Cambridge, pp 110–150

Shillington C, Verrell P (1997) Sexual strategies of a North American "tarantula" (Araneae, Theraphosidae). Ethology 103:588–598

Simon E (1903) Histoire Naturelle des Araignées. 2. Paris, Roret, pp 669–1080

Stradling DJ (1994) Distribution and behavioral ecology of an arboreal "tarantula" spider in Trinidad. Biotropica 26:84–97

Uetz GW, Stratton GE (1982) Acoustic communication and reproductive isolation in spiders. In: Witt P, Rovner J (eds) Spider communication: mechanisms and ecological significance. Princeton University Press, Princeton, pp 123–159

Varenberg M, Pugno NM, Gorb SN (2010) Spatulate structures in biological fibrillar adhesion. Soft Matter 6:3269–3272

Viera C, Japyassú HF, Santos A, Gonzaga MO (2007) Teias e forrageamento. In: Gonzaga MO, Santos AJ, Japyassu HF (eds) Ecologia e comportamento de aranhas. Editora Interciencia Ltda, Rio de Janeiro, pp 45–66

Wilder SM, Devito J, Persons MH, Rypstra AL (2005) The effects of moisture and heat on the efficacy of chemical cues used in predator detection by the wolf spider *Pardosa milvina* (Araneae, Lycosidae). J Arachnol 33:857–861

Wilson DM (1967) Stepping patterns in tarantula spiders. J Exp Biol 47:133–151

Wohlfart E, Wolff JO, Arz E, Gorb SN (2014) The whole is more than the sum of all its parts: collective effect of spider attachment organs. J Exp Biol 217:222–224

Wolff JO, Gorb SN (2012a) Comparative morphology of pretarsal scopulae in eleven spider families. Arthropod Struct Dev 41:419–433

Wolff JO, Gorb SN (2012b) The influence of humidity on the attachment ability of the spider *Philodromus dispar* (Araneae, Philodromidae). Proc R Soc B 279:139–143

Wolff JO, Gorb SN (2015) Adhesive foot pads: an adaptation to climbing? An ecological surveyin hunting spiders. Fortschr Zool 118:1–7

Wolff JO, Nentwig W, Gorb SN (2013) The great silk alternative: multiple co-evolution of web loss and sticky hairs in spiders. PLoS One 8:1–13

World Spider Catalog (2017) World Spider Catalog. Natural History Museum Bern, version 18.0. http://wsc.nmbe.ch. Accessed 8 Apr 2017

Yañez M, Locht A, Macías-Ordóñez R (1999) Courtship and mating behaviour of *Brachypelma klassi* (Araneae, Theraphosidae). J Arachnol 27:165–170

Chapter 3
Neotropical Arachnological Collections: Basic Tools for the Development of Spider Research

Antonio D. Brescovit, Oscar Francke, Álvaro Laborda, Griselda Montiel Parra, Cristina Scioscia, Miguel Simó, and Carlos Víquez Núñez

Abstract Natural history collections are essential tools for development in biological research. Since the nineteenth century, arachnological collections in the Neotropics have been collaborating to carry out research on spiders and human resource training. In many cases, they have been the starting point of several research fields and the first steps allowing the development of arachnological studies in many Latin American countries. An important part of the future production of arachnological knowledge is deposited in these collections. They preserve critical data about the natural history of Neotropical spiders in several areas of knowledge, such as taxonomy, systematics, ecology, ethology, and biogeography. The present chapter is focused on the historical, present, and future perspectives of Neotropical arachnological collections and their contribution to spider research. We also discuss the main role that araneological collections will play in the future of knowledge of Neotropical spider diversity and its conservation.

A.D. Brescovit (✉)
Laboratório Especial de Coleções Zoológicas, Instituto Butantan,
Av. Vital Brasil, 1500, CEP 05503-900 São Paulo, SP, Brazil
e-mail: antonio.brescovit@butantan.gov.br

O. Francke • G.M. Parra
Departamento de Zoología, Instituto de Biología, Universidad Nacional Autónoma de México, 3er Circuito exterior s/n, Ciudad Universitaria, Coyoacán CP 04510, Mexico
e-mail: offb@ib.unam.mx

Á. Laborda • M. Simó
Sección Entomología, Facultad de Ciencias, Universidad de la República,
Iguá 4225, CP 11400 Montevideo, Uruguay
e-mail: simo@fcien.edu.uy

C. Scioscia
División Aracnología, Museo Argentino de Ciencias Naturales "Bernardino Rivadavia" (CONICET), Av. Ángel Gallardo 470, C1405DJR Ciudad Autónoma de Buenos Aires, Argentina
e-mail: crisscio@yahoo.com.ar

C.V. Núñez
Independent Researcher, Heredia, Costa Rica
e-mail: cviquez1774@gmail.com

C. Viera, M.O. Gonzaga (eds.), *Behaviour and Ecology of Spiders*,
https://doi.org/10.1007/978-3-319-65717-2_3

Natural history collections represent the first step for biodiversity inventories. It is known that most of the living species in the world are undiscovered (Mora et al. 2011). Many of them are not registered, but another portion is deposited in scientific institutions, unstudied yet and waiting to be discovered. Collection data preserve critical information about distribution, phenology, vegetation, and ecology of organisms, with interest in several fields of research, such as taxonomy, systematics, ecology, ethology, and biogeography. As stated by Caetano and Aisenberg (2014), biological data outside repositories represent forgotten treasures about the natural history of life. The process of gathering and indexing information represents a powerful issue for understanding the diversity, ecology, and evolution of living species. Single-sample collection data have been used to estimate species richness through extrapolation, which makes it possible to make decisions regarding environmental conservation (Petersen and Meier 2003). But much more than that, metadata associated with the label of each specimen deposited in collections are a potential source for obtaining additional information about ecological interactions, or even specimens for supplying genetic or morphometric data (Schilthuizen et al. 2015).

The Neotropics is the most biodiverse region in the world, and as such it has served as a laboratory for many studies about the origins of biological diversity (Rull 2007). Spiders form a megadiverse group, with more than 45,000 species described (WSC 2017). In that region, the study of arachnids and particularly spiders began in the eighteenth century, with the voyages of European naturalists and the first arachnologists who collected and deposited specimens in collections in Europe (Platnick and Raven 2013). In the nineteenth century, natural history museums and spider collections were founded in Latin America, followed by the development of the first scientists who worked in this part of the world. After this, a cascade of researchers and the development of knowledge about Neotropical spiders has developed to date. The current scenario shows an important number of groups which work in many countries in the Neotropical region, most of them associated with arachnological collections. With the main objective of creating an inventory of the extant biodiversity, institutions developed several projects in natural areas of the Neotropics starting in the last decades of the twentieth century. They promoted a higher rate of increase in the number of specimens from different types of biota deposited in collections. From those, the Amazonian region is one of the most diverse Neotropical subregions, with more than 1,000 spider species described (Adis 2002); but other areas with high biodiversity and endemism remain understudied. Although some of them have been sampled, the specimens are conserved as large ecological collections waiting to be curated and become available for the scientific community to study. Obviously, such efforts demand the investment of space, funding, and staff in what frequently represents a bottleneck for advances in the knowledge of biodiversity. In a globalized world, the future perspectives of spider collections should be focused on archiving and sharing data in repositories for the international arachnological community. But this goes hand in hand with the fact that these institutions will need the necessary support to ensure their development and management.

This chapter refers to how representative natural history collections in Latin America have been contributing to the development of the study of Neotropical spiders, focusing on their history, specimens housed, and different areas of scientific productivity.

Conservation, Bites, and Smuggling: Facts of the Costa Rica Spider Collections

Naturalists who explored Costa Rica made the first contributions to the study of arachnids. The German physicians Karl Hoffmann and Alexander von Frantzius collected and sent thousands of specimens to the Royal Zoological Museum of Berlin. In 1869, Von Frantzius wrote a publication about a tarantula spider or *pica-caballo* (Spanish for "horse biter," which is a common name used in Costa Rica and other Central American countries). In that paper, the spider was misidentified as a *Mygale*, and 24 years later it was collected by Hoffmann; Ferdinand Karsch (1879) described it and named it as *Sphaerobothria hoffmanni* (Theraphosidae) (Hilje 2013) (Fig. 3.1a).

The scientific collections of the National Biodiversity Institute (INBio) became part of the National Museum of Costa Rica on March 27, 2015. After 25 years of continuous management of these collections, INBio was not in an economic position to take proper care of them. Collections were transferred to the Government of Costa Rica and lawfully received by the National Museum of Costa Rica, the current owner and custodian of the total valuable collections, including Insecta, Fungi, Plantae, Myriapoda, Onychophora and Arachnida.

The arachnological collection, which is physically separated from the insect collection, was established by Carlos Víquez around 1996. During early periods, part of the time was used to select and extract all the spiders that were mixed with other animals from different kinds of alcohol traps. In the 1980s, Dr. Carlos Valerio, a notable researcher in the field of arachnology in Costa Rica, donated a collection of books and reprints on various topics in arachnology, which helped to consolidate the foundations and knowledge that would be necessary to identify groups, families, and species in the collection. In 1995, INBio developed the fourth class for "parataxonomists", which lasted for 6 months, consisting of 16 people trained in various collection techniques and identification of different groups of Arachnids. These people were from the communities of Osa and La Amistad. This course was of great importance for the development of a formal collection; Dr. Gustavo Hormiga was invited as the main instructor, who trained the parataxonomists and one of us (CVN) in the Pittier research station in the National Park La Amistad Pacífico. Thanks to the contribution of Dr. Hormiga, the skilled parataxonomists continued to collect Arachnids in their respective areas of work, and already had some basis for recognizing the main groups and performing basic identifications. Today, the collection is housed in six cabinets, which are in a room with controlled temperature and

Fig. 3.1 Female of the species (**a**) *Sphaerobothria hoffmanni* and (**b**) *Phoneutria boliviensis*

humidity conditions. All the material is preserved in alcohol in concentrations varying between 80% and 95%. Now, the collection does not have any responsible curators or additional technical personnel working on it. The National Museum is in the process of improving the general conditions and is looking for ways to hire new staff to work full-time in this collection.

The collection houses 48,568 specimens, of which approximately 31,000 are sorted to class level, and the remaining specimens as follows: 499 specimens (2.74%) order Amblypygi, 14,261 specimens (78.31%) order Araneae, 1167 specimens (6.41%) order Opiliones, 561 specimens (3.08%) order Pseudoscorpiones, 123 specimens (0.68%) order Ricinulei, 96 specimens (0.53%) order Schizomida, 1434 specimens (7.87%) order Scorpiones, 57 specimens (0,31%) order Solifugae, 12 specimens (0.07%) order Uropygi; recently we found the first specimen of the order Palpigradi. These specimens represent 78 families, 248 genera and 483 species (approximately 46.9% of the species currently reported for Costa Rica).

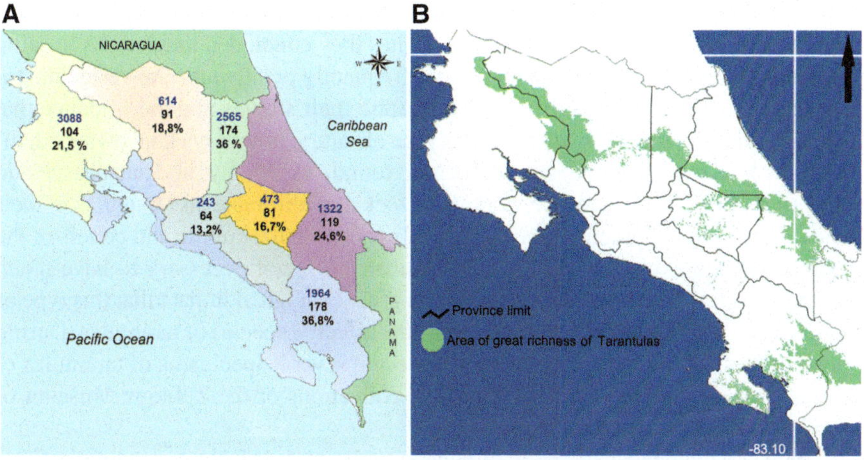

Fig. 3.2 (**a**) Map of Costa Rica showing databased records of specimens (in *blue*), species (in *black*, and percent). (**b**) Richness map for Tarantulas in Costa Rica

Guanacaste and Heredia are the provinces with the highest collecting effort, whereas the provinces of San Jose and Cartago have the lowest (Fig. 3.2a).

With regard to the material designated under the category of types, the collection houses 250 specimens, as follows: Holotypes (69), Paratypes (166), and Allotypes (15), representing the following orders: Amblypygi (one), Araneae (178), Opiliones (one), Pseudoscorpiones (one), Schizomida (56), and Uropygi (nine).

A total of 64 publications have used the scientific collections of the National Museum of Costa Rica, based on Santo Domingo de Heredia; 89.06% have been focused on aspects related to the description, taxonomy, and classification of species, while only 10.93% are dedicated to the description of the ecological aspects of the species. The total number of publications related to ecology have related to Costa Rica, whereas at the taxonomic level 33 international articles and 24 for Costa Rican fauna have been published. A very important aspect in the use of data of this collection is the creation of products for general use, which help people to come and learn more about the scientific research and the value of the collections.

Perhaps one of the most important uses of information from biological collections concerns health problems in Costa Rica. Some scientific articles have described cases where the utility of collections is clearly reflected in helping find solutions to problems related to human and animal health. Historical articles on spiders of Costa Rica are referenced, and we also show how new tools of research allow scientists to obtain new results. They also show that the collections provide the scientific bases to make decisions in biological and conservation aspects.

The first case concerns tarantulas, the family Theraphosidae, known in Costa Rica as "picacaballos" as explained above, which are feared due to their large size and hairy appearance (Valerio 1980). A series of myths have been developed with respect to the bite and the "orinada" (means peeing in Spanish) of these spiders. Valerio (1980) and Herrero and Odell (1988) studied the characteristics of the venom and digestive secretions of some species of tarantulas to determine the veracity of these myths. However, after those studies they concluded that lesions in cattle and horses attributed to these spiders were frequently positive for the vesicular stomatitis virus (Herrero and Odell 1988). Despite their studies and the evidence that freed spiders from being involved in this type of injury, nowadays land owners dedicated to livestock farming are frequently found, especially in Guanacaste and Puntarenas provinces, who hire workers to seek and eliminate all tarantulas in their land. These types of practices are very deeply rooted in the culture and beliefs of the people, and are extremely difficult to eliminate. It is not necessary to have much experience in conservation to determine the great ecological impact that this type of practice can have on natural populations of the affected species of tarantulas. During those studies, Drs. Valerio and Herrero collected several specimens of tarantulas of at least two new species, and reviewed the collections of the Zoology Museum of the University of Costa Rica.

Thanks to the latest generation of genetic tools, such as DNA barcoding or CO1, today it is possible to differentiate cryptic species in a more efficient way; the use of DNA from collections or directly from fresh tissues is a very important tool in the description and classification of species. The case of the species *Phoneutria boliviensis* Pickard-Cambridge 1897 (Ctenidae) (Fig. 3.1b) in Costa Rica might be a case where these genetic tools can reveal the existence of a misidentification (Víquez, in preparation). To be confident with regard to the correct identification of a species, it is critical to define other important ecological aspects such as behavior or even toxicity characteristics, which is impossible to do if we don't know that we are dealing with a complex of species. *Phoneutria* spiders live in most of the humid lowland of Costa Rica, and with the use of a combined new technology such as barcoding or ecological niche and potential distribution maps, it is also possible to get better results. Recently, some banana companies based in Costa Rica have had problems related to *Phoneutria* spiders, since occasionally these spiders manage to get past internal controls and get into packages of fruit ready to be exported, ending up in a supermarket or a house in Europe or the United States. When one of these spiders, or any other kind of spiders confused with a *Phoneutria* spider, is found, it can cause terror, because of the misinformation that surrounds them. People who are frightened after finding the spider file lawsuits against the companies that imported the bananas, causing heavy losses to the economy of a country like Costa Rica. The precise knowledge and good representation of specimens in collections help to prevent situations of this kind; often, good specimens of a collection can ensure a precise identification, and most of the time that is sufficient to dismiss a substantial demand.

The last case pertains to Black Widows (Theridiidae). In Costa Rica, there are at least three species of *Latrodectus* spiders, two of them introduced as invasive species [*Latrodectus geometricus* C. L. Koch, 1841 and *Latrodectus mactans* (Fabricius, 1775)], and the other is an undescribed native species (*Latrodectus* sp). The introduced species only inhabit the anthropized lands near buildings, houses, and other structures made by humans. In addition to this, there is only one record of these introduced spiders in coffee plantations. On the other hand, the native species frequents the natural dry forest areas in the Northern Pacific of Costa Rica, where it inhabits open areas with rocks (Víquez and Vásquez 2015). Samples of DNA barcoding of specimens from collections have confirmed that the *Latrodectus geometricus* are not significantly different from samples in other parts of the world, but apparently are most closely related to samples from the West Indies. For the species of the dry forest (*Latrodectus* sp), we know that there are similar specimens in collections from Mexico, Honduras, and Nicaragua; however, a comparative approach not has been made. Latrodectism in Central America has been published in recent years (Vasquez 2013; Víquez and Vasquez 2015). The cases were treated at the health center in the village of Samulali, municipality of San Ramon, Nicaragua, with the patients being subsequently transferred to the Hospital School Cesar Amador Molina. These cases were due to the bite of *Latrodectus* sp. in two children 12 and 13 years old, reported as moderate and severe respectively. People poisoning is a growing problem in the region; however, it tends to be underestimated by authorities. An improvement in collections and investment in research can be a preventive solution to dismiss future problems in human health.

Illegal animal smuggling is a global problem that directly and indirectly affects the Mesoamerican corridor and each country individually, including Costa Rica. Deeply understanding the characteristics of this market, how it is developing, who is involved and the methods used to smuggle specimens, will allow the implementation of strategies of mitigation and control, which will involve the application of national laws, the improvement of collections, and an increase in scientific research. A recent investigation (Víquez 2015) reveals an illicit market present in Costa Rica of more than 26 species of spiders, of which 70% are endemic, 15% restricted to only two countries (Costa Rica and Panama), and 15% are species of wide distribution in Central America, with only three species protected in CITES.

The lack of knowledge about existing fauna in the country, the inability to taxonomically identify one species from another, and the lack of national investment in research, collections, education, control, and punishment for acts of smuggling of wildlife, makes Costa Rica, and even more the Central American region, the ideal place to come to capture animals illegally!

If we analyze the distribution maps of species affected by the smuggling and overlay each of these maps, we can obtain a new map of species richness (Fig. 3.2b) that shows which parts of Costa Rica are mostly affected by these illegal practices. Exercises like this can be useful to plan control and conservation strategies of species.

Collections and scientific publications are not exempt from participation in this illegal market, while they still accept specimens that do not have real collecting permits or come from the market on the Internet (all specimens sold on the Internet that are from Costa Rica are illegal). When publishers, magazines, or bulletins accept articles where the legality of the specimens used is not checked, they are favoring and promoting illegal collecting and smuggling of specimens, and putting at risk the conservation of each of those species.

Spider Collections in Mexico: Discovering Their Diversity Before They Disappear

The Instituto de Biología (IB) of the Universidad Nacional Autónoma de Mexico (UNAM) was founded in 1929, and from its beginnings it had an important collection of chelicerates, including the "Colección de alacranes mexicanos de C. C. Hoffmann", as well as specimens from other orders of arachnids, many of them identified to species (Brailovsky et al. 1993). This collection was started in the 1970s and had representatives of all the extant orders of arachnids, plus 60 type specimens: scorpions (12), harvestmen (34), spiders (12) and pseudoscorpions (two) (Hoffmann 1993). Later, Dra. Anita Hoffmann, daughter of C. C. Hoffmann, donated her personal collection of arthropods to the IB, formalized by a legal contract signed on April 18, 1991. Dra. Tila Ma. Pérez Ortiz joined the IB in May of 1992 (Fig. 3.3a) and was named head of the Acarology Laboratory, where her duties

Fig. 3.3 (**a**) Dra.Tila Ma. Pérez Ortiz, Curator of the CNAN from 1991 to 2004. (**b**) Official donation of the Anita Hoffmann collection to the Instituto de Biología in 1992 (from left to right: Lic. Eduardo J. Vallejo Santin, Director of the Patrimonio Universitario UNAM, Dra. Anita Hoffmann Sandoval, Dr. José Sarukhan Kermes, Rector of UNAM, and Dr. Antonio Lot Helgueras, Director of the Instituto de Biología, UNAM)

Table 3.1 Arachnid taxa (excluding Acari) reported from Mexico and corresponding figures of those represented in the CNAN database

	Mexico			CNAN		
	Families	Genera	Species	Families	Genera	Species
Amblypygi	1	3	25	1	3	16
Araneae	67	453	2388	61	247	398
Opiliones	10	67	253	7	40	74
Palpigradi	1	1	4	1	1	4
Pseudoscorpiones	17	64	162	15	42	35
Ricinulei	1	1	16	1	1	14
Schizomida	2	6	43	2	6	18
Scorpiones	8	38	283	7	21	155
Solifugae	2	11	79	2	8	31
Uropygi	1	1	4	1	1	4
Total	**110**	**645**	**3257**	**98**	**370**	**749**

included curating the "A. Hoffmann" collection. Subsequently, through the kind support of entomologist/parasitologist Dionisio Pelaez, the IB acquired the scorpion collection of Mr. Jordi Juliá, which held hundreds of specimens from throughout Mexico. Additionally, the National Insect Collection turned over to Dra. Pérez the arachnids that they had, including part of the C. C. Hoffmann scorpion collection (Pérez 1993, Pérez et al. 2004). Thus, through the efforts of Dra. Pérez, the arachnid collection, including mites, grew considerably under her care.

The National Collection of Mites, "Colección Nacional de Ácaros" (CNAC), and the National Collection of Arachnids, "Colección Nacional de Arácnidos" (CNAN), were officially founded on November 4, 1997, the latter with Scientific Collection Register DFE.IN.041.1097 from the Instituto Nacional de Ecología (INE) and the Secretaría del Medio Ambiente y Recursos Naturales y Pesca (SEMARNAP). Dra. Pérez was named Curator of both collections. Dr. Oscar F. Francke, a scorpion taxonomist, joined the IB in September, 2004, and was named Curator of the CNAN, while Dra. Pérez remains as Curator of the CNAC. Currently, the curatorial staff is composed of Oscar F. Francke Ballvé, Ph. D. (Scorpion systematics) and Griselda Montiel Parra, M. Sc. (systematics of parasitic mites on arachnids). The CNAC has 52,000 specimens belonging to the ten non-acarine orders of arachnids. Most of the samples have been catalogued, even though some might be identified only to genus or family level; about 80% of the samples are catalogued, and the remaining 20% constitutes spiders and pseudoscorpions in need of sorting and identification. The number of species reported in Table 3.1 is therefore only an approximation, but includes 98 families, 370 genera, and 749 species. The catalogue was made public on March 9, 2016 and can be consulted at UNAM's "open data" portal (https://datosabiertos.unam.mx). The vast majority (90%) of the specimens are preserved in 80% ethanol and kept in glass vials and jars, depending on their size and the number of specimens in each lot. Only properly preserved pseudoscorpions and palpigrades are in permanent mounts in microscopic slides. Unidentified samples are in vials in 80% ethanol.

Starting with the donation of the Anita Hoffmann Collection, the proper conservation, protection, and documentation of type specimens became a priority for Dra. Pérez, and continues to this day (Fig. 3.3b). The most valuable and scientifically important subset of the CNAN contains 2096 specimens belonging to 246 species of all the orders of arachnids: there are 137 holotypes, 1891 paratypes, 51 allotypes, 11 topotypes, four lectotypes and two neotypes. A special cabinet holds the publication with the original description or original designation of each of the type specimens, duly recorded in a database in the program End Note.

Mexico has a very large biological diversity, primarily because within its territory there are two large biogeographic realms that converge: the Nearctic realm contributes with numerous species from temperate North America, and the Neotropical realm with elements from warmer Central and South America. Additionally, Mexico's complex geological history and the mountains of the various Sierras Madres create complex climatic patterns, allowing almost all known terrestrial ecosystems to exist (Sarukhán et al. 2009). Collecting efforts have targeted many of those ecosystems, extending from tropical rainforests, deciduous scrub forests, cloud forests, and temperate forests to deserts and sand dune areas. Urban and semi-urban areas have been the target of specific student projects. The CNAN holds specimens from the 31 political sub-divisions of the country, but most significantly it has samples of arachnid diversity in many protected areas such as Biosphere Reserves, National Parks, and National and State Protected Natural Areas. It is important to note that donations and interchanges with foreign colleagues have allowed the CNAN to obtain different arachnids from 32 countries. All the specimens are a source of information for taxonomic and systematic research, biogeography, ecology, behavior, environmental education, and conservation projects. Mexico has several species of scorpions and spiders of medical importance, and ranks first in the world in accidents from scorpion stings requiring medical attention. Thus, special efforts have focused on arachnids that have an impact on public health. Mexican tarantulas have been over-exploited for the pet trade, and studies at the CNAN have also focused on these endangered species.

The study of arachnid diversity in the Lacandona Rainforest has been of great importance. This region in the Ocosingo Municipality of Chiapas is one of the areas with the highest biodiversity in the country and in the humid tropics of the New World, located at 16° 40′ N and 91° 58′ W, with elevation ranging from 170 to 450 m. During a period of 4 years there were 14 collecting expeditions to the zone, each approximately 1 week long, with a variable number of participants (2–7), with variable experience in the field, and with various collecting techniques employed for the various habitats and micro-habitats used by different arachnids. A total of 17,667 arachnids were collected, representing ten of the 11 orders (Solifugae missing), 109 families, 235 genera, and 321 species. Several new species were collected, of which two spider species have been formally described (*Modisimus deltoroi* Valdez-Mondragón and Francke, 2009 (Pholcidae) and *Ochyrocera chiapas* Valdez-Mondragón, 2009 (Ochyroceratidae)), and others are undergoing description and illustration. This inventory is similar to the one reported from the "Reserva Duke" in Central Amazonia in Brazil (Adis et al. 2002), and it is the only project in Mexico to bring together Mexican arachnologists studying all the orders of arachnids.

Research done at the CNAN has resulted in a number of publications grouped as follows: (a) description of 138 species of arachnids new to science, from ten different orders (exc. Acari) (67 papers), (b) redescriptions of scorpions, spiders, and harvestmen (23 papers), (c) descriptions for the first time of males of spiders, scorpions, and pseudoscorpions (four papers), (d) chapters in books (four), and (e) miscellaneous contributions on arachnid diversity in certain environments, urban spiders, new country and state records, species delimitation methods, phylogenetics, etc. (36 papers).

Specimens from the CNAN have contributed in the production of 31 theses: 16 bachelor's degrees, 11 masters, and three doctorates. These include nine orders of arachnids: Amblypygi, Solifugae, and Uropygi (one), Araneae (15), Opiliones (one), Palpigradi (two), Schizomida (two), Scorpiones (seven), and Pseudoscorpiones (two). Finally, one thesis (López 2012) included a comparative analysis of abundance and richness of six orders of arachnids in slopes with different exposure in the same mountain (north facing, south facing, flat ridge) in deciduous thorn forest.

CNAN specimen data have been captured in three different databases: Biota: the Biodiversity Database Manager (ver. 2.04) (Colwell 2012), PostgreSQL (CCUD 2013), and Excel. Bibliographic references are also automated in EndNote X7.4; to date, it holds 3996 papers on the ten orders of arachnids. The collection also has a comprehensive coverage of specialized books.

The Instituto de Biología, UNAM, annually provides two budgets for the CNAN, one assigned to purchase materials and the second to support research. In addition, outside support has been obtained from the National Science Foundation (NSF, USA), Consejo Nacional de Ciencia y Tecnología (CONACyT, Mexico), and Instituto Bioclon (Mexico), which has been used to pay for equipment, field work, and student financial support.

Starting in 2010, the CNAN has participated in the Barcode of Life project (Red Temática de Código de Barras-CONACyT) or MexBOL, and a sub-set of recent collections is preserved according to molecular study specifications. Currently, it has 1171 refrigerated samples in 96% ethanol, of which 75% is already databased. Likewise, scanning electron microscope studies have increased, and the CNAN has a special collection of arachnids, or parts thereof, coated with gold in small plates; this collection has 234 plates from 92 species of harvestmen, spiders, schizomids, and ricinulids.

The CNAN website has information about its holdings, staff, students, publications, and field trips (http://www.ibiologia.unam.mx/html/Site_2/Welcome.html). There is also a web page with information on medically important arachnids from Mexico, their current and their potential distributions (http://www.ibiologia.unam.mx/html/mainframe.html).

The CNAN participates in the up-keep of a site for new species maintained since 2003 by the staff of the IB (http://www.unibio.unam.mx/especiesnuevas/). The staff and students are deeply committed to basic education, participating in classes, workshops, and conferences. Additionally, we provide assistance identifying arachnids brought in by the general public, particularly through the social networks Facebook (Aracnidos) and Twitter (@Aracnidos1).

The Arachnida Collection of the Butantan Institute: Before and After the Fire

The Butantan Institute was officially opened in 1901, and its origin was associated with the fight against the bubonic plague that around 1898 caused a severe epidemic on the São Paulo coast, at the port city of Santos. The Serumteráphico Institute was created in order to produce the cure against the plague; it was located at Butantan farm, situated west of São Paulo, and coordinated by Doctor Vital Brazil. In 1925, the official name was changed to Butantan Institute (IBSP), now linked to the Secretaria da Saúde do Estado de São Paulo. Since its foundation, the IBSP has concerned itself also with accidents involving scorpions and spiders. For such studies, initially, a collection of scorpions was received from the Paris Natural History Museum. On account of this initial exchange, other samples from various collections and suppliers began to arrive at the Institute, expanding this collection of dead animals and forming the first institutional collection.

However, unfortunately this material was not curated appropriately, as is the current preservation of collections today, and few of those older specimens can be found in the collection (Lucas 2014). The official collection of arachnids and millipedes at the IBSP was started only in 1924 and is today one of the largest in South America, even after a fire incident in October 2010 that destroyed part of the building which was shared by both collections.

In 1924, Jean Vellard (Fig. 3.4a) was appointed for the organization of the Museum and the systematics of arachnids (Oliveira 1980–81). He started the collection of spiders and scorpions, incorporating his private collection coming from the Natural History Museum in Paris, with material collected by him in Niterói, state of Rio de Janeiro (Lucas 2014), and Serra de Paranapiacaba, state of São Paulo (Brescovit et al. 2009).

The initial organization had no records in a registry book, but in 1925 this collection already had 250 cataloged lots. The records started only in 1926, at the initiative of Vellard, who organized the data of the first 400 spiders for the Araneae collection. This collection was represented by Mygalomorphae and Araneomorphae. In 1930, Vellard left the Butantan Institute after the retirement of Vital Brazil. The position held by Vellard was occupied by Alcides Prado, who initiated in 1932 the first records of the scorpion collection, giving preference to species of the genus *Tityus*, frequently used in the production of sera (Lucas 2014).

In February 1939, Wolfgang Bücherl (Fig. 3.4b), a biologist trained in Germany, was hired. His first projects were on systematics, biology, and the poison of chilopods (centipedes). From 1950 onwards, Bücherl reorganized the spider collection,

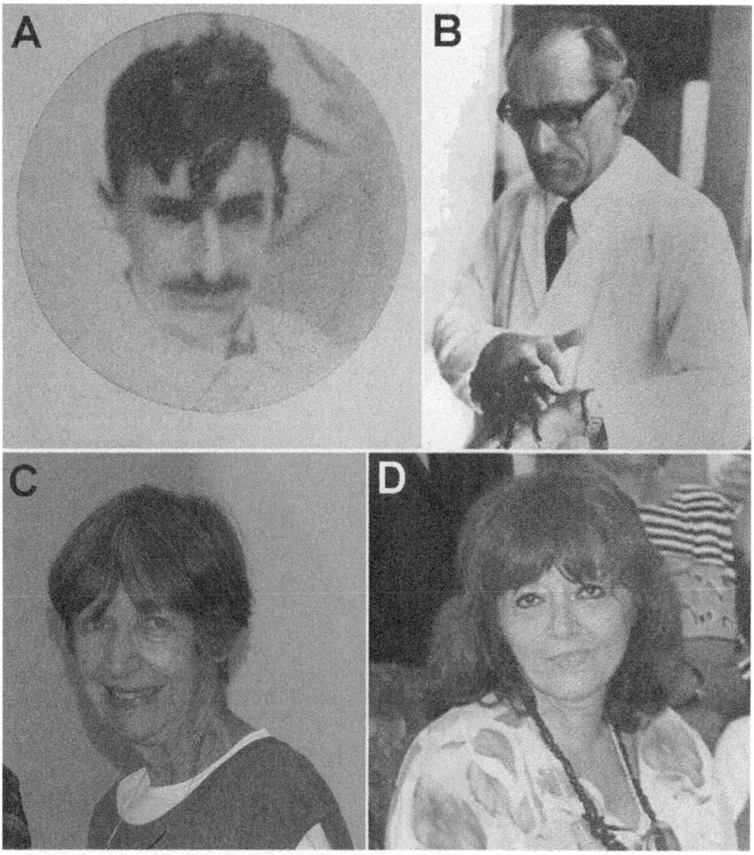

Fig. 3.4 (**a**) Jean Vellard, from the files of IBSP (Divisão de Extensão Cultural). (**b**) Wolfgang Bücherl, photo courtesy of Ilona Bücherl (daughter of Dr. Bücherl), also contained in IBSP files (Divisão de Extensão Cultural). (**c**) Sylvia M. Lucas, photo courtesy of Roberto H. Pinto Moraes. (**d**) Vera R. von Eickstedt, photo courtesy Eduardo Ramirez

dividing it into two sections; Araneomorphae and Mygalomorphae. He expanded the collections, organizing the material that came from the Animals Reception Sector. Bücherl retired in 1967, when he was replaced by the researcher Sylvia Marlene Lucas (Lucas 2014).

Under the direction of Sylvia Marlene Lucas (Mygalomorphae) (Fig. 3.4c) and with the collaboration of researcher Vera Regina von Eickstedt (Araneomorphae and Scorpions) (Fig. 3.4d), there was a momentum of growth of the IBSP collection which lasted until the 1990s. The largest contribution during this period occurred as a result of faunal redemptions during the construction of large hydroelectric plants in Brazil, such as UHE Tucuruí, UHE Balbina, and UHE Samuel in the Amazon, and several smaller ones in the state of São Paulo. In 1970, there was another reorganization in the IBSP, and the Laboratório de Zoologia Médica was renamed as the

Fig. 3.5 (**a**) Old building of the Laboratório de Artrópodes where the Arachnida collection was located. Photo courtesy of Rafael Indicatti. (**b**) Actual building of the Laboratório Especial de Coleções Zoológicas, where the IBSP collections are deposited. Photo courtesy of Roberto H. Pinto Moraes

Laboratório de Artrópodes, which remained in the building in conjunction with the Laboratório de Herpetologia until 2010 (Fig. 3.5a, b).

In 1994 four new researchers joined the Laboratório de Artropódes, expanding the work of gathering and diversifying the collection. In 2000, under the coordination of A. D. Brescovit, they obtained the Thematic Project Biota/FAPESP São Paulo (Brescovit et al. 2011), and for almost 2 years collections were performed in more than 50 areas of the Atlantic Forest and Cerrado in Brazil, increasing the collection with more than 150,000 specimens of arachnids.

In October 2010, a fire incident occurred in the building where the Herpetological and Arachnida collections were located, damaging some parts of this collection. After this incident, all collection trusteeship in IBSP was unified and zoological collections brought together in a single laboratory. As a result, on August 16 2011 the Laboratório Especial de Coleções Zoológicas (LECZ) was created, where all IBSP collections were gathered in the same space and under a Board of Curators. Soon after, in September 2013, 3 years after the fire, the building that today accommodates the zoological collections of IBSP was opened.

Today, the Arachnida collection of LECZ shelters representatives of all orders, but after the fire incident of October 2010, the material of Palpigradi, Ricinulei, Schizomida, and Solifugae was almost all lost. Today this collection has been totally revised, and the number of specimens included is approximately 203,000 lots and about 275,000 exemplars. As a result of the fire, 30% of the Arachnida acquisitions were lost. Among the most abundant orders in the now reduced collection are: Araneae — 80 families, about 950 genera, and about 1400 nominal species; Opiliones — 16 families, 137 genera, and 310 nominal species; and Scorpiones — 14 families, 70 genera, and 296 nominal species. In addition we still have in stock for inclusion in the collection about 40,000 specimens, of which 70% are spiders.

The type material is represented by Araneae — 240 types of holotypes and paratypes (several lost during the fire; inventory in progress); Opiliones — 91 types, 69 recovered and 22 lost in the fire (Coronato-Ribeiro et al. 2013); and Scorpiones — 35 types of holotypes and paratypes. The types of other orders were not inventoried before the fire, and today all may be lost.

The geographical representation (overall) of the specimens deposited in the collection is quite extensive, with examples from 62 countries, of which Brazil is the best represented today with over 185,000 lots (about 245,000 lots), followed by Peru with more than 700 lots, Chile with about 500 lots, and Germany with more than 350 lots. The representativeness of the collection is more concentrated in American countries, but we own or have received donated collections from South Africa, Yemen, and Malaysia totaling more than 100 lots. From Brazil, we have material from all states, especially from the states of São Paulo (with more than 85,000 lots), Bahia (18,000), Mato Grosso do Sul (12,000) and Minas Gerais (more than 10,000).

Today, 90% of the material of all orders in the collection has been scanned into Excel spreadsheets.

Due to the size, quantity, and diversity of the collection of material, today it is used for taxonomic work, inventories, ecological purposes, and even in biogeographical activities

From Brazil, we have relevant samples from the biomes Atlantic Forest, Cerrado, Caatinga, and Amazon rainforest. Many new species have been described from these areas, with emphasis on spiders and scorpions, both by IBSP researchers and researchers from other national and international institutions. We can also highlight numerous faunal inventories expanding the knowledge of these biomes, such as Brescovit et al. (2009), Nogueira et al. (2014), and Azevedo et al. (2016), to name but a few. A recent study showing the importance of the Opiliones collection of the IBSP are the biogeographic works published by DaSilva et al. (2015), who worked with areas of endemism in the Atlantic Forest.

It is also noteworthy that the IBSP collection now shelters the largest and most organized collection of Brazilian cave spiders, with more than 20,000 exemplars, of which 10% are troglomorphic representatives of which many have been described (Rheims and Brescovit 2004; Brescovit et al. 2012) or for which descriptions are in the process of being created (Brescovit and Sánchez-Ruiz 2016). In addition, other cave groups are also being targeted for work (Brescovit et al. 2016), contributing much to the enrichment of knowledge of cave fauna in the country.

We also highlight recent inventories in the South American countries Chile and Peru, where IBSP teams collected large samples in dry or desert regions of these countries, aiming at a specific fauna of these areas, such as representatives of the Sicariidae family, the genus *Sicarius* (Magalhaes et al. 2017) and Lycosidae, and the genus *Allocosa* (Brescovit and Taucare-Ríos 2013). Finally, we can define the collection of the IBSP today as one of the most important in Brazil, which has been the basis for studies by many arachnology professionals, has contributed significantly as an asset depository of representatives of the Brazilian fauna, and has fostered a high number of research projects with highly diverse material.

Arachnological Collections from Argentina: The Country with the Most Diverse Environments in the Neotropics

Arachnological collections in Argentina have their origins thanks to a widespread entomological tradition, since entomologists, in their frequent field trips, were the ones who deposited in their collections arachnids that had fallen into their nets.

The need for the creation of a Natural History Museum in the country, currently the Argentine Museum of Natural Sciences "Bernardino Rivadavia" (MACN-CONICET) was recognized in 1812; since then, collection of arachnological specimens has been carried out. Nevertheless, the first "formal" collection of arachnids from Argentina was that of the Museum of La Plata (FCNyM-UNLP, founded in 1884), created by the entomologist Carlos Bruch (1873–1943), and consolidated by Dr. Maximiliano Birabén (1893–1977) and his wife, Dra. María Hylton Scott, as a result of their numerous scientific excursions. This important material was sent to Brazil, between 1920 and 1950, to be studied by Dr. Cândido de Mello-Leitão, who described numerous species, and provided new records from Argentina. This collection of the Museum of La Plata, which houses the largest quantity of type specimens of the country, thus identified and organized by Mello-Leitão and Birabén, was used as a strong reference by the professors Rita Schiapelli (1906–1976) and Berta Gerschman de Pikelín (1905–1977) of the MACN of Buenos Aires (Fig. 3.6a) for identifying and organizing the arachnids they had been studying after they joined the Institution in 1929. In 1937, Schiapelli was named "responsible for the arachnid collection"; and in 1952, she was appointed as head of the newly created "Arachnology Section" within the Entomology Division. At that time, they started the inventory book and formally created the National Collection of Arachnology (today MACN-Ar), which is currently one of the largest and most organized Latin America collections. The subsequent incorporation into the present Arachnology

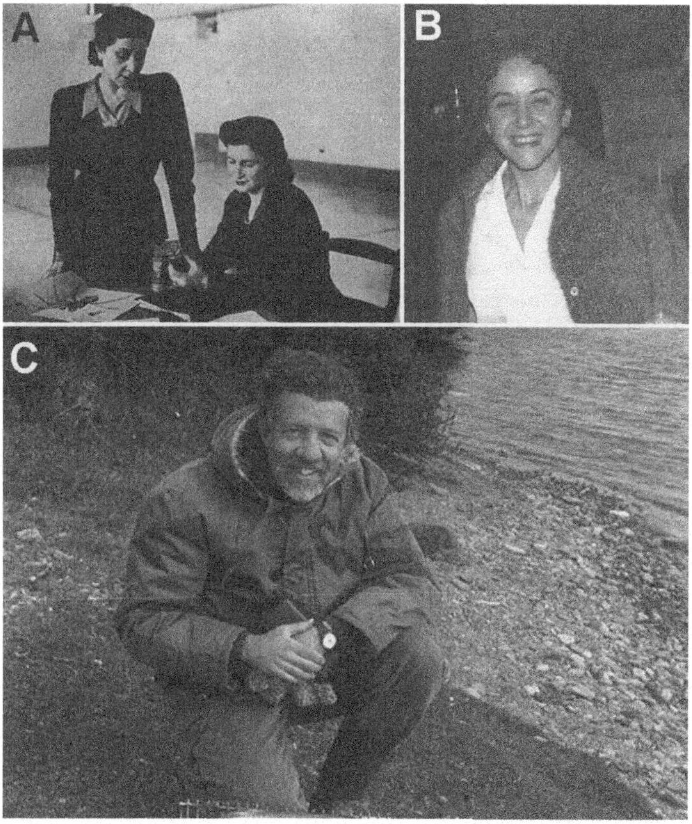

Fig. 3.6 (**a**) Rita D. Schiapelli (*left*) and Berta Gerschman de Pikelín (*right*). (**b**) María Elena Galiano. (**c**) Emilio Maury (Source: MACN https://sites.google.com/site/aracnologiamacn/nuestra-division)

Division of the work of María Elena Galiano (1928–2000) (Fig. 3.6b) and Emilio Maury (1940–1998) (Fig. 3.6c), tireless travelers and collectors, made the collection grow regularly. In this century, the staff in the division has increased significantly; researchers, technicians and fellowships contribute to the increase of the collection. In addition, thanks to the new trends of studies on biodiversity, systematized field trips, organized by specific areas, or through joint research projects between local and foreign researchers, the periodical provision of fresh, abundant, and relevant material from special groups or in specific areas of interest is assured.

In addition to these two important collections (FCNyM and MACN-Ar), other collections were appearing from other institutions of the country. Some of them having wide geographic representation, and others containing local or regional records, they all provide their data to the portal of the National System of Biological Information, from which they are then exported to the website of the Global Biodiversity Information Facility organization (GBIF). The records include taxonomic, biological, geographical, and temporal information, including collectors. The development of electronic catalogues has made data processing easier, and made it possible to work on large quantities of records for sharing information concerning specimens between websites on the Internet. At the same time, between 2004 and 2008, the International Barcode of Life Project (iBOL) arose and became established; Argentina, through the MACN, has had an outstanding role in the project as a regional node, being the center of barcoding activities for Argentina and the Southern Cone.

In decreasing order of numerical importance, the collections of Argentina are composed of specimens of Araneae, Scorpiones, Opiliones, Solifugae, Pseudoscorpiones, and Acari. A recent numerical estimation indicates the presence of about 150,000 batches of specimens, where "a batch" is a vial, which may contain one, few, or many specimens; 85% of this material belongs to the MACN-Ar collection. The type material is distributed between the collections from La Plata and Buenos Aires, and at a much lower percentage from the Cordoba collection, totaling about 1000 primary types and nearly 2000 paratypes.

The geographical representation of the collections is varied; some of them house specimens covering all the ecoregions of the country as well as specimens from many other countries, and others are more regional or restricted to the area of influence where they are located.

The National Collection of Arachnology (MACN-Ar)

This consists mainly of native specimens of spiders, scorpions, harvestmen, pseudoscorpions, solifuges, and mites from all provinces of the country, and including the Malvinas Islands. In much lesser numbers, there are specimens from Brazil, Chile, Uruguay, Paraguay, Bolivia, Peru, Ecuador, Venezuela, Panama, Guatemala, Haiti, Jamaica, Nicaragua, Cuba, Mexico, USA, South Africa, France, Belgium, Spain, Greece, England, Italy, Poland, Russia, Sweden, Germany, India, Nepal, Iran, Thailand, and Australia. It is important to highlight that in addition to the fixed specimens, this collection keeps numerous specimen stubs with samples prepared for scanning electron microscope (SEM); samples of webs, preserved by a method designed by researchers of the Arachnology Division; a photographic database of specimens and diagnostic structures obtained under different degrees of magnification (ca. 50,000 images); and samples of DNA to process through the iBOL project (Barcode of Life). In addition, for more than 60 years a breeding lab, originally created by María Elena Galiano, has been steadily maintained; and there is a live collection of Arachnids which normally are under study for the development of research projects in progress.

Arachnology and Miriapodology Section, Zoology of Invertebrates Division, Museum of La Plata, Faculty of Natural Sciences and Museum, National University of La Plata

This possesses spiders, scorpions, harvestmen, pseudoscorpions, solifuges, and mites from all over Argentina, as well as specimens from Uruguay, Bolivia, Brazil, Peru, Chile, Paraguay, Colombia, Ecuador, Venezuela, Surinam, Panama, Costa Rica, Mexico, USA, Canada, Italy, and Democratic Republic of Congo. This collection also has an annex repository in the Center of Parasitological Studies and Vectors (CEPAVE-CONICET), which preserves a great quantity of spiders from agroecosystems from diverse crops and regions of the country, as well as material from wild preserved areas.

Laboratory of Animal Diversity I (CDAI), Faculty of Exact, Physical, and Natural Sciences, National University of Córdoba

Started in the 1960s, this facility possesses spiders, scorpions, harvestmen, pseudoscorpions, and solifuges representative of almost the whole country, as well as from Brazil, Chile, Colombia, Paraguay, Peru, and Uruguay.

Institute of Invertebrates, Miguel Lillo Foundation, Tucumán

Started in the 1940s, this institute possesses spiders, scorpions and harvestmen, mainly from northern Argentina, neighboring countries, and Peru.

Entomological and Arachnological Collection: Argentine Research Institute of Arid Zones (IADIZA-CONICET), Mendoza

Started in the 1970s, this is a reference collection of the biodiversity from the arid ecosystems. It possesses spiders, scorpions, harvestmen, pseudoscorpions, and solifuges, from the biomes of the Mount, Patagonia, Puna, and Chaco, as well as from 11 provinces from Chile.

Collection of Arthropods of the National University of the Northeast (CARTROUNNE): Faculty of Exact and Natural Sciences and Survey, National University of the Northeast

Started in 2004, this collection possesses spiders, scorpions, harvestmen, and pseudoscorpions from the provinces of Formosa, Chaco, Corrientes, Santa Fe, and Misiones.

Entomological and Arachnological Collection: Municipal Museum of Natural Sciences "Lorenzo Scaglia" (MMPE), Mar del Plata, Province of Buenos Aires

Started in 1987, this collection possesses spiders, scorpions, harvestmen, pseudoscorpions, and solifuges, from different specific biotopes from the southeast of the province of Buenos Aires, as well as specimens from other Argentine provinces, Brazil, Chile, and Uruguay.

Laboratory of Invertebrate Zoology II (LZI), Department of Biology, Biochemistry and Pharmacy, National University of the South, Bahia Blanca, Province of Buenos Aires

Recently created, this is mainly composed of mygalomorph spiders from Argentine.

Institute for the Study of Invertebrate Biodiversity, Museum of Natural Sciences, Faculty of Natural Sciences, National University of Salta (IEBI-MCN-FCN-UNSa)

This collection deserves special comment because of the peculiarity of its material. Founded in 2007, it was established to deposit a large amount of material collected from different research projects related to the biodiversity of invertebrates in the North of Argentina, both actual and fossil fauna from natural and anthropized environments. The IEBI collection has invertebrates of natural and altered environments belonging to different ecoregions from Northern Argentina: Campos and Malezales, Espinal, Esteros del Ibera, Humid Chaco, Chaco Serrano, Yungas, Monte de Sierras and Bolsones, Prepuna-Puna, and Altos Andes (with lots of specimens collected from 4000 up to 6000 m a.s.l.). From disturbed environments, there are specimens from agro-ecosystems in the provinces of La Rioja, Tucumán, Salta, and Santa Fe including all kinds of crops; from urban environments of Salta; and from other Yungas environments affected by oil activity in Jujuy (Calilegua National Park). The material of several invasive species in the northwestern areas of Argentina is also recorded. The collection has more than 50,000 specimens identified to family, genus, and/or species/morphospecies level. All material is computerized, recorded in spreadsheets, and a database of digitized images of each species/morphospecies generated in the same IEBI through a Web application named IEBIData. Access to this Internet database is still restricted, and permission of the administrator is needed. Each species has a record of abundance, the locality with geographic coordinates, among other field data.

In Argentina, the diversity of climates and topographies has allowed the development of the most typical biomes, belonging to the Neotropical region and the Andean region. Morrone (2001, 2014) has redefined the biogeographic regions of Latin America and the Caribbean, on the basis of the distribution of endemic or characteristic taxa and, in many cases, using arachnid distributions for defining areas. According to these studies, Argentina is included in the following biogeographic regions, mentioning, where appropriate, the spider taxa used in their characterization:

1. Neotropical region: essentially, this comprises the tropics, from Northern Mexico to the center of Argentina, characterized by spiders: *Cerionesta*, *Hyetussa*, *Chira*, *Jollas* and *Rudra* (Salticidae), *Acanthoscurria* and *Cyriocosmus* (Theraphosidae).

    1.1 Brazilian sub-region is the largest area of the Neotropical region, extending through Southern and Central Mexico, Central America, and Northwestern South America.

    1.1.1 South Brazilian dominion, Amazonian forest, southwest of the Amazon river.

    1.1.1.1 Yungas province; extends along eastern slopes of the Andes, between 300 and 3500 m altitude, from Northern Peru to Northwestern Argentina.

    1.2 Chacoan sub-region: occupies Southeastern South America.

    1.2.1 Chacoan dominion; Northern and central Argentina, Southern Bolivia, Western and central Paraguay, Uruguay and central and Northeastern Brazil.

    1.2.1.1 Chaco province; extends from Southern Bolivia, Western Paraguay, Southern Brazil and North-Central Argentina, characterized by spiders: *Echemoides giganteus* Mello-Leitão, 1938, *E. mauryi* Platnick and Shadab, 1979 and *E. penicillatus* (Mello-Leitão, 1942) (Gnaphosidae), *Sumampattus pantherinus* (Mello-Leitão, 1942) (Salticidae), and *Cyriocosmus versicolor* (Simon, 1897) (Theraphosidae).

    1.2.1.2 Pampean province; occupies Central Western Argentina between 30 and 39° SL, Uruguay and Southern portion of the Brazilian state of Rio Grande do Sul, characterized by spiders: *Trachelopachys cingulipes* (Simon, 1886) (Trachelidae), *Echemoides argentinus* (Mello-Leitão, 1940) (Gnaphosidae), *Acanthogonatus tacuariensis* (Pérez-Miles and Capocasale, 1982), *Pycnotele auronitens* (Keyserling, 1891), *Stenoterommata crassistyla* Goloboff, 1995 and *S. tenuistyla* Goloboff, 1995 (Nemesiidae), *Homoeomma uruguayense* (Mello-Leitão, 1946) (Theraphosidae), and *Anelosimus misiones* Agnarsson, 2005 (Theridiidae).

1.2.2 Paraná dominion; Northeastern Argentina, Eastern Paraguay, Southern Brazil west of the Serra do Mar and toward central Rio Grande do Sul and Eastern Brazil, between 7 and 32° SL.

1.2.2.1 Paraná Forest province; extends across Southeastern Brazil, Northeastern Argentina, and Eastern Paraguay, characterized by spiders: *Rachias timbo* Goloboff, 1995, *Stenoterommata iguazu* Goloboff, 1995, and *S. uruguai* Goloboff, 1995 (Nemesiidae).

1.2.2.2 *Araucaria* Forest province; occupies Southern Brazil and Northeastern Argentina, between 600 and 1800 m altitude, characterized by spiders: *Anelosimus rabus* Levi, 1963 (Theridiidae).

1.3 South American transition zone; highlands of the Andes between Western Venezuela and Northern Chile and Central Western Argentina.

1.3.1 Monte province; extends along Central West of Argentina, approximately between 27 and 44° SL, from Salta to the Northeast of Chubut, characterized by spiders: *Echemoides balsa* Platnick and Shadab, 1979 (Gnaphosidae), *Acanthogonatus birabeni* Goloboff, 1995 (Nemesiidae), and *Sicarius rupestris* (Holmberg, 1881) (Sicariidae).

1.3.2 Puna province; comprises Bolivia, Northern Argentina and Chile, and Southern Peru, characterized by spiders: *Trachelopachys bidentatus* Tullgren, 1905, *T. machupicchu* Platnick, 1975, and *T. tarma* Platnick, 1975 (Trachelidae).

1.3.3 Prepuna province extends across Central and Northwestern Argentina, from Jujuy to Northern Mendoza.

2. Andean region; extends along the high mountain ranges of Venezuela, Colombia and Ecuador, through the coastal desert and the Puna of Peru, Bolivia, Northern Chile and Northern Argentina, up to the Argentinian–Chilean Patagonia.

2.1 Subantarctic sub-region; includes the Southern Andes, from them 37° SL until Cape Horn, including the archipelago of Southern Chile and Argentina, and Malvinas, South Georgia and Juan Fernandez Islands, characterized by spiders: *Acanthoceto cinerea* (Tullgren, 1901), *A. pichi* Ramírez, 1997, *Oxysoma* and *Ferrieria* (Anyphaenidae), *Scotinoecus* (Hexatelidae), and *Gnolus* (Mimetidae).

2.1.1 Maule Region province; situated in Southern Chile and Argentina, between the 34–37° SL, characterized by spiders: *Acanthoceto ladormida* Ramírez, 1997 (Anyphaenidae), *Scotinoecus cinereopilosus* (Simon 1889) (Hexatelidae), *Apodrassodes mercedes* Platnick and Shadab, 1983, *A. pucon* Platnick and Shadab, 1983 and *Echemoides malleco* Platnick and Shadab, 1979 (Gnaphosidae), *Mallecomigas* (Migidae), and *Acanthogonatus brunneus* (Nicolet, 1849), *A. hualpen* Goloboff, 1995, *A. mulchen* Goloboff, 1995, *A. nahuelbuta* Goloboff,

1995, *A. recinto* Goloboff, 1995, and *A. tolhuaca* Goloboff, 1995 (Nemesiidae).

2.1.2 Valdivian Forest province; located in Southern Chile and Argentina, to the south of Maule province, reaching 47° SL.

2.1.3 Magellanic Forest province; located in Southern Chile from 47° SL up to Cape Horn and Southern Argentina in small regions in the west of Santa Cruz and Tierra del Fuego, characterized by *Scotinoecus fasciatus* Tullgren, 1901 (Hexathelidae).

2.1.4 Magellanic Páramo province; located in Southern Chile and Argentina, limited to the East by the Magellanic forest province.

2.1.5 Malvinas Islands province; formed by the Argentine archipelago Malvinas Islands and South Georgia, situated in the South Atlantic Ocean, about 550 km from Tierra del Fuego.

2.2 Patagonian sub-region; extends along Southern Argentina, from central Mendoza, widening through Neuquén, Río Negro, Chubut, and Santa Cruz, up to the North of Tierra del Fuego; and reaches Chile in the Aisén and Magallanes provinces.

2.2.1 Central Patagonia province; located in Southernwest Argentina, from the Center of Mendoza up to the South of Santa Cruz

2.2.2 East Patagonia province; located in Southwest Argentina, this is a narrow strip along the austral Andes, widening towards the South up to Santa Cruz.

Although numerous surveys were conducted in recent decades across the country, several environments still remain without spider records. At the same time, many spider specimens are deposited, waiting to be studied. This highlights the value of the Argentinean arachnological collections for future studies in different fields of spider research.

Spider Collections in Uruguay: Understanding the assemblages at a Biological Crossroads

In Uruguay there are two arachnological collections, founded in the mid-twentieth century. One of them belongs to the Museo Nacional de Historia Natural (MNHN) from Montevideo. This institution, one of the oldest public organizations, was founded in 1837 and preserves the first spider collection of this country, which was created and curated for many years by Prof. Roberto Capocasale, containing exemplars deposited between 1940 and 2000. The other collection belongs to the Facultad de Ciencias (FCE), Universidad de la República. It was initiated by the first Uruguayan arachnologist, Pablo San Martín, who worked mainly in Scorpiones. Both collections preserve exemplars deposited over more than 60 years of arachnological research in this country.

The study of spiders in Uruguay began with R. Capocasale in the 1960s. At first, most records were collected by hand, mainly biased towards diurnal and ground species. Therefore, the oldest representatives are large- to medium-size spiders mainly collected under stones, trunks, or in environments associated with human activities such as buildings, roads, or crops. Although there are representatives from a great variety of localities, many of them came from localities in urban centers or near them. Post-graduate studies in spiders, and the beginning of research of rapid assessment in biodiversity using standardized methods, allowed the growth of the arachnological collections from the last decades of the twentieth century onwards. At this time, studies for conservation of natural areas included spiders in the surveys, and conducted an increased effort of curation and digitalization of collection data. The samples taken in these survey campaigns, from different places and seasons, provided useful phenological and reproductive data which produced numerous ecological and ethological publications about Uruguayan spiders. The First Meeting of Arachnologists from the Southern Cone, celebrated in Uruguay in 1997, represented an important step for the development and study of the material deposited in Uruguayan arachnological collections. Academic exchanges between arachnologists from 'Cone Sur' led to an increase in cooperative studies covering different disciplines such as taxonomy, systematics, ethology, and biogeography, increasing the identification of the material deposited and then improving the curation level of the collections.

As a result of many years of taxonomic research, Uruguayan mygalomorphs are mainly identified to species level (Montes de Oca and Pérez-Miles 2009), including type exemplars and representatives from most parts of Uruguay. As expected, Araneomorphae constitutes the most number of exemplars deposited, but only a few families are catalogued to species level. Conversely, most of them are classified only to family or morphospecies, but others are grouped as understudied ecological collections, which makes access for the research community more difficult. Many parts of the country have been covered in surveys, including areas with conservation interest, but a great part is waiting to be studied and identified. Up to now, digitalization is based on electronic worksheets, and a photographic database has been created for specimen identification. The last catalogue of Uruguayan spiders was published by Capocasale and Pereira (2003). Since then, a great amount of spiders have been deposited in the arachnological collection, certainly increasing the number of species known from the country. Taking into account the last 30 years, there has been a significant increase in the papers published, derived from study of the material deposited in both Uruguayan collections. Most of them are related to taxonomic and behavioral research.

Uruguayan collections preserve material from more than 30 graduate and postgraduate studies in the last two decades. Taxonomic, systematic, and biogeographic studies have been mainly focused on Theraphosidae. Uruguay has a large scientific production in ethology, publishing many articles on behavioral issues in various groups of arachnids, and many of the specimens deposited in Uruguayan collections come from experiments carried out in those investigations. The main lines of research in ethology are focused on three experimental models: theraphosids, wolf spiders, and sub-social spiders.

The specimens used in behavioral studies and deposited in the collections are the empirical evidence of the experiments, and therefore provide objective information that allows the replication of these experiments and the possibility to accept or refute the assumptions made by the ethologists. The collection of the Museo Nacional de Historia Natural from Montevideo contains specimens used in the investigation of the cryptic species *Lycosa thorelli* (Keyserling, 1877) and *Lycosa carbonelli* Costa and Capocasale, 1984 (Lycosidae). These species have very similar morphological and ecological characteristics, and the main isolation barrier is their sexual behavior (Costa and Capocasale 1984). The specimens used in behavioral experiments, deposited in FCE collection, made it possible to study morphological characteristics in the sexual organs to clearly differentiate these species and to support the separation previously observed on the basis of behavioral characteristics (Simó et al. 2002). The family Theraphosidae is widely distributed in Uruguay; they live mainly in hills and grasslands. The best-known species are *Grammostola anthracina* (C. L. Koch, 1842), *Grammostola quirogai* Montes de Oca et al. 2015, *Eupalaestrus weijenberghi* (Thorell, 1894), and *Acanthoscurria suina* Pocock, 1903. The research on these species covered several topics of reproductive biology, testing various hypotheses related to sexual selection, sexual communication, and reproductive isolation (Costa et al. 2015). Other aspects such as locomotion and adhesion have also been studied in this group (Pérez-Miles et al. 2015).

Behavioral research in wolf spiders in Uruguay has been focused on two species as the main models: *Schizocosa malitiosa* (Tullgren, 1905) and *Allocosa senex* (Mello-Leitão, 1945). *S. malitiosa* is a very common species in anthropic environments of the Uruguayan coast. It has been widely studied in basic and descriptive aspects of sexual behavior, agonistic behavior, sexual selection, sexual communication, and reproductive strategies (Costa 1975, Costa 2013). Studies of this species using different approaches have generated a thorough understanding of their behavior.

Allocosa senex is a burrowing wolf spider that inhabits the sandy coasts of Uruguay, northeastern Argentina, and southern Brazil (Simó et al. 2017). It presents a sex role reversal, an unusual characteristic in spiders (Aisenberg and Costa 2008). This behavioral characteristic makes it an interesting species, and its sexual behavior has been widely studied. Research in this species also covers other aspects of their biology such as foraging, cannibalism, habitat preference (Ghione et al. 2013, Jorge et al. 2015), and its usefulness as a bio-indicator. With regard to the sub-social spiders, the subject of study has been *Anelosimus vierae* Agnarsson, 2012; this species builds communal nests in native trees, and it is considered a good model for studying the evolution of sociality in spiders. In this species, various aspects of sexual behavior as well as development, social behavior, and maternal care have been studied (Viera et al. 2006).

Another topic that has promoted the study of the spiders deposited in Uruguayan collections is biogeography. Uruguay is situated at a biogeographic crossroads where biota belonging to the Pampa, Atlantic Forest, Paranaense Forest, Chaco

Fig. 3.7 (**a**) Cerro Arequita, Sierra de Minas, Lavalleja, (**b**) Quebrada de los Cuervos, Treinta y Tres. Two ecosystems with high spider diversity in Uruguay

province, and Espinal province converge (Grela and Brussa 2005; Simó et al. 2014; Morrone 2014). Furthermore, these areas are connected by numerous rivers, forests, grasslands, and hill elevations that generate this confluence zone. Studies performed on the Río Uruguay revealed the biological corridor condition of this fluvial course between the Paranaense Forest and areas of Pampa (Laborda et al. 2011). Spider surveys in hilly forest of Southern Uruguay such as Sierra de las Ánimas revealed past connections with similar areas in Argentina, through the Peripampasic Arc (Costa et al. 1991; Ferretti et al. 2012) (Fig. 3.7a). The Atlantic Ocean and Río de la Plata estuary constitute a connection with other coasts of Brazil and Argentine, allowing gene flow for the survival of numerous species adapted to live in sandy areas (Costa et al. 2006). In Northern Uruguay, the tabular hill ranges are relicts of connection with Cerrado and the Parana Forest (Simó et al. 2015). At the east of the country, Quebrada de los Cuervos (Fig. 3.7b) and Valle del Lunarejo represent relicts of Atlantic Forest biota in the country (Simó et al. 1994). However, environments with human impact such as urban habitats or agroecosystems have been studied, showing how synanthropic or invasive species interact with native representatives (Simó et al. 2011). Spider collections have provided data for field guides and books recently published about Uruguayan spiders. One thing appears to be critical for the future: scientific collections need financial support and human resources for their development; both are necessary for collections to continue with their condition as repositories of the natural heritage, and as bases for the production of knowledge in different fields of spider research.

Acknowledgements O.F. and G.M. give special thanks to Ubaldo Melo Samper Palacios for all his help and support with the PostgreSQL database and the electronic entering of thousands of specimen; to Carmen Loyola for the images, and to Ana Francke for improving the English language. A. D. B. thanks colleagues Rafael P. Indicatti and Marcelo R. Duarte for the photographs, Thamara A. Barros for assisting with numerical data, and Leandro Pedroso for the English translation. Funding agencies: CNPq (Process 301776/2004-0) and FAPESP (2011/50689-0). C.V. thanks Maris Briggs for time and assistance provided in the correction and improvement of English language in this work. C.S. is grateful for the curators of the Argentine Institutions other than MACN for providing information of the collections under their responsibility.

References

Adis J (2002) Taxonomical classification and biodiversity. In: Adis J (ed) Amazonian Arachnida and Myriapoda. Pensoft, Sofía, pp 13–15

Adis J, Bonaldo AB, Brescovit AD, Bertani R, Cokendolpher JC, Condé B, Kury AB, Lourenço WR, Manhert V, Pinto-da-Rocha R, Platnick NI, Reddell JR, Rheims CA, Rocha LS, Rowland JM, Weygoldt R, Woas S (2002) Arachnida at 'Reserva Ducke', Central Amazonia/Brazil. Amazonia 18:1–14

Aisenberg A, Costa FG (2008) Reproductive isolation and sex role reversal in two sympatric sand-dwelling wolf spiders of the genus *Allocosa*. Can J Zool 86:648–658

Azevedo R, Moura ES, Lopes AS, Carvalho LS, Dias SC, Brescovit AD (2016) Arachnids from Araripe Plateau, Ceará, Brazil. Check List 12:1–9

Brailovsky H, Beutelspacher CR, Zaragoza-Caballero S (1993) La Colección Entomológica del Instituto de Biología. In: Brailovsky H, Gómez B (Compls) Colecciones Biológicas Nacionales del Instituto de Biología. Colecciones Zoológicas. Instituto de Biología. UNAM, México, pp 67–100

Brescovit AD, Sánchez-Ruiz A (2016) Descriptions of two new genera of the spider family Caponiidae (Arachnida, Araneae) and an update of *Tisentnops* and *Taintnops* from Brazil and Chile. Zookeys 622:47–84

Brescovit AD, Taucare-Ríos A (2013) Description of the female of *Allocosa yurae* (Strand, 1908) (Araneae: Lycosidae, Allocosinae). Zootaxa 3647:495–498

Brescovit AD, Rheims CA, Indicatti RP (2009) Aranhas (Arachnida) de Paranapiacaba. In: Lopes MIMS, Kirizawa M, Melo MMRF (eds) Patrimônio da Reserva Biológica do Alto da Serra de Paranapiacaba. A antiga Estação Biológica do Alto da Serra. Instituto de Botânica, São Paulo, pp 503–527

Brescovit AD, de Oliveira U, Santos AJ (2011) Aranhas (Araneae, Arachnida) do Estado de São Paulo, Brasil: diversidade, esforço amostral e estado do conhecimento. Biota Neotrop 11:1–32

Brescovit AD, Ferreira RL, Silva MS, Rheims CA (2012) *Brasilomma* gen. nov., a new prodidomid genus from Brazil (Araneae, Prodidomidae). Zootaxa 3572:23–32

Brescovit AD, Magalhaes ILF, Cizauskas I (2016) Three new species of *Misionella* from northern Brazil (Araneae, Haplogynae, Filistatidae). Zookeys 589:71–96

Caetano DS, Aisenberg A (2014) Forgotten treasures: the fate of data in animal behaviour studies. Anim Behav 98:1–5

Capocasale RM, Pereira A (2003) Diversidad de la Biota Uruguaya. Araneae. An Mus Nac Hist Nat Antropol 10:1–32

CCUD (2013) Plataforma de digitalización de colecciones desarrollada por la CCUD. Documento Propiedad de la MA.CCUD/ODT/001/2013/A. Coordinación de Colecciones Universitarias Digitales, UNAM, México, p 7

Colwell RK (2012) Biota 2.04: the biodiversity database manager. Software and Manual. Sinauer Associates, Sunderland, p 880

Coronato-Ribeiro A, Pinto-da-Rocha R, Rheims CA (2013) Catalogue of Opiliones (Arachnida) types deposited in the Arachnida and Myriapoda collection of the Instituto Butantan, São Paulo, Brazil. Zootaxa 3637:501–520

Costa FG (1975) El comportamiento precopulatorio de *Lycosa malitiosa* Tullgren (Araneae: Lycosidae). Rev Bras Biol 35:359–368

Costa FG (2013) Interacciones entre machos de *Schizocosa malitiosa* (Araneae, Lycosidae) en contextos sexual y no sexual. Bol Soc Zool Urug 22:66–71

Costa FG, Curbelo B, Pérez-Miles F (2015) Long-term persistence and water resistance of female sex cues in the tarantula (Araneae: Theraphosidae). Arachnology 16(9):311–313

Costa F, Pérez-Miles F, Gudynas E, Prandi L, Capocasale RM (1991) Ecología de los arácnidos criptozoicos, excepto Ácaros, de Sierra de las Ánimas (Uruguay). Órdenes y Familias. Aracnología 13(15):1–41

Costa FG, Capocasale RM (1984) Lycosa carbonelli, sp. nov.; una etoespecie simpatrida, sibilina de Lycosa thorelli (Keyserling) (Araneae, Lycosidae). J Arachnol 11: 423–431

Costa FG, Simó M, Aisenberg A (2006) Composición y ecología de la fauna epígea de Marindia (Canelones, Uruguay) con especial énfasis en las arañas: un estudio de dos años con trampas de intercepción. In: Menafra R, Rodríguez-Gallego L, Scarabino F, Conde D (eds) Bases para la conservación y el manejo de la costa uruguaya. Vida Silvestre Uruguay, Montevideo, pp 427–436

DaSilva MB, Pinto-da-Rocha R, DeSouza AM (2015) A protocol for the delimitation of areas of endemism and the historical regionalization of the Brazilian Atlantic Rain Forest using harvestmen distribution data. Cladistics 31:692–705

Ferretti N, González A, Pérez-Miles F (2012) Historical biogeography of mygalomorph spiders from the peripampasic orogenic arc based on track analysis and PAE as a panbiogeographical tool. Syst Biodivers 10:179–193

Ghione S, Simó M, Aisenberg A, Costa FG (2013) *Allocosa brasiliensis* (Araneae, Lycosidae) as bio-indicator of coastal sand dunes in Uruguay. Bull Br Arach Soc 16:94–98

Grela I, Brussa C (2005) Novedades para la flora del Uruguay: nuevo registro de *Agarista* (Ericaceae). Acta Bot Brasil 19:511–514

Herrero MV, Odell GV (1988) Algunas características de los venenos y secreciones digestivas de *Aphonopelma seemanni* y *Sphaerobothria hoffmanii* (Araneae: Theraphosidae) de Costa Rica. Rev Biol Trop 36:241–246

Hilje L (2013) Los primeros exploradores de la entomofauna costarricense. Brenesia 80:65–88

Hoffmann A (1993) Las colecciones de artrópodos de A. Hoffmann. Cuadernos 19, Instituto de Biología, Universidad Nacional Autónoma de México, México, p 43

Jorge C, Laborda Á, Días MA, Aisenberg A, Simó M (2015) Habitat preference and effects of coastal fragmentation in the sand-dwelling spider *Allocosa brasiliensis* (Lycosidae, Allocosinae). Open J Anim Sci 05(03):309–324

Laborda A, Montes de Oca L, Useta G, Pérez-Miles F, Simó M (2011) Araneae, Deinopidae, *Deinopis amica* Schiapelli and Gerschman, 1957: first record for Uruguay and distribution msap. Check List 8:1301–1302

López PTG (2012) Variación espacial y temporal de la diversidad de arácnidos (Arthropoda: Arachnida) en el cerro de la Coronilla, Tepecoacuilco de Trujano, Guerrero, México. Tesis de Licenciatura, Facultad de Estudios Superiores-Iztacala, Universidad Nacional Autónoma de México, México, p 102

Lucas SM (2014) A coleção de Aracnídeos do Instituto Butantan. Resumos do Encontro Nacional sobre Coleções Biológicas e suas interfaces 1:29

Magalhaes ILF, Brescovit AD, Santos AJ (2017) Phylogeny of Sicariidae spiders (Araneae: Haplogynae), with a monograph on Neotropical *Sicarius*. Zool J Linnean Soc 179:767–864

Montes de Oca L, Pérez Miles F (2009) Las arañas Mygalomorphae del Uruguay: clave para familias, géneros y especies. Innotec 4:41–49
Mora C, Tittensor DP, Adl S, Simpson AGB, Worm B (2011) How many species are there on earth and in the ocean? PLoS Biol 9(8):e1001127
Morrone JJ (2001) Biogeografía de América Latina y el Caribe. M and T–Manuales and Tesis SEA, vol. 3. Zaragoza, p 148
Morrone JJ (2014) Biogeographical regionalisation of the Neotropical region. Zootaxa 3782:1–110
Nogueira AA, Venticinque EM, Brescovit AD, Lo-Man-Hung NF, Candiani DF (2014) List of species of spiders (Arachnida, Araneae) from the Pico da Neblina, state of Amazonas, Brazil. Check List 10:1044–1060
Oliveira JL (1980) Cronologia do Instituto Butantan (1888-1891). 1ª Parte: 1888-1945. Mem Inst Butantan 44(45):11–79
Pérez TM (1993) La Colección Acarológica del Instituto de Biología. In: Brailovsky H, Gómez B (Compls.). Colecciones Biológicas del Instituto de Biología. Colecciones Zoológicas Instituto de Biología. UNAM, México, pp 35–39
Pérez-Ortiz T, Castelo-Calvillo JL, Durán-Barrón CG, González-Santillán E, Montiel-Parra G, Villegas-Guzman GA (2004) Historia y aspectos generales de la Colección Nacional de Arácnidos (CNAN). In: Morales MMA, Ibarra M, Rivera AP, Stanford SC (eds) Entomología Mexicana, vol. III. Sociedad Mexicana de Entomología, A. C., México, pp 77–82
Pérez-Miles F, Perafán C, Santamaría L (2015) Tarantulas (Araneae: Theraphosidae) use different adhesive pads complementary during climbing on smooth surfaces: experimental approach in eightarboreal and burrowerspecies. Biol Open 4:1643–1648
Petersen FT, Meier R (2003) Testing species-richness estimation methods on single-sample collection data using the Danish Diptera. Biodivers Conserv 12:667–686
Platnick NI, Raven RJ (2013) Spider systematics: past and future. Zootaxa 3683:595–600
Rheims CA, Brescovit AD (2004) A new cave spider of the genus *Lygromma* Simon (Araneae, Prodidomidae) from Minas Gerais, Brazil. Rev Iber Aracnol 9:325–327
Rull V (2007) On the origin of present Neotropical biodiversity: a preliminary meta-analysis about speciation timing using molecular phylogenies. Orsis 22:105–119
Sarukhán J, Koleff P, Carabias J, Soberón J, Dirzo R, Llorente-Bousquets J, Halffter G, González R, March I, Mohar A, Anta S, Maza J (2009) Capital natural de México. Síntesis: Conocimiento actual, evaluación y perspectivas de sustentabilidad. Comisión Nacional para el Conocimiento y Uso de la Biodiversidad, México, p 100
Schilthuizen M, Vairappan C, Slade ME, Mann DJ, Miller J (2015) Specimens as primary data: museums and 'open science'. Trends Ecol Evol 30:237–238
Simó M, Pérez-Miles F, Ponce De León R, Achaval F, Meneghel M (1994) Relevamiento de fauna de la Quebrada de los Cuervos; Area Natural Protegida (Dpto. Treinta y Tres, Uruguay). Bol Soc Zool Uruguay 2:1–20
Simó M, Seguí R, Pérez-Miles F (2002) The copulatory organs of the cryptic species *Lycosa thorelli* and *Lycosa carbonelli* and their hybrid progeny, with notes on their taxonomy (Araneae, Lycosidae). J Arachnol 30(1):140–145
Simó M, Laborda A, Jorge C, Guerrero Antunez JC, Dias M, Castro M (2011) Introduction, distribution and habitats of the invasive spider *Badumna longinqua* in Uruguay with notes on its world dispersion (Araneae, Desidae). J Nat Hist 45:1637–1648
Simó M, Guerrero JC, Giuliani L, Castellano I, Acosta LE (2014) A predictive modeling approach to test distributional uniformity of Uruguayan harvestmen (Arachnida: Opiliones). Zool Stud 53:1–13
Simó M, Núñez M, Ojeda L, Laborda A, Queirolo D (2015) Knowing the biological linkage: spider composition and guilds in a hill range of northern Uruguay. Bol Soc Zool Uruguay 24:117–129
Simó M, Lise AA, Pompozzi G, Laborda A (2017) On the taxonomy of southern South American species of the wolf spider genus *Allocosa* (Araneae: Lycosidae: Allocosinae). Zootaxa 4216:261–278
Valerio CE (1980) Arañas Terafósidas de Costa Rica (Araneae, Theraphosidae). I *Sericopelma y Brachypelma*. Brenesia 18:259–288

Vásquez V (2013) Envenenamiento por araña viuda negra en un niño. Reporte de un caso. Rev Nicar Pediatr 1:26–28

Viera C, Ghione S, Costa FG (2006) Regurgitation among juveniles in the subsocial spider *Anelosimus studiosus* (Araneae, Theridiidae). J Arachnol 34:258–260

Víquez C (2015) Causas y efectos del contrabando de vida silvestre en los Arácnidos de Costa Rica. Informe final de Práctica Dirigida para optar por el título de Bachiller en Manejo de Recursos Naturales, p 89

Víquez C, Lozado V (2015) El Latrodectismo humano en Centroamérica, VI Congreso Científico de la Región de Azuero. Chitré, Herrera

World Spider Catalog (2017) World spider catalog. Natural History Museum Bern, online at http://wsc.nmbe.ch, version 17.0. Accessed 20th Feb 2017

Chapter 4
Habitat Selection and Dispersal

Dinesh Rao

Abstract Spiders can be roughly grouped into four guilds based on their manner of foraging: web builders, ambushers, wandering spiders, and refuge builders. These guilds have different needs that influence the choice of a site when focused on foraging, shelter, or reproduction. The choice of a place to settle is of fundamental importance to spiders, and this choice is mediated by a variety of factors. The physical attributes of the microhabitat also play an important role in influencing site occupancy. In this chapter, I introduce the term *Topophilia* to encapsulate these concepts. In most spiders, site selection may occur at two distinct stages: natal dispersal (when spiderlings leave the eggsac) or breeding dispersal (when adults or subadults search for the optimal location). The end result of dispersal is dispersion. Spiders can be solitary, in aggregations, colonial, or social; and these dispersion patterns imply different microhabitat needs. In this chapter, I review studies done on Neotropical spiders using the aforementioned classification as a guide to bringing out patterns of habitat selection and dispersal.

A spider's choice of a site to build its web or refuge is not a random decision; instead, it is predicated on a number of variables. Some of these variables, such as the point of attachment, are in the spider's control, but most are dependent on stochastic factors experienced in the microhabitat or decisions made by the preceding generation. Site settlement decisions, once made, are rarely permanent, especially for spiders that rebuild webs or relocate frequently. Prey abundance, frequency of damage to webs caused by inanimate objects such as falling twigs or leaves, and other factors can induce a change in site. Resistance to change is more likely for spiders that invest heavily in their webs or burrows, because of the energy cost incurred in site establishment. If a spider decides to abandon a site, it also incurs additional costs such as lost foraging opportunity or the ever-present threat of predation during movement, a time when the spider is most vulnerable. A constant tension exists between the decision to leave or stay, and this decision can be made

---

D. Rao (✉)
Inbioteca, Universidad Veracruzana, Xalapa, Veracruz, Mexico
e-mail: dinrao@gmail.com

C. Viera, M.O. Gonzaga (eds.), *Behaviour and Ecology of Spiders*,
https://doi.org/10.1007/978-3-319-65717-2_4

easier for the spider if site selection results in the occupation of a protected and food-rich spot.

Site selection, and consequently dispersion, is mainly driven by factors such as foraging needs, substrate availability, the presence of conspecifics, and physiological considerations. Animals seek out patches where they may encounter sufficiently profitable encounter rates with their preferred food source. One of the theoretical constructs for understanding the selection of a site is the optimal foraging theory, which suggests that organisms are more likely to choose a site if the expected gains from staying outweigh the costs such as predation risk or searching time. One of its central tenets is the marginal value theorem, which indicates that organisms must constantly sample their environment or have a near complete idea of their surroundings to make decisions regarding whether to stay or to leave.

However, the optimal foraging theory has been frequently criticized because of the difficulty in testing the concepts empirically. Because of the complexity of the natural world, there is limited evidence to indicate that spiders forage optimally. For example, in a study of the orb web spider *Gasteracantha fornicata*, spiders may not be able to assess the suitability of a site because of the high variability in numbers of prey intercepted (Edwards et al. 2009). Work with crab spiders by Morse (1993) showed that though spiders were more often found on high-value sites (e.g., the umbel of a flower), they did not use floral cues before settling. This work suggested that site selection was largely mediated by insect activity directly, which may vary according to the site or other factors such as seasonality (see Romero and Vasconcellos-Neto 2010).

The flip side of site selection is site desertion. According to optimal foraging considerations, a spider should move only when the benefit of moving is greater than the cost of relocation and cost of staying (e.g., because of a prey-poor site). Because the arrival of insects is a highly stochastic process, with a significant influence of wind, Janetos (1986) suggested that spiders such as sheet-web builders are more likely to stay, whereas orb spiders can relocate more often; this is attributed to a lower variance in prey arrival. Differences exist between species of orb-web builders, with spiders such as *Nephila* building large, semi-permanent structures that are repaired rather than recycled, whereas spiders such as *Leucauge* are known to build several webs in the same day.

Site desertion has been historically considered under the predictions of the marginal value theorem (Charnov 1976), which codifies the following idea: a forager should leave a site when the predator encounters less prey than is expected on average for that site. However, this idea rests on the assumption that the predator itself would contribute to a depletion in available prey in the patch. This has been called the giving-up density. The trap-building nature of many spiders make it difficult for the predators to actively deplete a site, especially since they are stationary and prey are mobile. In a review dealing with trap-building predators, Scharff et al. (2011) suggest that hunger levels may not be sufficient to prompt relocation, as a result of several factors such as the ability to modify the trap size (hungry spiders can build larger or stickier webs), the cost of predation pressure during relocation, and the inability to sample the site adequately before constructing a web. However, long-term starvation can prompt web neglect and possibly eventual relocation. In a series of experiments using the desert spider *Agelenopsis aperta*, spiders were given a choice of settling in habitats under conditions of varied temperature and prey availability (Riechert and Gillespie 1986). Spiders in this experiment did not settle in sites that had both preferred temperature and prey, but rather chose based on one of the variables. The authors hypothesized that this pattern was observed because temperature and prey availability are correlated in the desert, and hence the spider may make decisions based on either cue. But perhaps the critical factor here is that once settled, the spider can modify the thermal surroundings by changing the shape or the orientation of the web, or other common means of behavioral thermoregulation, rendering the site suitable.

Substrates that are of high quality in terms of availability of prey or web attachment support should be sought after by discerning spiders. In an experiment using artificial chicken wire substrates, Robinson (1981) demonstrated that increased diversity of attachment points led to an increase in colonization by spiders. However, under such circumstances, especially when high-quality sites are scarce, agonistic interactions between conspecifics is likely to occur, with resident spiders having to defend their territory from intruders. In a series of experiments with *Agelenopsis aperta*, Reichert and colleagues showed that spiders compete for high-quality sites, but the likelihood of prolonged contests is linked to the feeding history of the resident at the site (Riechert and Gillespie 1986). In the context of colonial spiders, even though a marked level of tolerance to conspecifics has been found, competition still occurs for favorable parts of the entire structure. The outer layer receives more prey but also more attention from predators, whereas the inner layer is more protected but with a lower likelihood of prey. Spiders therefore compete to occupy the middle layer. The prey disadvantage of being on the inner layer is somewhat mitigated by the 'ricochet effect' (Uetz 1989; Rao 2009), but competition to occupy the optimal position in the colony is intense (Rypstra 1979).

Abiotic factors such as temperature and humidity may also influence site selection and dispersion. Spiders that are in open areas under the sun may risk overheating, and therefore need to compensate in various ways, most notably by behavioral thermoregulation. This can include active changes in posture or orientation of the web, or passively through color morphs. For example, *Micrathena gracilis*, an orb-web spider, is known to orient its web to minimize the effect of sun rays, but only in open habitats as opposed to closed habitats (Biere and Uetz 1981). *Verrucosa arenata* is a colour polymorphic orb-web spider (Rao et al. 2015) that shows marked difference between the white and the yellow morphs in heating and cooling rates when exposed to an artificial light source (Rao and Mendoza-Cuenca 2016).

Topophilia

The choice of a site is contingent on multiple factors, namely substrate availability, potential prey, potential predators, abiotic factors such as humidity, precipitation, temperature, and presence of conspecifics, among others. Since substrate characteristics are of primary importance followed by subsequent site evaluation, there is a need for a single term to encapsulate the spider's choice. I propose to use the term 'topophilia', derived from the Greek (topos: place; philia: affinity to). This term was popularized in the 1970s by human geographer Yi-Fu Tuan in the context of human settlements (Tuan 2013). He defined it thus "Topophilia is the affective bond between people and place or setting", where the affective bond is informed by the perception, attitude, and worldview of the people in question. To look at this in terms of spiders, topophilia includes the behavioral as well as structural preferences of spiders, and with the acknowledgement that topophilia within a species is fluid (but constrained), and that it can change with changing environmental conditions or changes in ontogeny of the individual. In a sense, topophilia is conceptually similar to niche breadth, where the niche represents the range of tolerances and conditions required by an organism to survive in a given area; but topophilia places more emphasis on affinities, with particular weight placed on the role of microhabitat structure (Fig. 4.1). To elaborate, a spider might survive on a certain plant, but may prefer another.

In a sense, arachnologists already know this. Years of experience in the field gives us the ability to identify suitable location when in search of a particular species. We already know that if we are looking for *Theridion*, we need to look for folded leaves. It has to be a thin leaf and not a thick leaf. The fold is conical, usually with the tip in the uppermost part. If we are seeking a spider of the family Theridiosomatidae, we need to look in low-lying undergrowth close to water sources. Rolled-up leaves can indicate the presence of Salticid nests. We know that certain orb-web spiders prefer edges and clearings, and others prefer to build on tree trunks or low in the undergrowth. Topophilia is simply a formalization of this kind of natural history knowledge.

Traditionally, few studies have focused on site establishment — especially the process of choosing a site — perhaps because it is logistically challenging. Researchers generally find spiders by searching for already occupied spots, or conduct experiments under laboratory conditions where a large number of variables are controlled by necessity. Even the process of web building can be a long and tedious behavior to study, since typically spiders don't build webs from start to finish in one smooth sequence; there are various pauses in the construction that can last hours. To see the spider make choices would imply a time period that would extend beyond 24 h of continuous monitoring of individuals. In the case of dispersal studies, one generally finds already established sites and then extrapolates from there on. Special mention must be made of course to studies of ballooning, where it is possible to witness and measure the propensity of spiderlings to disperse, but measuring or recording the subsequent steps in the dispersal process is logistically challenging. In

Fig. 4.1 (**a**) Crab spider (Thomisidae) on a flower, awaiting pollinators. (**b**) Aggregation of *Nephila clavipes* (Araneidae), using conspecific webs for support. (**c**) *Metepeira* sp. (Araneidae), using a dried leaf as a refuge. (**d**) A Hersilid camouflaged on a tree trunk. (**e**) A tetragnathid, using a single grass stem as a base to build its web. (**f**) *Eustala* sp. (Araneidae), choosing a cactus scar as a refuge

some social and sub-social spiders, due to the limited dispersal range of the spiders, and given that subadults are also involved in the dispersal process, it may be easier to track the final settlement site of these individuals (perhaps by using marked spiders). However, despite these limitations, it is still possible to develop extensive understanding of the effects of microhabitat selection and dispersal in spiders. In my view, the ideal approach to these questions is to do a combination of field-based studies where a baseline is established and then do a laboratory- or large field-cage-based experiment in order to reduce the complexities of the study system. This approach may only work for certain systems, and it may well be that an idiosyncratic approach designed for the particular species in question will pay more dividends rather than a uniform approach.

Given the above-mentioned limitations in studying habitat selection and dispersal, it is not surprising that there are few studies that incorporate these aspects into studies of Neotropical spiders. From the review that I did to assemble material for this chapter, studies in Neotropical spiders can mostly be categorized based on the interests of the few groups spread across the continent. For the purposes of this chapter, I searched the Thomson–Reuters Web of Science database and the Google Scholar Database. I filtered the articles by choosing the following keywords: Neotropical, spiders, dispersal. I did a separate search for spiders with the various countries that make up the Neotropics and also used Spanish/Portuguese keywords. I filtered out taxonomic revisions and records of new species. Articles that survived included some information about site selection and/or dispersal. I only included articles where the work was done in the Neotropics, as defined by the tropical New World biogeographic region that extends south, east, and west from the central plateau of Mexico. I did not include any studies that were done north of Mexico, even though there are many species that overlap in range between the USA and Mexico. I have undoubtedly missed many papers, but I trust that I can provide a broad overview of the state of arachnological research in the Neotropics.

I have categorized this review into two main sections: habitat selection and dispersal. Habitat selection deals with where the spiders are found, and dispersal is how they get there and when. Within the section of habitat selection, I will focus on three levels of selection and utilization, largely in terms of association with plants:

1. *Macro level*: which includes habitats and environmental attributes at the larger scales such as forest types or managed ecosystems.
2. *Micro level*: which includes the process of selection and use of, for example, particular plants or types of plants.
3. *Nano level*: which includes selection and use of attributes within an individual plant.

However, there can be substantial overlap between these categories. In terms of abiotic factors, I will place particular emphasis on substrate suitability and water, either in the form of water bodies such as pools and rivers or in the form of precipitation.

With respect to dispersion, I use the following sense of the term: the spatial pattern of distribution of a species formed as a result of dispersal. In this context, the process of dispersal is discussed according to the status of a given species along the solitary–social continuum. Due to the paucity of studies in Neotropical spiders with regard to this process, I have mostly focused on aggregations, colonial spiders, and social/subsocial dispersal dynamics. This section is also informed by aspects of topophilia such as substrate choice.

Topophilia at the Macro Level

To a spider that is deciding where to settle, the world is not a homogeneous place. Sites that are adjacent to one another may yield completely different benefits in terms of prey availability and future reproductive success. Site selection is also intimately linked to the foraging mode of the spider. If we divide spiders into the now classic foraging guilds, it is obvious that different guilds occupy different positions in space. For example, orb-web builders are dependent on available substrate that allows them to construct a web that is 'floating' in space. Ground-burrowing spiders depend on the suitable soil structure that permits them to excavate their burrows. Even though these patterns are, at the finest level, a result of individual choices, they can still be appreciated in large scale biodiversity surveys (Flórez 1998). Sampling methods are frequently devised to take into account these preferences: for example, if you want to sample for Salticids in the understory, the most frequently recommended method is beating. The use of this technique for these spiders implies that microhabitat preferences have been considered in order to maximize sampling efficiency. Furthermore, low vegetation, which consists of a mix of microhabitats, can be a suitable habitat for multiple guilds of spiders, for example allowing for the co-existence of guilds such as Lycosids and Ctenids as well as the more typical web-building spiders (Flórez 1998).

Among the Ctenidae, *Phoneutria* spp. have been extensively studied in Neotropical ecosystems. There are species-level differences in habitat selection (Torres-Sánchez and Gasnier 2010). In a study comparing microhabitat preferences of two species, *P. reidyi* and *P. fera*, it was shown that *P. reidyi* was found in higher abundance in a swamp forest than in dense forest or heath forest areas, whereas *P. fera* was found in similar abundances in all three habitats.

The aforementioned study encapsulates many of the themes of this chapter: firstly, microhabitat preferences is intimately linked with not just plants as substrates but also plant species-specific associations. Secondly, there are microhabitat preferences that change according to the ontogeny of the species, i.e., juvenile preferences are different from those of adults. This is discussed in a subsequent section. Thirdly, there are inter- and intraspecific interactions that affect the choice of a site. And finally, abiotic factors, particularly water, play an important role in site selection, which in turn influences the distribution patterns of a species on a larger scale.

Gradients

Habitat selection can be highly influenced by both natural as well as artificial gradients in the landscape. Natural gradients are formed by changes in boundaries between different habitat types. The border between the habitats can be gradual, such as between scrub and dunes in coastal areas, or abrupt, such as at the edge of a stream. In natural areas, these changes in habitats can be seen in the form of forest gaps created by tree fall. These gaps allow in more light and fundamentally alter the composition of plant communities in the area and thereby influence the distribution of spiders. Bonaldo and Dias (2010) studied the effect of these gaps on spider species composition. They compared natural forest gaps with artificial gaps created by deforestation practices in Western Brazilian Amazonia, and found a significant difference in species richness between the gaps. Among specific families, the natural gaps had a higher abundance of Ctenids, whereas the artificial gaps were dominated by Lycosidae. They attributed the difference in species composition to the differential plant cover found within the gaps. Both these families can be used as environmental indicators of degradation, with Ctenids representing areas with little disturbance, and Lycosids representing areas with a high level of degradation. Another interesting result was that natural gaps harbored higher species richness than the surrounding matrix, which may be congruent with literature from other groups suggesting that there may be higher levels of species diversity in areas of intermediate disturbance (Connell 1978; but see Fox 2013 and Sheil and Burslem 2013 for a recent debate on the utility of the intermediate disturbance hypothesis). This last result was partially corroborated by another study done in an Atlantic forest remnant in Brazil (Peres et al. 2007), which looked at the difference in species composition between treefall gaps and mature forest. In this study, the authors found that there was a significant difference between the species composition of spiders in tree-fall gaps in comparison to the mature forest, but that there was no difference in overall numbers of species and individuals. Among the species that showed habitat preferences, there were more species among the group collected by hand in comparison to those collected by the pitfall sampling method. This suggests that certain guilds can be more attuned to habitat change than others. In a study done on the Barro Colorado Island in Panama, clearings and forest edge habitats supported higher densities of orb-web spiders than internal habitats (Lubin 1978). But this work adds support to the idea that even among orb-web spiders, there are distinct microhabitat preferences. Lubin states that *Nephila clavipes*, *Leucauge sp.*, and *Argiope argentata* can be considered as 'true clearing species' with distinct preferences for clearings and edges. This pattern of distribution is undoubtedly linked to prey availability, prey flight patterns, and behaviors in relation to the changing visibility of orb webs in different light habitats (Craig 1988).

As mentioned earlier, it is possible that Ctenids are good indicators of disturbance. Their value as environmental indicators lie in their absence. To test if different species of Ctenids responded similarly to anthropogenic change, four species were studied in fragmented forests and compared with those in contiguous forests (Rego et al. 2007). Of the four species studied, two of them — *C. amphora* and *C. villasboasi* — showed a marked decrease in population in forest fragments, whereas *C. manuara* and *C. crulsi* did not seem to be affected. The study also showed that another Ctenid, *Ancylometes rufus*, was only found in contiguous sites, and the authors suggest that this species may be restricted to larger areas due to the presence of flooded forests, which is an essential habitat for this species.

Soil

For two wandering spider species, *Ctenus crulsi* and *C. amphora*, their soil preferences were evaluated in a study done in Central Amazonia in Brazil (Portela et al. 2013). Here, the authors show that while *C. crulsi* preferred clayey soil to sand, *C. amphora* did not have any such preferences. This preference is presumably linked to prey availability and suitability.

However, just the fact that spiders occupy certain niches is not always indicative of their preferences; topophilia emphasizes affinities to a certain space. In a study done on sand-dwelling Lycosids in Uruguay, male *Allocosa brasiliensis* spiders were selected from either fine sand habitats or coarse sand habitats and then given the choice of building their burrows in fine or coarse sand. These spiders significantly preferred building burrows in coarse sand (Albin et al. in prep). These results suggest that site selection can be constrained by other factors (e.g., predation pressure, prey availability) subsequent to structural preferences. Another fine example of niche expansion is seen in the Theridiosomatid spider *Wendilgarda galapagenisis*. Although other members of this genus build webs attached to the planar surface of water, in the Cocos Islands these spiders have moved away from water and into open habitats, and in addition build a variety of different webforms (Eberhard 1989). Niche expansion in this case can be attributed to probable relaxation of predation pressure in these islands, but this remains to be tested.

Agroecosystems

The modern landscape is a mosaic of remnant forests and managed ecosystems changed due to human land use. However, within managed ecosystems, there exists a gradient in types of habitat. Some agro-ecosystems, for example shade coffee plantations, have been known to retain a large amount of spider diversity and may play an important role in preserving spider populations (Pinkus-Rendón et al. 2006). In these managed ecosystems, vegetation cover plays a critical role. In a study looking at spider diversity along a tropical habitat gradient with 18 habitat types, Pinkus-Rendón et al. (2006) showed that ground-dwelling spider diversity was significantly correlated with tree cover during the rainy season but not during the other seasons, emphasizing the importance of abiotic variables and seasonality in determining species distributions. This study showed a direct link between habitat complexity, as measured by canopy cover and abundance of different plant forms (herbs, shrubs, trees, and epiphytes) and spider diversity, irrespective of the season. This measure of habitat complexity, however, is on a large scale, and spans several distinct habitats. The relationship between management regimes and spider diversity is, however, not constant.

Spider diversity in coffee plantations was studied by comparing two different management regimes [conventional coffee plantation and organic (shade coffee)] (Pinkus Rendón et al. 2006). There was no relationship between management regime and spider diversity. However, when analyzed according to seasons, the conventional site harbored higher diversity than the organic site in the dry season but not in the wet season. Web-building spiders and hunting spiders were generally more dominant in the conventional site. The authors attribute this effect to intense bird predation (untested) at shaded (organic) sites, and to the fact that they only studied the system for a year. However, in this study only spiders on coffee bushes were sampled, and other areas in the coffee plantation such as in soil or on trees were not tested. It is possible that there exists a high rate of disturbance due to human activity among the coffee bushes, leading to the patterns thus described. The authors note that in most cases there was no difference between the sites in terms of spider diversity. Clearly, there need to be further studies looking at the agro-ecosystem as a whole.

Epiphytes

An alternative approach to looking at coffee ecosystems was put forward by Méndez-Castro and Rao (2014). In this study, the authors compared the diversity of epiphyte-dwelling spiders in remnant cloud forest with those in shade coffee plantations, arguing that the existence of epiphytes in both ecosystems allows for a natural point of comparison that is hard to achieve in other similar studies (Méndez-Castro and Rao 2014). They reported that spider species richness was highly correlated to the type of the epiphyte, with bromeliads harboring more species than other types of epiphytes such as ferns, orchids, and others. Their study also showed that the shade coffee plantations had more spider diversity than the cloud forest remnant, suggesting that the natural population in the canopy has been augmented by more species that exploit this habitat.

In recent years, epiphyte-dwelling spiders have become a focus of study, thanks largely to several studies in Brazil and Central America. Microhabitat preference of these spiders is discussed in detail in the following section, but there are a couple of studies that look at epiphyte preferences over a larger scale. Epiphytes can be broadly categorized into two types based on their water-holding capacity: those that retain rain water and those that do not (Romero 2006). With respect to the geographic range of epiphyte-dwelling spiders, for the Salticid group, some species are highly associated with particular species of epiphytes and other species are more generalist; and this pattern is seen across a wide range of habitats. Microhabitat preference can be attributed to the regional availability of bromeliads. Furthermore, Salticids that showed a preference for water-retaining epiphytes were not found on the non-water-retaining epiphytes (Romero 2006).

Other studies in the Neotropical canopy have identified Salticids as one of the major group of spiders that live in epiphytes (e.g., Méndez-Castro and Rao 2014). However, in a study of the canopy in Costa Rica, araneids and linyphiids were the dominant groups (Yanoviak et al. 2002). This can be partly attributed to the sampling method used, since the authors collected by means of visual search and hand-collecting. In canopy studies, collection of arthropods by chemical fogging has long been the favored technique, but this method leaves much to be desired in terms of efficiency for the spider community. There is a high possibility of spiders clinging on to epiphytic plants even after fogging, and furthermore due to the inverse conical shape of most epiphytes, fogged spiders are more likely to fall into the center of the plants and thereby be under-sampled. The more exhaustive and time-consuming (but more accurate) option is to extract bagged epiphytes (at the tree) and disassemble them in a laboratory set-up; this method has the further advantage that spiders that live among the roots such as the Oonopids can be adequately sampled. Furthermore, Yanoviak et al. (2002) showed that in primary forests, spider abundance and richness was higher at the understory level than in the crowns. This result is also not in accordance with Méndez-Castro and Rao (2014), where there was a positive effect of height on spider species richness. Interestingly, Yanoviak et al. show that there may be family level differences in spatial distribution, suggesting once again that spiders actively partition the available microhabitat in accordance with their needs.

Water

The importance of the presence of water has been highlighted by several studies in this review. In two species of wandering spiders of the family Miturgidae, *Syspira longipes* and *S. tigrina*, there was evidence of niche differentiation (Nieto-Castañeda and Jiménez-Jiménez 2009). *S. longipes* was restricted to cooler areas that had high relative humidity during July and January, and to the areas that were warmer and had lower relative humidity during October and May. This study was done in Baja California Sur, Mexico. Even though these two species are very similar in habit, they do segregate temporally and spatially. In cases where there is a direct effect of water, such as in spiders that live on aquatic plants, spatial segregation is also seen. In a study looking at species richness in aquatic plants in Brazil, it was shown that vertical structure of the plants was an important factor affecting the abundance, richness, and guild composition of spiders (Cunha et al. 2012). This is possibly directly linked to vegetation complexity, since the number of available substrates increases with plant complexity and thus leads to a occupation of these areas, especially by web-building spiders.

Spiders can adapt to conditions of constant water presence by using plants or rocks to occupy areas that are normally inaccessible. However, in situations when there is seasonal flooding, the scenario is very different. A safe and productive habitat during the dry season can be fatal in the rainy season due to flood water. The retreat and expansion of swamp areas offer special challenges to mostly sessile spiders. Migration by spiders up trees to escape the floodwaters is a common phenomenon (Hénaut et al. 2014), but this brings a different set of challenges, such as predation pressure from different guilds of predators. A study in the floodplain ecosystem in Amazonia showed that spider species richness was higher in a non-flooded ('terra firme') site than an inundation site (Höfer 1997). This preference was particularly striking in soil- and litter-inhabiting species such as mygalomorphs and Zodariids. Wandering spiders of the families Corinnidae, Ctenidae, Lycosidae, and

Pisauridae have been observed moving away from the waterline in a steady process. Ballooning is yet another method by which spiders can disperse away from the advancing flood water.

Topophilia at a Micro Scale

Apart from selecting sites at a larger scale, individual spiders also sample the environment at a fine scale. Spiders prefer certain parts of plants over others depending on their foraging mode. Differences in preferences can be seen especially in studies that look at vertical stratification in spider communities. In a study looking at diversity of spider in aquatic macrophytes, Raizer and Amaral (2001) showed that while comparing species richness, plants with the greatest structural complexity showed the highest number of unique spider species. Once again, preference was determined by guilds. In particular while comparing two plant species, *Echinodorus paniculatus* and *Salvinia auriculata*, these differences were made clear. *E. paniculatus* has a greater height (above water) and lower leaf and branch density which favored web-building spiders; whereas *S. auriculata*, which had a lower height but higher branch density, promoted wandering spiders.

The height of the plant has also been shown to be important for spider diversity for web-building spiders on land as well (Greenstone 1984). Interestingly, in this study done in Costa Rica, Greenstone found that prey availability did not significantly predict species diversity, suggesting that there may be some sort of hierarchical approach to site selection. The first critical step may be substrate availability and then prey availability. Web spiders can move if a site is not productive, despite the cost of movement. Alternatively, if a site on the whole is highly productive in terms of potential prey, then prey availability is of lesser importance in site selection. In contrast to the study above, the vertical distribution of spiders in oak tree habitat was studied by Vanegas et al. (2012). They found a distinct segregation of microhabitat by web type: small sheet webs in the leaf litter and orb webs higher off the ground. They divided the trees into height classes, and found that the spider composition (species richness) varied accordingly. The lowest height class was significantly different from the other classes, suggesting that there may be a threshold where the microhabitat is no longer suitable for some guilds and very suitable for others. In these studies, foraging guilds are the usual way to discriminate between species.

However, in a study done on habitat use by wandering spiders of the families Trechaleidae and Ctenidae, the spiders were categorized based on microhabitat use and adhesion capabilities (Lapinski and Tschapka 2013). There were three subguilds recorded comprising eight species, namely (1) two semi-aquatic species with low adhesion ability, (2) three forest-floor-dwelling species with good adhesion ability, and (3) three vegetation-dwelling species showing very good adhesion ability. This study showed that the spiders were segregated based on the use of the habitat as well as microhabitat, with *Ancylometes bogotensis* and *Trechalea tirimbina* particularly associated with proximity to water. With respect to adhesion ability, the three *Ctenus* species were suggested to be particularly adapted for vegetal surfaces. The segregation based on microhabitats in terms of proximity to water and vegetation may also be reflected in significantly lower water loss rates and desiccation susceptibility of vegetation-dwelling spiders (Lapinski and Tschapka 2014). And finally, as an example of morphological changes reflecting the behavioral data, the three subguilds differed in ecomorphological traits. The forest-ground dwellers had smaller claw tufts relative to body mass than the vegetation dwellers, which consistently showed the best adhesion performance (Lapinski et al. 2015).

Bromeliads: A Special Focus

Bromeliads are a very interesting case study on topophilia in spiders for two reasons. Firstly, the epiphytes of this family are abundant in the Neotropics, and over the years there has been considerable interest in the characteristics of the spider–bromeliad association. Secondly, bromeliads create special microhabitats possessing characteristics that are clearly delineated from the surrounding ecosystems, and which have often been referred to as treetop islands (Richardson 1999). There is considerable evidence that bromeliads serve not only as temporary refuges for spiders, but also for spiders that spend their whole lives in and around epiphytes, and this can be inferred because of a study that observed spider egg sacs, silken retreats, female spiders that were carrying eggs, and nurseries with spiderlings as well as molts (Méndez-Castro and Rao 2014). Almost every part of the bromeliad can be used by spiders either for foraging, structural necessities, or even for the water retained in large tank-bromeliad species. Though there are certain families such as the Salticidae that are overrepresented in collections from bromeliads, they are by no means the only families that use the plants. There have been records of up to 26 families recorded in epiphytes (Méndez-Castro and Rao 2014). Such a diversity of species implies a diversity of foraging guilds, and these result in a very efficient niche exploitation and segregation mechanism.

Bromeliads may function as diversity amplifiers in spiders. To test this, Gonçalves-Souza et al. compared the spider species composition sampled from bromeliads, ground vegetation, and shrubs (Gonçalves-Souza et al. 2010a). Although the vegetation habitat had the highest richness, the bromeliads were responsible for a 41% increase in richness due to a large number of exclusive fauna. The authors attribute the extraordinary success of bromeliads as spider habitats to the availability of spatial niches with the following variables: foliar axils, foliar blades, space between leaves, dry and green leaves, and central and peripheral tanks.

However, another study by the same group showed that spider species composition was not related to the architectural complexity of the bromeliads (Gonçalves-Souza et al. 2010b). This counterintuitive result can be partly explained by a differential response by web-building and hunting spiders. As the architectural complexity of the bromeliads (measured as leaf width, leaf length, and number of leaves) increased, web spiders were less likely to be found. The authors attribute this to reduced space available for the construction of orb webs. In a deviation from other bromeliad studies, this study also looked at terrestrial bromeliads, i.e., not only epiphytic bromeliads. In this case, one would assume that the conditions that cause the creation of a unique microhabitat at the canopy level are very different at the ground level.

However, the vertical distribution of bromeliads did not influence species composition in total, but there was a significant effect when considering the guilds. Hunting spiders were more likely to be associated with the bromeliad species that show greater vertical distribution range, and this is attributed to the fact that many of these spiders use the bromeliads as a 'permanent site', unlike web builders which may be transient. Of all the architectural variables, the number of leaves showed a strong positive correlation with web spider abundance but a negative correlation with web spider richness, suggesting that some web spider species are better at exploiting the bromeliad habitat than others when there are more leaves. To put it in the context of topophilia, some web spiders may have a stronger affinity to the bromeliads, or have a favorable calculus in relation to the costs and benefits of utilizing this site. In contrast, the number of leaves had a positive correlation with both hunting spider abundance as well as richness. This study is a nice example of the central thesis of this chapter: that topophilia is deeply and sometimes idiosyncratically bound with the foraging modes of spiders.

Species Specialization

The two studies discussed above leads us to the conclusion that studies that look at topophilia at a broader scale, for example at the scale of diversity and richness, are perhaps too coarse to reveal the fine details of microhabitat preference. It may be more relevant to look at topophilia from the perspective of a single species. Salticids are particularly suitable for studying these interactions because of the frequent reports of this family being represented in high abundance among epiphytes and particularly bromeliads. Some species of jumping spiders are exclusively found on bromeliads, implying a high degree of spider–plant association, which is quite rare in spiders (Romero et al. 2007; Romero and Vasconcellos-Neto 2004a). As a counterpoint, in another study this sort of strict plant association was not seen in a Theraphosid spider irrespective of the differences in plant architecture (as measured by leaf surface area and leaf number) or the thermal properties of the water in the phytotelmata (Dias and Brescovit 2004).

Omena and Romero (2008) looked at fine-scale microhabitat specialization in a Salticid species *Psecas chapoda* with respect to bromeliads. They evaluated the presence of this spider in three bromeliad species, with reference to the effect of plant architecture (as measured by leaf length, width, and number), presence of *phytotelmata* (i.e., standing water held by plants in miniature pools), and leaf spines (a putative protective function) (Omena and Romero 2008). They found that the microhabitat specialization in this spider is related to plant architecture rather than plant taxonomy. In particular, the presence of *phytotelmata* played a very important role in determining the spider's presence. The authors suggest that since *P. chapoda* evolved in regions in where tank-bromeliads are rare, they do not use the presence of standing water in a similar way as other Salticids. *P. chapoda* also discriminate between different stages of the same bromeliad (Romero and Vasconcellos-Neto 2005a). *Bromelia balansae*, a preferred host for this spider, changes in architecture as it blooms, switching from a 3-dimensional conical shape to a 2-dimensional flattened shape. In an experiment to look at the effect of change in plant architecture, spiders were more likely to colonize the plants that retained a 3-dimensional shape as well as less likely to colonize a bromeliad with dry leaves in the funnel (Romero and Vasconcellos-Neto 2005b).

These studies show that the life history of spiders and Salticids in particular are intimately bound with their host plants. A subsequent chapter in this book explores these interactions in more detail.

Topophilia at a Nano Scale

As mentioned in the last section, on fine-scale habitat discrimination, spiders are capable of discriminating and showing affinity to different plant structures within the same plant. In the following section, I focus on topophilia at a 'nano scale'. To put this concept in perspective, one has to consider the fine-scaled partitioning of a given plant based on different spatial and structural attributes such as the presence of flowers, spines, accumulation of dried leaves, predation pressure, mutualism, and finally the potential prey visitation rates. The special relationship between spiders and plants will be discussed in detail in subsequent chapters (Chapters 7 and 8), and so I will here only illustrate topophilia in broad strokes. One of the most interesting interactions is seen in the orb-web spiders *Eustala illicita* and *E. oblonga* which are found in the acacia trees (*Acacia collinsii*) that are patrolled by *Pseudomyrmex* ants in Panama (Hesselberg and Triana, 2010). The ant–plant relationship is well studied: the ants defend the plant against herbivores, and in return consume sugar and Beltian bodies and use nest space. It is thought that the spiders exploit this already existing mutualism to occupy an unused niche. The webs extend from the tree and therefore the spiders are out of physical contact with the ants, especially since they are nocturnal. The spiders rest in a refuge during the day close to the ant nests, and the mechanism for their defense against ant predation is not known.

In the case of Oxyopid spiders, there is substantial evidence that they prefer plants with sticky glandular trichomes, since the sticky hairs function as insect traps (Morais-Filho and Romero 2008, 2010; Vasconcellos-Neto et al. 2007). Though it would be reasonable to assume that the Oxyopids may prefer the reproductive structure of the plant, because this would attract potential insect prey in the form of pollinators, the association between the plant and the spider is driven by the glandular trichomes (Morais-Filho and Romero 2008). Some Thomisids also show a preference for plants with glandular trichomes (Romero and Vasconcellos-Neto 2004b). In the case of Thomisids that wait on or around flowers for prey, a study showed that they do not appear to use color for selecting their foraging sites, but do remain on color-matched flowers for longer times (Peixoto et al. 2012). However, this study did not take into account the colors of the flower or the spider from the point of view of potential prey. Many Thomisids as well as flowers reflect light in the UV wavelength (which can be detected by insects) and therefore these results have to be viewed with caution.

The relationship between the preferred plant species and a spider can also extend to modifications made by the spider. In this case — which can be referred to as an extended topophilia — spiders such as *Cupiennius sp.*, which prefer plants such as bromeliads for their retreats, can also use Zingiber and Aracaceae, but modify the plants to make them suitable substrates (Barth et al. 1988). Spiders can use silk to close up the open areas of the plants, or turn leaves into protective tubes by bending and rolling them and further affixing them with silk (Barth et al. 1988). These plants have the added function of being suitable substrates for vibratory communication between conspecifics as well as from prey.

Though plants make up the overwhelming majority of substrates for spiders in terms of selection, there are a few studies that emphasize the importance of hetero-specific webs as points of attachment. The interspecific association between the Uloborid *Philoponella vicina* and the Tengellid *Tengella radiata* is a case in point (Fincke 1981). Here, the Uloborids actively choose to build in *T. radiata* webs. This association seems to be commensal in nature, and the Uloborids that were associ-ated to *T. radiata* webs persisted at a site for longer, and there was significantly greater prey capture. The association depends on the host spider's presence, since Uloborids associated with abandoned *T. radiata* webs soon left the site. This asso-ciation may benefit the Uloborids in terms of enhanced prey capture through the ricochet effect (Rao 2009), or protection from predators.

Webs of other species can also be fruitful patches for kleptoparasitic spiders of the *Argyrodes* group (Theridiidae). The host web size is a good predictor of klepto-parasitic load, and this is enhanced when the host webs are clustered (Agnarsson 2003).

Ontogeny and Topophilia

Up to now I have been considering mostly adult spiders. However, topophilia is not uniform between the males and females of a species, nor between different stages in the ontogeny of the species. This discrimination is particularly prominent in the study done on site selection by *Peucetia flavia* (Oxyopidae). Juveniles were more likely to occur on the lower regions of the crown of a glandular plant (Morais-Filho and Romero 2008). In *Misumenops argenteus* (Thomisidae), adult females were more likely to use the reproductive branches, and juveniles were found in flower-heads with a greater frequency for the following reasons: possibility of camouflage, shelter provided by the flowers, and the availability of specific prey (Romero and Vasconcellos-Neto 2004a). In Ctenids, there was a clear effect of change in microhabitat preference with ontogeny (Torres-Sánchez and Gasnier 2010). Subadults and adults of *Phoneutria reidyi* were less likely to be found on the ground and were correlated with the presence of palms, and this pattern was not encountered in *P. fera*. The choice of spiderlings to use vegetation substrates in both species is probably driven by protection from predators. The study suggests that *P. reidyi* adults stay in the vegetation in order to avoid predation by *P. fera*.

Dispersal and Dispersion

From a topophilia perspective, dispersal is the process by which spiders, which have well-defined needs and motivations, arrive at a suitable site. Many animals can 'sample' the environment before deciding to settle, but in spiders the process is complicated by two factors.

Firstly, spiders usually disperse as spiderlings, and usually through a process known as ballooning where an individual releases a single strand that is caught by the wind till the spider is launched into the air. The dependence on wind conditions implies that spiders do not have control over where they land. Ballooning is not restricted to web spiders; it has also been recorded in Ctenizids such as *Ummidia* (Eberhard 2006). They may subsequently move to refine the choice of the landing site, but it is worth keeping in mind that dispersal in spiders is a risky process since the spiders are subject to predation pressure, lost foraging opportunity, and high energetic costs of relocation (Jakob et al. 2001).

Secondly, a highly suitable site in terms of substrate may still be unsuitable in terms of potential prey. The spider essentially has to make a bet that this site is a good one before investing in establishing a site, either through the construction of a web or a refuge. There is the possibility that spiders can evaluate the potential prey just by receiving airborne vibrations, but this has not been rigorously tested as far as I know. There are anecdotal accounts that it is easier to get orb-web spiders to build webs in captivity if flying prey are released in the cage beforehand. A bad choice in site can lead to further costs of relocation, predation during relocation, and also lost foraging opportunity. Therefore, it is likely that a spider would use plant attributes such as flowers as cues (Morse 1993).

A useful cue to a suitable site is the presence of conspecific silk, a phenomenon known as 'sericophily'. An example of this has been shown in the subsocial spider *Anelosimus baeza*, where dispersing subadult spiders showed a strong preference to settle in sites that contained conspecific silk (Rao and Aceves-Aparicio 2012). Another cue of site suitability, and perhaps the most commonly used one, is to settle in sites that have a proven record of profitability, such as being close to or at the mother's site. This has been termed "natal philopatry" and has been studied in the context of social spiders, but is by no means restricted to this group. For example, in the Theraphosid *Brachypelma vagans*, spiders are rarely observed far from the burrow, and there is a strong clustering in terms of spatial patterns of dispersion (Reichling 2000).

Clustering can occur in orb spider populations as well. Normally, these are referred to as aggregations, with the definition that spiders build webs very close in space to each other, and they may or may not have webs that connect. Spiders that almost always connect webs are called colonial spiders. In both aggregations as well as colonial spiders, the individual spider maintains its own territory. In subsocial and social spiders, the webs are connected to form a large irregular mass, but there is no distinct demarcation of individual territory. Most social and subsocial spiders build webs, but there are a few exceptions such as the social crab spiders in Australia (Evans 1998).

Vertical stratification is seen in colonial spider groups. In *Leucauge sp.*, there was size-dependent stratification: larger spiders were found higher in the group, and lower spiders also maintained a size-based hierarchy (Salomon et al. 2010). It is probable that prey availability is enhanced at higher levels of the group. But another advantage in building higher is that these spiders may use substrates that have greater architectural stability, thereby ensuring that they are less prone to disturbances. This pattern, of larger spiders building higher than juveniles, has also been reported in *Metepeira gressa* (Viera 2003).

In Costa Rica, *Metabus gravidus* build interconnected orb webs over streams (Buskirk 1975), and they favor the narrow sections over the broader sections. They show a distinct preference for river sectors with a gentle slope and a steady current, and they show an affinity to areas where there are prominences jutting out from the shoreline. Very rarely are spiders seen to build solitary webs away from the river. In this case, the dispersion and resulting aggregation in these spiders is driven by topophilia. There are abiotic affinities associated with the riverine system, and the connecting up of webs allows the spiders to access prey and spaces that would be impossible to access in other ways.

Social and Subsocial Spiders

Dispersal in Neotropical spiders has been mostly studied in social and subsocial spiders, and hence this section will focus on these groups. Among the spiders that occupy the sociality continuum, the genus *Anelosimus* is perhaps the best studied, especially in the Americas. Dispersal tendency (as measured by the propensity of individuals to bridge with or without stimulation) and dispersal ability (in males) was negatively correlated with the degree of sociality in seven *Anelosimus* species (Corcobado et al. 2012). The authors suggest that the loss of dispersal ability in social species contributes to the maintenance of the social systems in these spiders. It is interesting to note the differential role played by males in this study. The bridging speed decreased in males with an increase in sociality level, suggesting that the males were not motivated to disperse by the need to find females, since they were found within the same colony. Interestingly, the study found a significant difference in male dispersal ability (bridging speed) between social and subsocial spiders.

In a study looking at dispersal and dispersion in the subsocial spider *Anelosimus baeza* (Aceves-Aparicio et al. in review), males were more likely to disperse during the night than the day. This study followed the fates of dispersal and source colonies in a single tree for the entire breeding season. Females establish dispersal webs and males seek them out, but this process of occupancy of dispersal webs is highly dynamic across the season. In general, the occupancy of dispersal webs declines over the season, but there are fluctuations in both female 'site fidelity' and male visitation rates. The authors propose that in subsocial spiders, the dispersal sites act as stepping stones towards eventual colony formation, but could also serve as temporary staging areas before return to the original source colony. From a topophilia perspective, this study is a nice illustration of site choice. The dispersal webs were found on the opposite side of the tree with respect to the colony webs. Furthermore, dispersing females that were spatially and temporally proximate, i.e., spiders that were close to each other in space and time (of initiation of web construction) showed similar trends in occupancy over the season, suggesting that similar microhabitat conditions influence similar levels of site fidelity. These conditions could include average prey interception rates, predation pressures, temperature/wind fluctuations, and protection.

The central issue of dispersal is in understanding the reasons or the factors that drive individuals to leave the colony. There are many hypotheses, among them competition for resources, inbreeding depression, condition (nutritional status) of the individual, and sexually differentiated dispersal. In a study done on a subsocial spider, *Anelosimus vierae* (Ferreira-Ojeda et al. in review), it was shown that both the size of the colony as well as the amount of food available influenced dispersal. In particular, dispersal was more commonly recorded in large colonies and in colonies which experienced food scarcity.

Colony Foundation It has been demonstrated that in subsocial species, dispersal web sites decline over the season; i.e., not every dispersal web converts into a colony (Aceves-Aparicio et al. in review). The survival rates of colonies are generally higher. For example, *A. baeza* spiders prefer to settle in sites that show previous occupation (Rao and Aceves-Aparicio 2012). In *Anelosimus eximius*, colony foundation may occur through three processes, perhaps even simultaneously: through budding, active migration of gravid females, or passive emigration of a part of the colony (both adults and juveniles) (Pasquet and Krafft 1989). This study found colonies mainly distributed along roads, which indicates a preference for edge habitats, and also that colonies formed "colony complexes" which consisted of a large colony surrounded by small colonies. These observations further indicate that there is limited dispersal, which results in clustered or aggregated dispersion. Another study also recorded the preference of *A. eximius* for borders (Venticinque et al. 1993), but colonies found in the forest interior had a significantly longer life span. This leads us to speculate that new colony formation is perhaps opportunistic, occurring only in times of food scarcity. Alternatively, since colonies on the edges are more prone to disturbance and experience higher microclimate fluctuations, these colonies may be more likely to fail over time. This study is a larger-scale perspective of the findings reported in *A. baeza* (Aceves-Aparicio et al. in review). This pattern is also seen across social and subsocial spiders. The more social spiders which form longer-lived colonies are found in the interior of the forest, while the less social species tend to be found in forest-edge habitats (Purcell et al. 2012).

Two interrelated factors have been shown to be of vital importance in understanding colony dynamics. The first is prey availability and the second is seasonal variations with emphasis on precipitation rates. Social spiders have been shown to be absent in areas of high latitudes and high elevations (Avilés et al. 2007; Yip et al. 2008). These areas contain fewer insects of the largest size classes (Guevara and Avilés 2007). The biomass intake per capita peaks at intermediate colony sizes, which explains why dispersal is seen in the larger colonies (Yip et al. 2008). Rainfall in particular has also thought to be a factor influencing the survival of social colonies. Social species occur in areas of higher annual rainfall (Majer et al. 2013), and it has been speculated that heavy and frequent rain may influence the failure of smaller nests.

Conclusions and Future Research

Site selection and dispersal in spiders are vital factors that contribute to their survivorship. In this chapter, I have reviewed the studies that have contributed to our state of knowledge in various species spread across the Neotropics. To summarize, I looked at site establishment in spiders across three broad levels of habitats, and focused on the special relationship that spiders have with their plant substrate. Subsequently, I looked at a few studies that explored dispersal and dispersion, specifically in spiders that are along the solitary–sociality continuum.

It is worthwhile looking at spider communities through the lens of topophilia. The choice of a site leads us to explore other facets of a spider's behavior, and allows us to discover the strength of the spider–substrate relationship. This information can further funnel into future research questions. For example, we know that *Leucauge* species prefer edges and are often found in disturbed habitats. We can now ask questions such as: what are the characteristics of web-building behavior that permit the occupation of 'new' habitats? Lycosids are a common species that occupy pasture lands. What are the features of this habitat that benefit Lycosids to such an extent?

Furthermore, a topophilic consideration can lead us to evaluate the true availability of space and site. Even if the substrate is identical to another, there are unseen preferences that allow a spider to occupy one over the other. This in turn has ramifications on behavior such as courtship and reproduction.

While substantial work has been done on these questions, there are still many areas of research that need attention. The location of egg sacs is one area that needs further information. The same preferences that drive site establishment are further refined for eggsac placement, and are more driven by site characteristics. Some orb spiders bury their eggsacs, and the choice of web-site location may be dependent on the soil texture. Other questions such as sex-biased differentiation of site selection and dispersal are known only for a few groups of spiders. Changes in site preferences with ontogeny is another area that needs further investigation.

In conclusion, this chapter aimed to give a perspective of research in the Neotropics over a range of species and habitats. The integrating of behavioral studies with spatial information, and especially by looking at the level of the individual, can lead to a better understanding of a spider's sense of place.

References

Aceves-Aparicio A, Macías-Ordóñez R, Tapia-McClung H, Rao D (in review) Subsocial spiders in space and time: fine scale approach to dynamics of dispersal

Agnarsson I (2003) Spider webs as habitat patches — the distribution of kleptoparasites (*Argyrodes*, Theridiidae) among host webs (*Nephila*, Tetragnathidae). J Arachnol 31:344–349

Albin A, Simó M, Aisenberg A (in prep) Substrate choice by burrowing males of *Allocosa brasiliensis* (Araneae: Lycosidae)

Avilés L, Agnarsson I, Salazar PA, Purcell J, Iturralde G, Yip EC, Powers KS, Bukowski TC (2007) Altitudinal patterns of spider sociality and the biology of a new midelevation social *Anelosimus* species in Ecuador. Am Nat 170:783–792

Barth FG, Seyfarth EA, Bleckmann H, Schüch W (1988) Spiders of the genus *Cupiennius* Simon 1891 (Araneae, Ctenidae). I. Range distribution, dwelling plants, and climatic characteristics of the habitats. Oecologia 77:187–193

Biere J, Uetz G (1981) Web orientation in the spider *Micrathena gracilis* (Araneae: Araneidae). Ecology 62:336–344

Bonaldo AB, Dias SC (2010) A structured inventory of spiders (Arachnida, Araneae) in natural and artificial forest gaps at Porto Urucu, Western Brazilian Amazonia. Acta Amaz 40:357–372

Buskirk RE (1975) Coloniality, activity patterns and feeding in a tropical orb-weaving spider. Ecology 56:1314–1328

Charnov EL (1976) Optimal foraging, the marginal value theorem. Theor Popul Biol 9:129–136

Connell JH (1978) Diversity in tropical rain forests and coral reefs. Science 199:1302–1310

Corcobado G, Rodríguez-Gironés MA, Moya-Laraño J, Avilés L (2012) Sociality level correlates with dispersal ability in spiders. Funct Ecol 26:794–803

Craig C (1988) Insect perception of spider orb webs in three light habitats. Funct Ecol 2:277–282

Cunha ER, Thomaz SM, Mormul RP, Cafofo EG, Bonaldo AB (2012) Macrophyte structural complexity influences spider assemblage attributes in wetlands. Wetlands 32:369–377

Dias SC, Brescovit AD (2004) Microhabitat selection and co-occurrence of *Pachistopelma rufonigrum* Pocock (Araneae, Theraphosidae) and *Nothroctenus fuxico* sp. nov. (Araneae, Ctenidae) in tank bromeliads from Serra de Itabaiana, Sergipe, Brazil. Rev Bras Zool 21:789–796

Eberhard WG (1989) Niche expansion in the spider *Wendilgarda galapagensis* (Araneae, Theridiosomatidae) on Cocos Island. Rev Biol Trop 37:163–168

Eberhard W (2006) Dispersal by Ummidia spiderlings (Araneae, Ctenizidae): ancient roots of aerial webs and orientation? J Arachnol 34:254–257

Edwards W, Whytlaw PA, Congdon BC (2009) Is optimal foraging a realistic expectation in orb web spiders? Ecol Entomol 34:527–534

Evans TA (1998) Factors influencing the evolution of social behaviour in Australian crab spiders (Araneae: Thomisidae). Biol J Linn Soc 63:205–219

Ferreira-Ojeda L, Rao D, Viera C (in review). Factors that influence the dispersal of the subsocial spider *Anelosimus vierae* (Theridiidae)
Fincke OM (1981) An association between two Neotropical spiders (Araneae: Uloboridae and Tengellidae). Biotropica 13:301–307
Flórez E (1998) Estructura de comunidades de arañas (Araneae) en el departamento del Valle, suroccidente de Colombia. Caldasia 20:173–192
Fox JW (2013) The intermediate disturbance hypothesis should be abandoned. Trends Ecol Evol 28:86–92
Gonçalves-Souza T, Brescovit AD, Rossa-Feres DC, Romero GQ (2010a) Bromeliads as biodiversity amplifiers and habitat segregation of spider communities in a Neotropical rainforest. J Arachnol 38:270–279
Gonçalves-Souza T, Almeida-Neto M, Romero GQ (2010b) Bromeliad architectural complexity and vertical distribution predict spider abundance and richness. Austral Ecol 36:476–484
Greenstone MH (1984) Determinants of web spider species diversity: vegetation structural diversity vs. prey availability. Oecologia 62:299–304
Guevara J, Avilés L (2007) Multiple techniques confirm elevational differences in insect size that may influence spider sociality. Ecology 88:2015–2023
Hénaut Y, Corbara B, Pélozuelo L, Azémar F, Céréghino R, Herault B, Dejean A (2014) A tank bromeliad favors spider presence in a Neotropical inundated forest. PLoS One 9(12):e114592–e114513
Hesselberg T, Triana E (2010) The web of the acacia orb-spider Eustala illicita (Araneae: Araneidae) with notes on its natural history. J Arachnol 38(1):21–26
Höfer H (1997) The spider communities. In: Junk WJ (ed) The Central Amazon floodplain. Springer-Verlag, Berlin/Heidelberg
Jakob EM, Porter AH, Uetz GW (2001) Site fidelity and the costs of movement among territories: an example from colonial web-building spiders. Can J Zool 79:2094–2100
Janetos AC (1986) Web-site selection: are we asking the right questions. In: Shear WA (ed) Spiders: webs, behavior and evolution. Stanford University Press, Stanford, pp 9–22
Lapinski W, Tschapka M (2013) Habitat use in an assemblage of central American wandering spiders. J Arachnol 41:151–159
Lapinski W, Tschapka M (2014) Desiccation resistance reflects patterns of microhabitat choice in a central American assemblage of wandering spiders. J Exp Biol 217:2789–2795
Lapinski W, Walther P, Tschapka M (2015) Morphology reflects microhabitat preferences in an assemblage of Neotropical wandering spiders. Zoomorphology 134:219–236
Lubin YD (1978) Seasonal abundance and diversity of web-building spiders in relation to habitat structure on Barro Colorado Island, Panama. J Arachnol 6:31–51
Majer M, Agnarsson I, Svenning JC, Bilde T (2013) Social spiders of the genus Anelosimus occur in wetter, more productive environments than non-social species. Naturwissenschaften 100:1031–1040
Méndez-Castro FE, Rao D (2014) Spider diversity in epiphytes: can shade coffee plantations promote the conservation of cloud forest assemblages? Biodivers Conserv 23:2561–2577
Morais-Filho JC, Romero GQ (2008) Microhabitat use by *Peucetia flava* (Oxyopidae) on the glandular plant *Rhyncanthera dichotoma* (Melastomataceae). J Arachnol 36:374–378
Morais-Filho JC, Romero GQ (2010) Plant glandular trichomes mediate protective mutualism in a spider–plant system. Ecol Entomol 35:485–494
Morse DH (1993) Choosing hunting sites with little information: patch-choice responses of crab spiders to distant cues. Behav Ecol 4:61–65
Nieto-Castañeda IG, Jiménez-Jiménez ML (2009) Possible niche differentiation of two desert wandering spiders of the genus *Syspira* (Araneae: Miturgidae). J Arachnol 37:299–305
Omena PM, Romero GQ (2008) Fine-scale microhabitat selection in a bromeliad-dwelling jumping spider (Salticidae). Biol J Linn Soc 94:653–662
Pasquet A, Krafft B (1989) Colony distribution of the social spider Anelosimus Eximius (Araneae, Theridiidae) in French Guiana. Insect Soc 36:173–182

Peixoto PEC, Souza JC, Schramm JE Jr (2012) To be or not to be … a flower? A test of possible cues influencing hunting site selection in subadult females of the crab spider *Epicadus heterogaster* (Guerin 1812) (Araneae: Thomisidae). Stud Neotropical Fauna Environ 47:73–79

Peres MCL, Silva JMC, Brescovit AD (2007) The influence of treefall gaps on the distribution of web-building and ground hunter spiders in an Atlantic Forest remnant, northeastern Brazil. Stud Neotropical Fauna Environ 42:49–60

Pinkus Rendón MA, Ibarra-Núnez G, Parra-Tabla V, García-Ballinas JÁ, Hénaut Y (2006) Spider diversity in coffee plantations with different management in Southeast Mexico. J Arachnol 34:104–112

Pinkus-Rendón MA, Leon-Cortes JL, Ibarra-Nunez G (2006) Spider diversity in a tropical habitat gradient in Chiapas, Mexico. Divers Distrib 12:61–69

Portela E, Willemart RH, Gasnier TR (2013) Soil type preference and the coexistence of two species of wandering spiders (*Ctenus amphora* and *C. crulsi*: Ctenidae) in a rainforest in Central Amazonia. J Arachnol 41:85–87

Purcell J, Vasconcellos-Neto J, Gonzaga MO, Fletcher JA, Avilés L (2012) Spatio-temporal differentiation and sociality in spiders. PLoS One 7:e34592–11

Raizer J, Amaral M (2001) Does the structural complexity of aquatic macrophytes explain the diversity of associated spider assemblages? J Arachnol 29:227–237

Rao D (2009) Experimental evidence for the amelioration of shadow competition in an orb-web spider through the "ricochet" effect. Ethology 115:691–697

Rao D, Aceves-Aparicio A (2012) Notes on the ecology and behavior of a subsocial spider Anelosimus Baeza (Araneae: Theridiidae) in Mexico. J Arachnol 40:325–331

Rao D, Mendoza-Cuenca L (2016) The effect of colour polymorphism on thermoregulation in an orb web spider. Sci Nat 103:63

Rao D, Castañeda-Barbosa E, Nuñez-Beverido N, Díaz-Fleischer F (2015) Foraging benefits in a colour polymorphic Neotropical orb web spider. Ethology 121:187–195

Rego FNAA, Venticinque EM, Brescovit AD (2007) Effects of forest fragmentation on four *Ctenus* spider populations (Araneae: Ctenidae) in central Amazonia, Brazil. Stud Neotropical Fauna Environ 42:137–144

Reichling SB (2000) Group dispersal in juvenile *Brachypelma vagans* (Araneae, Theraphosidae). J Arachnol 28:248–250

Richardson BA (1999) The bromeliad microcosm and the assessment of faunal diversity in a Neotropical Forest. Biotropica 31:321–336

Riechert SE, Gillespie RG (1986) Habitat choice and utilization in web-building spiders. In: Spiders: webs, behavior and evolution. Stanford University Press, Stanford, pp 23–48

Robinson JV (1981) The effect of architectural variation in habitat on a spider community: an experimental field study. Ecology 62:73–80

Romero GQ (2006) Geographic range, habitats, and host plants of bromeliad-living jumping spiders (Salticidae). Biotropica 38:522–530

Romero G, Vasconcellos-Neto J (2004a) Spatial distribution patterns of jumping spiders associated with terrestrial bromeliads. Biotropica 36:596–601

Romero GQ, Vasconcellos-Neto J (2004b) Foraging by the flower-dwelling spider, *Misumenops argenteus* (Thomisidae), at high prey density sites. J Nat Hist 38:1287–1296

Romero GQ, Vasconcellos-Neto J (2005a) Spatial distribution and microhabitat preference of *Psecas chapoda* (Peckham and Peckham) (Araneae, Salticidae). J Arachnol 33:124–134

Romero G, Vasconcellos-Neto J (2005b) The effects of plant structure on the spatial and microspatial distribution of a bromeliad-living jumping spider (Salticidae). J Anim Ecol 74:12–21

Romero GQ, Vasconcellos-Neto J (2010) Foraging by the flower-dwelling spider, *Misumenops argenteus* (Thomisidae), at high prey density sites. J Nat Hist 38:1287–1296

Romero GQ, Santos AJ, Wienskoski EH (2007) Association of two new Coryphasia species (Araneae, Salticidae) with tank-bromeliads in southeastern Brazil: habitats and patterns of host plant use. J Arachnol 35:181–192

Rypstra AL (1979) Foraging flocks of spiders. Behav Ecol Sociobiol 5:291–300

Salomon M, Sponarski C, Larocque A, Avilés L (2010) Social organization of the colonial spider *Leucauge* sp. in the Neotropics: vertical stratification within colonies. J Arachnol 38:446–451
Scharf I, Lubin Y, Ovadia O (2011) Foraging decisions and behavioural flexibility in trap-building predators: a review. Biol Rev Camb Philos Soc 86(3):626–639, http://doi.org/10.1111/j.1469-185X.2010.00163.x
Sheil D, Burslem DFRP (2013) Defining and defending Connell's intermediate disturbance hypothesis: a response to Fox. Trends Ecol Evol 28:571–572
Torres-Sánchez MP, Gasnier TR (2010) Patterns of abundance, habitat use and body size structure of *Phoneutria reidyi* and *P. fera* (Araneae: Ctenidae) in a central Amazonian rainforest. J Arachnol 38:433–440
Tuan YF (2013) Topophilia: a study of environmental perceptions, attitudes, and values. Columbia University Press, New York
Uetz G (1989) The "ricochet effect" and prey capture in colonial spiders. Oecologia 81:154–159
Vanegas S, Fagua G, Flórez E (2012) Distribución vertical de arañas asociadas a *Quercus humboldtii* y *Clusia* spp. en el santuario de fauna y flora Iguaque, Colombia. Acta Biolo Colomb 17:635–656
Vasconcellos-Neto J, Romero GQ, Santos AJ, Dippenaar-Schoeman AS (2007) Associations of spiders of the genus *Peucetia* (Oxyopidae) with plants bearing glandular hairs. Biotropica 39:221–226
Venticinque EM, Fowler HG, Silva CA (1993) Modes and frequencies of colonization and its relation to extinctions, habitat and seasonality in the social spider *Anelosimus eximius* in the Amazon (Araneidae: Theridiidae). Psyche 100:35–42
Viera C (2003) Spatial and temporal variability in webs of *Metepeira gressa* (Keyserling, 1892) (Araneae, Araneidae): a year field study. An Biol 25:13–20
Yanoviak S, Kragh G, Nadkarni N (2002) Spider assemblages in costa Rican cloud forests: vertical stratification and effects of forest level and forest age. Stud Neotropical Fauna Environ 38:145–154
Yip EC, Powers KS, Avilés L (2008) Cooperative capture of large prey solves scaling challenge faced by spider societies. Proc Natl Acad Sci USA 105:11818–11822

Chapter 5
Dispersal Strategies, Genetic Diversity, and Distribution of Two Wolf Spiders (Araneae, Lycosidae): Potential Bio-Indicators of Ecosystem Health of Coastal Dune Habitats of South America

Leticia Bidegaray-Batista, Miquel Arnedo, Ana Carlozzi, Carolina Jorge, Patricio Pliscoff, Rodrigo Postiglioni, Miguel Simó, and Anita Aisenberg

Abstract Dispersal strategies are essential for species survival. Animals need to move to search for food, to locate potential sexual partners, to find refuge and escape from predators, and to avoid inbreeding and local competition for resources. The degree of plasticity of those traits will determine the ability of the species or population to respond successfully to changes in the environment, which is particularly important in species with a restricted habitat. *Allocosa marindia* and *Allocosa senex* are two nocturnal lycosids that construct burrows along the sandy coasts of Argentina, Brazil, and Uruguay. Both species show a reversal in the typical sex roles and size

L. Bidegaray-Batista
Laboratorio de Etología, Ecología y Evolución, Instituto de Investigaciones Biológicas Clemente Estable, Montevideo, Uruguay

Departamento de Biodiversidad y Genética, Instituto de Investigaciones Biológicas Clemente Estable, Montevideo, Uruguay

M. Arnedo
Biodiversity Research Institute UB, Departament de Biologia Evolutiva, Ecologia i Ciències Ambientals, Universitat de Barcelona, Barcelona, Spain

A. Carlozzi
Laboratorio de Etología, Ecología y Evolución, Instituto de Investigaciones Biológicas Clemente Estable, Montevideo, Uruguay

Consejo de Educación Secundaria, Montevideo, Uruguay

C. Jorge
Instituto Superior de Estudios Forestales, CENUR NORESTE, Sede Tacuarembó, Universidad de la República, Tacuarembó, Uruguay

P. Pliscoff
Instituto de Geografía, Facultad de Historia, Geografía y Ciencia Política, Pontificia Universidad Católica de Chile y Departamento de Ecología, Facultad de Ciencias Biológicas, Pontificia Universidad Católica de Chile, Santiago, Chile

C. Viera, M.O. Gonzaga (eds.), *Behaviour and Ecology of Spiders*,
https://doi.org/10.1007/978-3-319-65717-2_5

dimorphism expected in spiders: females are the mobile and courting sex, and males are larger than females. *A. marindia* and *A. senex* are strictly associated to coastal sand dunes with scarce native vegetation. During recent decades, the South American coastline has been reduced and disturbed due to urbanism and touristic activities, leading to the isolation of populations of *Allocosa* species. In the present chapter, we review the available information about the natural history of *Allocosa* species, providing data about their mechanisms of dispersal, distribution, genetic diversity, and spatial patterns of genetic variation, as well as their role as biological indicators for the coastlines of Southern South America. We integrate information provided by dispersal behavior data, genetic data, and GIS (geographic information systems) and SDMs (species distribution models) tools, and discuss the predictive maps of distribution for each species and their possible fate under a global-warming scenario.

Dispersal and Conservation in Spiders

Locomotion is fundamental over the lifetime of organisms. Individuals move to search for food, to mate, to avoid inbreeding and competition for resources, to escape from predators, and to track refugia under unfavorable environmental conditions (Nathan et al. 2008; Peterson 2009; Bonte et al. 2012). The movement of individuals depends on both intrinsic factors (e.g., physiological, morphological, and cognitive) that account for why, how, and where they move, and extrinsic factors (e.g., biotic interactions and climatic fluctuations) (Nathan et al. 2008). When locomotion involves displacement of individuals from the place of birth to other locations where they reproduce, this behavior is referred as dispersal (Nathan et al. 2008; Matthysen 2012). Dispersal over time and space contributes to the determination of species distributions as well as their spatial and genetic structures; thus, dispersion becomes a key trait in the evolutionary history of a species (Mathias et al. 2001; Nathan et al. 2008; Bonte et al. 2012; Gillespie et al. 2012).

The dispersal ability of individuals has also implications for species conservation, specifically by ensuring gene flow among populations, avoiding inbreeding depression, and facilitating the colonization of new habitats during unfavorable environmental conditions (e.g., global warming) (Frankham et al. 2004; Peterson 2009; Travis et al. 2013). On the other hand, the vagility of species may be directly or indirectly affected by processes such as habitat fragmentation and habitat loss because of urbanism, touristic activities, deforestation, and climatic oscillations, among other factors (Travis and Dytham 1999; Bonte et al. 2012; Travis et al. 2013).

R. Postiglioni • A. Aisenberg (✉)
Laboratorio de Etología, Ecología y Evolución, Instituto de Investigaciones Biológicas Clemente Estable, Montevideo, Uruguay
e-mail: anita.aisenberg@gmail.com

M. Simó
Sección Entomología, Facultad de Ciencias, Universidad de la República, Iguá 4225, CP 11400, Montevideo, Uruguay

The effect of climate change on dispersal ability has received increasing attention because climatic variables define in part the ecological niche and geographic area available for a species (Sexton et al. 2009). The classical approach of delimiting potential distributions based solely on the correlative responses of their presence to a set of environmental variables (Guisan and Zimmermann 2000) has been recently revised to incorporate dispersal as an essential attribute to determine more dynamic distribution models, based on a greater realism in the movement ability of organisms (Gallien et al. 2010). The incorporation of dispersal to define the range of a species has led to the design of new models that more accurately reflect the characteristics of the geographical landscape (e.g., corridors and geographic barriers) where the organisms move, which in turn allows the prioritization of spatially explicit conservation areas (Engler et al. 2009). These new approaches incorporate connectivity as an essential criterion to ensure movement through the landscape at different spatial scales (Engler et al. 2009). This is particularly relevant for both species with high and species with low vagility, because it is possible to define connectivity models to ensure the movement from broad to very small spatial scales (Bellard et al. 2012). Distribution models that incorporate spatially explicit dispersion rules with connectivity designs are fundamental in order to generate more realistic analysis at the demographic level of organisms. From a conservation perspective, distribution models that incorporate dispersion data contribute to the development of strategies to mitigate the global change effects on biota, especially habitat loss and climate change (Zurell et al. 2016).

Molecular tools provide relevant information on conservation planning by assessing the genetic diversity, size, and connectivity (i.e., gene flow) of populations, and delimiting evolutionary significant lineages for conservation (Moritz 1994, 1995; Frankham et al. 2002). Genetic data reveal the structure of populations over time and space, providing indirect evidence of species dispersal abilities. Different patterns are expected depending on the vagility of species. For example, good dispersers are expected to exhibit weaker geographic structure and more chances to track refuges and colonize new habitats under unfavorable environmental conditions, hence experiencing larger shifts in range than low-vagility species (McPeek and Holt 1992; Papadopoulou et al. 2009; Bidegaray-Batista et al. 2016).

Dispersal among spiders is mainly conducted by ambulatory locomotion or aerial displacements by means of silk threads, commonly referred as 'ballooning' (Decae 1987; Bonte 2013). The former dispersal strategy involves short-distance movements in which spiders walk the necessary distance for foraging or mating. Aerial dispersal, on the other hand, implies both short- and long-distance movements, allowing spiders to colonize new and remote habitats and quickly shift their distribution range during unfavorable conditions (Gillespie et al. 2012; Bonte 2013). Hence, studies of dispersal mechanisms and ballooning propensity may help to explain population structure and provide information on the response of spider species to climatic change and habitat loss.

Spiders are good indicators of human impact and ecosystem health (Kremen et al. 1993; Wise 1995). However, because of their great diversity and abundance, and difficult taxonomy, spiders have been rarely included in monitoring programs of the conservation status of particular habitats (Cardoso 2008; Crespo et al. 2014). The development of rapid biodiversity assessment programs specifically tailored to study spider communities (Coddington et al. 1991; Cardoso 2009; Cardoso et al. 2016; Emerson et al. 2016; Malumbres-Olarte et al. 2017), has helped to overcome the inherent limitation in sampling megadiverse groups, and has paved the way to investigate spider diversity patterns across multiple regions in a comparative framework (Cardoso et al. 2011b). In addition, the use of DNA barcoding approaches may further help to ameliorate the burden and difficulty of taxonomic identification in large bio-inventory studies (Smith et al. 2005; Emerson et al. 2016). Semi-quantitative biodiversity assessment protocols and DNA barcoding tools have been successfully combined to infer regional patterns of species diversity and abundance in spiders of the Panamanian cloud forest (Labarque 2012) and the Spanish Network of Nationals Parks (Arnedo and collaborators, ongoing research).

Spiders are largely underrepresented in the IUCN red list of threatened species. Less than 0.5% of the approximately 46,000 spider species (World Spider Catalog 2016) have been evaluated following to the IUCN criteria, of which only 133 have been catalogued as threatened species (IUCN 2017). Cardoso et al. (2011a) have highlighted the difficulties in applying the IUCN criteria to spiders and other invertebrates, and have justified the need for revising and adapting the criteria to these organisms. The assessment of the conservation status of the Maderian endemic wolf spider *Hogna ingens* (Blackwall 1857) provides a nice example of how spiders can be accommodated to the IUCN criteria (Crespo et al. 2014). The relevance of including spiders in the red list has been nicely exemplified by the linyphiid spider *Nothophantes horridus* Merrett & Stevens, 1995 from southwest England (World Spider Catalog 2016). The inclusion of the spider in the global Red List of Threatened Species as critically endangered resulted in the dismissal of a proposal to build a housing development in one of the few known localities of this spider. The recent creation of the Spider and Scorpion Specialist Group within the IUCN will certainly help to increase the number of spiders evaluated and eventually included in the red list, leading to the future protection and recovery of numerous arachnid species.

In the age of the sixth mass extinction, it is essential to assess species and ecosystems conservation status to predict their fate in the face of global change and ensure their preservation for the next generations. As discussed above, a complete understanding of the health of the ecosystem requires the integration of the information provided by ecology, behavior, and genetics. Determining spider dispersal abilities, identifying the climatic factors governing their distribution, and revealing their population structure and patterns of gene flow are unavoidable steps towards assessing their conservation status and developing conservation plans. In the next sections of the present chapter, we will summarize and integrate the results obtained in different multidisciplinary studies conducted on two sand-dwelling wolf spiders: *Allocosa marindia* Simó et al. 2017 and *A. senex* (Mello-Leitão 1945). Recently, *A. marindia* was proposed as a new species, including the Uruguayan specimens considered previously as *A. alticeps* (Mello-Leitão 1944), whereas *A. senex* which remained as a junior synonym of *A. brasiliensis* (Petrunkevitch 1910) was revalidated (Simó et al. 2017). These spiders exhibit extraordinary morphological, ecological, and behavioral adaptations to live in the sandy coasts of Southern South America, where they are confined in a severely reduced and transformed habitat.

Neotropical Sand-Dwelling Wolf Spiders with Atypical Behavioral Roles

Lycosidae, commonly called wolf spiders, is one of the most specious spider family, with approximately 2400 species currently recognized (World Spider Catalog 2016). Lycosids are medium- to large-sized spiders that usually show moderate levels of sexual size dimorphism (Walker and Rypstra 2003). As commonly seen in spiders, females are larger than males (Vollrath and Parker 1992; Hormiga et al. 2000; Moya-Laraño et al. 2002; Foellmer and Moya-Laraño 2007). Males are usually the wandering sex that searches and competes for potential mates, and initiates courtship (Foelix 2011; Huber 2005; Schneider and Andrade 2011). On the other hand, females are more sedentary and choosy, and sexual cannibalism on males may occur during courtship or after mating (Moya-Laraño et al. 2003; Wise 2006). Interestingly, cases of reversal in the sex roles, in which females turn into the roving sex that searches and courtship males, while males are selecting potential mates, have been documented in several animal groups such as birds, fish, and insects, among others (Gwynne 1991; Andersson 1994; Bonduriansky 2001). Although male mate assessment and female request of mating has been described for many spiders (Robinson and Robinson 1980; Costa and Pérez-Miles 2002; Rypstra et al. 2003; Schulte et al. 2010) and sex role reversal seems to be relatively common in insects, hypotheses related to changes in traditional roles have rarely been tested in spiders (Knoflach 1998; Aisenberg 2014).

The Neotropical sand-dwelling wolf spiders *A. marindia* and *A. senex* are the first two identified cases of sex role reversal in spiders, giving us the opportunity to study the ecological, behavioral, and phylogenetic factors driving the origin and maintenance of these atypical behaviors.

Allocosa marindia and *A. senex* are two burrowing wolf spiders that inhabit the sandy coasts of rivers, lakes, and the seashore of Southern South America (see Fig. 5.1) (Capocasale 1990). *A. senex* is larger than *A. marindia* and, unlike most spiders, in both species males are larger than females (*A. marindia*: females 2.9 ± 0.3 mm, males 3.3 ± 0.5 mm, *A. senex*: females 4.6 ± 0.5 mm, males 5.8 ± 0.6 mm) (Aisenberg 2014). The species are sympatric but not syntopic because they show consistent differences in their microhabitat (Costa et al. 2006; Aisenberg et al. 2011c): *A. senex* is associated to open dunes with scarce native psammophilic vegetation, while *A. marindia* is found in areas with higher abundance of vegetation, which can include exotic plant species (see Fig. 5.1). Both spiders construct silk-lined burrows in the sand, where they stay during daylight and winter (Costa 1995; Costa et al. 2006). During the summer nights, these spiders become active and leave their burrows to forage and search for mates. Male burrows in both species are much longer than those corresponding to females, which are simple and short silk tubes (Aisenberg et al. 2007; Aisenberg and Costa 2008; Albín et al. 2015).

Studies on *A. marindia* and *A. senex* have revealed reversal in the typical sex roles expected for spiders (Aisenberg et al. 2007; Aisenberg and Costa 2008; Aisenberg 2014). Males emit volatile sex pheromones that aid females in locating

Fig. 5.1 (**a**) Female of *Allocosa marindia* (Photo: Marcelo Casacuberta). (**b**) Male of *Allocosa senex* (Photo: Marcelo Casacuberta). (**c**) Typical habitat of *A. marindia*, Parque del Plata Beach, Department of Montevideo, Uruguay (Photo: Anita Aisenberg). (**d**) Typical habitat of *A. senex*, La Serena Beach, Department of Rocha, Uruguay (Photo: Marcelo Casacuberta)

them inside their burrows (Aisenberg et al. 2010a). Mating occurs inside the male burrow and involves several mounts and dismounts (Aisenberg et al. 2007; Aisenberg and Costa 2008). When mating ends, the male leaves the burrow and the female stays in. He closes the burrow entrance from the outside, while the female cooperates by laying silk from the inside. The male leaves only after the burrow entrance is completely closed and camouflaged (Aisenberg et al. 2007, Aisenberg and Costa 2008). The female lays the egg sac inside the male burrow and leaves after spiderling emergence and for further dispersal (Postiglioni et al. 2008).

The donation of the burrow has implications for male fitness because mating opportunities are reduced until the construction of a new burrow. In *A. marindia* and *A. senex*, both sexes are selective when taking mating decisions, in agreement with the high reproductive investment on mate searching, courtship, and egg laying by females, and in burrow donation by males (Aisenberg 2014). Females prefer males with longer burrows (Aisenberg et al. 2007, Aisenberg and Costa 2008), and rejected males can respond by lengthening their burrows to maximize future mating opportunities (Carballo et al. 2017). On the other hand, males prefer virgin females in good body condition (Aisenberg and González 2011; Aisenberg et al. 2011a). In addition, only in *A. senex*, rejected females can be expelled from male burrows or attacked and cannibalized (Aisenberg et al. 2011a).

The drivers of sex role reversal in these particular spiders are still a matter of investigation. Based on the available information in other animal groups, sex role reversal occurs when there is high male reproductive investment that limits mating opportunities in this sex, and makes them a scarce resource for which females have to compete (Gwynne 1991). Nevertheless, high male reproductive cost alone seems not to be a sufficient condition to induce sex role reversal (Gwynne 1991; Bonduriansky 2001). Therefore, additional behavioral, ecological, and phylogenetic characteristics of the species may be required to drive the evolution of these unique strategies.

Despite the presence of sex role reversal in both sand-dwelling wolf spiders, they show marked differences in life history and reproductive traits. *A. senex* life span is longer than *A. marindia* (adults of *A. senex* can survive two reproductive seasons, while *A. marindia* adults survive only one) (Aisenberg and Costa 2008). Moreover, breeding under laboratory conditions has revealed a strong female bias in the sex ratio in *A. marindia*, but not in *A. senex* (Aisenberg and Costa 2008), which agrees with field observations (Costa et al. 2006; Aisenberg et al. 2011b). The presence of *Wolbachia*, an endosymbiont bacterium that can cause female-biased sex ratios in arthropods (Werren et al. 2008; Goodacre and Martin 2012, 2013), in certain populations of *A. marindia* recently reported by Lerette et al. (2015) could explain the aforementioned observations.

High male reproductive investment and sex role reversal in arthropods have been linked to species inhabiting harsh and unpredictable environments (Karlsson et al. 1997; Lorch 2002). Furthermore, these changes in female and male roles can affect only certain populations and they can occur very fast, according to the behavioral plasticity of the species (Gwynne and Simmons 1990). Because of the wide temperature and rainfall variation and changes in prey availability, the sandy coasts of South America could be considered harsh and temporal habitats (Aisenberg et al. 2011c).

Allocosa as Bio-Indicator of Coastal Sand Dunes in Uruguay

The presence and abundance of bio-indicator species provide evidence on ecosystem health (Kremen et al. 1993; Wise 1995; Maelfait and Hendrickx 1998). Spiders have been used as biological indicators, for example to evaluate the stress induced by heavy metal pollution, soil quality, or the conservation status of natural habitats (Maelfait and Hendrickx 1998; Hartley et al. 2008; Uehara-Prado et al. 2009; Ghione et al. 2013).

Allocosa marindia and *A. senex* are two of the most frequent arachnids inhabiting the sandy coasts of Uruguay (Costa et al. 2006), and are considered good bio-indicators of the quality of this habitat (Ghione et al. 2013). Organisms living in sandy coastal habitats have evolved several morphological and behavioral adaptations to cope with the dramatic variations in temperature, solar radiation, and humidity conditions in these habitats (McLachlan 1991; Stanley et al. 2013; Aisenberg 2014). The body of both *A. marindia* and *A. senex* is densely covered by setae, which provides protection against variation in humidity and temperature

(Dondale 1986). The lateral posterior spinnerets are longer compared to other wolf spiders of similar size (Dondale 1986), probably as an adaptation for more efficient silk deposition during the construction of the burrows. Male pedipalps are highly elongated and present conspicuous macrosetae, which facilitate digging in the sand (Aisenberg et al. 2010b).

The 740-km stretch of coast of the Río de la Plata estuary and the Atlantic Ocean of Southern and Eastern Uruguay was traditionally characterized by broad extensions of sandy areas with fixed and mobile dunes, and scarce psammophilic vegetation (Costa et al. 2006). During the last century, most of these areas have been dramatically fragmented and reduced due to agriculture, cattle raising, urbanization, and tourism (Lercari and Defeo 2003; Costa et al. 2006; Goso and Muzio 2006; Panario and Gutiérrez 2006; Ríos et al. 2010; Aisenberg et al. 2011c). Moreover, the forestation with exotic plants has increased erosion and profoundly changed the physiognomy of the coast. The growth of urban centers along the coast interferes with the dynamics of the dunes, while the extraction of sand for construction has reduced the dune lines, generating in many cases artificial lagoons.

The drastic reduction and transformation of South American sandy coasts due to human activity has had a profound impact on the native fauna of these areas. During recent years, plans for recovering the dune line in some areas along the Southern Uruguayan coast have been implemented, including the installation of wooden barriers that retain the sand (Panario and Gutiérrez 2006). Although the restoration of dunes is crucial for the conservation of both *Allocosa* species, this alone is not enough. Jorge et al. (2015) observed that *A. senex* only colonizes restored dunes if they have native psammophile plant species such as *Panicum racemosum* (P. Beauv.) (Poaciae), *Senecio crassiflorus* (Poir.) D. C. (Asteraceae) and *Hydrocotyle bonariensis* Lam. (Apiaceae). This native vegetation could provide the ideal conditions for digging burrows in the open sand, while it would also protect the burrow against variable climatic conditions.

Bonte et al. (2006a) indicated that spiders inhabiting fragmented areas in coastal sand dunes show larger body sizes and longer generation times, characteristics that make them more vulnerable to extinction. Furthermore, habitat loss and fragmentation reduces the possibility that specialized species such as *A. marindia* and *A. senex* may successfully disperse by ballooning, mainly because the probability of successful landing in the sandy areas is limited. This pressure could lead to increased ambulatory dispersal, as has been reported for *Pardosa monticola* (Clerck 1757), another wolf spider inhabiting sand dunes at the Northern Hemisphere (Bonte et al. 2002). Natural corridors between sandy patches should be designed or restored to ensure connectivity between isolated populations of *Allocosa* and other obligate sand-dwellers.

The establishment of conservation plans at the Uruguayan Southern coast is urgently needed to preserve the coastal dynamics and the native biota, and avoid the introduction of exotic species. In this context, *A. marindia* and *A. senex* represent promising bio-indicators of the human impact on the Uruguayan coastline. Unfortunately, their sensitivity to the perturbations in their peculiar habitat is threatening the viability of their populations, and both *Allocosa* have been declared as priority species for conservation in Uruguay (Ghione et al. 2017).

Ambulatory Dispersal in *A. senex*

Ambulatory locomotion is essential for foraging, mate searching, and for avoiding predation and competition (inter and intraspecific). The efficiency of this dispersal strategy depends on the habitat structure (Bonte et al. 2003b; Bonte 2013). Baker (2007) proposed that when the habitat of an animal is fragmented into small and disconnected patches, there is a tendency to reduced mobility between them.

Ambulatory dispersal has been little studied in wolf spiders (but see Kiss and Samu 2000, Bonte et al. 2003b, Samu et al. 2003, Fernández-Montraveta and Cuadrado 2008). In general, wolf spiders are wandering or ambusher hunters that do not use silk webs to capture prey (Foelix 2011). Lycosids can move freely on different substrates, such as soil, rocks, sand, leaf-litter, and vegetation, and in some cases can even walk on water surfaces (Suter et al. 2003, Stratton et al. 2004, Suter 2013, Albín et al. 2017). Most lycosids are active during the night (Shook 1978; Ortega-Escobar 2002; Schonewolf et al. 2006; Aisenberg 2014), but there are few examples of species active during daylight or both daylight and night (Ford 1978; Marshall et al. 2002; González Pérez 2015). Ambulatory dispersal can occur individually or it can involve a group of related individuals. A distinctive characteristic of wolf spiders is that after spiderlings emerge from their egg sac, the mother carries them on their back for some days until they molt and disperse (Foelix 2011).

Jorge et al. (2015) conducted a capture–mark–recapture study in two beaches with different degree of shoreline modification in Uruguay to investigate terrestrial dispersal in *A. senex*. The species presented higher abundance, both in density of burrows and walking individuals, in less fragmented beaches with dunes with scarce psammophile vegetation (coverage between 25% and 50%). The species was absent in areas with exotic plants or with substrate covered by abundant leaf-litter, which probably concurred with trails frequently trampled by people walking down to the beach during the summer. Moreover, the occurrence of invasive vegetation including deciduous bushes or trees reduces the availability of substrate for digging burrows, and probably affects communication among individuals of *A. senex*. A similar negative association between density and leaf litter has also been reported in the burrowing wolf spider *Geolycosa xera* McCrone, 1963, which inhabits the sandy coasts of Florida (Marshall 1995).

The authors recorded displacements of 0.5–115 m from their capture sites for *A. senex* females and 2.6–54 m for males (Jorge et al. 2015). Females were more frequently recaptured and moved over greater distances than males, fulfilling the expectations of higher mobility in females due to the sex role reversal in this species (Aisenberg et al. 2007). Jorge et al. (2015) also found that in small patches the number of recaptures was higher, suggesting that the spiders were not able to disperse to new patches. Similar findings have been reported in dune ecosystems of Bulgaria, where the spider distributions seem to be shaped, among other things, by the size and connectivity between suitable habitat remnants (Bonte et al. 2004).

Aerial Dispersal

Spiders are able of airborne dispersal by means of silk threads that function as parachutes, a behavior that has been referred as ballooning. The use of silk for aerial dispersal has always captured the attention of humans. It was first documented in the seventeenth century (Cantabrigian 1969 in Bell et al. 2005). Ballooning has also inspired Latin American literature, as in the story named "Babas del diablo" (*"The devil's slimes"*) from the Argentinean writer Julio Cortázar (1959).

From a mechanical perspective, ballooning implies the secretion of silk threads released from the spinnerets that are dragged by air currents, generating a sufficient pulling force for lifting off the spider body and initiate flying. Because body weight is an obvious limiting factor, ballooning is frequently restricted to juveniles or adults of tiny species as linyphiids (Greenstone et al. 1987; Weyman et al. 2002). Ballooning involves two different yet non-exclusive pre-ballooning behaviors: (1) to climb to a high position and start releasing silk threads, raising the abdomen and stretching the legs in a display known as "tip-toeing," and (2) to climb to a high spot and drop hanging from a silk dragline, known as "dropping on dragline," then the spider body is tilted upwards by the wind and the silk line breaks near the attaching point (Coyle 1983; Decae 1987; Eberhard 1987; Bell et al. 2005; Ferretti et al. 2013).

Ballooning is a semi-controlled dispersal strategy. Although spiders may decide when to start the stereotypical behavior for aerial dispersal, they have little or no control over the flight duration or the landing place (Decae 1987; Bonte 2013). In this regard, the advantages of ballooning are counterbalanced by the settlement costs due to the risk of landing in an unsuitable habitat (Bonte 2013). Data gathered from ships, airplanes, and mountain-top traps have revealed that the traveling distances achieved with this aeronautic behavior can reach hundreds of kilometers, and thousands of meters of height (Weyman 1993; Bell et al. 2005). Interestingly, a recent study has shown that ballooning is associated with the ability to survive landings on water (Hayashi et al. 2015). This ability allows individuals to disperse repeatedly among landmasses separated by water bodies such as rivers, lakes, and seas (e.g., Kuntner and Agnarsson 2011).

Ballooning ability differs between spider families; it is widespread among araneomorphs but has been rarely documented in mygalomorphs (Decae 1987; Bell et al. 2005; Coddington 2005; Ferretti et al. 2013). It has been documented in 17 spider families, but most frequently occurs in Linyphiidae, Araneidae, Salticidae, Theridiidae, Thomisidae and Lycosidae (Bell et al. 2005). Lycosidae is among the RTA (retrolateral tibial apophysis) clade families with the higher number of ballooning reports (Bell et al. 2005).

Environmental Conditions and Ballooning

Available data has revealed an interesting trend between microclimatic factors such as temperature, wind speed, humidity, and light, and ballooning behavior (Richter 1970; Vugts and Van Wingerden 1976; Bishop 1990; Greenstone 1990; Weyman 1993; Duffey 1998; Reynolds et al. 2007). These interactions are largely expected, since spiders possess many sensorial organs that provide such climatic information (Weyman 1993; Bell et al. 2005; Foelix 2011). Temperature seems to be positively correlated with ballooning (Yeargan 1975; Vugts and Van Wingerden 1976; Bishop 1990; Weyman 1993; Bell et al. 2005). For example, Bonte and collaborators (2003a) found that under laboratory conditions there is a negative relationship between temperature and ballooning latency. Ballooning requires upward currents that the spider can sense when it starts releasing silk lines during pre-ballooning behavior (Suter 1999). There is a general consensus that ballooning activity declines when the mean wind speed exceeds 3 m^{-1} (Vugts and Van Wingerden 1976; Bishop 1990; Greenstone 1990; Weyman 1993; Reynolds et al. 2007). In wolf spiders, ballooning is related to the interaction between the patch size where the spiderlings were captured and wind speed. Juveniles of *Pardosa purbeckensis* F.O. Pickard-Cambridge, 1895 from larger patches showed higher ballooning frequencies under high wind currents (Bonte et al. 2007). The combined effect of warm temperatures and low winds, on the other hand, seems to result in longer ballooning distances (Reynolds et al. 2007). Unlike temperature and wind, the effects of humidity on ballooning are not yet conclusive. Bishop (1990) found a negative relation between humidity and ballooning, and proposed that since humid air is less dense than dry air, humidity could reduce wind dragging. Conversely, Vugts and Van Wingerden (1976) observed ballooning behavior under high humidity levels. Ballooning is more frequently observed during daylight, probably due to the presence of upward warm air currents (Weyman 1993, Bell et al. 2005). A positive relationship has been observed between both light-hours and the intensity of sunlight, and ballooning (Vugts and Van Wingerden 1976; Greenstone 1990).

Food Shortage, Habitat Features, and Ballooning

The shortage of feeding resources is a major ecological pressure driving dispersal. Legel and Van Wingerden (1980, in Weyman 1993) found a positive relation between ballooning frequency and the lack of food. Similarly, Mestre and Bonte (2012) reported a positive relationship between long-distance dispersal and both individual and mother-induced food deprivation. Interestingly, well-fed linyphiids showed lower latencies of ballooning in comparison with other food-deprived groups (Bonte et al. 2003a).

The relationship between habitat stability (i.e., temporary vs long-term stable habitats) and ballooning has also been a matter of investigation (Southwood 1962).

Greenstone (1982) found that wolf spider species inhabiting a more unpredictable habitat (in terms of meteorological conditions, availability of prey, and refuges, among other factors) showed higher ballooning frequencies compared to wolf spiders living in a more predictable habitat. Several studies support the former observation (see Bell et al. 2005). However, Richter (1970) identified habitat availability rather than predictability as the main factor driving ballooning in eight species of wolf spiders. Specialization, on the other hand, seems to be negatively related with ballooning frequency, which is generally higher in generalist spiders (Bonte et al. 2003c). The same seems to hold for fragmented areas, where ballooning is less frequent that in continuous landscapes (Bonte et al. 2006b).

Ballooning at the Coastal Sand Dunes of Uruguay

We have conduced laboratory and field experiments to investigate ballooning behavior in sand-dwelling *Allocosa* (Postiglioni et al., in press; Carlozzi et al., in press). In these studies, the juveniles of the species *Schizocosa malitiosa* (Tullgren 1905) are used as positive control, since this species performs ballooning and individuals can be found on grasslands at neighboring sites in the sand dunes. Moreover, similarly to *Allocosa* but unlike most spiders, *S. malitiosa* performs ballooning during the day but also at night. In agreement with Greenstone (1982), field studies in the two *Allocosa* species which inhabit the highly unpredictable coastal sand dune habitat showed higher ballooning frequencies compared to *S. malitosa*, a species from more stable and modified areas (Postiglioni et al., in press). Diurnal ballooning in *A. senex* was positively related with temperature, wind speed, and humidity, but negatively related with atmospheric pressure. However, both temperature and atmospheric pressure were negatively related with daylight ballooning in *A. marindia*. On the other hand, only in *A. marindia* during nocturnal trials, ballooning was negatively related to atmospheric pressure but positively related to humidity. The differences between *Allocosa* species could be related with predatory avoidance strategies, since intra-guild predation mainly due to *A. senex* predating on *A. marindia* has been reported (Aisenberg et al. 2009). Finally, the contrasting results between ballooning frequencies and climatic factors suggest a complex relationship between these phenomena, and that optimal conditions for taking-off depend on different contexts (i.e., day and night).

Genetic Diversity and Species Distribution Models in *Allocosa*

Genetic data provides essential information for inferring the phylogenetic relationships of species and populations and the timeline of their diversification, allowing the identification of the abiotic factors (e.g., climatic changes and geological events) and/or biological factors (e.g., dispersion) that shape genetic diversity. Moreover, genetic data provide fine-scale information to identify evolutionary significant lineages for conservation and to assess the size, structure, and connectivity (or gene flow) of populations (Moritz 1994; Frankham et al. 2004). On the other hand, genetic data provide limited information about the changes of species distribution ranges throughout time, and none about the environmental variables that shape these distributions. Identifying the environmental constraints on species distributions is fundamental to understand how species distributions are affected by environmental changes. Technological and methodological advances in geographic information systems for modeling species distributions offer the opportunity to identify such key environmental variables. Several studies have recently shown that the integration of species distribution models with genetic data are very useful in understanding patterns and in inferring processes that generate and shape the genetic and geographic distributions of populations and species, as well as for the prediction of future changes in the distribution patterns (Carstens and Richards 2007; Carnaval et al. 2009; Chan et al. 2011; Marske et al. 2011; Bidegaray-Batista et al. 2016). In this context, the following sections summarize ongoing phylogeographic, demographic, and species distribution modelling studies in *A. marindia* and *A. senex* aimed at shedding light on the processes that shaped their distributions. We further discuss how current global warming may affect the future distribution of these species.

Phylogeography, Demography, and Lineage Ages

Phylogeography aims at inferring the patterns of genetic variation of populations and closely related species in a geographical context, in order to understand the factors that generated and shaped population structure and species divergences (Avise et al. 1987; Avise 1998). Generally, co-occurring species sharing ecological requirements and similar ages are expected to show similar patterns of genetic variation because they may have endured the same historical and biogeographic events (Berminham and Moritz 1998; Avise 2000; Arbogast and Kenagy 2001; Hewitt 2004). Conversely, incongruent phylogeographic patterns in co-distributed species may indicate either differences in the species-specific traits (i.e., behavior, dispersal ability, and habitat preferences, among others) (Irwin 2002; Garrick et al. 2008; Papadopoulou and Knowles 2016) or a different timing of origin in the region.

Because of the ecological, evolutionary, and biogeographic similarities, we should expect similar phylogeographic patterns in *A. marindia* and *A. senex*. Ongoing studies in our research team based on the mitochondrial gene cytochrome c oxidase subunit I (*cox1*) have revealed similar coalescent times of *A. marindia* and *A. senex*, which trace back to the Pleistocene (~1 Ma) (Bidegaray-Batista et al., unpublished data). The two species also show similar nucleotide diversity[1]

[1] Nucleotide diversity (π_n): the average number of nucleotide differences per site between two sequences.

(π_n = 0.004), but the haplotype diversity[2] is higher for *A. senex* (h = 0.79) than for *A. marindia* (h = 0.48) (Bidegaray-Batista et al. unpublished data). None of the species show population genetic differentiation, and the distribution of the mitochondrial haplotypes[3] is not geographically structured, showing a star-like haplotype network around a very frequent and widespread mitochondrial haplotype, which is probably the ancestral haplotype of the species (Fig. 5.2) (Bidegaray-Batista et al. unpublished data; Postiglioni 2015). Star-like networks may be indicative of recent demographic expansion (Templeton et al. 1995), and this demographic signal was statistically confirmed by several tests in *A. senex* but not in *A. marindia*. The recent demographic expansion scenario revealed by the mitochondrial data in *A. senex* was further confirmed with nuclear data (i.e., AFLP, amplified fragment length polymorphism) (Postiglioni 2015). This expansion could be related to regional climatic oscillations and geologic changes during the Pleistocene (i.e., formation of sandy areas during cold and dry periods of the upper Pleistocene) (Clapperton 1993; Carignano 1999; Mon and Gutiérrez 2009; Turchetto-Zolet et al. 2013). Moreover, the absence of genetic differentiation among populations (and high connectivity) in both species could be related with their aerial dispersal abilities. Ephemeral environments, at both historical and geological time scales, favor strategies involving high dispersal rates, increasing the probability of finding new suitable places (Southwood 1977, 1988; Travis and Dytham 1999). This hypothesis has also been proposed to explain the lack of genetic structure observed in other sandy coast and riparian wolf spiders inhabiting highly dynamic environments (Boulton et al. 1998; Bonte et al. 2003b; Lambeets et al. 2009).

Interestingly, in both species the haplotypes most distantly related to the ancestral haplotype are found in the easternmost sandy coasts of the Rocha Department in Uruguay (*A. senex* hap3, hap8 and hap10, and *A. marindia* hap11) (see Fig. 5.2), where they coexist with other more frequent and widespread haplotypes. The highly divergent haplotypes could be the remnants of isolated populations that survived Pleistocene glaciations close to where they inhabit today, and subsequently mixed with newcomers following population expansions. Because of the highest genetic diversity found, the coast of the Rocha Department should be considered in future conservation planning.

Our results reveal similar phylogeographic patterns in *A. marindia* and *A. senex*, probably due to similar responses to the same past geological and environmental changes. However, *A. marindia* differ in its lower genetic diversity, and the lack of demographic expansion processes. This difference could simply reflect the inability to detect subtle changes due to the limited amount of data used, which highlights the need for increasing the geographic, taxonomic, and molecular marker sampling, and the use of a more powerful statistical framework (Knowles 2009; Papadopoulou and Knowles 2016). Alternatively, the reported differences in species-specific traits

[2] Haplotype diversity (h): the average number of alleles per locus.

[3] Haplotype or allele: a unique combination of genetic markers present in a sample.

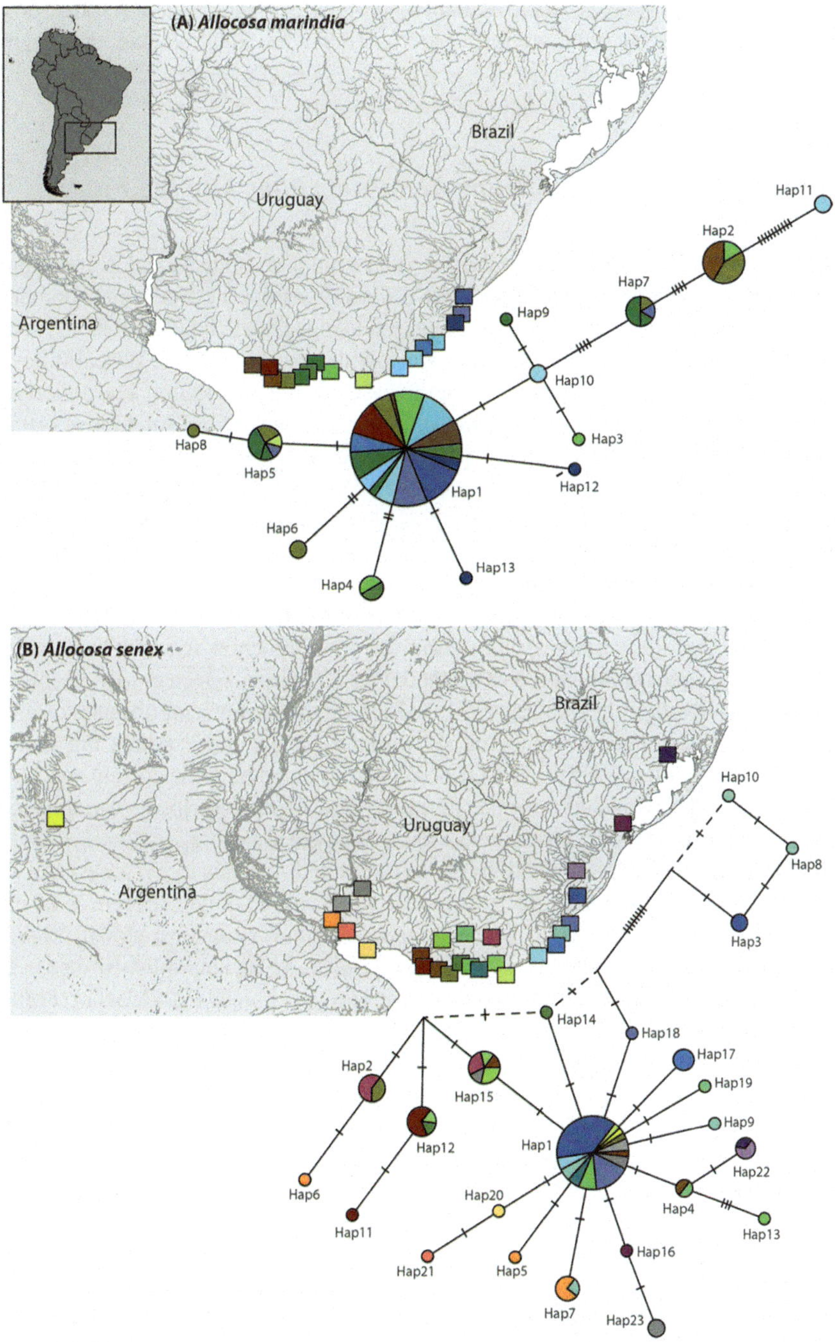

Fig. 5.2 Statistical parsimony network of the cox1 gene of *A. marindia* (**a**) and *A. senex* (**b**). The area of the pie plots is proportional to the number of individuals found for each haplotype. Dashes indicate the number of mutational differences between haplotypes, and discontinuous lines indicate alternative network connections. Haplotypes' colors match those of the localities in the map

between the two species, such as habitat preferences and behavior traits, could explain the different demographic patterns (Knowles 2009, Papadopoulou and Knowles 2016).

Species Distribution Modeling and Projections into the Future

"Species distribution models" (SDMs) or "niche models" (Guisan and Zimmermann 2000) are correlative models based on the relationships between environmental variables and known occurrences of a species, using statistical techniques. By using this approach, it is possible to define areas with greater or fewer probabilities of detecting the occurrence of a species in the past, present, and future. The underlying concept of SDMs is the ecological niche concept developed by Hutchinson (1957), which refers to an n-dimensional hyper volume defined by the combination of environmental conditions (factors) in which the species can survive. This definition is very important for the conceptualization of SDMs, allowing the geographic space to relate directly through the notion of a multidimensional space of variables where a species lives or potentially can do so.

SDMs are generated using two sources of information: presence data of species/ecosystems and the variables that define the environmental space, which will be distributed in geographic space. The modeling technique selected will establish a relationship between the geographical position on the presence or absence of information and the range of values of all variables where these points are located (Guisan and Thuiller 2005). The SDMs can be projected onto the current space, but they can also be projected onto the future or past, using descriptive variables corresponding to the time frame that we want to analyze.

SDMs have been constructed for *A. marindia* and *A. senex* to: (i) assess the current distribution of both sympatric species, which have subtle differences in their microhabitat preference, and (ii) predict their distribution under a global-warming scenario onto the next 70 years (Bidegaray-Batista et al. unpublished data). The SDMs were constructed with the maximum entropy technique (MAXENT) (Phillips et al. 2006; Elith et al. 2011), using unique localities of species presences and a set of six uncorrelated bioclimatic variables for the study area (temperature and precipitation seasonality, precipitation of driest and warmest quarter, minimum temperature of coldest month, and temperature annual range). The SDMs were projected onto the current and onto the future surface in the year 2080 using two RCP (report representative concentration pathways) scenarios (2.6 and 8.5) from the HADGEM2-ES global circulation model (GCM) presented by the IPCC5 (Intergovernmental Panel on Climate Change, Fifth Assessment Report). The results are presented in Fig. 5.3, where high environmental suitability is shown in the color red, and lower probabilities of occurrence are shown in green (Fig. 5.3a–d). The SDMs show that the current distribution for both species is sympatric. However, we can detect some differences between the species, where the distribution of *A. senex* is markedly more extensive and continuous (Fig. 5.3a). When future distribution is predicted for the year 2080, the distribution of both species shifts southwards and reduces (Fig. 5.3b, c, e, f). *A. marindia* will be most affected under climate change scenarios, tending to disappear under the worst scenario of CO_2 emission (RCP 8.5). These results call for the urgent implementation of a monitoring program and the development of conservation strategies for both species (with a greater concern in the case of *A. marindia*) and the supralittoral coastal ecosystem that they inhabit.

Conclusions and Future Directions

The sand-dwelling wolf spiders inhabiting the Uruguayan sandy coastal habitats — *A. marindia* and *A. Senex* — provide an excellent model system to investigate multiple aspects of the evolutionary process and ecological adaptations. Moreover, these two highly adapted species have been proposed as bio-indicators of the health of the costal dune ecosystems of South America, a disappearing habitat, highly threatened by human activities. Unfortunately, the two species are also of conservation concern, and predictive species distribution models under current global warming trends reveal a very pessimistic scenario for the future of the two species.

Allocosa marindia and *A. senex* are unique among spiders in showing a reversal in the traditional sex roles. Available data suggests that the evolution of this behavior may be linked to species dwelling in temporary and harsh environments, such as the coastal sand dunes inhabited by the *Allocosa* species. The availability of two species exhibiting sex role reversal living in sympatry and sharing similar ecological requirements and ways of life offers the opportunity to investigate the drivers for the evolution and the maintenance of this bizarre behavior in a comparative framework. Unfortunately, little is known about the phylogenetic relationships of these species, and there are even doubts about their generic assignment and species distribution boundaries. To date, the most inclusive molecular phylogenetic analyses conducted on the family Lycosidae did not include any *Allocosa* representative (Murphy et al. 2006). *Allocosa*, as many other lycosid genera, is in need of a profound taxonomic revision. With some exceptions mostly restricted to the Western Palearctic and a few Nearctic and African species, lycosid species seem to be misplaced at the generic level (Piacentini and Grismado 2009). *A. marindia* and *A. senex* were once included in a more inclusive genera, later considered junior synonymies of the world-wide genus *Allocosa* (Capocasale 1990), which currently includes approximately 140 species (World Spider Catalog 2016). A major effort should be done to infer a multi-locus phylogeny of a thorough sample of South American representatives of the genus *Allocosa* and their relatives (i.e., Allocosinae). A well-supported phylogeny will help to delimit the boundaries of the different species and will provide a phylogenetic framework to test hypothesis about the evolution of sex role reversal and the adaptations to sandy habitats. Specifically, it will allow testing the phylogenetic correlation between the adaptation to sand dune habitats and sex role reversal to confirm or reject whether inhabiting a highly dynamic and harsh habitat is involved in the evolution of these behaviors.

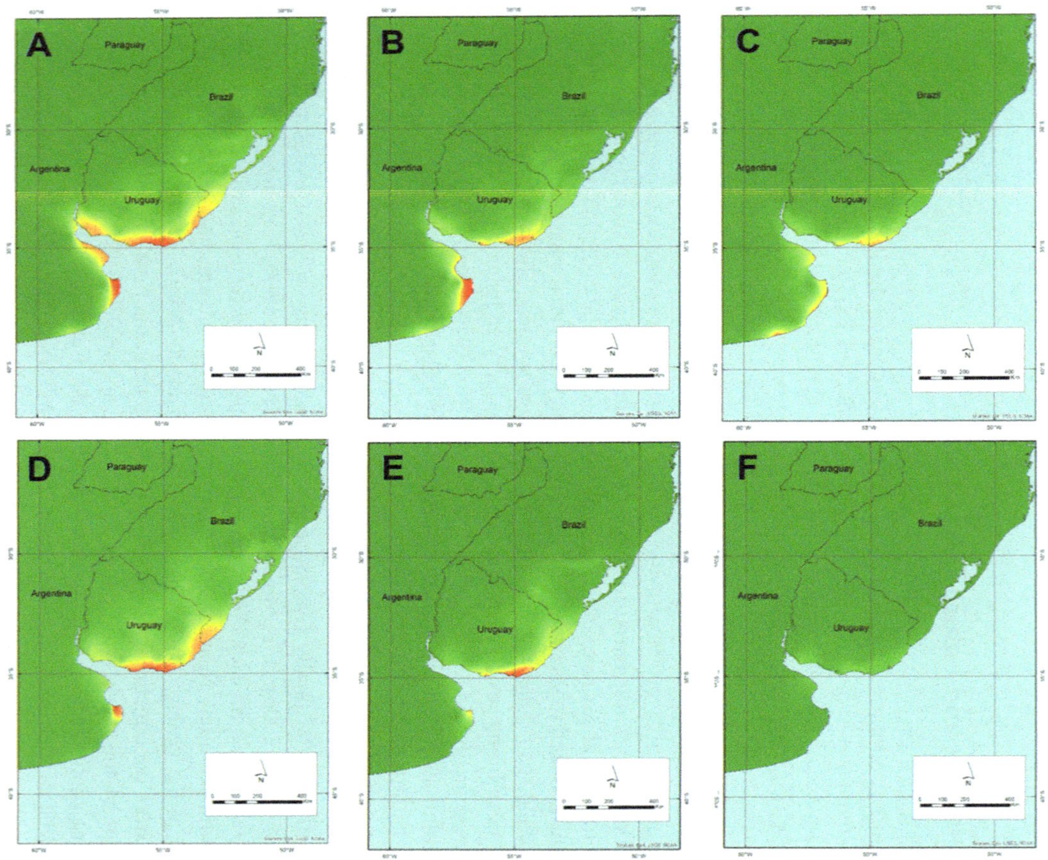

Fig. 5.3 Maps of predicted species occurrences in the present and in the future (2080 year) using two IPCC5 RCP scenarios (2.6 and 8.5) from HADGEM2-ES GCM. High probabilities of presence are shown in red and lower probabilities in green. (**a**, **d**) correspond to maps for the present for *A. senex* and *A. marindia* respectively, (**b**, **e**) correspond to maps for the future (RCP 2.6) for *A. senex* and *A. marindia* respectively, (**c**, **f**) correspond to maps for the future (RCP 8.5) for *A. senex* and *A. marindia*, respectively

Preliminary mtDNA population data, as well as genomic scanning trough AFLP for *A. senex*, have revealed the lack of structuring in the population of the Uruguayan *Allocosa* species. These results suggest high levels of gene flow among localities, which is most likely the result of the high dispersal ability of these spiders (i.e., ballooning). In this regard, South American sand-dwelling *Allocosa* species may be better considered as metapopulations (Hanski and Gaggiotti 2004; Hanski 2009). The recurrent dispersal into temporary patches in a heterogeneous landscape would guarantee the persistence of the species in places where otherwise they would go extinct (Alexander et al. 2012). Alternatively, the apparent lack of population structure may be the result of the lack of resolution of the molecular markers used to date. The present-day population structuring of the two species could be extremely subtle, with very shallow divergence that will require a large number of unlinked loci across the genome to be effectively detected. Next-generation sequencing methods would facilitate the fast and cost-effective generation of thousands of genome-wide genetic markers for non-model organisms, such as wolf spiders (Davey et al. 2011). Future studies on the population dynamics of *Allocosa* species will require the generation and analysis of such kind of molecular data.

According to species distribution models, these sand-dwelling wolf spiders may be brought on the verge of extinction if current trends of global warming persist in the forthcoming years. Important as they are to increase the sensitivity of governments and public and private agencies committed with the conservation of biodiversity, these predictions have not yet fully incorporated dispersal as a parameter to reflect the accessibility of the species to future suitable environments (Holloway et al. 2016). Ongoing research on environmental and intrinsic factors affecting ambulatory dispersal and ballooning in Uruguayan *Allocosa* will provide the necessary information to integrate into future predictive models the potential constraints to successful dispersal. This holistic approach will improve our understanding of connectivity and its integration into successful conservation strategies (Vasudev et al. 2015).

As in many other areas of coastal South America, human activity has drastically reduced and transformed the sandy coast along the Río de la Plata and the Atlantic Ocean coasts of Uruguay (Lercari and Defeo 2003; Goso and Muzio 2006; Costa et al. 2006; Aisenberg et al. 2011c), which has had a profound impact on the native fauna of these areas. Because of their narrow ecological requirements and high habitat fidelity, the two *Allocosa* species have been shown to be good indicators of the ecosystem quality of sand-dune habitats. Monitoring of these species populations through space and time will provide accurate information on the degree of impact of human disturbance and the success of the recovery plans designed to improve the quality of the sand dune ecosystems of Uruguay. Promoting public awareness on the factors that threaten sand dune ecosystems and on the important role of *Allocosa* spiders as bio-indicators of these fragile habitats will reinforce ongoing and spark new initiatives directed to protect and recover this unique habitat.

Acknowledgements We are grateful to two anonymous reviewers who improved the final version of the chapter. AA, LB, and MS acknowledge financial support by Sistema Nacional de Investigación (ANII) and Programa de Desarrollo de las Ciencias Básicas, UdelaR, Uruguay. LB acknowledges financial support by Project FCE_3_2013_1_100607 (ANII), MAA to project CGL2012-36863 from the Spanish Ministry of Economy and Competitivity and 2014SGR1604 from the Catalan Government, PP to FONDECYT Iniciación N° 11140357, and AA to Project FCE-2-2011-1-6023 (ANII).

References

Aisenberg A (2014) Adventurous females and demanding males: sex role reversal in a neotropical spider. In: Macedo RH, Machado G (eds) Sexual selection: perspectives and models from the Neotropics. Elsevier, Boston, pp 163–182

Aisenberg A, Costa FG (2008) Reproductive isolation and sex-role reversal in two sympatric sand-dwelling wolf spiders of the genus *Allocosa*. Can J Zool 86:648–658

Aisenberg A, González M (2011) Male mate choice in *Allocosa alticeps* (Araneae: Lycosidae), a sand-dwelling spider with sex role reversal. J Arachnol 39:444–448

Aisenberg A, Viera C, Costa FG (2007) Daring females, devoted males, and reversed sexual size dimorphism in the sand-dwelling spider *Allocosa brasiliensis* (Araneae, Lycosidae). Behav Ecol Sociobiol 62:29–35

Aisenberg A, González M, Laborda A, Postiglioni R, Simó M (2009) Reversed cannibalism, foraging, and surface activities of *Allocosa alticeps* and *Allocosa brasiliensis*: two wolf spiders from coastal sand dunes. J Arachnol 37:135–138

Aisenberg A, Baruffaldi L, González M (2010a) Behavioural evidence of male volatile pheromones in the sex-role reversed wolf spiders *Allocosa brasiliensis* and *Allocosa alticeps*. Naturwissenschaften 97:63–70

Aisenberg A, Costa FG, González M, Postiglioni R, Pérez-Miles F (2010b) Sexual dimorphism in chelicerae, forelegs and palpal traits in two burrowing wolf spiders (Araneae: Lycosidae) with sex-role reversal. J Nat Hist 44:1189–1202

Aisenberg A, Costa FG, González M (2011a) Male sexual cannibalism in a sand-dwelling wolf spider with sex role reversal. Biol J Linn Soc 103:68–75

Aisenberg A, González M, Laborda A, Postiglioni R, Simó M (2011b) Spatial distribution, burrow depth and temperature: implications for the sexual strategies in two *Allocosa* wolf spiders. Stud Neotropical Fauna Environ 46:147–152

Aisenberg A, Simó M, Jorge C (2011c) Spider as a model towards the conservation of coastal sand dunes in Uruguay. In: Murphy JA (ed) Sand dunes: conservation, shapes/types and desertification. NOVA Science Publishers, New York, pp 75–93

Albín A, Carballo M, Stanley E, Aisenberg A, Simó M (2017) Nocturnal surface activity and behavioural responses to floods in Allocosa senex (Araneae, Lycosidae). Arachnology 17(4):1–7

Albín A, Simó M, Aisenberg A (2015) Characterisation of burrow architecture under natural conditions in the sand-dwelling wolf spider *Allocosa brasiliensis*. J Nat Hist 50:201–209

Alexander HM, Foster BL, Ballantyne F, Collins CD, Antonovics J, Holt RD (2012) Metapopulations and metacommunities: combining spatial and temporal perspectives in plant ecology. J Ecol 100:88–103

Andersson MB (1994) Sexual selection. Princeton University Press, Princeton

Arbogast BS, Kenagy G (2001) Comparative phylogeography as an integrative approach to historical biogeography. J Biogeogr 28:819–825

Avise JC (1998) The history and purview of phylogeography: a personal reflection. Mol Ecol 7:371–379

Avise JC (2000) Phylogeography: the history and formation of species. Harvard University Press, Cambridge, MA

Avise JC, Arnold J, Ball RM, Bermingham E, Lamb T, Neigel JE, Reeb CA, Saunders NC (1987) Intraspecific phylogeography: the mitochondrial DNA bridge between population genetics and systematics. Annu Rev Ecol Syst 18:489–522
Baker L (2007) Effect of corridors on the movement behavior of the jumping spider *Phidippus princeps* (Araneae, Salticidae). Can J Zool 85:802–808
Bell JR, Bohan DA, Shaw EM, Weyman GS (2005) Ballooning dispersal using silk: world fauna, phylogenies, genetics and models. Bull Entomol Res 95:69–114
Bellard C, Bertelsmeier C, Leadley P, Thuiller W, Courchamp F (2012) Impacts of climate change on the future of biodiversity. Ecol Lett 15:365–377
Berminham E, Moritz C (1998) Comparative phylogeography: concepts and applications. Mol Ecol 7:367–369
Bidegaray-Batista L, Sánchez-Gracia A, Santulli G, Maiorano L, Guisan A, Vogler AP, Arnedo MA (2016) Imprints of multiple glacial refugia in the Pyrenees revealed by phylogeography and paleodistribution modelling of an endemic spider. Mol Ecol 25:2046–2064
Bishop L (1990) Meteorological aspects of spider ballooning. Environ Entomol 19:1381–1387
Bonduriansky R (2001) The evolution of male mate choice in insects: a synthesis of ideas and evidence. Biol Rev 76:305–339
Bonte D (2013) Cost-benefit balance of dispersal and the evolution of conditional dispersal strategies in spiders. In: Nentwig W (ed) Spider Ecophysiology. Springer, Heidelberg/New York/Dordrecht/London, pp 67–78
Bonte D, Baert L, Maelfait JP (2002) Spider assemblage structure and stability in a heterogeneous coastal dune system (Belgium). J Arachnol 30:331–343
Bonte D, Deblauwe I, Maelfait J-P (2003a) Environmental and genetic background of tiptoe-initiating behaviour in the dwarf spider *Erigone atra*. Anim Behav 66:169–174
Bonte D, Lens L, Maelfait J-P, Hoffmann M, Kuijken E (2003b) Patch quality and connectivity influence spatial dynamics in a dune wolfspider. Oecologia 135:227–233
Bonte D, Vandenbroecke N, Lens L, Maelfait J-P (2003c) Low propensity for aerial dispersal in specialist spiders from fragmented landscapes. P Roy Soc Lond B Bio 270:1601–1607
Bonte D, Baert L, Lens L, Maelfait JP (2004) Effects of aerial dispersal, habitat specialisation, and landscape structure on spider distribution across fragmented grey dunes. Ecography 27:343–349
Bonte D, Lens L, Maelfait J-P (2006a) Sand dynamics in coastal dune landscapes constrain diversity and life-history characteristics of spiders. J Appl Ecol 43:735–747
Bonte D, Borre JV, Lens L, Maelfait J-P (2006b) Geographical variation in wolf spider dispersal behaviour is related to landscape structure. Anim Behav 72:655–662
Bonte D, Bossuyt B, Lens L (2007) Aerial dispersal plasticity under different wind velocities in a salt marsh wolf spider. Behav Ecol 18:438–443
Bonte D, Van Dyck H, Bullock JM, Coulon A, Delgado M, Gibbs M, Lehouck V, Matthysen E, Mustin K, Saastamoinen M (2012) Costs of dispersal. Biol Rev 87:290–312
Boulton AM, Ramírez MG, Blair CP (1998) Genetic structure in a coastal dune spider (*Geolycosa pikei*) on Long Island, New York Barrier Islands. Biol J Linn Soc 64:69–82
Capocasale RM (1990) Las especies de la subfamilia Hippasinae de America del Sur (Araneae, Lycosidae). J Arachnol 18:131–141
Cardoso P (2008) Biodiversity and conservation of Iberian spiders: past, present and future. Bull Soc Entomol Arag 42:487–492
Cardoso P (2009) Standardization and optimization of arthropod inventories—the case of Iberian spiders. Biodivers Conserv 18:3949–3962
Cardoso P, Borges PA, Triantis KA, Ferrández MA, Martín JL (2011a) Adapting the IUCN red list criteria for invertebrates. Biol Conserv 144:2432–2440
Cardoso P, Pekár S, Jocqué R, Coddington JA (2011b) Global patterns of guild composition and functional diversity of spiders. PLoS One 6:e21710
Cardoso P, Carvalho JC, Crespo LC, Arnedo MA (2016) Optimal inventorying and monitoring of taxon, phylogenetic and functional diversity. bioRxiv. https://doi.org/10.1101/060400
Carignano CA (1999) Late Pleistocene to recent climate change in Códoba Province, Argentina: geomorphological evidence. Quat Int 57:117–134

Carballo M, Baldenegro F, Bollatti F, Peretti AV, Aisenberg A (2017) No pain, no gain: male plasticity in burrow digging according to female rejection in a sand-dwelling wolf spider. Behav Process 140:174–180
Carlozzi A, Bidegaray-Batista L, González-Bergonzoni I, Aisenberg A (In press) Flying sand-dwelling spiders: aerial dispersal in Allocosa marindia and A. senex (Araneae, Lycosidae). J Arachnol
Carnaval AC, Hickerson MJ, Haddad CFB, Rodrigues MT, Moritz C (2009) Stability predicts genetic diversity in the Brazilian Atlantic forest hotspot. Science 323:785–789
Carstens BC, Richards CL (2007) Integrating coalescent and ecological niche modeling in comparative phylogeography. Evolution 61:1439–1454
Chan LM, Brown JL, Yoder AD (2011) Integrating statistical genetic and geospatial methods brings new power to phylogeography. Mol Phylogenet Evol 59:523–537
Clapperton CM (1993) Nature of environmental changes in South America at the last glacial maximum. Palaeogeogr Palaeoclimatol Palaeoecol 101:189–208
Coddington JA (2005) Phylogeny and classification of spiders. In: Ubick D, Paquin P, Cushing PE, Roth V (eds) Spiders of North America: an identification manual. American Arachnological Society, Columbia, pp 18–24
Coddington JA, Griswold CE, DS D, Peñaranda E, Larcher SF (1991) Designing and testing sampling protocols to estimate biodiversity in tropical ecosystems. In: Dudley EC (ed) The Unity of evolutionary biology: proceedings of the fourth international congress of systematics and evolutionary biology. Dioscorides Press, Portland, pp 44–60
Cortázar J (1959) Las armas secretas. Editorial Sudamericana, Buenos Aires
Costa FG (1995) Ecología y actividad diaria de las arañas de la arena *Allocosa* spp. (Araneae, Lycosidae) en Marindia, localidad costera del sur del Uruguay. Rev Bras Biol 55:457–466
Costa FG, Pérez-Miles F (2002) Reproductive biology of Uruguayan theraphosids (Araneae, Mygalomorphae). J Arachnol 30:571–587
Costa FG, Simó M, Aisenberg A, Menafra R, Rodríguez-Gallego L, Scarabino F, Conde D (2006) Composición y ecología de la fauna epígea de Marindia (Canelones, Uruguay) con especial énfasis en las arañas: un estudio de dos años con trampas de intercepción. In: Menafra R, Rodríguez-Gallego L, Scarabino F, Conde D (eds) Bases para la conservación y el manejo de la costa uruguaya. Vida Silvestre Uruguay, Montevideo, pp 427–436
Coyle FA (1983) Aerial dispersal by mygalomorph spiderlings (Araneae, Mygalomorphae). J Arachnol 11:283–286
Crespo LC, Silva I, Borges PAV, Cardoso P (2014) Assessing the conservation status of the strict endemic Desertas wolf spider, *Hogna ingens* (Araneae, Lycosidae). J Nat Conserv 22:516–524
Davey JW, Hohenlohe PA, Etter PD, Boone JQ, Catchen JM, Blaxter ML (2011) Genome-wide genetic marker discovery and genotyping using next-generation sequencing. Nat Rev Genet 12:499–510
Decae AE (1987) Dispersal: ballooning and other mechanisms. In: Ecophysiology of spiders. Springer, Berlin, pp 348–356
Dondale CD (1986) The subfamilies of wolf spiders (Araneae, Lycosidae). In: Barrientos JA (ed) X Congreso Internacional de Aracnología. Imprenta Juvenil, S.A., Jaca, pp 337–332
Duffey E (1998) Aerial dispersal in spiders. In: Proceedings of the 17th European Colloquium of Arachnology, pp 187–191
Eberhard WG (1987) How spiders initiate airborne lines. J Arachnol 15:1–9
Elith J, Phillips SJ, Hastie T, Dudík M, Chee YE, Yates CJ (2011) A statistical explanation of MaxEnt for ecologists. Divers Distrib 17:43–57
Emerson BC, Casquet J, López H, Cardoso P, Borges PA, Mollaret N, Oromí P, Strasberg D, Thébaud C (2016) A combined field survey and molecular identification protocol for comparing forest arthropod biodiversity across spatial scales. Mol Ecol Resour. https://doi.org/10.1111/1755-0998.12617
Engler R, Randin CF, Vittoz P, Czáka T, Beniston M, Zimmermann NE, Guisan A (2009) Predicting future distributions of mountain plants under climate change: does dispersal capacity matter? Ecography 32:34–45

Fernández-Montraveta C, Cuadrado M (2008) Microhabitat selection in the potentially endangered wolf spider *Donacosa merlini* (Araneae, Lycosidae): implications for spider conservation. Can J Zool 86:1280–1288
Ferretti N, Pompozzi G, Copperi S, Schwerdt L (2013) Aerial dispersal by *Actinopus* spiderlings (Araneae: Actinopodidae). J Arachnol 41:407–408
Foelix RF (2011) Biology of spiders. Oxford University Press, New York
Foellmer MW, Moya-Laraño J (2007) Sexual size dimorphism in spiders: patterns and processes. In: Fairbairn DJ, Blanckenhorn WU, Székely T (eds) Sex, size and gender roles: evolutionary studies of sexual size dimorphism. Oxford University Press, Oxford, pp 71–81
Ford MJ (1978) Locomotory activity and the predation strategy of the of the wolf-spider *Pardosa amentata* (Clerck)(Lycosidae). Anim Behav 26:31–35
Frankham R, Ballou JD, Briscoe DA (2002) Introduction to conservation genetics. Cambridge University Press, Cambridge, UK
Frankham R, Ballou JD, Briscoe DA (2004) A primer of conservation genetics. Cambridge University Press, Cambridge, UK/New York
Gallien L, Munkemuller T, Albert CH, Boulangeat I, Thuiller W (2010) Predicting potential distributions of invasive species: where to go from here? Divers Distrib 16:331–342
Garrick RC, Rowell DM, Simmons CS, Hillis DM, Sunnucks P (2008) Fine-scale phylogeographic congruence despite demographic incongruence in two low-mobility saproxylic springtails. Evolution 62:1103–1118
Ghione S, Coelho L, Costa FG, García LF, González M, Jorge C, Laborda Á, Montes de Oca L, Pérez-Miles F, Postiglioni R, Simó M, Toscano-Gadea C, Viera C, Aisenberg A (2017) Arácnidos prioritarios para la conservación en Uruguay. Bol Soc Zool Urug 26(1):1–8
Ghione S, Simó M, Aisenberg A, Costa FG (2013) *Allocosa brasiliensis* (Araneae, Lycosidae) as a bioindicator of coastal sand dunes in Uruguay. Arachnology 16:94–98
Gillespie RG, Baldwin BG, Waters JM, Fraser CI, Nikula R, Roderick GK (2012) Long-distance dispersal: a framework for hypothesis testing. Trends Ecol Evol 27:47–56
González Pérez MdlM (2015) Aspectos reproductivos de *Aglaoctenus lagotis*: estudio interpoblacional de una araña lobo sedentaria de gran variabilidad fenotípica. Ph.D. dissertation, Universidad Nacional de Córdoba, Córdoba
Goodacre SL, Martin OY (2012) Modification of insect and arachnid behaviours by vertically transmitted endosymbionts: infections as drivers of behavioural change and evolutionary novelty. Insects 3:246–261
Goodacre SL, Martin OY (2013) Endosymbiont infections in spiders. In: Spider ecophysiology. Springer, Berlin, pp 93–105
Goso C, Muzio R (2006) Geología de la costa uruguaya y sus recursos minerales asociados. In: Menafra R, Rodríguez-Gallego L, Scarabino F, Conde D (eds) Bases para la conservación y el manejo de la costa uruguaya. Vida Silvestre Uruguay, Montevideo, pp 9–19
Greenstone MH (1982) Ballooning frequency and habitat predictability in two wolf spider species (Lycosidae: *Pardosa*). Fla Entomol 65:83–89
Greenstone M (1990) Meteorological determinants of spider ballooning: the roles of thermals vs. the vertical windspeed gradient in becoming airborne. Oecologia 84:164–168
Greenstone MH, Morgan CE, Hultsch A-L, Farrow RA, Dowse J (1987) Ballooning spiders in Missouri, USA, and New South Wales, Australia: family and mass distributions. J Arachnol 15:163–170
Guisan A, Thuiller W (2005) Predicting species distribution: offering more than simple habitat models. Ecol Lett 8:993–1009
Guisan A, Zimmermann NE (2000) Predictive habitat distribution models in ecology. Ecol Model 135:147–186
Gwynne DT (1991) Sexual competition among females: what causes courtship-role reversal? Trends Ecol Evol 6:118–121
Gwynne DT, Simmons LW (1990) Experimental reversal of courtship roles in an insect. Nature 346:172–174

Hanski I (2009) Incorporating the spatial configuration of the habitat into ecology and evolutionary biology. In: Cantrell S, Cosner C, Ruan S (eds) Spatial ecology. CRC Press, London, pp 167–188

Hanski I, Gaggiotti OE (2004) Ecology, genetics, and evolution of metapopulations. Elsevier Academic Press, Burlington

Hartley W, Uffindell L, Plumb A, Rawlinson H, Putwain P, Dickinson N (2008) Assessing biological indicators for remediated anthropogenic urban soils. Sci Total Environ 405:358–369

Hayashi M, Bakkali M, Hyde A, Goodacre SL (2015) Sail or sink: novel behavioural adaptations on water in aerially dispersing species. Evol Biol 15:118

Hewitt GM (2004) The structure of biodiversity — insights from molecular phylogeography. Front Zool 1:1–16

Holloway P, Miller JA, Gillings S (2016) Incorporating movement in species distribution models: how do simulations of dispersal affect the accuracy and uncertainty of projections? Int J Geogr Inf Sci 30:2050–2074

Hormiga G, Scharff N, Coddington JA (2000) The phylogenetic basis of sexual size dimorphism in orb-weaving spiders (Araneae, Orbiculariae). Syst Biol 49:435–462

Huber BA (2005) Sexual selection research on spiders: progress and biases. Biol Rev 80:363–385

Hutchinson G (1957) Concluding remarks: Cold Spring Harbor Symposia on Quantitative Biology. Yale University, New Haven

Irwin DE (2002) Phylogeographic breaks without geographic barriers to gene flow. Evolution 56:2383–2394

IUCN (2017) The IUCN Red List of Threatened Species. Version 2017-2. http://www.iucnredlist.org. Downloaded on 14 September 2017

Jorge C, Laborda Á, Alves Días M, Aisenberg A, Simó M (2015) Habitat preference and effects of coastal fragmentation in the sand-dwelling spider *Allocosa brasiliensis* (Lycosidae, Allocosinae). Open J Anim Sci 5:309–324

Karlsson B, Leimar O, Wiklund C (1997) Unpredictable environments, nuptial gifts and the evolution of sexual size dimorphism in insects: an experiment. Proc R Soc B 264:475–479

Kiss B, Samu F (2000) Evaluation of population densities of the common wolf spider *Pardosa agrestis* (Araneae: Lycosidae) in Hungarian alfalfa fields using mark-recapture. Eur J Entomol 97:191–196

Knoflach B (1998) Mating in *Theridion varians* Hahn and related species (Araneae: Theridiidae). J Nat Hist 32:545–604

Knowles LL (2009) Statistical phylogeography. Annu Rev Ecol Evol Syst 40:593–612

Kremen C, Colwel RK, Erwin TL, Murphy DD, Noss RF, Sanjayan MA (1993) Terrestrial arthropod assemblages: their use in conservation planning. Conserv Biol 7:796–808

Kuntner M, Agnarsson I (2011) Phylogeography of a successful aerial disperser: the golden orb spider Nephila on Indian Ocean islands. Evol Biol 11:119

Labarque FM (2012) Evaluación de identificaciones taxonómicas mediante código de barras del ADN en un grupo tropical megadiverso. Ph.D. dissertation, Universidad de Buenos Aires, Buenos Aires

Lambeets K, Breyne P, Bonte D (2009) Spatial genetic variation of a riparian wolf spider *Pardosa agricola* (Thorell, 1856) on lowland river banks: the importance of functional connectivity in linear spatial systems. Biol Conserv 143:660–668

Legel GJ, Van Wingerden WKRE (1980) Experiments on the influence of food and crowding on the aeronautic dispersal of Erygone arctica (White, 1852) (Araneae, Linyphiidae). In: Gruber J (ed) Proceedings, 8th International Arachnological Congress. Vienna, pp 97–102

Lercari D, Defeo O (2003) Variation of a sandy beach macrobenthic community along a human-induced environmental gradient. Estuar Coast Shelf Sci 58:17–24

Lerette S, Aisenberg A, Costa FG, Sotelo-Silveira JR, Bidegaray-Batista L (2015) Detección de la bacteria endosimbionte *Wolbachia* en la araña lobo *Allocosa alticeps*. Bol Soc Zool Uruguay 24:171–177

Lorch PD (2002) Understanding reversals in the relative strength of sexual selection on males and females: a role for sperm competition? Am Nat 159:645–657
Maelfait J-P, Hendrickx F (1997/1998) Spiders as bio-indicators of anthropogenic stress in natural and semi-natural habitats in Flanders (Belgium): some recent developments. In: Selden P (ed) Proceedings of the 17th European colloquium of arachnology, Edinburgh, pp 293–300
Malumbres-Olarte J, Scharff N, Pape T, Coddington JA, Cardoso P (2017) Gauging megadiversity with optimized and standardized sampling protocols: a case for tropical forest spiders. Ecol Evol 7:494–506
Marshall SD (1995) Natural history, activity patterns, and relocation rates of a burrowing wolf spider: *Geolycosa xera* archboldi (Araneae, Lycosidae). J Arachnol 23:65–70
Marshall SD, Pavuk DM, Rypstra AL (2002) A comparative study of phenology and daily activity patterns in the wolf spiders *Pardosa milvina* and *Hogna helluo* in soybean agroecosystems in southwestern Ohio (Araneae, Lycosidae). J Arachnol 30:503–510
Marske KA, Leschen RAB, Buckley TR (2011) Reconciling phylogeography and ecological niche models for New Zealand beetles: looking beyond glacial refugia. Mol Phylogent Evol 59:89–112
Mathias A, Kisdi É, Olivieri I (2001) Divergent evolution of dispersal in a heterogeneous landscape. Evolution 55:246–259
Matthysen E (2012) Multicausality of dispersal: a review. In: Clobert J, Baguette M, Benton TG, Bullock JM (eds) Dispersal ecology and evolution. Oxford University Press, Oxford, UK, pp 3–18
McLachlan A (1991) Ecology of coastal dune fauna. J Arid Environ 21:229–243
McPeek MA, Holt RD (1992) The evolution of dispersal in spatially and temporally varying environments. Am Nat 140:1010–1027
Mestre L, Bonte D (2012) Food stress during juvenile and maternal development shapes natal and breeding dispersal in a spider. Behav Ecol 23:759–764
Mon R, Gutiérrez AA (2009) The Mar Chiquita Lake: an indicator of intraplate deformation in the central plain of Argentina. Geomorphology 111:111–122
Moritz C (1994) Defining "evolutionarily significant units" for conservation. Trends Ecol Evol 9:373–375
Moritz C (1995) Uses of molecular phylogenies for conservation. Philos Trans R Soc Lond B 349:113–118
Moya-Laraño J, Halaj J, Wise DH (2002) Climbing to reach females: Romeo should be small. Evolution 56:420–425
Moya-Laraño J, Pascual J, Wise DH (2003) Mating patterns in late-maturing female Mediterranean tarantulas may reflect the costs and benefits of sexual cannibalism. Anim Behav 66:469–476
Murphy NP, Framenau VW, Donnellan SC, Harvey MS, Park YC, Austin AD (2006) Phylogenetic reconstruction of the wolf spiders (Araneae: Lycosidae) using sequences from the 12S rRNA, 28S rRNA, and NADH1 genes: implications for classification, biogeography, and the evolution of web building behavior. Mol Phylogenet Evol 38:583–602
Nathan R, Getz WM, Revilla E, Holyoak M, Kadmon R, Saltz D, Smouse PE (2008) A movement ecology paradigm for unifying organismal movement research. Proc Natl Acad Sci U S A 105:19052–19059
Ortega-Escobar J (2002) Evidence that the wolf-spider *Lycosa tarantula* (Araneae, Lycosidae) needs visual input for path integration. J Arachnol 30:481–486
Panario D, Gutiérrez O (2006) Dinámica y fuentes de sedimentos de las playas uruguayas. In: Menafra R, Rodríguez-Gallego L, Scarabino F, Conde D (eds) Bases para la conservación y el manejo de la costa uruguaya. Vida Silvestre Uruguay, Montevideo
Papadopoulou A, Knowles LL (2016) Toward a paradigm shift in comparative phylogeography driven by trait-based hypotheses. Proc Natl Acad Sci USA 113:8018–8024
Papadopoulou A, Anastasiou I, Keskin B, Vogler AP (2009) Comparative phylogeography of tenebrionid beetles in the Aegean archipelago: the effect of dispersal ability and habitat preference. Mol Ecol 18:2503–2517

Peterson AT (2009) Phylogeography is not enough: the need for multiple lines of evidence. Front Biogeogr 1:19–25

Phillips SJ, Anderson RP, Schapire RE (2006) Maximum entropy modeling of species geographic distributions. Ecol Model 190:231–259

Piacentini LN, Grismado CJ (2009) *Lobizon* and *Navira*, two new genera of wolf spiders from Argentina (Araneae: Lycosidae). Zootaxa 2195:1–33

Postiglioni R (2015) Estructuración genética y variación morfológica en ambientes fluviales y oceánico-estuarinos en la araña *Allocosa brasiliensis* (Lycosidae) del sur de Uruguay. M.Sc. dissertation, Universidad de la República, Montevideo

Postiglioni R, González M, Aisenberg A (2008) Permanencia en la cueva masculina y producción de ootecas en dos arañas lobo de los arenales costeros. In: Proceedings IX Jornadas de Zoología del Uruguay, Montevideo, p 145

Postiglioni R, Aisenberg A, Carlozzi A, Bidegaray-Batista L (In press) The dark side of ballooning: nocturnal aerial dispersal in wolf spiders from the South American coastline. Arachnology

Reynolds AM, Bohan DA, Bell JR (2007) Ballooning dispersal in arthropod taxa: conditions at take-off. Biol Lett 3:237–240

Richter CJ (1970) Aerial dispersal in relation to habitat in eight wolf spider species (*Pardosa*, Araneae, Lycosidae). Oecologia 5:200–214

Ríos M, Bartesaghi L, Piñeiro V, Garay A, Mai P, Delfino L, Masciadri S, Alonso-Paz E, Bassagoda MJ, Soutullo A (2010) Caracterización y distribución espacial del bosque y matorral psamófilo. Avaiable from Grupo Guayubira Informe No 23. Proyecto Fortalecimiento del Proceso de Implementación del Sistema Nacional de Áreas Protegidas del Uruguay. SNAP Ecoplata, Montevideon

Robinson MH, Robinson B (1980) Comparative studies of the courtship and mating behavior of tropical araneid spiders. Pac Ins Monogr 36:1–218

Rypstra AL, Wieg C, Walker SE, Persons MH (2003) Mutual mate assessment in wolf spiders: differences in the cues used by males and females. Ethology 109:315–325

Samu F, Sziranyi A, Kiss B (2003) Foraging in agricultural fields: local 'sit-and-move' strategy scales up to risk-averse habitat use in a wolf spider. Anim Behav 66:939–947

Schneider JM, Andrade MCB (2011) Mating behaviour and sexual selection. In: Herberstein ME (ed) Spider behaviour: flexibility and versatility. Cambridge University Press, Cambridge, pp 215–274

Schonewolf KW, Bell R, Rypstra AL, Persons MH (2006) Field evidence of an airborne enemy-avoidance kairomone in wolf spiders. J Chem Ecol 32:1565–1576

Schulte KF, Uhl G, Schneider JM (2010) Mate choice in males with one-shot genitalia: limited importance of female fecundity. Anim Behav 80:699–706

Sexton JP, McIntyre PJ, Angert AL, Rice KJ (2009) Evolution and ecology of species range limits. Annu Rev Ecol Evol Syst 40:415–436

Shook RS (1978) Ecology of the wolf spider, *Lycosa carolinensis* Walckenaer (Araneae, Lycosidae) in a desert community. J Arachnol 6:53–64

Simó M, Lise AA, Pompozzi G, Laborda A (2017) On the taxonomy of southern South American species of the wolf spider genus *Allocosa* (Araneae: Lycosidae: Allocosinae). Zootaxa 4216(3):261–278

Smith MA, Fisher BL, Hebert PD (2005) DNA barcoding for effective biodiversity assessment of a hyperdiverse arthropod group: the ants of Madagascar. Philos Trans R Soc B 360:1825–1834

Southwood TRE (1962) Migration of terrestrial arthropods in relation to habitat. Biol Rev 37:171–211

Southwood T (1977) Habitat, the templet for ecological strategies? J Anim Ecol 46:337–365

Southwood T (1988) Tactics, strategies and templets. Oikos 52:3–18

Stanley E, Toscano-Gadea C, Aisenberg A (2013) Spider hawk in sand dunes: *Anoplius bicinctus* (Hymenoptera: Pompilidae), a parasitoid wasp of the sex-role reversed spider *Allocosa brasiliensis* (Araneae: Lycosidae). J Insect Behav 26:514–524

Stratton GE, Suter RB, Miller PR (2004) Evolution of water surface locomotion by spiders: a comparative approach. Biol J Linn Soc 81:63–78
Suter RB (1999) An aerial lottery: the physics of ballooning in a chaotic atmosphere. J Arachnol 27:281–293
Suter RB (2013) Spider locomotion on the water surface: biomechanics and diversity. J Arachnol 41:93–101
Suter RB, Stratton G, Miller PR (2003) Water surface locomotion by spiders: distinct gaits in diverse families. J Arachnol 31:428–432
Templeton AR, Routman E, Phillips CA (1995) Separating population structure from population history: a cladistic analysis of the geographical distribution of mitochondrial DNA haplotypes in the Tiger Salamander, *Ambystoma tigrinum*. Genetics 140:767–782
Travis JM, Dytham C (1999) Habitat persistence, habitat availability and the evolution of dispersal. Proc R Soc Lond B 266:723–728
Travis JM, Delgado M, Bocedi G, Baguette M, Bartón K, Bonte D, Boulangeat I, Hodgson JA, Kubisch A, Penteriani V (2013) Dispersal and species' responses to climate change. Oikos 122:1532–1540
Turchetto-Zolet A, Pinheiro F, Salgueiro F, Palma-Silva C (2013) Phylogeographical patterns shed light on evolutionary process in South America. Mol Ecol 22:1193–1213
Uehara-Prado M, de Oliveira FJ, de Moura BA, Machado G, Santos AJ, Vaz-de-Mello FZ, Freitas AVL (2009) Selecting terrestrial arthropods as indicators of small-scale disturbance: a first approach in the Brazilian Atlantic Forest. Biol Conserv 142:1220–1228
Vasudev D, Fletcher RJ, Goswami VR, Krishnadas M (2015) From dispersal constraints to landscape connectivity: lessons from species distribution modeling. Ecography 38:967–978
Vollrath F, Parker GA (1992) Sexual dimorphism and distorted sex ratios in spiders. Nature 360:156–159
Vugts HF, Van Wingerden WKRE (1976) Meteorological aspects of aeronautic behaviour of spiders. Oikos 27:433–444
Walker SE, Rypstra AL (2003) Sexual dimorphism and the differential mortality model: is behaviour related to survival? Biol J Linn Soc 78:97–103
Werren JH, Baldo L, Clark ME (2008) Wolbachia: master manipulators of invertebrate biology. Nat Rev Microbiol 6:741–751
Weyman GS (1993) A review of the possible causative factors and significance of ballooning in spiders. Ethol Ecol Evol 5:279–291
Weyman G, Sunderland K, Jepson P (2002) A review of the evolution and mechanisms of ballooning by spiders inhabiting arable farmland. Ethol Ecol Evol 14:307–326
Wise DH (1995) Spiders in ecological webs. Cambridge University Press, New York
Wise DH (2006) Cannibalism, food limitation, intraspecific competition, and the regulation of spider populations. Annu Rev Entomol 51:441–465
World Spider Catalog (2016) Natural History Museum Bern. http://wsc.nmbe.ch. Accessed 18 Oct 2016
Yeargan KV (1975) Factors influencing the aerial dispersal of spiders (Arachnida: Araneida). J Kans Entomol Soc 48:403–408
Zurell D, Thuiller W, Pagel J, Cabral JS, Münkemüller T, Gravel D, Dullinger S, Normand S, Schiffers KH, Moore KA (2016) Benchmarking novel approaches for modelling species range dynamics. Glob Chang Biol 22:2651–2664

Chapter6
Webs: Diversity, Structure and Function

Sean J. Blamires, Shichang Zhang, and I-Min Tso

Abstract Web building has been such a highly successful foraging innovation among spiders that the vast majority of extant spiders are web builders. The structure of spider webs varies substantially between species, and web building has even been lost completely in some clades. Examples of different web forms include the classic orb webs, which may be orientated vertical to the ground or horizontal, sheet webs, and cobwebs, which consist of three-dimensional meshwork and ascending sticky threads for support and capture of prey. The architecture of webs may also vary within clades and even within species. This may be a consequence of: (i) individuals adapting their web structures to the environment; e.g., larger webs are built in areas where more space is available, (ii) spiders varying their webs to tune its performance, e.g., when spiders are exposed to different prey, or (iii) silk expression constraints, e.g., when on diets lacking certain nutrients. We review the literature, focusing on contributions from the Neotropical region, showing that spider webs vary in structure and function at multiple levels and so must be considered a dynamic, variable, extended phenotype of its builder. Webs accordingly depict the foraging, mating, and defensive strategies, and physiological status, of the spider.

S.J. Blamires (✉)
Evolution & Ecology Research Centre, School of Biological,
Earth & Environmental Sciences, The University of New South Wales,
Sydney 2052, Australia

Department of Life Science, Tunghai University, Taichung 40704, Taiwan
e-mail: s.blamires@unsw.edu.au

S. Zhang
Department of Life Science, Tunghai University, Taichung 40704, Taiwan

I.-M. Tso
Department of Life Science, Tunghai University, Taichung 40704, Taiwan

Center for Tropical Ecology & Biodiversity, Tunghai University,
Taichung 40704, Taiwan

C. Viera, M.O. Gonzaga (eds.), *Behaviour and Ecology of Spiders*,
https://doi.org/10.1007/978-3-319-65717-2_6

Spiders of the infraorder Araneomorphae, which are often misleadingly (as not all members build webs) called web-building or true spiders (Turnbull 1973), comprise 93.9% of all extant spider species, and thus represent by far the most diverse spider group. The webs of Araneomorph spiders are highly distinguishable and found in almost every ecosystem on Earth, but they are particularly abundant in the Neotropics.

The primarily function of spider webs is to capture prey. However, they can serve other functions, including acting as a sensory system, a courtship and/or mating platform, thermoregulatory platform, and antipredator barrier. Despite their ubiquitousness across environments, the ecological and structural importance of spider webs within ecosystems is poorly known. This is partly because the ecological, evolutionary, and biophysical aspects of webs for individual spiders, populations, and species are largely unexplored. A significant portion of what is known about spider web diversity, evolution, ecology, and building behaviours is a consequence of over a century of detailed observations on Neotropical spiders.

Probably the most readily recognizable form of spider web is the orb web. It was once thought to be the pinnacle of spider web evolution. New molecular evidence has nonetheless suggested two alternative scenarios (Bond et al. 2014; Fernández et al. 2014): (1) that the orb web evolved earlier than originally postulated and may represent the ancestral form of all spider webs, or (2) the orb web has had multiple independent origins. The latter hypothesis would conform with hypotheses formed prior to the advancement of more methodical cladistic analyses (e.g., Lehtinen 1967) but seems to be the less likely. We do not weigh into the debate regarding the origins of orb webs herein. Rather, we explore how studies using Neotropical spiders have assisted the development of the various hypotheses over the years.

In addition to our understanding of the evolutionary trajectories of spider webs, our understanding of the function of spider webs has been significantly enhanced of late (see Blackledge et al. 2011; Harmer et al. 2011). Still lacking, however, is a clear understanding of the genetic underpinnings of spider webs and web building. Our knowledge of the genetic underpinnings of various spider silks has expanded of late (Prosdocimi et al. 2011; Clarke et al. 2015; Collin et al. 2016), and this knowledge might provide insights into the genetic influences over web building and web functionality. Here, we overview spider web and silk diversity and variability, with specific reference to the plenitude of work on Neotropical spiders.

Neotropical Araneomorph Diversity

We know that up to half of all genera of orb-web spiders are found in the Neotropics. Brazil alone has arguably the world's greatest web-building spider diversity, containing up to 72 of the 112 known families of Araneomorphs (World Spider Catalog 2016). Such an impressive diversity qualifies the Neotropics as a spider biodiversity hotspot and highlights the importance of on-going research in the region (see Chap. 1).

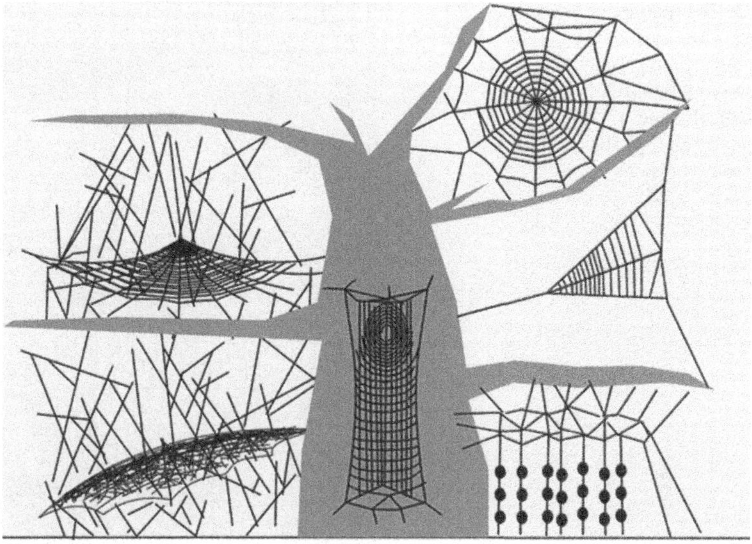

Fig. 6.1 Examples of the diversity of Neotropical spider web architectures, showing (clockwise from the upper left-hand side) two-dimensional Araneid planar horizontal orb webs, an example of a derived orb web, Theridiid cobwebs with gumfooted threads, three-dimensional sheet webs, and three-dimensional orb webs of *Cyrtophora* and *Mecynogea*. An example of a 'ladder web' is shown on the trunk of the tree

A wide variety of web types and foraging modes are found among Neotropical Araneomorphs. These include the use of silken aerial and ground webs, snares and trip lines, sit-and-wait and cursorial foraging. Among the webs, a diversity of web architectures are found (depicted in Fig. 6.1), from two-dimensional planar horizontal or vertical orb webs, to three-dimensional sheet webs and cobwebs, to elongated two dimensional 'ladder webs', highly modified webs, and webs comprising but a few capture threads.

Some Examples of Well-Known Araneomorph Groups

Synotaxidae

The Synotaxidae are a family of spiders found in South America, Australia, and New Zealand (Griswold et al. 1998; Agnarsson 2003a). Spiders of this group were once thought to belong to the Theridiidae and to be related to the Argyrodinids (Forster et al. 1990; Griswold et al. 1998; Agnarsson 2003a). Nevertheless, upon detailed assessments of leg and abdomen micro-characteristics, spiders in the group were designated to an independent family (Forster et al. 1990; Agnarsson 2003a). A diagnostic characteristic of the Synotaxidae is the unique 'chicken-wire' web

Fig. 6.2 The unique 'chicken-wire' web of the Synotaxidae, comprising irregularly meshed silk with vertically and horizontally aligned sticky silk threads (From http://tolweb.org/Synotaxidae/93137)

comprising irregularly meshed silk with vertically and horizontally aligned sticky silk threads (Eberhard 1977, 1995; Griswold et al. 1998; Dimitrov et al. 2017) (Fig. 6.2). The web building, eggsac production and prey-wrapping behaviours in *Synotaxus ecuadorensis* are well described and seem to be typical of the group (Eberhard 1995; Barrantes and Eberhard 2007).

RTA Clade

The retrolateral tibial apophysis (RTA) is a megadiverse clade containing almost half of all Araneomorph spiders. The majority of species in the RTA clade do not build webs. The latest genomic analyses have nevertheless placed the clade sister to the web-building Deinopoidae (Bond et al. 2014; Fernández et al. 2014). This suggests that the RTA clade represent a group that lost the ability to build webs. The few RTA spiders that build webs today are thus interesting subjects for testing hypotheses about the costs and benefits of web building.

The Neotropics are rich in RTA clade spiders (Santos and Brescovit 2001; Santos 2007; Silva et al. 2008). Significantly, the Neotropics has good representations of web-building RTA species, including species of web-building pisaurids (Pisauridae) and wolf spiders (Lycosidae). Accordingly, the region seems to be a hotbed for studies testing hypotheses about the evolutionary significance of web building (Stefani et al. 2001; Macrini et al. 2015).

Fig. 6.3 Web of *Aglaoctenus lagotis* (Lycosidae) from Serra do Japi, SP, Brazil. (Photo: M.O. Gonzaga)

The genus *Aglaoctenus* (Lycosidae), for example, is widely distributed across South America (Uruguay to Colombia) and found in a variety of different environments (Santos and Brescovit 2001; Piacentini 2011; González et al. 2015a, b). This is particularly interesting because the genus includes species that build tube-shaped webs and display sub-social behaviour (Macrini et al. 2015). Spiders of the genus belong to an ancient subfamily of wolf spiders (Sosippinae). Accordingly, it may be hypothesized that the modern wolf spider had at some time jettisoned web building (Murphy et al. 2006). The webs of *Aglaoctenus* spp. are comprised of a series of attachments, a meshed sheet, and a funnel in which the spider can retreat (Stefani and Del-Claro 2015) (Fig. 6.3). The tube-shaped webs are built low to the ground, and may be a modification of a web resembling that of some ecribellate spiders (Murphy et al. 2006).

A preliminary silk gene expression analysis for *Aglaoctenus lagotis* and *A. oblongus* from Uruguay shows the presence of the *MaSp2* protein (spidroin) gene (Blamires, unpublished data). This finding, if confirmed, is interesting because the MaSp2 protein was thought to appear first in the ecribellate spiders (Hinman and Lewis 1992; Blackledge et al. 2012; Blamires et al. 2017). Clearly more comparative genetic and behavioural analyses are needed to reveal the evolutionary origins of the various silk proteins, the RTA spider webs, cribellate spider webs, and modern ecribellate spider webs.

Uloboridae

Neotropical spiders of the family Uloboridae usually spin horizontal orb webs containing a fibrous dry ecribellate adhesive silk (see section on 'Spider silks') (Lehtinen 1967; Eberhard 1972; Opell 1987). Modifications of the usual horizontally aligned orb webs, such as the highly reduced webs of *Polenecia* spp. (Wiehle 1927), *Hyptiotes* spp. (Peters 1938; Opell 1982), and *Miagrammopes* (Pickard-Cambridge 1904; Lubin et al. 1978; Opell 1987), are well described (see also Nascimento and Gonzaga 2015, Santos and Gonzaga 2017 for variations).

There are numerous publications on the web-building behaviours of Neotropical Uloborids thanks to almost 50 years of meticulous observations by Eberhard (Eberhard 1972, 1973, 1990a; Lubin et al. 1978; Opell and Eberhard 1984; Eberhard and Barrantes 2015). Significantly meticulous behavioural observations of Neotropical Uloborids were among the first to document variability in web building and web architectures across environments and contexts (Eberhard 1990a). These studies have been integral to our understanding of the evolution of different Araneomorph web forms (Eberhard 1990b; Eberhard 2014; Hormiga and Griswold 2014; Eberhard and Barrantes 2015).

Theridiidae

Sociality has evolved four times within the theridiid clade, suggesting the clade contains characters and behaviours that promote sociality (Agnarsson 2002; Agnarsson et al. 2006, 2007). Members of the genus *Anelosimus* have evolved a unique sociality, one with no hierarchy but the sharing of brood care, prey capture, and web construction. As a consequence their behaviours and communal webs are exceptionally well studied (Nentwig and Christenson 1986; Agnarsson 2002; Whitehouse and Lubin 2005; Yip et al. 2008) (see Chap. 13).

Theridiids of the genus *Argyrodes* have attracted similar interest because of their kleptoparasitic lifestyle (Agnarsson 2002; Su and Smith 2014). Interestingly, phylogenetic analyses of both sociality and kleptoparasitism show similar origins, and both may have diverged from some form of maternal care (Agnarsson 2002). Unlike sociality, kleptoparasitism arose only once in the Theridiidae (Agnarsson 2002). Neotropical species of the genus *Argyrodes* inhabit the webs of larger spiders, most commonly *Nephila clavipes*, and steal food from the host's web. While *Argyrodes* kleptoparasites have a distribution beyond the Neotropics, the Neotropical species have been most extensively studied (Vollrath 1979; Agnarsson 2002, 2004, 2011). From these studies we know much about the behaviours of *Aryrodes* spp. and the type of interactions they have with their host spider and its web (Vollrath 1979; Higgins and Buskirk 1998; Agnarsson 2002, 2003b, 2011). These studies by and large suggest that *Argyrodes* negatively influences the fitness of the host (Vollrath 1979; Higgins and Buskirk 1998). However, it seems that under certain ecological circumstances some *Argyrodes* can benefit the host spider (Elgar 1994; Peng et al. 2013).

The cobwebs and combfoot webs made by theridiid spiders are well-known. These include the webs built by Neotropical *Latrodectus* spp., *Steatoda* spp. and *Theridion* spp. These spiders all construct three-dimensional 'irregular sheet' webs (Benjamin and Zschokke 2002). The cobweb typically has four additional components: (i) a tangled retreat, a small 'pocket' consisting of supporting tangle threads and a non-sticky sheet where the spider hides, (ii) supporting threads, which form a large inaccessible tangle, (iii) gumfooted threads which vertically descend from the tangled retreat and interact with prey crawling below the web, and (iv) gumfooted gluey silk

Fig. 6.4 (**a**) Diagram of a cobweb showing the architectural components: the retreat consisting of supporting tangle threads and a non-sticky sheet (*a*), *b* vertically descending gumfooted threads, and *c* gumfoot sticky droplets deposited at the base of the gumfooted threads. (**b**) Web of *Achaearanea tingo* (Theridiidae) from Parque Estadual Intervales, Ribeirão Grande, SP, Brazil (Photo: M.O. Gonzaga)

droplets at the base of the ascending threads, which adhere to the crawling prey (Fig. 6.4). Distinctive web patterns in Neotropical theridiids, however, have been previously described. This is the case of the webs of *Helvibis longicauda* and *Chrysso intervales*, which are entirely composed of viscid silk lines (Gonzaga et al. 2006).

Araneidae

Spiders in this family are abundant in Neotropical ecosystems (Bonaldo et al. 2007; Baldissera et al. 2008). The depth of spider research in the region means that there is a dearth of information on the diversity, behaviours, ecology, and phenology of Neotropical Araneid spiders. There is information, however, on several aspects of their biology in the Neotropics, such as foraging strategies (e.g., Moura et al. 2016; Rito et al. 2016), habitat selection (e.g., Messas et al. 2014; Souza et al. 2015), web structures and architecture (e.g., Eberhard 1988b, 2008, 2014; Xavier et al. 2017), sexual selection and parental care (e.g., Moura et al. 2017; Moura and Gonzaga 2007), defences against natural enemies (e.g., Eberhard 2003; Gonzaga and Vasconcellos-Neto 2005; Magalhaes and Santos 2012), among others.

Nephila clavipes is one of the most well-known of all Neotropical orb-web Araneids due to its size, abundance, and ubiquitousness throughout the Neotropics. The species builds a large (often several metres in width and height), tightly meshed, two-dimensional orb web (often with an accompanying three-dimensional 'barrier web' that may contain stacked prey carcasses, Higgins 1992) across high canopy gaps in the forest, where they potentially capture almost all insects prey that flies through the corridor (Vasconcellos-Neto and Lewinsohn 1984). They have also been known to capture and consume small bats and birds (Brooks 2012; Nyffeler and Knornschild 2013) and are a common host of *Argyrodes* kleptoparasites (Vollrath 1979). Like other species in the genus (Blamires et al. 2010), *N. clavipes* often aggregates its web. The behaviours and ecology of *N. clavipes* are so well-studied that it serves as a model for a plenitude of web and silk studies, and behavioural phenomena, such as centrally placed foraging, sexual cannibalism, and sexual-size dimorphism (Herberstein and Hebets 2013).

Tetragnathidae

Spiders of the genus *Leucauge* are very common in Neotropical forests and are identified by their distinctive silver, black, green, and/or yellow body colouration (Tso et al. 2006) and their sub-canopy, horizontally aligned two-dimensional orb webs (Eberhard 1990a, b; Hénaut et al. 2006; Briceno and Eberhard 2011). They may be solitary, aggregated, or subsocial (Eberhard 1990b; Salomon et al. 2010).

The web-building behaviours of *L. mariana*, for example, have been the subject of close observation as they make precision movements to produce webs with highly consistent spacing between spiral threads despite evidently constructing the web 'blindly' (Eberhard 1987, 1988a, b; Briceno and Eberhard 2011). The extra reach of their extensively long legs I and II are thought to facilitate rapid grasping of the radial threads and precise placement of the spiral threads during construction (Briceno and Eberhard 2011).

Spider Silks

Araneomorphs have evolved a 'toolkit' of silks, with each silk having particular properties for particular uses (Blackledge and Hayashi 2006). We will overview the work on silks used predominantly in webs, namely the major ampullate (MA), aciniform, and the cribellate and ecribellate gluey silks. While minor ampullate, pyriform, and some other silks (e.g., pseudoflagelliform and ribbon silks of filistatids) may function within webs, there is scant documentation of their functions within Neotropical spider webs.

Major Ampullate Silk

The longstanding model organism for studies of MA silk proteins, protein structures, and physical properties is *Nephila clavipes*. This is partly because its large body size makes it amenable for keeping in the laboratory and repeatedly extracting large quantities of silk, which is essential for most chemical assays, and partly because the pioneering work was done using this species (Lewis 1992).

The interest in MA silk properties emanates from its unique combination of high strength, extensibility, and toughness (Lewis 1992; Craig 2003; Harmer et al. 2011; Blamires et al. 2017). Accordingly, there are a multitude of potential commercial implications associated with understanding the mechanisms by which MA silk performs. Ecologically, the properties of MA silk enable it to function efficiently as a safety line, a propagator of tactile information, as a supporting frame for the web, and as impact-absorbing radial threads in orb webs (Osaki 1996; Sensenig et al. 2012; Blamires et al. 2017).

Phylogenetic analyses show that a stiff MA silk precursor was secreted by the major ampullate glands prior to the evolution of aerial web building (Garb et al. 2007, 2010; Prosdocimi et al. 2011). MA silk appeared in the first webs as moorings and supports (Denny 1976; Craig 2003). The proline-rich, highly elastic protein MaSp2 was first incorporated into MA silk by orb-web-building spiders (Hinman and Lewis 1992; Blackledge et al. 2012). Its high extensibility meant that incorporating radial threads into orb webs remarkably improved the web's ability to absorb the kinetic energy imparted by impacting prey (Denny 1976; Blackledge et al. 2012; Sensenig et al. 2012). Furthermore, the inclusion of the MaSp2 protein induced MA silks to shrink upon exposure to water, a phenomena called supercontraction (Boutry and Blackledge 2010). Supercontraction made it possible for MA silk to become stretched in humid environments, and for the web to tighten when the humidity fell (Boutry and Blackledge 2010, 2013). A finding of the *MaSp2* gene in the Uruguayan wolf spider (RTA clade) *Aglaoctenus lagotis*, and the prsence of a MaSp2-like protein in cribellate silks (Piorkowski and Blackledge 2017), suggests that *MaSp2* may have evolved prior to the building of two-dimensional, vertically aligned ecribellate orb webs.

Aciniform Silk

Aciniform silk is used by most Araneomorph spiders for wrapping and immobilizing prey, building sperm webs, as an egg-case liner, and for web decorations in *Argiope argentata* (Blackledge and Hayashi 2006). Aciniform silk, like major ampullate silk, is comprised of a core of spidroins (Hayashi et al. 2004). The aciniform spidroin, AcSp1, of *Latrodectus* spp. contains poly-alanine motifs similar to major ampullate silk, albeit containing fewer glycine motifs (Ayoub et al. 2013). Aciniform silk of the widely distributed *Argiope trifasciata* is thought to be as tough as major ampullate silk (Hayashi et al. 2004). This high toughness seems to be functionally important for protecting eggs (Blackledge and Hayashi 2006; Hsia et al. 2011). Nevertheless, when aciniform silk is used as web decorations by *Argiope* spp. it plays no role in the web's mechanical performance (Herberstein et al. 2000a; Blackledge and Hayashi 2006). It is probably utilized because it reflects ultraviolet light more strongly than the web silks, and so is useful as a silk-based signal to predators, prey, or perhaps both (Cheng et al. 2010).

Cribellate Silk

Spiders from the Deinopidae and Uloboridae families produce dry cribellar silks made of thousands of silk fibrils that surround supporting axial fibres (Opell and Bond 2000). These spiders produce their cribellar silks by drawing fibrils from spigots on the cribellum using the calamistrum, a comb of setae on the metatarsus of the fourth legs (Hawthorn and Opell 2003). The fine fibrils combine with thicker supporting strands to produce the characteristic wooliness of the silk (Eberhard and Pereira 1993; Hawthorn and Opell 2003).

Cribellate silk are used as capture threads in the horizontal orb webs of Uloborids. These wool-like threads are relatively stiff and inelastic compared to those of ecribellate orb webs (Lubin 1986; Kohler and Vollrath 1995; Blackledge and Hayashi 2006). They adhere to insects that get caught in the web via van der Waals and hygroscopic attractive forces between the thread and insect cuticle (Hawthorn and Opell 2003; Sahni et al. 2014a). Opell (1994a, b), examined the properties of the spiral threads of Neotropical *Miagrammopes* spp., *Hyptiotes* spp., and *Uloborus* spp., and found differences in stickiness across species. Web architecture appeared to be a major factor correlating with stickiness, with the more reduced webs of *Miagrammopes* and *Hyptiotes* being the stickiest.

Ecribellate Spiral Threads

The spiral threads of the derived Araneoid spiders, including orb weavers and cobweb builders, are made from either pseudoflagelliform or flagelliform silk coated with viscid aggregate silks (Townley and Tillinghast 2013; Sahni et al. 2014a, b). The flagelliform and aggregate silk genes and their orthologues were first identified and described for the Neotropical orb web spider *Nephila clavipes* (Hayashi and Lewis 1998; Choresh et al. 2009; Collin et al. 2016).

The viscid coated threads are produced by a triad of two aggregate-secreting and one flagelliform-secreting spigots located on the posterior lateral spinneret (Sahni et al. 2014a). The aggregate silk coating the flagelliform thread is highly hygroscopic and absorbs atmospheric moisture immediately upon spinning, thus swelling before transfiguring into regularly distributed ellipsoid droplets along the thread that resemble beads along a string (Edmonds and Vollrath 1992; Sahni et al. 2014a; Townley et al. 2006).

The stickiness of the viscid spiral threads is due to the presence of glycoproteins (Opell and Hendricks 2010; Townley and Tillinghast 2013; Sahni et al. 2014a, b). The aggregate silk additionally contains inorganic salts, as well as organics salts, amino acids (e.g., glycine), neurotransmitters and saturated fatty acids (Vollrath et al. 1990). The viscid threads are produced more quickly and are less visible to insects than cribellate sticky silks (Craig 1986; Opell 1996, 1998). This advantage may have driven the immense diversity of ecribellate spiders that exists today.

Other Silks

A defining feature of Araneomorphs is their ability to spin multiple task-specific silks (Vollrath and Knight 2001; Blackledge and Hayashi 2006; Blamires et al. 2017). Silks utilized in addition to major ampullate, aciniform, and different kinds of sticky silks include tubuliform silk, which has high stiffness and low toughness (Blackledge and Hayashi 2006) and is used to form the inner coating of eggsacs (Hu et al. 2005; Tian and Lewis 2005; Gnesa et al. 2012). A phylogenetic examination across the major Araneomorph families (including Deinopidae, Uloboridae, Theridiidae, Araneidae) shows the *TuSp1* gene to be, despite its property differences, monophyletic and closely related to all of the other spidroin genes (Tian and Lewis 2005; Garb et al. 2010).

Minor ampullate (MiA) silk is used as a temporary capture spiral in orb webs, for prey-wrapping by theridiids, and as a component of dragline threads by most spiders (Vollrath and Knight 2001; La Mattina et al. 2008; Blamires et al. 2017). While most of the mechanical properties of minor ampullate silk are comparable with major ampullate silk (Blackledge and Hayashi 2006), it does not, unlike major ampullate silk, supercontract in water (Blamires et al. 2017). The spidroin-coding genes of minor ampullate (*MiSp1* and *MiSp2*) spidroins have been identified in *Nephila clavipes* and seem to be *MaSp1* orthologues (Colgin and Lewis 1998).

Pyriform silks cement the dragline and web frame to substrates, and glue silk threads together during web construction (Perry et al. 2010; Wolff et al. 2015). The spidroin PySp1 is manufactured exclusively in the pyriform gland of black widow spiders (Blasingame et al. 2009). A comparative PCR analysis of the *Argiope trifasciata*, *Nephila clavipes*, and *Nephilingis cruentata* PySp1 protein found two amino acid motifs unique to the spidroin: (1) a motif where every other amino acid is proline, and (2) a glutamine-rich motif of 6–8 amino acids (Perry et al. 2010). Nevertheless, relatively little is known of the properties of this spidroin.

A ribbon-like prey-wrapping silk is found in among spiders from the family Filistatidae (Eberhard and Pereira 1993). There is conflicting evidence whether or not these silks are cribellar threads that have been smoothed as a consequence of a division in the cribellum or whether it is composed of individually rippled tubuliform fibres (Eberhard and Pereira 1993; Lopardo and Ramirez 2007).

Web Function

The primary function of spider webs is to catch flying insect prey. Prey capture by a spider web involves three actions — prey interception, stopping, and retention (Eberhard 2014; Zaera et al. 2014), with the different web silks modified in a way within webs to perform one or more of these steps (Blackledge and Hayashi 2006; Blamires et al. 2017).

The architectural components of all two-dimensional orb webs are: (i) the capture surface area, (ii) the width of the spaces between the sticky spirals or mesh size, (iii) the number of radii that traverse the spiral thread, and (iv) the length or pattern of any decorations (stabilimenta) added to the web (Fig. 6.5).

Uloborid Orb Webs

Uloborid orb webs tend to be aligned horizontal to the ground. The reduced webs of *Miagrammopes* spp. and *Hyptiotes* spp., however, are aligned vertical to the ground, which might be facilitated by their greater cribellate thread stickiness (Opell 1994a,

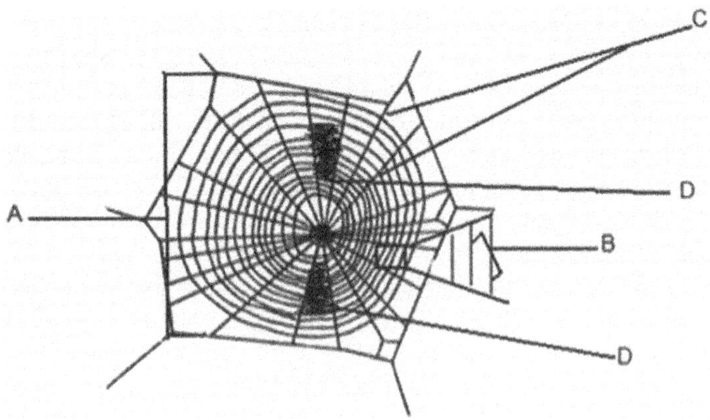

Fig. 6.5 Diagram of a vertical two-dimensional orb web, showing the architectural components: (*A*) capture surface area, (*B*) space between the sticky spirals (mesh size), (*C*) radii traversing the spiral thread and (*D*) decoration (stabilimenta)

b). Recently Santos and Gonzaga (2017) described a new Uloborid genus (*Uaitemuri*) from Southeastern Brazil which also builds vertical orb webs. The horizontal orientation seems to reduce the prey interception rates of webs (Bishop and Connolly 1992) but, considering prey retention is entirely attributable to benign Van der Waals forces (Hawthorn and Opell 2003), it might be utilized out of necessity to retain prey in the web.

Uloborids, such as *Philoponella undulata*, often aggregate their webs or build webs in close proximity to the webs of other spiders (Finke 1981). When multiple webs are found in close proximity, insects that bounce off or fly through any one web are likely to have their velocity reduced, thereby rendering them more easily caught by a nearby web (Uetz 1989; Yip et al. 2008; Blamires et al. 2010). The more webs in the vicinity, the more likely it is that an insect will eventually be caught. This phenomenon is described as the 'ricochet effect' and has been proposed as having an evolutionary benefit associated with spider web aggregations (Uetz 1989).

Deinopid Net-Casting Webs

The Neotropical net-casting spiders *Deinopis spinosa* and *D. longipes* (Deinopidae) produces cribellate capture threads, and they have a unique prey-capturing method. They position themselves on vegetation and spin a rectangular-shaped capture net of cribellate silk (Fig. 6.6), releasing the net as an insect moves beneath. The net surrounds the insect, which becomes entangled in the woolly cribellate silk (Opell

Fig. 6.6 The cribellate capture net of *Deinopis* sp., showing the spider in position so to release the net as an insect moves beneath.

1994a; Getty and Coyler 1996). Large flying insects, such as moths, are caught by flicking the cribellate silk backwards (Getty and Coyler 1996). The silk reflects ultraviolet light (Craig et al. 1994) which might be utilized to lure prey toward the net.

Theridiid Cobwebs

The gumfoot threads of a cobweb extend downward from the tangled retreat (see Fig. 6.4). The gumfoot glue droplets at the base of the thread adhere to prey crawling along the ground, and when the prey struggle, the thread is released from the substrate. Upon release from its pyriform attachment, a gumfoot thread transmits vibrational stimuli toward the cobweb to inform the spider that prey has been captured (Peters 1987). Viscid globules have been found within the tangled retreat of *Achaeranea tesselata* webs (Barrantes and Weng 2006). Nevertheless, the function of these globules remains unclear. Their small size suggests that they are of little value in prey retention, but this function should not be ruled out (Benjamin et al. 2002; Barrantes and Weng 2006).

Cobweb Function Compared to Orb Webs

In both spiral and gumfoot threads, the glue coalesces under surface forces into droplets that disperse along the axial thread (Opell and Hendricks 2010; Sahni et al. 2011; Blamires et al. 2014a) (Fig. 6.7). The glues of both types of thread are comprised of an aqueous solution of glycoproteins and low molecular weight organic and inorganic salts (Sahni et al. 2014a, b). Gumfoot glue, however, contains additional water-soluble peptides (Hu et al. 2007; Sahni et al. 2011, 2014b). While orb web axial threads are comprised of extensible flagelliform silk (Peters 1987), gumfoot axial threads consist of stiffer major ampullate gland (MA) threads (Sahni et al. 2014b). The mechanical properties of spiral and gumfoot threads differ, most likely as a consequence of the different properties of the respective axial silks (Peters 1987) although differences in the biochemistry of the glues may also play a role (Blamires et al. 2014a).

Araneidae

Vertically Aligned Orb Webs

Orb webs aligned vertical to the ground, such as those spun by *Neoscona*, *Argiope*, *Nephila*, *Araneus*, and *Larinia*, appear to be adapted for the capture of high-kinetic-energy prey (Kohler and Vollrath 1995; Harmer et al. 2011; Sensenig et al. 2012). The radial threads are the threads that play the greatest role in stopping prey, as the energy absorption capacity of their silks is an order of magnitude greater than that of the

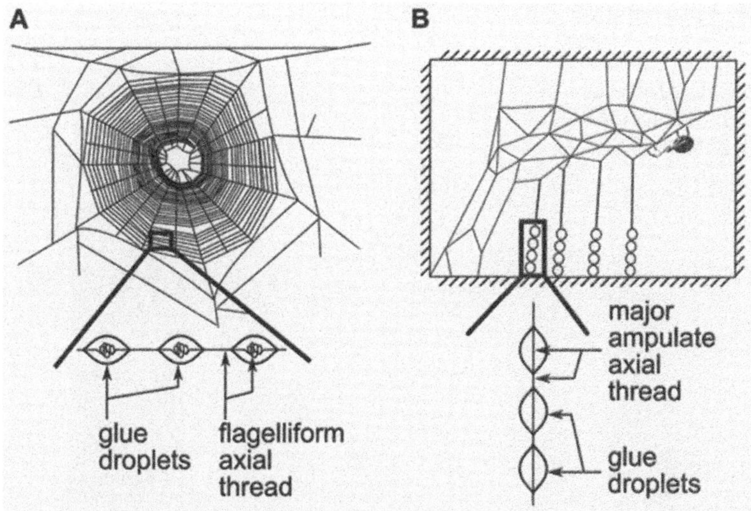

Fig. 6.7 The gluey silks and underlying axial threads of (**a**) orb webs and (**b**) cobwebs, showing that the glues of both are comprised of an aqueous gluey silk, comprising a solution of and low molecular weight organic and inorganic salts and glycoproteins surrounding flagelliform (orb-web spiral threads) or major ampullate (cobweb gumfoot threads)

viscid silks (Vollrath 1994). The initial softness and ultimate strength of radial threads provides the inelastic absorption for high-kinetic-energy interception (Denny 1976; Craig 1987; Harmer et al. 2011; Sensenig et al. 2012). The energy absorbed depends where on the web's surface area, the force applied, and the angle of interception (Craig 1987). Covering the flagelliform threads with aqueous aggregate glue causes the flagelliform silk to plasticise and become highly extensible. This enables the kinetic energy of impacting prey to be imparted onto the web, reducing the probability of the prey bouncing off or flying through the web (Boutry and Blackledge 2013).

The primary function of the sticky spiral threads nonetheless is to retain prey (Sahni et al. 2014b). Some insects, e.g., flies, are better retained by webs with cribellate silk while others, e.g., bugs and beetles, are better retained by webs with aggregate silk (Vollrath 1994). Other insects, e.g., moths, have low adhesion to both (Vollrath 1994; Opell and Schwend 2007). The spiral silk's ability to maintain tension when stretched and relaxed in rapid succession is a critical feature of its performance. (Zhou and Zhang 2005).

Modified Orb Webs

Neotropical spiders of the genus *Scoloderus* build an elongated (up to 1 m long but 0.2 m wide) web with the hub located toward the bottom (Eberhard 1975) (Fig. 6.8). This is in contrast with other Araneids that build elongated orb webs, such as the

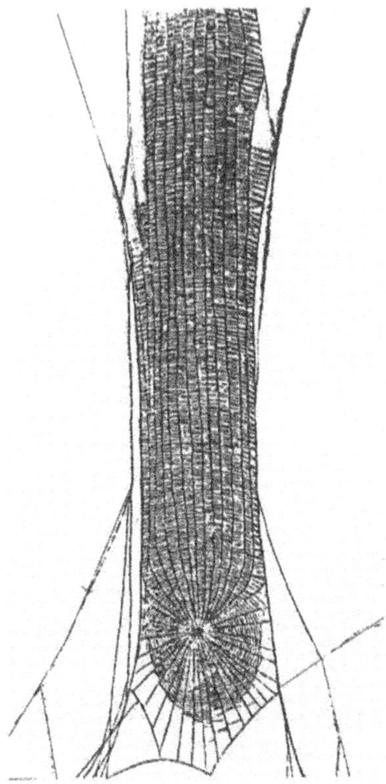

Fig. 6.8 A diagram of the elongated (up to 1 m long) webs of spiders from the Neotropical spider genus *Scoloderus* (Reproduced from Eberhard (1975))

Australian ladder-web spiders, *Telaprocera* spp. which build elongated webs with the hub positioned at the top of it (Harmer et al. 2011). The reason the hub of *Scoloderus* webs are at the bottom is because they specialize in capturing moths. When a moth strikes the web it sheds scales in its struggle and rather than breaking free tumbles down the web (Stowe 1978). The extreme elongation of the web means the moths will continue to tumble until they finally adhere to the spirals near the hub (Eisner et al. 1964), thus enabling the spider to rapidly attack its prey (Stowe 1978).

Cyrtophora citricola is native to Africa and the Middle-East (Lubin 1974) but has been recently documented living in colonies in Columbia, the Dominican Republic, Cuba, and Florida (Levi 1997; Pulido 2002; Alayón 2003). They build a web containing a horizontally aligned prey-catching orb that lacks sticky silks (Levi 1997; Eberhard 1990a; Blamires et al. 2013), onto which it adds a three-dimensional silk barrier structure that extends vertically upward to ~1 m (Lubin 1974, 1980; Berry 1987; Blamires et al. 2013). The spiders position themselves on the underside of the orb and catch the prey that fall into it from above upon striking the barrier structure's threads.

Much of what we know about the function of their distinctive 'tent web' comes from examination of the New World congener *Cyrtophora moluccensis* (Lubin 1980;

Berry 1987; Blamires et al. 2013, 2014b). The architecture of these 'tent webs' fundamentally differs to those of closely related orb-web spiders, such as *Argiope* spp. (Eberhard 1990a; Blackledge et al. 2011). It has been supposed that the barrier structure is used to support the horizontal orb (Berry 1987) but there is no empirical evidence for this function. Most likely it provides protection, and enables the spiders to more effectively capture large-bodied prey, such as moths, which might fly through regular orb webs or fail to adhere to the sticky spirals (Blamires et al. 2013).

Horizontally Aligned Webs

Webs built in horizontal alignment to the ground, such as those of *Leucauge spp.* are generally not in the flight path of flying insects so intercept less prey (Bishop and Connolly 1992). Accordingly, they appear to be less effective than webs built aligned vertical to the ground (Craig 1987; Blackledge et al. 2011). Nonetheless, horizontally aligned webs appear to have some advantages over vertically aligned web, including protection from inadvertent damage from wind or flying animals, the exploitation of low-lying prey habitats or prey flight patterns not exploited by vertical webs, ability to better cope with the spider's weight, and freedom from the need to absorb enormous amounts of kinetic energy, since insects are rarely caught in full flight, and tensioning of the web by supercontraction (Craig 1987; Blackledge et al. 2011; Bishop and Connolly 1992; Boutry and Blackledge 2013). There is nevertheless currently no definitive evidence of an advantage for building horizontally aligned webs over vertically aligned webs. In fact, a study that manipulated *Leucauge regnyi* webs into various alignments found vertically aligned webs to catch fewer prey less efficiently than horizontal webs (Bishop and Connolly 1992).

Web Plasticity

Orb Webs

The architectural components of orb webs are the most readily identifiable and quantifiable among spider webs. Measurements used to quantify orb-web architectural variability (i.e., plasticity) include counting the number of radial threads and any decorations (stabilimenta), measuring the widths and heights of the arc encapsulated by spirals, and calculating parameters such as mesh height and capture area using various formulae (Herberstein and Tso 2000; Blackledge and Gillespie 2002; Tso et al. 2007). Environmental factors that can be ascribed as influential over the relative number and size of orb-web architectural components include: (i) the presence of predators, (ii) ambient temperature, wind, and light levels, and (iii) prey availability, prey types, and prey nutrient value (Heiling and Herberstein 2000;

Boutry and Blamires 2013). Change in web architecture has also been documented across ontogeny in orb-web spiders (Hesselberg 2010; Kuntner et al. 2010; Escalante 2013).

Observations of Neotropical Araneids, Uloborids and Tetragnathids were the first to document variability in web building and web architectures across environments and ontogeny (Eberhard 1990a, b). Subsequent studies have expanded to ecribellate orb web spiders, such as *Nephila clavipes* and *Eustala illicita* (Hesselberg 2010; Hesselberg and Triana 2010). An orb-web-building spider generally follows the following steps sequentially in building an orb web: (i) exploration, (ii) frames construction, (iii) building of a proto hub, (iv) radii construction, (v) the construction of a spiral scaffold, and (vi) the construction of the spiral threads. *Nephila clavipes* differs from other orb web spiders in that they do not remove the spiral scaffolding upon placement of the spiral thread (Eberhard 1990b). Variations in behaviours at any one of these steps will result in measurable variation to the web architectural components (Eberhard 1990a; Zschokke and Vollrath 1995). Decorations are added to completed orb webs by Neotropical *Cyclosa* spp. using detritus, with the size and shape of the detritus decoration dependent upon the availability of prey remains (Gonzaga and Vasconcellos-Neto 2012). Other plastic behaviours associated with web building include variability in predatory behaviours. For instance, the order, frequency, and timing of prey-attacking behaviours can vary across spider ontogeny (Castanho and Oliveira 1997; Japyassu and Caires 2008).

Web plasticity may influence web structural variation across the spider phylogeny (Blackledge et al. 2009, 2011; Kuntner et al. 2010). Accordingly, some clades might exhibit relatively minor variability in web forms across species, environments, and ontogeny. While others, e.g., Theridiidae, might exhibit exceptional variability in web forms across species, environments, and ontogeny (Forster and Forster 1985; Eberhard et al. 2008).

Diet-Induced Web Plasticity

Satiated spiders build webs with smaller capture areas than starved spiders (Sandoval 1994; Herberstein et al. 2000b). The capture area of a web is often associated with a reduction in mesh height (Sandoval 1994; Tso et al. 2007). If the spider expects to feed on a homogeneous prey type it will not change mesh height, but it will invest in longer spiral threads to accommodate the web size change (Sandoval 1994). Spiders might also alter mesh height to tune the web for catching a specific type of prey (Sandoval 1994; Blackledge and Zevenbergen 2006; Blamires et al. 2011). Whether spiders adjust their web architectures under different diets because of constraints placed on the production of certain silks or to adjust the performance of the web is difficult to ascertain, because web and silk properties strongly co-vary across diets (Tso et al. 2007; Blamires et al. 2015, 2016).

Cobwebs

Cobwebs are the specialized three-dimensional webs of spiders of Theridiidae (Blackledge et al. 2005; Eberhard et al. 2008; Boutry and Blamires 2013). A number of studies have examined cobweb architectural plasticity (Blackledge and Zevenbergen 2007; Salomon 2007; Zevenbergen et al. 2008). As with orb webs, spatial constraints, hunger, and the type of prey consumed influence cobweb architectures (Jorger and Eberhard 2006; Blackledge and Zevenbergen 2007; Zevenbergen et al. 2008). For instance, satiated *Lactodectus hesperus* build webs with fewer gumfooted threads, with each thread containing fewer sticky droplets, than do those of starved *L. hesperus* (Blackledge and Zevenbergen 2007). The sizes of the individual gumfoot droplets nevertheless do not differ between satiated and starved spiders (Blamires et al. 2014a). Temperature and light influence cobweb architecture, most likely by affecting the spider's ability to spin silk or to build webs (Lubin et al. 1991; Blackledge and Zevenbergen 2007; Zevenbergen et al. 2008).

Plasticity in Other Webs

The Pholcid *Physocyclus globosus* builds a finely meshed, irregular, domed sheet web below a veiled tangle web. While the web form does not differ substantially, the web-building behaviours vary significantly across ontogeny in this species. Juveniles lay 91% of their threads within the sheet chamber, while adult females lay 55% and adult males 41%; adult males more often utilize exploration threads (Escalante 2013). Variations in behaviour by different individuals have been shown to result in variable web architectures (Madrigal-Brenes and Barrantes 2009). Female web-building wolf spiders of the genus *Aglaocetenus* may abandon web building at certain times of year or in certain locations, or vary their web building, i.e., shorten the attachment threads or deposit more fine threads without any apparent function, in the presence of male conspecifics or other spiders, e.g., Uloborids (González et al. 2014).

Conclusion

The diversity of spider webs and web-building spiders from the Neotropics has been well described owing to over a century of detailed observations. We have given an overview of some examples of the multitude of different web forms found in the Neotropics, highlighting some striking web forms and their structural and functional variability. Since the Neotropics contains up to half of all genera of orb-web spiders, the spider web forms and the various modifications thereof described herein will be a close representation of overall spider web diversity.

A spider's web is an extended phenotype depicting its foraging, mating, and defensive strategies, and physiological status. Spiders exhibiting web plasticity can continue to build functional webs across highly variable environments. The seasonality and geography of the Neotropical region, and the array of spider predators and prey that can be found there, have undoubtedly shaped the unique diversity of the region's spiders.

References

Agnarsson I (2002) Sharing a web: on the relation of sociality and kleptoparasitism in theridiid spiders (Theridiidae, Araneae). J Arachnol 30:181–188

Agnarsson I (2003a) The phylogenetic placement and circumscription of the genus *Synotaxus* (Araneae: Synotaxidae), a new species from Guyana, and notes on theridioid phylogeny. Invertebr Syst 17:719–734

Agnarsson I (2003b) Spider webs as habitat patches — the distribution of kleptoparasites (*Argyrodes*, Theridiidae) among host webs (*Nephila*, Tetragnathidae). J Arachnol 31:344–349

Agnarsson I (2004) Morphological phylogeny of cobweb spiders and their relatives (Araneae, Araneoidea, Theridiidae). Zool J Linnean Soc 141:447–626

Agnarsson I (2011) Habitat patch size and isolation as predictors of occupancy and number of argyrodine spider kleptoparasites in *Nephila* webs. Naturwissenschaften 98:163–167

Agnarsson I, Avilés L, Coddington JA, Maddison WP (2006) Sociality in Theridiid spiders: repeated origins of an evolutionary dead end. Evolution 60:2342–2351

Agnarsson I, Maddison WP, Avilés L (2007) The phylogeny of the social *Anelosimus* spiders (Araneae: Theridiidae) inferred from six molecular loci and morphology. Mol Phylogenet Evol 43:833–851

Alayón GG (2003) *Cyrtophora citricola* (Araneae: Araneidae), registro neuvo de araña para Cuba. Cocuyo 13:14

Ayoub NA, Garb JE, Kuelbs A, Hayashi CY (2013) Ancient properties of spider silks revealed by the complete gene sequence of the prey-wrapping silk protein (AcSp1). Mol Biol Evol 30:589–601

Baldissera R, Ganade G, Brescovit AD, Hartz SM (2008) Landscape mosaic of *Araucaria* forest and forest monoculture influencing understory spider assemblages in southern Brazil. Austral Ecol 33:45–54

Barrantes G, Eberhard WG (2007) The evolution of prey-wrapping behaviour in spiders. J Nat Hist 41:1631–1658

Barrantes G, Weng JL (2006) Viscid globules in webs of the spider *Achaearanea tesselata* (Araneae: Theridiidae). J Arachnol 34:480–482

Benjamin SP, Duggelin M, Zschokke S (2002) Fine structure of sheet-webs of *Linyphia triangularis* (Clerck) and *Microlinyphia pusilla* (Sundevall), with remarks on the presence of viscid silk. Acta Zool 83:49–59

Benjamin SP, Zschokke S (2002) Untangling the tangle web: web construction behaviours of the gumfooted spider *Steatoda triangulos* and comments on phylogenetic implications (Araneae: Theridiidae). J Insect Behav 15:791–809

Berry JW (1987) Notes on the life history and behavior of the communal spider *Cyrtophora moluccensis* (Doleschall) (Araneae, Araneidae) in Yap, Carolina Islands. J Arachnol 15:309–319

Bishop L, Connolly SR (1992) Web orientation, thermoregulation, and prey capture efficiency in a tropical forest spider. J Arachnol 20:173–178

Blackledge TA, Gillespie RG (2002) Estimation of capture areas of spider webs in relation to web asymmetry. J Arachnol 30:70–77

Blackledge TA, Hayashi CY (2006) Silken toolkits: biomechanics of silk fibers spun by the orb web spider *Argiope argentata* (Fabricius 1775). J Exp Biol 209:2452–2461
Blackledge TA, Zevenbergen JM (2006) Mesh width influences prey retention in spider orb webs. Ethology 112:1194–1201
Blackledge TA, Zevenbergen JM (2007) Condition-dependent spider web architecture in the western black widow, *Latrodectus hesperus*. Anim Behav 73:855–864
Blackledge TA, Summers AP, Hayashi CY (2005) Gumfooted lines in black widow cobwebs and the mechanical properties of spider capture silk. Zoology 108:41–46
Blackledge TA, Scharff N, Coddington JA, Szuts T, Wenzel JW, Hayashi CY, Agnarsson I (2009) Reconstructing web evolution and spider diversification in the molecular era. Proc Natl Acad Sci USA 106:5229–5234
Blackledge TA, Kuntner M, Agnarsson I (2011) The form and function of spider orb webs: evolution from silk to ecosystems. Adv Insect Physiol 41:175–262
Blackledge TA, Perez-Riguiero J, Plaza GR, Perea B, Navarro A, Guinea GV, Elices M (2012) Sequential origin in the high performance properties of orb spider dragline silk. Sci Rep 2:782
Blamires SJ, Lee YH, Chang CM, Lin IT, Cheg JA, Lin TY, Tso IM (2010) Multiple structures interactively influence prey capture efficiency in spider orb webs. Anim Behav 80:947–953
Blamires SJ, Chao IC, Liao CP, Tso IM (2011) Multiple prey cues induce foraging flexibility in a trap-building predator. Anim Behav 81:955–961
Blamires SJ, Hou C, Chen LF, Liao CP, Tso IM (2013) Three-dimensional barricading of a predatory trap reduces predation and enhances prey capture. Behav Ecol Sociobiol 67:709–714
Blamires SJ, Sahni V, Dhinojwala A, Blackledge TA, Tso IM (2014a) Nutrient deprivation induces property variations in spider gluey silk. PLoS One 9:e88487
Blamires SJ, Hou C, Chen LF, Liao CP, Tso IM (2014b) A predator's body coloration enhances its foraging profitability by day and night. Behav Ecol Sociobiol 68:1253–1260
Blamires SJ, Piorkowski D, Chuang A, Tseng YH, Toft S, Tso IM (2015) Can differential nutrient extraction explain property variations in a predatory trap? R Soc Open Sci 2:140479
Blamires SJ, Tseng YH, Wu CL, Toft S, Raubenheimer D, Tso IM (2016) Spider silk and web performance landscapes across nutrient space. Sci Rep 6:26383
Blamires SJ, Blackledge TA, Tso IM (2017) Physico-chemical property variation in spider silks: ecology, evolution and synthetic production. Annu Rev Entomol 62:443–460
Blasingame E, Tuton-Blasingame T, Falick AM, Zhao L, Fong J, Vaidyanathan V, Vispers A, Guerts P, Hu XW, LaMattina C, Viera C (2009) Pyriform spidroin 1, a novel member of the silk gene family that anchors dragline silk fibers in attachment discs of the black widow spider, *Latrodectus hesperus*. J Biol Chem 284:29097–29108
Bonaldo AB, Marques MAL, Pinto-da-Rocha R, Gardner T (2007) Species richness and community structure of arboreal spider assemblages in fragments of three vegetational types at Banhado Grande wet plain, Gravataí River, Rio Grande do Sul, Brazil. Iheringia 97:143–151
Bond JE, Garrison NL, Hamilton CA, Godwin RL, Hedin M, Agnarsson I (2014) Phylogenomics resolves a spider backbone phylogeny and rejects a prevailing paradigm for orb web evolution. Curr Biol 24:1765–1771
Boutry C, Blackledge TA (2010) Evolution of supercontraction in spider silk: structure–function relationship from tarantulas to orb-weavers. J Exp Biol 213:3505–3514
Boutry C, Blackledge TA (2013) Wet webs work better: humidity, supercontraction and the performance of spider orb webs. J Exp Biol 216:3606–3610
Boutry C, Blamires SJ (2013) Plasticity in spider webs and silk: an overview of current evidence. In: Santerre M (ed) Spiders: morphology behavior and geographic distribution. Nova, New York, pp 1–46
Briceno RD, Eberhard WG (2011) The hub as a launching platform: rapid movements of the spider *Leucauge mariana* (Araneae: Tetragnathidae) as it turns to attack prey. J Arachnol 39:102–112
Brooks DM (2012) Birds caught in spider webs: a synthesis of patterns. Wilson J Ornithol 124:345–353

Castanho LM, Oliveira PS (1997) Biology and behaviour of the neotropical ant-mimicking spider *Aphantochilus rogersi* (Araneae: Aphantochilidae): nesting, maternal care and ontogeny of ant-hunting techniques. J Zool 242:643–650
Cheng RC, Yang EC, Lin EP, Herberstein ME, Tso IM (2010) Insect form vision as one potential shaping force of spider web decoration design. J Exp Biol 213:759–768
Choresh O, Bayarmagnai B, Lewis RV (2009) Spider web glue: two proteins expressed from opposite strands of the same DNA sequence. Biomacromolecules 10:2852–2856
Clarke TH, Garb JE, Hayashi VY, Arensberger P, Ayoub NA (2015) Spider transcriptomes identify ancient large-scale gene duplication event potentially important in silk gland evolution. Genome Biol Evol 7:1856–1870
Colgin MA, Lewis RV (1998) Spider minor ampullate silk proteins contain new repetitive sequences and highly conserved non-silk like "spacer regions". Protein Sci 7:667–672
Collin MA, Clarke TH, Ayoub NA, Hayashi CY (2016) Evidence from multiple species that spider silk glue component ASG2 is a spidroin. Sci Rep 6:21589
Craig CL (1986) Orb-web visibility: the influence of insect flight behaviour and visual physiology on the evolution of web designs within the Araneoidea. Anim Behav 34:54–68
Craig CL (1987) The ecological and evolutionary interdependence between web architecture and web silk spun by orb web weaving spiders. Biol J Linn Soc 30:135–162
Craig CL (2003) Spiderwebs and silk: tracing evolution from molecules to genes to phenotypes. Oxford University Press, Oxford
Craig CL, Bernard GD, Coddington JA (1994) Evolutionary shifts in the spectral properties of spider silks. Evolution 48:287–296
Denny MW (1976) The physical properties of spider's silk and their role in the design of orb webs. J Exp Biol 65:483–506
Dimitrov D, Benavides LR, Arnedo MA, Giribet G, Griswold CE, Scharff N, Hormiga G (2017) Rounding up the usual suspects: a standard target-gene approach for resolving the interfamilial phylogenetic relationships of ecribellate orb-weaving spiders with a new family–rank classification (Araneae, Araneoidea). Cladistics 33:221–250. https://doi.org/10.1111/cla.12165
Eberhard WG (1972) The web of *Uloborus diversus* (Araneae : Uloboridae). J Zool 166:417–465
Eberhard WG (1973) Stabilimenta on the webs of *Uloborus diversus* (Araneae: Uloboridae) and other spiders. J Zool 171:367–384
Eberhard WG (1975) The 'inverted ladder' orb web of *Scoloderus* sp. and the intermediate orb of Eustala (?) sp. Araneae : Araneidae. J Nat Hist 9:93–106
Eberhard WG (1977) Rectangular orb-webs of *Synotaxus* (Araneae: Theridiidae). J Nat Hist 11:501–507
Eberhard WG (1987) Effects of gravity on temporary spiral construction by *Leucauge mariana* (Araneae: Araneidae). J Ethol 5:29–36
Eberhard WG (1988a) Memory of distances and directions moved as cues during temporary spiral construction in the spider *Leucauge mariana* (A., Araneidae). J Insect Behav 1:51–66
Eberhard WG (1988b) Behavioral flexibility in orb web construction: effects of supplies of different silk glands and spider size and weight. J Arachnol 16:295–302
Eberhard WG (1990a) Function and phylogeny of spider webs. Annu Rev Ecol Syst 21:341–372
Eberhard WG (1990b) Early stages of orb construction by *Philoponella vicina*, *Leucauge mariana*, and *Nephila clavipes* (Araneae, Uloboridae and Tetragnathidae), and their phylogenetic implications. J Arachnol 18:205–234
Eberhard WG (1995) The web and building behavior of *Synotaxus ecuadorensis* (Araneae, Synotaxidae). J Arachnol 23:25–30
Eberhard WG (2003) Substitution of silk stabilimenta for egg sacs by *Allocyclosa bifurca* (Araneae: Araneidae) suggests that silk stabilimenta function as camouflage devices. Behaviour 140:847–868
Eberhard WG (2008) *Araneus expletus* (Araneae, Araneidae): another stabilimentum that does not function to attract prey. J Arachnol 36:191–194
Eberhard WG (2014) A new view of orb webs: multiple trap designs in a single structure. Biol J Linn Soc 111:437–449

Eberhard WG, Barrantes G (2015) Cues guiding uloborid construction behavior support orb web monophyly. J Arachnol 43:371–387

Eberhard WG, Pereira F (1993) Ultrastructire of cribellate silk of nine species in eight families and possible taxonomic implications (Araneae: Amaurobiidae, Deinopidae, Desidae, Dictynidae, Filistatidae, Hypochilidae, Stiphidiidae, Tengellidae). J Arachnol 21:161–174

Eberhard WG, Agnarsson I, Levi HW (2008) Web forms and the phylogeny of theridiid spiders (Araneae: Theridiidae): chaos from order. Syst Biodivers 6:1–61

Edmonds DT, Vollrath F (1992) The contribution of atmospheric water vapour to the formation and efficiency of a spider's capture web. Proc R Soci B 248:145–148

Eisner T, Alsop R, Ettershank G (1964) Adhesiveness of spider silk. Science 146:1058–1061

Elgar MA (1994) Experimental evidence of a mutualistic association between two web-building spiders. J Anim Ecol 63:880–886

Escalante I (2013) Ontogenetic and sexual differences in exploration and web construction in the spider *Physocyclus globosus* (Araneae: Pholcidae). Arachnology 16:61–68

Fernández R, Hormiga G, Giribet G (2014) Phylogenomic analysis of spiders reveals nonmonophyly of orb weavers. Curr Biol 24:1772–1777

Finke OM (1981) An association between two Neotropical spiders (Araneae: Uloboridae and Tengellidae). Biotropica 13:301–307

Forster LM, Forster RR (1985) A derivative of the orb web and its evolutionary significance. N Z J Zool 12:455–465

Forster RR, Platnick NI, Coddington JA (1990) A proposal and review of the spider family Synotaxidae (Araneae, Araneoidea), with notes on theridiid interrelationships. Bull Am Mus Nat Hist 193:1–116

Garb JE, DiMauro T, Lewis RV, Hayashi CY (2007) Expansion and intragenic homogenization of spider silk genes since the Triassic: evidence from Mygalomorphae (tarantulas and their kin) spidroins. Mol Biol Evol 24:2454–2464

Garb JE, Ayoub NA, Hayashi CY (2010) Untangling spider silk evolution with spidroin terminal domains. BMC Evol Biol 10:243

Getty RM, Coyler FA (1996) Observations on prey capture and anti-predator behaviors of ogre-faced spiders (Dienopis) in southern Costa Rica (Araneae, Dienopidae). J Arachnol 24:93–100

Gnesa E, Hsia Y, Yarger JL, Weber W, Lin-Cereghino J, Lin-Cereghino G, Tang S, Agari K, Viera C (2012) Conserved C-terminal domain of spider tubuliform spidroin 1 contributes to extensibility in synthetic fibers. Biomacromolecules 13:304–312

Gonzaga MO, Vasconcellos-Neto J (2005) Testing the functions of detritus stabilimenta in webs of *Cyclosa fililineata* and *Cyclosa morretes* (Araneae : Araneidae): do they attract prey or reduce the risk of predation? Ethology 111:479–491

Gonzaga MO, Vasconcellos-Neto J (2012) Variation in the stabilimenta of *Cyclosa fililineata* Hingston, 1932, and *Cyclosa morretes* Levi, 1999 (Araneae: Araneidae), in southeastern Brazil. Psyche 2012:396594

Gonzaga MO, Leiner NO, Santos AJ (2006) On the stick cobwebs of two theridiid spiders (Araneae: Theridiidae). J Nat Hist 40:293–306

González M, Costa FG, Peretti AV (2014) Strong phenological differences between two populations of a Neotropical funnel-web wolf spider. J Nat Hist 48:2183–2197

González M, Peretti AV, Costa FG (2015a) Reproductive isolation between two populations of *Aglaoctenus lagotis*, a funnel-web wolf spider. Biol J Linn Soc 114:646–658

González M, Costa FG, Peretti AV (2015b) Funnel-web construction and estimated immune costs in *Aglaoctenus lagotis* (Araneae: Lycosidae). J Arachnol 43:158–167

Griswold CE, Coddington JA, Hormiga G, Scharff N (1998) Phylogeny of the orb-web building spiders (Araneae, Orbiculariae: Deinopoidea, Araneoidea). Zool J Linnean Soc 123:1–99

Harmer ATM, Blackledge TA, Madin JS, Herberstein ME (2011) High-performance spider webs: integrating biomechanics, ecology and behaviour. J R Soc Interface 8:457–471

Hawthorn AC, Opell BD (2003) Van der Waals and hygroscopic forces of adhesion generated by spider capture threads. J Exp Biol 206:3905–3911

Hayashi CY, Lewis RV (1998) Evidence from flagelliform silk cDNA for the structural basis of elasticity and modular nature of spider silks. J Mol Biol 275:773–784
Hayashi CY, Blackledge TA, Lewis RV (2004) Molecular and mechanical characterization of aciniform silk: uniformity of iterated sequence modules in a novel member of the spider silk fibroin gene family. Mol Biol Evol 21:1950–1959
Heiling AM, Herberstein ME (2000) Interpretations of orb-web variability: a review of past and current ideas. Ekologia 19:97–106
Hénaut Y, Garcia-Ballinas JA, Alauzet C (2006) Variations in web construction in *Leucauge venusta* (Araneae, Tetragnathidae). J Arachnol 34:234–240
Herberstein ME, Hebets EA (2013) Behaviour: why are spiders good models for research? In: Penny D (ed) Spider research in the 21st century: trends and perspectives. SIRI Scientific Publishing, Manchester, pp 230–250
Herberstein ME, Tso IM (2000) Evaluation of formulae to estimate the capture area and mesh height of orb webs (Araneoidea, Araneae). J Arachnol 28:180–184
Herberstein ME, Craig CL, Coddington JA, Elgar MA (2000a) The functional significance of silk decorations of orb-web spiders: a critical review of the empirical evidence. Biol Rev 75:649–669
Herberstein ME, Craig CL, Elgar MA (2000b) Foraging strategies and feeding regimes: web and decoration investment in *Argiope keyserlingi* Karsch (Araneae: Araneidae). Evol Ecol Res 2:69–80
Hesselberg T (2010) Ontogenetic changes in web design in two orb-web spiders. Ethology 116:535–545
Hesselberg T, Triana E (2010) The web of the acacia orb-spider *Eustala illicita* (Araneae: Araneidae) with notes on its natural history. J Arachnol 38:21–26
Higgins LE (1992) Developmental changes in barrier web structure under different levels of predation risk in *Nephila clavipes* (Araneae: Tetragnathidae). J Insect Behav 5:635–655
Higgins LE, Buskirk RE (1998) Spider-web kleptoparasites as a model for studying producer-consumer interactions. Behav Ecol 9:384–387
Hinman MB, Lewis RV (1992) Isolation of a clone encoding a second dragline silk fibroin: *Nephila clavipes* dragline silk is a two-protein fiber. J Biol Chem 267:19320–19324
Hormiga GC, Griswold E (2014) Systematics, phylogeny, and evolution of orb-weaving spiders. Annu Rev Entomol 59:487–512
Hsia Y, Gnesa E, Jeffrey F, Tang S, Viera C (2011) Spider silk composites and applications. In: Cuppoletti J (ed) Metal ceramic and polymeric composites for various uses. Intech, Rijeka, pp 303–324
Hu XW, Lawrence BD, Kohler K, Falick AM, Moore AMF, McMullen E, Jones PR, Viera C (2005) Araneoid egg case silk: a fibroin with novel ensemble repeat units from the black widow spider, *Latrodectus hesperus*. Biochemistry 44:10020–10027
Hu XW, Yuan J, Wang X, Vasanthavada K, Falick AM, Jones PR, La Mattina C, Viera C (2007) Analysis of aqueous glue coating proteins on the silk fibers of the cob weaver, *Latrodectus hesperus*. Biochemistry 46:3294–3303
Japyassu HF, Caires RA (2008) Hunting tactics in a cobweb spider (Araneae-Theridiidae) and the evolution of behavioral plasticity. J Insect Behav 21:258–284
Jorger KM, Eberhard WG (2006) Web construction and modification by *Achaearanea tesselata* (Araeae, Theridiidae). J Arachnol 34:511–523
Kohler T, Vollrath F (1995) Thread biomechanics in the two orb-weaving spiders *Araneus diadematus* (Araneae, Araneidae) and *Uloborus walckenaerius* (Araneae, Uloboridae). J Exp Zool 271:1–17
Kuntner M, Kralj-Fiser S, Gregorič M (2010) Ladder webs in orb-web spiders: ontogenetic and evolutionary patterns in Nephilidae. Biol J Linn Soc 99:849–866
La Mattina C, Reza R, Hu XW, Falick AM, Vasanthavada K, McNary S, Yee R, Viera C (2008) Spider minor ampullate silk proteins are constituents of prey wrapping silk in the cob weaver *Latrodectus hesperus*. Biochemistry 47:4692–4700

Lehtinen PT (1967) Classification of the cribellate spiders and some allied families, with notes on the evolution of the suborder Araneomorpha. Ann Zool Fenn 4:199–468
Levi HW (1997) The American orb weavers of the genera *Mecynogea, Manogea, Kapogea,* and *Cyrtophora* (Araneae: Araneidae). Bull Mus Comp Zool 155:215–255
Lewis RV (1992) Spider silk: the unraveling of a mystery. Acc Chem Res 25:392–398
Lopardo L, Ramirez MJ (2007) The combing of cribellar silk by the prithine *Misionella mendensis,* with notes on other Filistatid spiders (Araneae: Filistatidae). Am Mus Novit 3563:1–14
Lubin YD (1974) Adaptative advantages and the evolution of colony formation in *Cyrtophora* (Araneae, Araneidae). Zool J Linnean Soc 54:321–339
Lubin YD (1980) The predatory behavior of *Crytophora* (Araneae: Araneidae). J Arachnol 8:159–185
Lubin YD (1986) Web building and prey capture in the Ulobidae. In: Shear WA (ed) Spiders: Webs Behavior and Evolution. Stanford University Press, Stanford, pp 132–171
Lubin YD, Eberhard WG, Montgomery GG (1978) Webs of *Miagrammopes* (Araneae , Uloboridae) in the Neotropics. Psyche 85:1–23
Lubin Y, Kotzman M, Ellner S (1991) Ontogenetic and seasonal changes in webs and websites of a desert widow spider. J Arachnol 19:40–48
Macrini CMT, Peres EA, Solferini VN (2015) Cryptic diversity of *Agloactenus lagotis* (Araneae, Lycosidae) in the Brazilian Atlantic rainforest: evidence from microsatellite and mitochondrial DNA sequence data. J Appl Biol Biotechol 3:009–014
Madrigal-Brenes R, Barrantes G (2009) Construction and function of the web of *Tidarren sisyphoides* (Araneae: Theridiidae). J Arachnol 37:306–311
Magalhaes I, Santos AJ (2012) Phylogenetic analysis of *Micrathena* and *Chaetacis* spiders (Araneae: Araneidae) reveals multiple origins of extreme sexual size dimorphism and long abdominal spines. Zool J Linn Soc-Lond 166:14–53
Messas YF, Souza HS, Gonzaga MO, Vasconcellos-Neto J (2014) Spatial distribution and substrate selection by the orb-weaver spider *Eustala perfida* Mello-Leitão, 1947 (Araneae: Araneidae). J Nat Hist 48:2645–2660
Moura RR, Gonzaga MO (2007) Temporal variation in size-assortative mating and male mate choice in a spider with amphisexual care. Sci Nat 104:28. https://doi.org/10.1007/s00114-017-1448-6
Moura RR, Leal LC, Kloss TG (2016) Does nutritional status constrain adoption of more costly and less risky behaviour in an Amazonian shelter-building spider? J Nat Hist 50:2829–2837
Moura RR, Vasconcellos-Neto J, Gonzaga MO (2017) Extended male care in *Manogea porracea* (Araneae: Araneidae): the exceptional case of a spider with amphisexual care. Anim Behav 123:1–9
Murphy NP, Framenau VW, Donnellan SC, Harvey MS, Park YC, Austin AD (2006) Phylogenetic reconstruction of the wolf spiders (Araneae: Lycosidae) using sequences from the 12S rRNA, 28S rRNA, and NADH1 genes: implications for classification, biogeography, and the evolution of web building behavior. Mol Phylogenet Evol 38:583–602
Nascimento AL, Gonzaga MO (2015) Maternal defensive behaviors of *Uloborus* sp. (Aranea, Uloboridae): behavioral repertoire and influence of clutch size and female size on female aggressiveness. Acta Ethol 19:33–41
Nentwig W, Christenson TE (1986) Natural history of the non-solitary sheetweaving spider *Anelosimus jucundus* (Araneae: Theridiidae). Zool J Linnean Soc 87:27–35
Nyffeler M, Knornschild M (2013) Bat predation by spiders. PLoS One 8:e58120
Opell BD (1982) Post-hatching development and web production of *Hyptiotes cavatus* (Hentz) (Araneae, Uloboridae). J Arachnol 10:185–191
Opell BD (1987) The new species *Philoponella herediae* and its modified orb-web (Araneae , Uloboridae). J Arachnol 15:59–63
Opell BD (1994a) Factors governing the stickiness of cribellar prey capture threads in the spider family Uloboridae. J Morphol 222:111–119
Opell BD (1994b) Increased stickiness of prey capture threads accompanying web reduction in the spider family Uloboridae. Funct Ecol 8:85–90

Opell BD (1996) Functional similarities of spider webs with diverse architectures. Am Nat 148:630–648
Opell BD (1998) Economics of spider orb-webs: the benefits of producing adhesive capture threads and of recycling silk. Funct Ecol 12:613–624
Opell BD, Bond JE (2000) Capture thread extensibility of orb-weaving spiders: testing punctuated and associative explanations of character evolution. Biol J Linn Soc 70:107–120
Opell BD, Eberhard WG (1984) Resting postures of orb-weaving uloborid spiders (Araneae,Uloboridae). J Arachnol 11:369–376
Opell BD, Hendricks ML (2010) The role of granules within viscous capture threads of orb-weaving spiders. J Exp Biol 213:339–346
Opell BD, Schwend HS (2007) The effect of insect surface features on the adhesion of viscous capture threads spun by orb-weaving spiders. J Exp Biol 210:2352–2360
Osaki S (1996) Spider silk as a mechanical lifeline. Nature 384:419
Peng P, Blamires SJ, Agnarsson I, Lin HC, Tso IM (2013) A colour-mediated mutualism between two arthropod predators. Curr Biol 23:172–176
Perry DJ, Bittencourt D, Liberels-Stilberg J, Rech EL, Lewis RV (2010) Pyriform spider silk sequences reveal unique repetitive elements. Biomacromolecules 11:3000–3006
Peters HM (1938) Ober das Netz der Dreieckspinne, *Hyptiotes paradoxus*. Zool Anz 121:49–59
Peters HW (1987) Fine structure and function of capture threads. In: Nentwig W (ed) Ecophysiology of spiders. Springer-Verlag, Berlin, pp 187–202
Piacentini L (2011) Three new species and new records in the wolf spider subfamily Sosippinae from Argentina (Araneae: Lycosidae). Zootaxa 3018:27–49
Pickard-Cambridge FO (1904) Arachnida — Araneida and Opiliones. In: Biologia Centralia-Americana. Zoology, London, pp 465–560
Piorkowski D, Blackledge TA (2017) Punctuated evolution of viscid silk in orb web spiders supported by mechanical behavior of wet cribellate silk. Sci Nat 6: doi:10.1007/s00114-017-1489-x
Prosdocimi F, Bittencourt D, Rodrigues da Silva F, Kirst M, Motta PC, Rech EL (2011) Spinning gland transcriptomics from two main clades of spiders (order: Araneae) — insights on their molecular, anatomical and behavioral evolution. PLoS One 6:e21634
Pulido FJI (2002) Manejo de la araña del Mediterraneo o araña parda enredadora. Instituto Colombiano Agropecuario report. 5p
Rito KF, Hanashiro FTT, Peixoto EC, Gonzaga MO (2016) Optimal foraging or predator avoidance?: why does the Amazon spider *Hingstepeira foliscens* (Araneae: Araneidae) adopt alternative foraging behaviors? Zoologia 33, e20150147
Sahni V, Blackledge TA, Dhinojwala A (2011) Changes in the adhesive properties of spider aggregate glue during the evolution of cobwebs. Sci Rep 1:41
Sahni V, Dhinojwala A, Opell BD, Blackledge TA (2014a) Prey capture adhesives produced by orb-weaving spiders. In: Asakura T, Miller T (eds) Biotechnology of silk. Springer, Dordrecht, pp 203–217
Sahni V, Miyoshi T, Chen K, Jain D, Blamires SJ, Blackledge TA, Dhinojwala A (2014b) Direct solvation of glycoproteins by salts in spider silk glues increases adhesion and helps to explain the evolution of modern spider orb webs. Biomacromolecules 15:1225–1232
Salomon M (2007) Western black widow spiders express state-dependent web-building strategies tailored to the presence of neighbours. Anim Behav 73:865–875
Salomon M, Sponarski C, Larocque A, Aviles L (2010) Social organization of the colonial spider *Leucauge* sp in the Neotropics: vertical stratification within colonies. J Arachnol 38:446–451
Sandoval CP (1994) Plasticity in web design in the spider *Parawixia bistriata*: a response to variable prey type. Funct Ecol 8:701–707
Santos AJ (2007) A revision of the Neotropical nursery-web spider genus *Architis* (Araneae: Pisauridae). Zootaxa 1578:1–40
Santos AJ, Brescovit AD (2001) A revision of the South American spider genus *Aglaoctenus* Tullgren, 1905 (Araneae, Lycosidae, Sosippinae). Andrias 15:75–90

Santos AJ, Gonzaga MO (2017) Systematics and natural history of *Uaitemuri*, a new genus of the orb-weaving spider family Uloboridae (Araneae: Deinopoidea) from south-eastern Brazil. Zool J Linnean Soc 180:155–174

Sensenig A, Lorentz KA, Kelly SP, Blackledge TA (2012) Spider orb webs rely on radial threads to absorb prey kinetic energy. J R Soc Interface 9:1880–1891

Silva ELC, Lise A, Carico JE (2008) Revision of the Neotropical spider genus *Enna* (Araneae, Lycosoidea, Trechaleidae). J Arachnol 36:76–110

Souza HS, Messas YF, Gonzaga MO, Vasconcellos-Neto J (2015) Substrate selection and spatial segregation by two congeneric species of *Eustala* (Araneae: Araneidae) in southeastern Brazil. J Arachnol 43:59–66

Stefani V, Del-Claro K (2015) The effects of forest fragmentation on the population ecology and natural history of a funnel-web spider. J Nat Hist 49:211–223

Stefani V, Del-Claro K, Silva LA, Guimarães B, Tizo-Pedroso E (2001) Mating behaviour and maternal care in the tropical savanna funnel-web spider *Aglaoctenus lagotis* Holmberg (Araneae: Lycosidae). J Nat Hist 45:1119–1129

Stowe MK (1978) Observations of two nocturnal orbweavers that build specialized webs: *Scoloderus cordatus* and *Wixia ectypa* (Araneae: Araneidae). J Arachnol 6:141–146

Su YC, Smith D (2014) Evolution of host use, group-living and foraging behaviours in kleptoparasitic spiders: molecular phylogeny of the Argyrodinae (Araneae : Theridiidae). Invertebr Syst 28:415–431

Tian M, Lewis RV (2005) Molecular characterization and evolutionary study of spider tubuliform (eggcase) silk protein. Biochemistry 44:8006–8012

Townley MA, Tillinghast EK (2013) Aggregate silk gland secretions of Araneoid spiders. In: Nentwig W (ed) Spider ecophysiology. Springer, Berlin, pp 283–302

Townley MA, Tillinghast EK, Neefus CD (2006) Changes in composition of spider orb web sticky droplets with starvation and web removal and synthesis of sticky droplet compounds. J Exp Biol 209:1463–1486

Tso IM, Liao CP, Huang RP, Yang EC (2006) Function of being colorful in web spiders: attracting prey or camouflaging oneself? Behav Ecol 17:606–613

Tso IM, Chiang SY, Blackledge TA (2007) Does the giant wood spider *Nephila pilipes* respond to prey variation by altering web or silk properties? Ethology 113:324–333

Turnbull AL (1973) Ecology of the true spiders (Araneomorphae). Annu Rev Entomol 18:305–348

Uetz GW (1989) The 'ricochet effect' and prey capture in colonial spiders. Oecologia 81:154–159

Vasconcellos-Neto J, Lewinsohn TM (1984) Discrimination and release of unpalatable butterflies by *Nephila clavipes*, a Neotropical orb-weaving spider. Ecol Entomol 9:337–344

Vollrath F (1979) Behaviour of the kleptoparasitic spider *Argyrodes elevatus* (Araneae, theridiidae). Anim Behav 27:515–521

Vollrath F (1994) General properties of some spider silks. In: Kapaln DL, Adams WW, Farmer B, Viney C (eds) Silk polymers: materials science and biotechnology. American Chemical Society, Washington, D.C, pp 17–28

Vollrath F, Knight DP (2001) Liquid crystalline spinning of spider silk. Nature 410:541–548

Vollrath F, Fairbrother WJ, Williams RJP, Tillinghast EK, Bernstein DT, Gallagher KS, Townley MA (1990) Compounds in the droplets of the orb spider's viscid spiral. Nature 345:526–528

Whitehouse MEA, Lubin YD (2005) The functions of societies and the evolution of group living: spider societies as a test case. Biol Rev 80:347–361

Wiehle H (1927) Beitrage zur Kenntnis Uloboriden. Z Morph Okol Tiere 8:468–537

Wolff JO, Grawe I, Writh M, Karstedt A, Gorb SN (2015) Spider's super-glue: thread anchors are composite adhesives with synergistic hierarchical organization. Soft Matter 11:2394–2403

World Spider Catalog (2016) Natural History Museum Bern, version 16.5. Available from http://research.amnh.org/entomology/spiders/catalog/index. Accessed 28 Mar 2017

Xavier GM, Moura RR, Gonzaga MO (2017) Orb web architecture of *Wixia abdominalis* O. Pickard-Cambridge, 1882 (Araneae: Araneidae): intra-orb variation of web components. J Arachnol (in press) 45: 160–165

Yip EC, Powers KS, Aviles L (2008) Cooperative capture of large prey solved scaling challenge faced by spider societies. Proc Natl Acad Sci USA 105:11818–11822

Zaera R, Solar A, Teus J (2014) Uncovering changes in spider orb-web topology owing to aerodynamic effects. J R Soc Interface 11:20140484

Zevenbergen JM, Schneider NK, Blackledge TA (2008) Fine dining or fortress? Functional shifts in spider web architecture by the western black widow *Latrodectus hesperus*. Anim Behav 76:823–829

Zhou H, Zhang Y (2005) Hierarchical chain model of spider capture silk elasticity. Phys Rev Lett 94:028104

Zschokke S, Vollrath F (1995) Web construction patterns in a range of orb-weaving spiders (Araneae). Eur J Entomol 92:523–541

Chapter 7
Spider–Plant Interactions: An Ecological Approach

João Vasconcellos-Neto, Yuri Fanchini Messas, Hebert da Silva Souza, German Antonio Villanueva-Bonila, and Gustavo Quevedo Romero

Abstract Spiders are among the most common animals in diverse terrestrial environments, and display a variety of lifestyles and foraging modes. This chapter represents an overview of our knowledge of spider–plant interactions. Spiders are strongly influenced by plant architecture, rather than being randomly distributed in the vegetation; structures such as rosette-shaped clusters of leaves or glandular trichomes are particularly common in plants that have associations with spiders. Spiders derive benefits from plants such as shelter and access to insect prey. In turn, they can protect plants against herbivory. However, they may also consume or deter pollinators, imposing a cost that can exceed their benefit to the plant. Specific spider–plant associations are mutualistic if spiders provide protective or nutritional benefits, thus improving plant fitness, and if plants provide shelter and suitable foraging sites to spiders. We examine several case studies of spiders living in association with plants, and describe spatial/temporal adaptations in spider–plant relationships.

Plant vegetation represents a heterogeneous complex of environments (Morse et al. 1985; Sugihara and May 1990; Scheuring 1991), and the animals associated with it must have morphological, physiological, and behavioral characteristics that facilitate their feeding, survival, and reproduction in this context, especially predators such as spiders (Foelix 2011). Plants can directly benefit spiders by providing substrates for web building and hunting (Wise 1993), attracting prey (Morse 1999; Schmalhofer 2001), and offering suitable microclimates (Riechert and Tracy 1975;

J. Vasconcellos-Neto (✉) • G.Q. Romero
Depto de Biologia Animal, IB, UNICAMP, Campinas, Brazil
e-mail: jvascont@gmail.com

Y.F. Messas • H. da Silva Souza
Programa de Pós-Graduação em Biologia Animal, IB, UNICAMP, Campinas, Brazil

G.A. Villanueva-Bonila
Depto de Biologia Animal, IB, UNICAMP, Campinas, Brazil

Programa de Pós-Graduação em Biologia Animal, IB, UNICAMP, Campinas, Brazil

C. Viera, M.O. Gonzaga (eds.), *Behaviour and Ecology of Spiders*,
https://doi.org/10.1007/978-3-319-65717-2_7

Whitney 2004), whereas spiders can benefit plants by consuming or deterring herbivores and, in some cases, by providing nutritional resources to plants.

Spiders are among the most common animals in terrestrial environments, and they inhabit areas ranging from the hottest, most arid deserts to the deepest caverns and the highest, coldest mountains. They have the seventh largest number of species of any animal group, with 46,650 described species (Coddington and Levi 1991; World Spider Catalog 2017). They present a wide variety of lifestyles and behaviors (Foelix 2011). Many species are dispersed by the wind and can achieve great heights in the atmosphere (Turnbull 1973; Suter 1999). Spiders are among the most common arthropods that inhabit vegetation (e.g., Wise 1993), although they generally do not consume plant tissue, and are useful in studies that investigate how the habitat structure affects the community of arthropod predators (Gunnarsson 1990, 1992; Romero and Vasconcellos-Neto 2005a; Sanders et al. 2008).

Spiders capture prey using a variety of different foraging strategies. Some spiders are active hunters, such as the jumping spiders (Salticidae), which pursue their prey, whereas others remain motionless on vegetation, especially flowers, waiting for an insect to approach. Spiders of the family Thomisidae are typical hunters of the sit-and-wait type. Several other foraging modes exist along a continuum between the active-hunter and sit-and-wait strategies. In addition, many species build webs to capture prey (Romero and Vasconcellos-Neto 2007a, b). Several studies have sought to determine which foraging strategies cause the greatest indirect effects within terrestrial trophic cascades (Preisser et al. 2007).

Because spiders are predators, they can profoundly affect the dynamics of prey populations and the structure of prey communities (Wise 1993; Romero and Vasconcellos-Neto 2007a). In addition, as they often consume insect herbivores, the presence of spiders on plants can decrease herbivory on these plants (review in Romero and Vasconcellos-Neto 2007a). In fact, spiders are excellent biological control agents of pests in agroecosystems (review in Romero 2007). Spiders are among the most abundant and diverse arthropods in vegetation, but studies focusing on their interactions with plants are relatively scarce. In fact, only a few studies have reported specific associations between spiders and plants. Moreover, few studies have used an integrative approach to address the mutualistic relationships between spiders and plants. Spiders are often used as models of predators to answer questions related to the dynamics and structures of food webs (review in Romero and Vasconcellos-Neto 2007a). However, we lack a complete understanding of how morphological and structural aspects of plants can benefit spiders and how changes in the architecture of plants affect the composition and distribution of spiders (Halaj et al. 1998; Souza and Martins 2005; Souza 2007; Diniz 2011). A few studies have reported that spiders benefit from plants by obtaining alternative food resources, such as nectar and pollen (Romero and Vasconcellos-Neto 2007b; Meehan et al. 2009; Nahas et al. 2012; Stefani et al. 2015).

In this chapter we explore the associations between spiders and plants, covering topics such as defense, foraging, and reproduction, as well as providing recent evidence of facultative mutualistic interactions between spiders and plants. Several species of spiders are exclusively associated with plants that have certain types of architecture, which benefit them in many ways. In return for the benefits that they receive, spiders can remove herbivores and even nourish their host plants with feces and prey carrion. Spiders can have mixed effects on flowers: if they capture herbivorous insects that consume parts of the flower or the whole flower, spiders can benefit plants and even increase their reproductive success, but if they capture or expel the insects that pollinate flowers, their presence on the plant can impose a cost.

Guilds of Spiders Associated with Plants

The term *guild* was applied by plant and animal ecologists to describe trophic groups called Genossenschaften (Schimper 1903) or Syntrophia (Balogh and Loksa 1956). Modern usage of the term *guild* was formalized in a study of avian niche exploitation patterns as "a group of species that exploit the same class of environmental resources in a similar way" (Root 1967), and this concept was later extended to the arthropod fauna of collards (Root 1973). Thus, a guild comprises potentially competing species and is a fundamental aspect of ecological communities (Uetz et al. 1999). Since the term was coined, the guild concept has been applied to numerous animal and plant communities (Hawkins and MacMahon 1989; Simberloff and Dayan 1991).

Spiders may be classified into guilds according to the different strategies they use to capture their prey. Scientists have proposed different numbers of spider guilds based on their ecological and foraging characteristics: 2 (Uetz 1977), 3 (Nyffeler 1982), 4 (Young and Edwards 1990), 8 (Riechert and Lockley 1984), and 11 (Post and Riechert 1977). A commonly used classification of the different foraging strategies was proposed by Uetz et al. (1999). They performed quantitative analyses of ecological characteristics of families and suggested eight guilds based on hunting strategies: (1) stalkers (e.g., Salticidae and Oxyopidae), (2) ambushers (Thomisidae and Pisauridae), (3) foliage runners (Anyphaenidae and Clubionidae), (4) Ground Runners (Lycosidae and Gnaphosidae), (5) funnel web-builders (Agelenidae and Amaurobiidae), (6) wandering sheet/tangle weavers (Linyphiidae), (7) orb weavers (Araneidae, Tetragnathidae, and Uloboridae), and (8) 3D web builders (Theridiidae and Pholcidae). Höfer and Brescovit (2001) proposed a classification that assigned different families to 12 guilds, and Dias et al. (2010) refined these categories by creating subdivisions within certain families, since different sub-groups or genera of the same family fit better in different guilds, which resulted in 11 groups.

The families of spiders that make up the guilds of the stalkers, ambushers, and foliage runners are generally the most common inhabitants of vegetation (Uetz et al. 1999; Höfer and Brescovit 2001; Romero and Vasconcellos-Neto 2007a; Mohsin et al. 2010; Cardoso et al. 2011). In an extensive study, Nentwig et al. (1993) recorded many spider species associated with flowers, leaves, and trunks of various plant species in Panama, and all the spiders observed belonged to these three guilds.

Up to 70% of the spiders found in the flowers of *Lantana camara* (Verbenaceae) were Thomisidae, and more than 90% of the spiders collected in these flowers hunted by ambushing or stalking. These flowers were also occupied by spiders belonging to the families Salticidae, Anyphaenidae, Oxyopidae, Pisauridae, and Clubionidae. In contrast, 46% of the spiders on *Palicourea guianensis* (Rubiaceae) flowers belonged to the family Salticidae. In *Rhynchospora nervosa* (Cyperaceae) flowers, Nentwig et al. (1993) observed a large number of spiders belonging to the families Salticidae, Thomisidae, Oxyopidae, and Clubionidae. They also reported that wandering spiders associated with leaves mainly belonged to the families Salticidae, Pisauridae, and Anyphaenidae; Salticidae and Pisauridae spiders occurred preferentially in flat and xeromorphic leaves, and Anyphaenidae occurred in leaves with trichomes. According to Nentwig et al. (1993), the most common spiders on tree trunks belonged to the family Salticidae, including approximately half of all the spiders sampled.

As the spiders belonging to the stalker, ambush, and foliage runner guilds do not build webs but live in constant contact with the vegetation, they often have closer relationships with this type of substrate than do web builders. In addition to using plants directly for foraging, they use them for shelter and breeding habitat. Therefore, the spiders that belong to these guilds are the main predators in tri-trophic interactions and the main control agents of insect herbivores (Romero and Vasconcellos-Neto 2007b; Romero 2007).

Plant Architecture, Species Richness, and Diversity of Spiders

Understanding the patterns of species richness and abundance, as well as the processes that promote and maintain them, is a central theme in ecological studies (Gonçalves-Souza et al. 2011; Brown 2014). In terrestrial ecosystems, the habitat heterogeneity hypo (MacArthur and MacArthur 1961; Pianka 1966) indicates that complex environments are the predominant determinant of animal diversity (Tews et al. 2004). This hypothesis is supported by several studies of different taxonomic groups and different environments (e.g., Souza 1999; Halaj et al. 2000; Langellotto and Denno 2004; Tews et al. 2004). Vegetation is one element that provides structural diversity to habitat, as different patterns of branching and the modular organization of plants can provide a wide range of architectural arrangements (Bell et al. 1979; Küppers 1989; Bell 1991). Numerous studies have found that the architecture of plants is a major factor in determining the diversity of fauna associated with the vegetation, especially among the arthropod community (e.g., Halaj et al. 2000; Hatley and MacMahon 1980; Lawton 1983; Souza and Martins 2005; Woodcock et al. 2007; Ribas et al. 2011).

For example, the species diversity of birds depends more on the architectural diversity of the vegetation than on the taxonomic diversity of plants (MacArthur and MacArthur 1961). In a shinnery oak ecosystem, lower frequency and abundance of rodent species were recorded in open spaces with no vegetation than in the densely vegetated areas around the oaks (Cramer and Willig 2002). Among arthropods, the diversity of beetles and phytophagous arthropods were explained more by the architectural diversity of vegetation than by the diversity of the plant community (Brose 2003; Woodcock et al. 2007).

Although a strong correlation exists between the increasing architectural complexity of the vegetation and the diversity and abundance of species, studies of the influence of the architectural complexity of plants are biased toward vertebrates, particularly birds, which cover a third of the studies and represent less than 1% of animal diversity (Tews et al. 2004). Moreover, the concept of architectural complexity is difficult to generalize, as the operating variables of architecture, as well as the definition of habitat architecture, vary from author to author (McCoy and Bell 1991; Tews et al. 2004). In contrast, taxonomic groups that represent a large portion of overall animal diversity (e.g., arthropods) have been little studied. For arthropods associated with vegetation, a single plant is the whole habitat, so even small changes in its architecture can have consequences on the community structure and on the foraging efficiency of arthropods (Price et al. 1980; Tews et al. 2004).

Several studies have examined the influence of the architectural characteristics of plants on the abundance and diversity of arthropods, particularly spiders (Riechert and Gillespie 1986; Gunnarsson 1996). This influence is related to the various vegetative parts of the plant (e.g., needles, branches, leaves) and to the presence of reproductive structures that can provide, for example, a large variety of shelters, favorable microclimate conditions, anchoring points for prey capture webs, and opportunities to use different foraging methods (Greenstone 1984; Riechert and Gillespie 1986; Uetz 1991; Dennis et al. 1998).

Inflorescences

Some studies showed a high number of spider species inhabiting plants with inflorescences. These structures provide favorable microclimatic conditions and shelter against possible predators. In addition, inflorescences can attract different types of prey, representing a remarkable benefit for spiders. Structural features in inflorescences, such as number and size of flowers and leaves, arrangement in space, and branch size can vary among plant species and among the inflorescences of the same plant at different phenological stages (e.g., open flowers, flower buds).

Souza and Módena (2004) compared the differences in abundance and size distribution of ambush spiders (Thomisidae) and active hunting spiders (Salticidae, Oxyopidae, Clubionidae, and Anyphaenidae) in different types of inflorescences in *Melanthera latifolia*, *Conyza bonariensis*, and *Eupatorium hecatanthum* (all belonging to the family Asteraceae). The researchers recorded the architectural features, including the number of inflorescences, the inflorescence branch length, and the size and openness of flowers. *M. latifolia* had larger (6.5 cm) and more open flowers than the other two species (*C. bonariensis* = 1.30 cm and *E. hecatanthum* = 3.47 cm), but it showed fewer flowers per inflorescence. The most spiders were recorded in *M. latifolia*, followed by *E. hecatanthum* and then *C. bonariensis*. Evidently, bigger and more open flowers attract more insects (e.g., Bell 1985; Cohen and Shmida 1993), which become potential prey for spiders. Overall, smaller spiders inhabited *M. latifolia* than *E. hecatanthum* and *C. bonariensis*. Although larger flowers might attract larger insects (Dafni et al. 1997) and thus potentially feed larger spiders, studies indicate that larger spiders on the vegetative branches of plants are more vulnerable to predation by birds (Waldorf 1976; Askenmo et al. 1977). In fact, *M. latifolia* has the lowest number of flowers, which indicates fewer possible retreats against predators.

In the system *Peucetia viridans* (Oxyopidae) and *Croton ciliatoglandulifer* (Euphorbiaceae), Jiménez-Salinas and Corcuera (2008) found that *P. viridans* is most abundant in plants with the highest number of inflorescences and greater vegetation cover. The researchers experimentally manipulated the architecture of the plants by removing the inflorescences of some of the plants. More spiders were observed in control plants (from which inflorescences were not removed) than in treatment plants (from which all inflorescences were removed). In addition, more spiders were found in plants with the highest number of inflorescences, and more spiders were found in male flowers than in female flowers. The researchers argued that because male flowers produce pollen and, in some cases, more nectar, they attract more insects and provide more resources for the spiders.

Later, in the same spider–plant system, Corcuera et al. (2010) experimentally placed artificial inflorescences on *C. ciliatoglandulifer* to evaluate their effect on the distribution of the spider *P. viridans*. The treatments included (1) 15 plants from which all the inflorescences were removed, (2) 15 plants whose natural inflorescences were replaced by artificial ones, and (3) 15 control plants whose natural inflorescences were not changed. More spiders were recorded in plants with natural and artificial inflorescences than in plants from which the inflorescences were removed. However, the abundance of spiders was similar between the control treatment plants and the artificial inflorescence plants. Possibly, *P. viridans* does not use the scents of flowers as a sign of an available and appropriate microhabitat, but relies only on the inflorescence architecture (e.g., the flower's shape and size). These studies confirm the positive effect of the inflorescences on the distribution of some groups of spiders. These reproductive structures may provide different resources (e.g., shelter, prey attractants) to the spiders associated with them.

Spacing Between the Branches and Leaf Density

Few experimental studies have manipulated plant architectural complexity, and those that did so have often confused the effects of the architecture with the effects of area. In addition, several studies were restricted to a particular plant species or genus (Hatley and MacMahon 1980), limiting the ability to extend the results related to increasing architectural complexity of the plant and the abundance and diversity of spiders.

Souza and Martins (2005) compared the abundance of spiders in seven species of plants, which were grouped into structural complexity levels. This level of complexity was measured by the leaf density index, which was the number of leaves per branch divided by the estimated volume of the branch. The plant species selected for the study were *Baccharis dracunculifolia* DC. (Asteraceae), *Bidens gardneri* Baker (Asteraceae), *Chromolaena laevigata* (Lam.) King and H. Rob. (Asteraceae), *Diplusodon virgatus* Pohl (Lythraceae), *Microlicia euphorbioides* Mart. (Melastomataceae), *Microlicia fasciculata* Mart. ex Naud. (Melastomataceae), and *Vochysia tucanorum* Mart. (Vochysiaceae). As expected, the average number of spiders per branch was positively correlated with the structural complexity index. The highest number of spiders was recorded in *B. dracunculifolia* (which had the highest value complexity index), whereas *D. virgatus* and *B. gardneri* had intermediate and low leaf densities, respectively.

Subsequently, to isolate the effects of leaf density from those of biomass, the researchers experimentally manipulated the density of the branches' leaves, replacing natural branches with artificial ones. Artificial branches were made with wire and plastic models containing 24 leaves per branch. The leaf model was made of cloth to avoid the effects of chemical components and to replicate the texture of natural branches. For this experiment, three plant species were used: *B. dracunculifolia*, *D. virgatus*, and *M. fasciculata*. Two treatments were used per plant species: artificial branches with high leaf density and branches with low leaf density. The average number of spiders that colonized the artificial branches was similar among the three plant species. However, more spiders colonized the branches with high leaf density. The researchers suggest that the architecture of the branches has a significant effect on the abundance of spiders, regardless of biomass, leaf surface area, and branch texture. This effect was also reported by other studies (Hatley and MacMahon 1980; Gunnarsson 1990, 1992). However, in all these studies the effects of biomass and/or surface area available for colonization by spiders were not isolated from the effects of the branches' architecture.

Diniz (2011 and unpublished data) record changes in the composition of spiders after modifying the spacing of the branches of experimental plants. Closed architectures (closely spaced vegetative branches) favored runner spiders, presumably because they offered a greater number of shelters. This "daytime shelter" resource apparently led runner spiders to accumulate in dense vegetation. Halaj et al. (2000) and Hatley and MacMahon (1980) also reported that closed plant architecture favored non-weaver spiders. In contrast, open architecture (widely spaced branches) strongly favored weaver spiders. Larger open areas are advantageous for the construction of webs, especially orb webs, which need space for expansion and require few attachment points (Shear 1986). Thus, a more open architecture provides adequate space as well as sufficient anchorage points for the web (Diniz 2011 and unpublished data). Hatley and MacMahon (1980) also recorded that these spiders preferred open architecture (greater spacing between the branches and between the leaves of the branches) in Utah, near the entrance to the "Green Canyon."

Density of Needles

The coniferous forests of northern and central Europe are severely affected by air pollution (Schulze 1989). This contamination results in accelerated loss of leaves (needles) and has been interpreted as a symptom of environmental stress (Sundberg and Gunnarsson 1994). The loss of leaves affects the architecture of the trees, making the branches more exposed (lower coverage) and directly affecting the fauna associated with these types of trees. Gunnarsson (1988) compared the abundance and distribution of spider sizes on branches with different needle densities on *Picea abies* (L.) Karst. (Pinaceae), a predominant conifer in the forests of southwest Sweden. There were greater numbers of large spiders on the branches with high needle density, whereas both the number of small spiders and the total number of spiders did not significantly differ between branches with dense and sparse needles. Two possible explanations are that the microclimate conditions in the branches with low needle density are more favorable, and that predation by birds is lower on branches with high needle density. Predation by birds is one of the highest causes of mortality in spiders associated with fir (*P. abies*) (Gunnarsson 1983), and branches with more thorns can provide better shelter from potential predators. Another important finding was the positive relationship among families of weaver spiders, such as Linyphiidae, with sparsely needled branches, contrary to the results for raptorial spiders (e.g., Thomisidae), which were reported more frequently in branches with high needle density.

Later, Gunnarsson also conducted experimental field and laboratory studies by manipulating the density of needles on the branches to investigate the effect on the abundance and size distribution of spiders. In both studies (Gunnarsson 1990, 1992), a positive correlation was recorded between the density of needles and the abundance of spiders, supporting the hypothesis that these structures play an important role in the survival of spiders. In several experiments, there was no correlation between the size of spiders (bigger or smaller) and the density of needles on the branches. A positive correlation was found, however, between larger spiders and density of branches (not needles). In a later experiment, Gunnarsson (1996) isolated the effect of predation by birds from the effect of the needle density. Again, spider abundance of spiders was correlated with higher needle density on the branches. In this case, a significant decrease in the abundance of spiders in treatments in which the predation by birds was not isolated indicated that these predators exert strong pressure on spiders. However, the effect of the needle density on the distribution of the spiders' sizes is not clear, as smaller spiders appear to be affected by the change in density of needles.

As spiders are generalist predators and influence the balance of the populations of other arthropods (Wise 1993), changes in the architectural complexity of vegetation may culminate in top-down effects on the community of spiders, which would affect the populations of phytophagous arthropods and primary productivity as well (Denno et al. 2005; Sanders 2007).

Use of Specific Parts of Plants by Spiders

The different structures of plants (e.g., tree trunks, branches, leaves, flowers), comprise their structural complexity, and, as we have seen, are determinants of community distribution of spiders on vegetation. Most studies that address this topic have focused on the relationship between measures of structural complexity and the diversity of spiders. Only a few studies have examined the specific use of certain parts of plants and specific associations between spider species and plants (Fig. 7.1).

Fig. 7.1 Use of different plant structures by spiders. Flowers: (**a**) Thomisidae on *Rubus rosifolius* (Rosaceae). (**b**) *Epicadus heterogaster* (Thomisidae) on *Trixis antimenorrhoea* (Asteraceae). (**c**) Adult male and female of Anyphaenidae using flowers as refuge. (**d**) *Misumenops argenteus* (Thomisidae) feeding on *Pseudoscada erruca* (Ithomiinae) on *Trichogoniopsis adenantha* (Asteraceae). *Leaves*: (**e**) Refuge of Anyphaenidae, (**f**) Spiderlings on folded leaves. (**g**) Salticidae feeding on extrafloral nectary. (**h**) Adult female of Anyphanidae protecting its eggsac. Camouflage on stems: (**i**) *Ariamnes* sp. (Theridiidae). (**j**) Araneidae. (**k–l**) *Senolocus* sp. (Senoculidae) (Photographs: **a–c**, **f–l** Yuri Fanchini Messas; **d**, **e** João Vasconcellos-Neto)

Spiders of the family Thomisidae are typical sit-and-wait predators, hunting prey that visit flowers (Morse 1981). These spiders choose locations on the flowers so that they are not perceived or even increase insects' attraction to the flowers (Heiling et al. 2006). The associations between these spiders and flowers are well known, but little is known about specific associations or with specific groups of plants, let alone whether they represent specific examples of mutualism.

Spiders use the leaves on plants, as well as dried leaves on the ground, for foraging, shelter, and reproduction (Foelix 2011). There is little information about the specific use of certain plant species. Some studies have reported that spiders, such as those in the family Anyphaenidae, use leaves with specific formats belonging to certain species of plants for mating, nest shelter, or eggsac protection (Foelix 2011; Zanatta 2013; Zanatta et al. 2016). In some of these cases, leaf characteristics (e.g., the presence of trichomes) are essential to their use as shelter for the ovisac (see examples below).

Certain types of branches or stems can be used for foraging and/or protection, including camouflage (Messas et al. 2014; Souza et al. 2015). The set of characteristics belonging to a plant helps determine the composition of spider species associated with it or not associated with it.

Spiders that Feed on Pollen and Plant Fluids

Most spiders are considered obligatory carnivores that feed almost exclusively on insects and other arthropods (Wise 1993; Foelix 2011). However, recent studies have examined the possibility of vegetation as a direct food source, and some spiders appear to be true omnivores, as they can feed on nectar and/or pollen under certain environmental conditions (e.g. Taylor 2004; Eggs and Sanders 2013; Suetsugu et al. 2014; Nyffeler 2016). In a recent review, Nyffeler et al. (2016) recorded more than 60 species belonging to ten families of spiders that feed directly on plant products (e.g., pollen, nectar, Beltian corpuscles) in natural conditions. These families include non-weaver (Anyphaenidae, Clubionidae, Eutichuridae, Oxyopidae, Salticidae, Thomisidae, and Trachelidae) and weaver (Araneidae, Linyphiidae, and Theridiidae) spiders. A species of the Salticidae family was observed feeding on Mullerian corpuscles in trees of the genus *Cecropia* in the Serra do Japi, Jundiaí, São Paulo State, Brazil (JVN personal observation). Laboratory studies indicate that other families of spiders can also feed on plant products in nature.

Smith and Mommsen (1984) performed one of the first extensive studies on spiders that feed on pollen. These researchers reported that newly emerged *Araneus diadematus* (Araneidae) reared in the laboratory had longer life expectancies and produced more silk web when they fed on pollen than did newly emerged spiders that fed on aphids and spore fungi. The consumption of pollen could be adaptive, because during the time when the young spiders emerge (spring), few or no insects are available, but ample pollen is produced by dominant woody plants in temperate regions (e.g., *Pinus*) and dispersed by wind. Vogelei and Greissl (1989) tested the survival of *Thomisus onustus* (Thomisidae) spiderlings that were given no food (control), *Erigeron annuus* pollen, *Bellis perennis* pollen (Asteraceae), artificial nectar (30% sucrose solution), or *Drosophila melanogaster*. The control group of spiders survived an average 21 days. The spiders that were fed pollen survived for 35–49 days, depending on the plant species, and those fed artificial nectar survived for 130 days. However, only individuals who were fed with flies were able to develop normally, showing normal molting (ecdysis), and only they survived to the end of the experiment (>250 days).

Some wandering spiders, such as *Hibana velox*, *Hibana similaris* (Anyphaenidae) *Cheiracanthium mildei* (Miturgidae), and *Trachelas similis* (Corinnidae), were observed by Taylor and Foster (1996) feeding on both floral nectaries and on extra-floral nectaries of various plant species in several locations in Costa Rica and Florida. The researchers reported that there is evidence that *Myrmarachne foenisex* (Salticidae), a spider associated with ants, feeds on the exudate of coccidia (Coccidae). To test the role of nectar in the longevity of newly emerged *H. velox* spiders, the researchers provided young spiders with either water or 25% sucrose solution. The spiders in the sucrose group lived twice as long as did those in the control group.

Pollard et al. (1995) observed *Misumenoides formosipes* (Thomisidae) males feeding on nectar from extra-floral nectaries of some plant species. To determine whether these individuals consumed nectar as a source of water or energy, the researchers developed double-choice experiments, introducing small amounts of water vs 30% sucrose solution in experimental arenas, and found that the spiders preferred the sucrose. Even individuals that had consumed their fill of water ingested the sucrose solution. Males that drank only water died sooner than those that fed on nectar. The researchers suggest that as the males of this spider species are much smaller than females, they can become more dehydrated; therefore, feeding on the nectar of extra-floral nectaries may be an adaptive behavior.

In nature, Jackson et al. (2001) observed 31 species of Salticidae feeding on floral nectar. In the laboratory, they tested the preference of 90 species of Salticidae for distilled water vs 30% sucrose solution. All of the species selected and remained longer on the sucrose solution, indicating that nectarivory must be a common habit in that family. The researchers suggest that feeding on nectar may be advantageous because the fluids are rich in amino acids, lipids, vitamins, and minerals and because feeding on a flower involves no risk of injuries, unlike the capture of prey. The use of nectar by spiders can also benefit the plants. In fact, Ruhren and Handel (1999) showed that the presence of *Eris* sp. and *Metaphidippus* sp. (Salticidae) increased the production of fruits and seeds of the plant *Chamaecrista nictitans* (Caesalpineaceae). As in some species of ants, these spiders feed on the nectar of extra-floral nectaries. Meehan et al. (2009) recorded in the field and in the laboratory, using stable isotopes, that the main diet of the *Bagheera kiplingi* spider, a Neotropical salticid, is the Beltian corpuscles of the plant *Vachellia* sp. (Fabaceae). *B. kiplingi* presented concentrations of ^{15}N and ^{13}C isotopes in an intermediate range between ants that feed on the plant and other spiders that do not feed on plant products, confirming that this salticid consumes plant products.

Taylor and Bradley (2009) also showed the importance of nectar in the diets of the non-weaver spiders *C. mildei* (Miturgidae) and *H. velox* (Anyphaenidae). The researchers tested the importance of extrafloral nectaries to the survival, molting, and nighttime activity of these spiders. Tests of survival and molting were conducted in plastic terrariums containing a single spider newly emerged from the ovisac. To test the effects on survival, *H. velox* spiders were assigned to receive 25% sucrose, nectar from *Terminalia catappa* (Combretaceae), or water. A second experimental group was assigned to receive 69% sucrose, nectar, or water. To test the effects on molting, a single adult *Drosophila* was provided to a *C. mildei* spider on alternate days until the spider molted. *T. catappa* nectar was provided on the days when the *Drosophila* was not provided. In a second experiment, nectar was not provided (although water was given as a control). To test the effects on nighttime activity, newly emerged spiders received either water and nectar or water alone (control). To quantify the nocturnal activity, the number of spiders in each group that ran continuously during the night for at least 1 min was recorded. The spiders were filmed at 10-min intervals, for a total of 54 observation periods, and both species of spiders were used.

In the survival test, the spiders that received nectar or either sucrose concentration survived significantly longer than those that received water. There were no differences in survival between nectar and sucrose treatments. In the molting test, the number of spiders that molted was significantly higher in the nectar group. Finally, in the nighttime activity test, spiders ran more when their diets included nectar. These results suggest that nectar can be a source of energy for spiders, especially during periods when prey are scarce, since survival and molt rates were significantly higher when the nectar was provided. The researchers noted that the sugars obtained from the nectar supplied much of the energy demands of locomotion, freeing up the proteins contained in reserves for use in growth and/or new deposition of cuticle (Dalingwater 1987).

These results, however, are not easily generalizable. Carrel et al. (2000) found that *Frontinella pyramitela* (Linyphiidae) individuals gained weight when fed *D. melanogaster* but lost weight when fed pine pollen, suggesting that polinivory is restricted to particular groups of spiders and/or conditions of food scarcity.

Non-weaver spiders are not the only species that feed on nectar or pollen. Weaver spiders (Araneidae) may dismantle and eat their webs at regular intervals, which enables them to recycle the silk proteins efficiently. As the webs are not just a trap for potential prey, but also a trap for air plankton, spiders may also feed on adhered spores and pollen when they are recycling the webs. Eggs and Sanders (2013) tested the percentage of pollen in the diet of the orb-weaver spiders *Aculepeira ceropegia* and *Araneus diadematus* (Araneidae) in the presence of other food sources (insects). Their experiment included ten spiders that were fed fruit flies (*Drosophila*) and pollen from *Betula pendula* (Betulaceae) that was adhered to their webs, and ten spiders that were fed exclusively *Drosophila*. An analysis of stable isotopes in the body tissues of the spiders was performed. The results indicated that about 25% of the diet of spiders was composed of pollen and the other 75% was composed of flying insects, mainly small Diptera and Hymenoptera. The amount of pollen consumed was similar in laboratory and field observations (10–40% of the diet). Therefore, orb-weaver spiders actively feed on pollen, together with insects, to obtain the essential nutrients they need, at least during the early stages of life. The consumption of pollen by adult spiders decreased in the natural environment when insects became more abundant (during the summer season). The researchers suggested that this group of spiders be classified as omnivores, rather than as strict predators, as both carnivory and herbivory occur during important life stages of these orb-weaver spiders.

Spider–Plant Specific Associations

Arthropod–plant interactions have been studied extensively in some groups, for example, phytophagous insects, which live exclusively on vegetation and show highly specific relationships with their host plants (Schoonhoven et al. 1998). However, few studies have demonstrated this kind of association in spiders, despite the incredible diversity of spiders and their habitats. Some spider species belonging to the families Oxyopidae, Thomisidae, Salticidae, Araneidae, Ctenidae, Corinnidae, Selenopidae, and Theraphosidae have been shown to live strictly associated with a particular plant species or groups of plants that share morphological characteristics (e.g., glandular trichomes, rosettes, tree bark containing specific structures) (see review in Romero 2006; Messas et al. 2014).

Few studies have explored the reasons that spiders have specific associations with plants. Generally speaking, plants can provide suitable sites for protection, foraging, and reproduction. Recently, Hormiga and Scharff (2014) described a new species of Linyphiidae, *Laetesia raveni* (Araneae) collected in New South Wales and Queensland (Australia). This new linyphiid species seems to build its webs almost exclusively on two plant species, *Calamus muelleri* Wendland (Arecaceae) and *Solanum inaequilaterum* Domin (Solanaceae), both densely covered with thorns. The abundant thorns may protect the spiders from certain predators. Another unusual and little explored association involves the thomisid *Synema obscuripes* and the carnivorous plant *Nepenthes madagascariensis* (Nepenthaceae) (Rembold et al. 2012). This species spends its entire life cycle within the plant pitchers, structures that provide food (the pitchers attract insects) and shelter against predators (the pitchers secrete a liquid containing digestive enzymes).

In Central America, at least seven species of wandering spiders belonging to the *Cupiennius* genus (Ctenidae) are known for their intimate associations with certain plant groups (Barth et al. 1988a, review in Romero 2006). These ctenid spiders live exclusively on plants of the families Musaceae and/or Bromeliaceae, sheltering in them during the day and using them for ecdysis, courtship, and foraging at night. For example, *Cupiennius salei* lives in certain species of bromeliads (Barth and Seyfarth 1979; Barth et al. 1988a) and exchanges vibratory signals through the leaves of these plants (Barth et al. 1988b; Baurecht and Barth 1992). Other spider species, such as *Pachistopelma rufonigrum* (Theraphosidae; Santos et al. 2002), *Nothroctenus fuxico* (Ctenidae; Dias and Brescovit 2003, 2004), and various species of jumping spiders (Salticidae; see review in Romero 2006), are exclusively associated with tank bromeliads in several countries. Some corinnid species use bromeliads as habitats, in coastal and inland areas of Brazil (Cotgrave et al. 1993; Dias et al. 2000; Mestre et al. 2001; Araújo et al. 2007; Gonçalves-Souza et al. 2010). According to Gonçalves-Souza et al. (2010), of the five species of Corinnidae found in three types of habitat, four occurred only on bromeliads. However, only a few studies have demonstrated exclusivity with regard to Corinnidae and bromeliads. The first study demonstrating this specific association was Piccoli (2011), who reported that the spider *Corinna* sp. nov., described later by Rodrigues and Bonaldo (2014) as *Corinna demersa*, is exclusively associated with *Quesnelia arvensis* (Bromeliaceae) in *restinga* forests. The spider constructs a shelter in the axils of leaves or along the central tank and dives into the water when threatened.

Spiders that are associated with plants necessarily have adaptations that facilitate their relationships with host plants. These adaptations are usually related to spiders' sensory systems, allowing them to discern specific plant species through visual, olfactory, and tactile stimuli (review in Romero and Vasconcellos-Neto 2007b). The spider *Misumena vatia* (Thomisidae), for example, when presented with a choice of differently colored artificial flowers, more often chose yellow flowers (Greco and Kevan 1994). The morphological characteristics of plants also affect spiders' selection of habitat. In a field study, Morse (1990) demonstrated that *M. vatia* lays its eggsacs preferably in plants of the genus *Asclepias* (Apocynaceae). Leaf characteristics (e.g., flexibility, large size, high density of trichomes) appear to be fundamental factors determining the selection of this plant as an oviposition site. A social species, *Diaea* (Thomisidae), chooses *Eucalyptus* leaves, also using morphological leaf characteristics to recognize them (Evans 1997). However, this species selects smaller leaves than does *M. vatia*, because it is too small to handle the large leaves.

Another way that Thomisidae spiders find their foraging sites is through volatile substances. Heiling et al. (2004) offered crab-spiders (*Thomisus spectabilis*) and floral visitors (*Apis mellifera*, Hymenoptera) both flowers with natural scents and those from which the scents had been removed. Both species preferred the flowers with the scents. The spider and the bee favored different visual characteristics of flowers (size and reflectance). Krell and Krämer (1998) showed that the volatile eugenol [2-methoxy-4-(2-propenyl) phenol], a component of the floral fragrance found in plants of different families around the world, attracts the cogeneric spiders *Thomisus daradioides* and *T. blandus* (Thomisidae). The researchers suggest that the spiders are able to memorize common chemical compounds in flowers and use them as cues to locate their foraging sites, which in turn are highly visited by pollinators.

Lyssomanes viridis is a translucent green jumping spider that perches on the abaxial surface of leaves. This salticid has a chemically mediated preference for, and higher hatching success on, the sweet gum *Liquidambar styraciflua* L. (Altingiaceae) during the summer. Compared with other sympatric species, the sweet gum leaves presents a potent broad-spectrum antimicrobial volatile compound, notably the monoterpene terpinen-4-ol, a well-studied antimicrobial agent known from tea tree oil. This chemical compound could protect the spider eggs against microbes, promoting higher hatchings (Tedore and Johnsen 2015).

The contact, but not airborne, with chemical cues of this plant species are attractive to *L. viridis*. These spiders overwinter predominately on leaves of a broadleaf evergreen species, the American holly *Ilex opaca* Aiton (Aquifoliaceae), so must migrate to American holly in the autumn, and back to sweet gum in the spring once its leaves have re-emerged (Tedore and Johnsen 2015). Posteriorly, these same authors (2016) expected that *L. viridis* might use leaf shape to find sweet gum, and green coloration to detect American holly. However, their results suggest that *L. viridis* does not attend to the leaves color or shape, but does have a visually mediated preference for a particular level of ambient illumination and possibly perceived leaf brightness. In experimental conditions, spiders overtake any potential preference for leaf size. Importantly, if they had not controlled for the effect of leaf size on the ambient illumination in the area surrounding the leaf, they would have concluded that *L. viridis* was capable of judging the relative sizes of leaves using other parameters, like relative area or linear dimension. The authors conclude that ambient illumination was the most important factor in determining which leaf the spiders settled during their experiments.

Some studies have demonstrated intimate associations between spiders and plant species or groups of plants that share features in common. The best-known associations involve interactions between species of the genus *Peucetia* (Oxyopidae) and plants with grandular trichomes, jumping spiders (Salticidae) and Bromeliaceae plants, and araneids of the genus *Eustala* and tree species or dry vegetation structures. Some of these studies will be discussed later.

Associations Between Spiders and Plants with Glandular Trichomes

Several plant species from different taxa have glandular trichomes on the surfaces of their leaves and stems. These structures possibly arose as a direct defense against herbivores and pathogens (Duffey 1986). Enhancing the plants' defense systems, some spiders belonging to the families Thomisidae and Oxyopidae forage and reproduce preferentially on plants containing this type of glandular structure.

Spiders of the genus *Peucetia* (Oxyopidae) do not construct webs, weaving only silk threads leading to the branches, leaves, or flowers of the plants in which they live. Females lay their eggsacs under leaves and remain near them for several days, until the emergence of the spiderlings. Some species belonging to this genus are commonly found on plants that have glandular trichomes. In a literature review coupled with over 30 years of field observations, Vasconcellos-Neto et al. (2007) showed that ten species of *Peucetia* occur in association with more than 55 species of plants that have these trichomes. The plant species more frequently used by these spiders belong to the families Solanaceae, Asteraceae, and Melastomataceae. Whereas the species *Peucetia flava*, *Peucetia rubrolineata*, *Peucetia longipalpis*, and *P. viridans* occur in the Americas (Brazil, Colombia, Panama, Mexico, and the United States), the oxyopids *Peucetia arabica*, *Peucetia crucifera*, *Peucetia maculifera*, *Peucetia nicolae*, *Peucetia transvaalica*, and *Peucetia viridis* occur in the Old World, including Spain and some parts of Africa (Fig. 7.2).

In the municipality of Sumaré, São Paulo (Brazil), *P. rubrolineata* occurred more frequently on Solanaceae species that contained leaves with a high density of glandular trichomes (Vasconcellos-Neto et al. 2007). In other regions, *P. flava* and *P. rubrolineata* were exclusively associated with plants containing these structures (Morais-Filho and Romero 2008, 2009; Vasconcellos-Neto et al. 2007).

Why do several species of the genus *Peucetia* specialize in plants containing glandular trichomes? Vasconcellos-Neto et al. (2007) suggested that this specialization may have evolved as a result of the adhesive nature of glandular trichomes, which hold small insects captive or hinder the movements of larger arthropods. In fact, many arthropod predators, such as insects belonging to the family Miridae, also have specific associations with plants containing glandular trichomes (Anderson and Midgley 2003; Sugiura and Yamazaki 2006; Romero and Vasconcellos-Neto 2004b) and capture prey that become adhered to these plant structures. Therefore, trichomes make it easier to capture prey, enabling predators to save the energy they would have expended in the capture and subjugation of prey.

Do spiders actively select plants with glandular trichomes, and does this behavior really benefit the spiders? To answer this question, Romero et al. (2008a) conducted field experiments using *P. rubrolineata* and *P. flava* in Serra do Japi, in

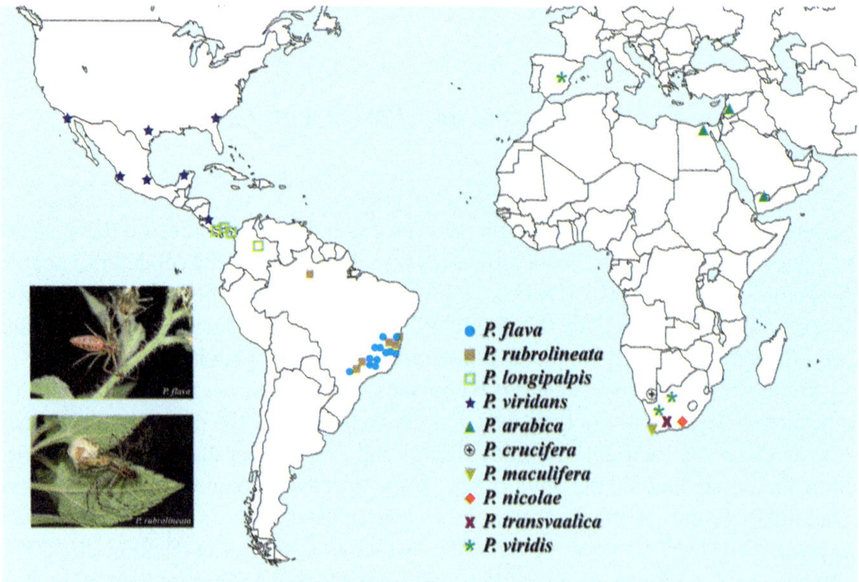

Fig. 7.2 Distribution map of *Peucetia* (Oxyopidae) associated to plants containing glandular trichomes (Photographs: Yuri Fanchini Messas)

Jundiaí, São Paulo (Brazil). The researchers estimated the residence time of these spiders in plants with glandular trichomes (*Trichogoniopsis adenantha*, Asteraceae) and without glandular trichomes (*Melissa officinalis*, Lamiaceae, and *Lantana camara*, Verbenaceae). Both spider species remained significantly longer on plants with trichomes (approximately 60 h on average) than on plants without these structures (a few minutes). In a second experiment, the researchers placed dead vestigial *Drosophila* flies (with atrophied wings) on *T. adenantha* plants (which have glandular trichomes) and on *M. officinalis* (which do not have trichomes). The plants were placed individually in exclusion cages and a spider was maintained on each plant. The spider's biomass was estimated at the beginning of the experiment and 6 days later. Spiders on plants with glandular trichomes did not lose or gain biomass, whereas spiders on plants without trichomes lost biomass. The findings indicate that *Peucetia* spiders consume dead organisms attached to trichomes and therefore act as scavengers on these structures. Detection and recognition of dead prey on vegetation are not common behaviors among spiders. The dead prey adhered to the trichomes probably provide extra nutrients during periods of live food scarcity.

In a complementary study, Morais-Filho and Romero (2010) used razor blades to remove the glandular trichomes from some *Rhynchanthera dichotoma* (Melastomataceae) plants. As in the study by Romero et al. (2008a), *P. flava* spiders stayed longer on plants with intact trichomes than on those whose trichomes had been removed. Next, the researchers released 30 vestigial winged *Drosophila* flies on each plant in a sample that included both plants with intact glandular trichomes and those from which the trichomes had been removed. More flies adhered to plants with intact trichomes. The study confirmed the hypothesis that arthropods can become adhered to these plant structures.

As shown here, *Peucetia* can consume dead insects that are attached to trichomes. In nature, do these spiders consume more live or dead insects? Morais-Filho and Romero (2010) enriched vestigial *Drosophila* with large amounts of nitrogen-15 isotope (^{15}N, see the procedure in Romero et al. 2006), to identify how much of the enriched prey was transferred to the spiders. The treatment groups included plants with dead enriched flies and plants with live enriched flies. Spiders were introduced to both treatments, and after a few days were collected for isotopic measurements of their body tissues. The spiders that preyed on dead and live flies showed similar amounts of the nitrogen-15 isotope, indicating that the consumption of live prey and decomposing insects on trichomes occurs in similar proportions.

Among the Thomisidae spiders, *Misumenops argenteus* was also found on plants with glandular trichomes in Serra do Japi (Romero and Vasconcellos-Neto 2004b). This spider occurred more frequently in *T. adenantha* and *Hyptis suaveolens* (Laniaceae), both containing trichomes, than in other plants available in the study area that do not present these structures. *T. adenantha* blooms all year and can thus attract potential prey throughout the life cycle of the spider. The glandular trichomes hinder the movement of ants and Chironomidae mosquitoes, which constitute up to 21% of the diet of *M. argenteus* (Romero and Vasconcellos-Neto 2003). According to Romero and Vasconcellos-Neto (2004a), these characteristics of the spider–plant interaction may all contribute to make the relationship beneficial to both species.

Specializations of Spiders for Bromeliads or Similar Plants

A wide diversity of aquatic and terrestrial arthropods inhabit plants belonging to the family Bromeliaceae, especially in Neotropical environments (Benzing 2000). Some spider species occur preferentially on bromeliads, and some present morphological features (e.g., dorsoventrally flat body) that facilitate their colonization of these plants. Associations between spiders and bromeliads and/or similar plants (e.g., plants that have leaves arranged as rosettes) have been described for the spider families Anyphaenidae (Brescovit 1993), Araneidae (Figueira and Vasconcellos-Neto 1991), Corinnidae (Piccoli 2011), Ctenidae (Barth et al. 1988a, b; Dias and Brescovit 2004), Salticidae (Young and Lockley 1989; Maddison 1996; Rossa-Feres et al. 2000; Frank et al. 2004; Romero and Vasconcellos-Neto 2005a, b, c; Romero 2006; Romero et al. 2007), Trechaleidae (Brescovit and do Oliveira 1994), and Theraphosidae (Dias and Brescovit 2004). The jumping spider *Pelegrina tillandsiae* (Salticidae) was recorded on Spanish moss (*Tillandsia usneoides*, Bromeliaceae) in the southeastern United States (Romero 2006).

Among these, the most studied associations involve Salticidae species. So far, nine species of jumping spiders associated with bromeliads in South America have been reported. The studies showing those associations were conducted in several countries, including Brazil, Bolivia, Argentina, and Paraguay, in areas containing different vegetation types, including *cerrado* regions, semi-deciduous forests, vegetation of coastal dunes, *restingas*, rocky outcrops (inselbergs), chacos, seasonal forests, dense rainforests, and tropical montane forests (Rossa-Feres et al. 2000; Romero and Vasconcellos-Neto 2004c, 2005a, b, c; Romero 2006). These studies showed that spiders use bromeliads as foraging sites and breeding, nursery, and shelter sites where they can avoid predators and adverse weather conditions.

Some of these species are specialists that are associated almost exclusively with one type of bromeliad (e.g., *Psecas chapoda* and *Bromelia balansae*) over a large geographical area (Fig. 7.3). In contrast, other species are generalists (other *Psecas* sp., *Coryphasia* spp., *Eustiromastix nativo*, *Uspachus* sp. new) that have been found inhabiting as many as eight species of bromeliads. The specialists occur in phytogeographical regions such as the *cerrado* and semi-deciduous forests, where one species of bromeliad (*B. balansae*) typically dominates, while general spiders usually live in areas with high species richness and diversity of bromeliads, such as in the rain forest (Romero 2006). Bromeliad species that occur in rainforests share morphological characteristics (e.g., broad leaf, presence of a tank) that are attractive to spiders.

So far, the spider–bromeliad association that has been studied the most involves *Psecas chapoda* (Salticidae) and *Bromelia balansae* (Bromeliaceae). This spider occurs almost exclusively on *B. balansae* in various regions of *cerrado* and semi-deciduous forest in Brazil, Bolivia, and Paraguay (Rossa-Feres et al. 2000; Romero and Vasconcellos-Neto 2005a, b, c). The spider uses the bromeliad throughout its reproductive cycle, from courtship and mating to the deposition of eggsacs and population recruitment of young spiders. Each female can lay up to two eggsacs,

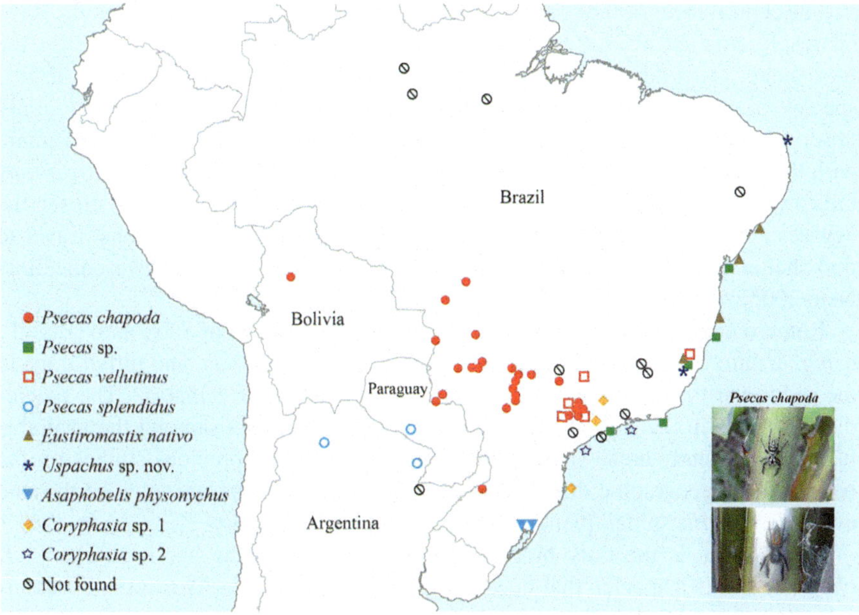

Fig. 7.3 Distribution map of Salticidae species associated to bromeliads on South America (Photographs: Gustavo Quevedo Romero)

always in the middle region and on the concave side of the leaves, covering them with a silk sheet woven over the leaf edges (Rossa-Feres et al. 2000).

Romero and Vasconcellos-Neto (2005c) found that *P. chapoda* occur more frequently in open areas (fields) than within the forest. The distribution may be related to the blocking of the rosettes' central base (which are used by the spiders as shelter) by dry leaves that fall from the trees. To test this hypothesis, Romero and Vasconcellos-Neto (2005a) conducted a field experiment in open areas containing two treatments (bromeliads with and without dry leaves in rosettes) and found that plants containing leaves were less often colonized by the spiders. However, in a similar experiment carried out within the forest, the spiders did not occupy bromeliads that lacked dry leaves. As the abundance of prey (insects) was significantly higher in open areas than in the forest, the researchers suggested that both the presence of dry leaves and the availability of prey affect the spatial distribution of *P. chapoda.*

The spiders of this species rarely occur in flowering plants and may occupy up to 90% of the plants that lack inflorescences (Romero and Vasconcellos-Neto 2005b, c). The leaves of bromeliads close to the ground decline when the plant blooms, and this structural modification exposes the flowers to pollinators (e.g., hummingbirds). Romero and Vasconcellos-Neto (2005a) showed that this change affects spiders by modifying their shelter and the nesting sites available inside the rosette, exposing jumping spiders to abiotic (e.g., severe weather conditions) and biotic (e.g., natural enemies) factors.

The central layer of the rosette is occupied by the majority (approximately 70%) of freshly emerged *P. chapoda*. The center may be preferred because it offers the best shelter from desiccation and/or cannibalism, which is very common in this species of spider (G. Q. Romero unpublished data). Females are commonly observed in external layers of the plant, but they build their eggsacs and remain with them in the inner layers, where young spiders find shelter. This behavior may indicate the existence of maternal care against cannibalism (i.e., the closer the eggsacs are to the center of the bromeliad, the less the spiderlings must travel to find shelter and the greater their chances of survival) (Romero and Vasconcellos-Neto 2005c).

Romero and Vasconcellos-Neto (2005b) collected spiders in São Paulo (Brazil), using an entomological umbrella, a visual search of vegetation, and pitfall traps in the soil to verify whether *P. chapoda* lives exclusively on *B. balansae*. The species did not occur in the soil and was found only in *B. balansae* among the available plants in the study area. These results, in addition to the previous studies on the behavior and geographic distribution of *P. chapoda* on *B. balansae,* were the first evidence that this spider–plant association could be obligatory.

B. balansae is the only bromeliad species found in the habitat range of *P. chapoda* and is a species that does not accumulate water in its rosette. To determine whether the selection of bromeliads by *P. chapoda* is species-specific, Omena and Romero (2008) planted blocks of bromeliads in the field, each containing three species: *B. balansae*, *Aechmea distichantha,* and *Aechmea blanchetiana. A. distichantha* has a leaf architecture that is similar to *B. balansae*, but retains water in its central tank, while *A. blanchetiana* has much wider leaves than the other two species and also accumulates water in the rosette. Spiders colonized *B. balansae* and *A. distichantha* equally, but occupied *A. blanchetiana* less frequently. Therefore, *P. chapoda* selects plants according to their architecture (long, narrow leaves) and is not species-specific. When an observer approaches *P. chapoda* spiders on their host plant (*B. balansae*), spiders flee to the base of the leaves. However, spiders that colonized bromeliads which accumulate water in the tank could not escape to the base of the rosettes. Interestingly, jumping spiders inhabiting tank bromeliads in other geographic regions (along the Brazilian coast) can dive into the tanks to escape from predators (Romero and Vasconcellos-Neto 2004c).

The study by Omena and Romero (2008) demonstrated that *P. chapoda* selects plants with specific architectures. However, the mechanisms by which the spiders detect, identify, and evaluate the plants remain unclear. Jumping spiders have good vision and, therefore, may be able to use visual cues to choose plants. To test this hypothesis, Omena and Romero (2010) offered spiders a choice of four plants: *Agave augustifolia* (which has a rosette similar to a bromeliad), *Euterpe oleracea* (a palm), *Croton floribundus* (dicotyledonous with large leaves), and *Delonix regia* (dicotyledonous with small folioles). Almost all spiders chose the agave, which has similar architecture to a bromeliad, showing that these spiders select plants with similar architectural features by using visual cues. In a similar experiment, Omena and Romero (2010) offered *P. chapoda* spiders a choice of four bromeliads: *B. balansae*, *A. distichantha*, *A. blanchetiana*, and *Aechmea fasciata*. The first two species have long and narrow leaves, while the others have short and broad leaves. The spiders more frequently chose the bromeliad with narrow leaves, indicating that they use fine details to choose their microhabitats. In order to eliminate the possible effects of color and scent, the researchers used life-size black-and-white photographs of the bromeliad species used in the previous experiment. Interestingly, spiders still chose photographs of bromeliads with long, narrow leaves. These two studies showed that spiders use visual cues to choose bromeliads, relying mainly on the plant architecture and not on coloration.

Three other salticid species were recorded living in bromeliads in the coastal regions of Brazil (Romero and Vasconcellos-Neto 2004c; Santos and Romero 2004). The jumping spiders *E. nativo* and *Psecas* sp. are associated with bromeliads in two different types of vegetation in Linhares (ES): native grasslands (a plant formation similar to *restingas*) and mussunungas, a low forest growing on sandy soils that is typically found in the northern region of this state. *E. nativo* was also found on bromeliads in a *restinga* region in the city of Trancoso (BA). Another species, *Uspachus* sp., also occurs in native grasslands in Linhares and is more frequent in dune areas in Natal (RN). Romero and Vasconcellos-Neto (2004c) suggest that these three spider species are associated specifically with plants of the Bromeliaceae family, as they were not found on other plants. They also point out that members of this plant family have a highly complex architecture and provide favorable microhabitat for jumping spiders.

As with *P. chapoda*, characteristics related to the physical structure of bromeliads and the environment can affect habitat selection by jumping spiders. *E. nativo* occurs preferentially on large bromeliads in two different regions (Linhares and Trancoso). Larger bromeliads have increased sheltering capacity and are more likely to be visited by insects (offering higher availability of prey) due to the larger surface area; thus, they are considered better quality foraging sites (Romero and Vasconcellos-Neto 2004c).

In contrast with *E. nativo*, which occurred more frequently in bromeliads in open areas (native grassland), *Psecas* sp. mostly occupied bromeliads from adjacent forests in Linhares. In Trancoso, even in the absence of *Psecas* sp., *E. nativo* occurred only in open areas (*restingas*), indicating that this pattern of distribution reflects the habitat and/or microhabitat choice, rather than being due to interspecific competition between these two species of spiders. Romero and Vasconcellos-Neto (2004c) concluded that *E. nativo* first selects the habitat and then chooses the microhabitat.

In other regions of Brazil and Argentina, five other species of Salticidae (*Psecas vellutinus*, *P. splendidus*, *Coryphasia* sp. 1 and sp. 2, and *Asaphibelis physonychus*) were observed specifically associated with bromeliads. The biology and natural history of these species remain unknown (Romero 2006).

Associations of Spiders with Arboreal Plants

Arboreal plant species provide a high diversity of microhabitats due to their huge biomass (large surfaces) and high structural complexity (Draney et al. 2014). Among these microhabitats, tree trunks can provide concavities, cracks, epiphytes (e.g., other vegetables, moss, and lichen), fissures, and patches of loose bark (Szinetár and Horváth 2005; Michel and Winter 2009; Messas et al. 2014). Due to this variation, the bark of a particular tree can have its own microclimate (Nicolai 1986, 1989) and that resource can significantly affect the distribution of species in tree trunks (Prinzing 2001, 2005).

Spiders may use the tree trunk as an exclusive, facultative, or occasional habitat (Wunderlich 1982). The spider *Neriene radiata* (Linyphiidae), for example, sometimes occurs on tree trunks, but it prefers the more stable environment offered by the bark to the understory environment (Herberstein 1998). Species that live exclusively associated with tree trunks generally present behavioral (e.g., seeking shelter under tree bark), morphological (e.g., flattened body), physiological (e.g., camouflage), and phenological adaptations to the environment in which they live (Szinetár and Horváth 2005). Bark-dwelling spiders (e.g., *Telaprocera*; Harmer 2009, Harmer and Herberstein 2009) and *Eustala perfida* (Messas et al. 2014) can construct vertically long webs, called ladder-webs, whose shape is probably due to the horizontal space limitation caused by the trunks or to a specialization for specific prey (e.g., moths; Harmer and Herberstein 2010). Bark-dwellers occur more frequently in trunks that exhibit surfaces with specific characteristics (e.g., *E. perfida*; Messas et al. 2014). These spiders usually select microhabitats containing essential characteristics such as shelter, high prey availability, and anchorage points for web construction (Herberstein 1998; Harmer 2009; Draney et al. 2014; Messas et al. 2014). The structural characteristics of the bark invite a wide variety of potential prey to spiders (Horvath et al. 2005).

Messas et al. (2014) investigated the spatial distribution and habitat selection of *E. perfida* (Araneidae), a spider that presents chromatic polymorphism, with colors ranging from green, red, white, and black. The study was conducted in Serra do Japi, a semi-deciduous rainforest located in São Paulo state (Brazil), with altitudinal variation from 700 to 1300 m. To verify the spatial distribution of this species, the researchers delimited plots on the edge and in the interior of the forest at different altitudes (basal, intermediate, and high) and performed a visual search for spiders on vegetation. The spiders were not found at the edge or on shrubby and herbaceous vegetation; instead, they occurred strictly on tree trunks inside the forest. Therefore, the species is an exclusively bark-dwelling spider.

Subsequently, Messas et al. (2014) proposed that *E. perfida* prefers trunks containing specific characteristics. They measured structural attributes of the trunks within the plots, characterizing each trunk according to texture (smooth or rough bark), size (diameter at breast height), and the presence of features such as lichens, mosses, and concavities. The characteristics of over 3000 tree trunks were evaluated in an analysis of use by spiders. *E. perfida* was found in different tree species, both native and introduced, indicating that the spider did not require a unique host plant species. *E. perfida* occurred most frequently on trees with rough trunks and mosses, lichens, and/or concavities (Fig. 7.4a). These structures provide insertion points for the construction of orb webs. In addition, more spiders were found on trunks with larger diameters at intermediate and lower regions of the mountain. Larger trunks have more surface area for web construction and are more common in these low-altitude areas. In contrast, the highest region of the mountain consists of a semideciduous rainforest that is typical for the altitude, with thinner trees and, coincidentally, fewer spiders.

These studies demonstrated that *E. perfida* occurs in narrowly defined environments, determined by the type of vegetation (large trees) and tree trunks that share the same structural characteristics. Furthermore, the species presents chromatic polymorphism (at least from a human's point of view) that is similar to the colors found in the bark or in elements of the trunk, such as mosses and lichens. This adaptation is probably due to the pressure exerted by visually oriented predators such as birds and hymenopteran parasitoids. The researchers observed some spiders whose bodies contained larvae of the koinobiont ectoparasitoid *Acrotaphus tibialis* (Hymenoptera, Ichneumonidae), but the parasitism rate was extremely low (Messas et al. in preparation). In fact, during the study period, few *E. perfida* predation events by other animals were observed, mostly involving other spiders (e.g., *Gelanor* sp., *Argyrodes* sp., and a species of Salticidae). This indicates that the camouflage in this species is effective, but further studies should be conducted to determine how predators perceive the coloration of *E. perfida.*

In another study conducted in Serra do Japi, Villanueva-Bonilla (2015) investigated habitat selection by the wall crab spider *Selenops cocheleti* (Selenopidae), which lives on tree trunks and presents a dorsoventrally flattened body. Unlike *E. perfida*, this selenopid is strongly associated with trees that have desquamative stems and smooth texture. This preference is related to the spiders' use of cracks as shelter, since the flat body of *S. cocheleti* enables the spider to shelter in areas between the bark and the tree trunk. Furthermore, this spider species prefers Myrtaceae plants to other plants in the study area (Fig. 7.4b). Nevertheless, as with *E. perfida* (Messas et al. 2014), the species does not show specificity for a single plant species, but for a set of trees that share structural characteristics.

Associations of Spiders and Dry Structures of Vegetation

Spiders are commonly found living on shrubby and herbaceous vegetation. Within the group of orb-weaver spiders, some species of the Araneidae family are associated with specific plants (Hesselberg and Triana 2010) or with plants that share characteristics in common, such as density and architecture of branches that enable the construction of orb webs (Turnbull 1973).

Souza et al. (2015) investigated the spatial distribution and habitat selection of two sympatric and cogeneric species of orb-weaver spiders, *Eustala taquara* and *Eustala sagana* in Serra do Japi. Both species have chromatic polymorphism, with

Fig. 7.4 (**a**) Adult female of *Eustala perfida* (Araneidae). (**b**) Expected and observed frequencies of *E. perfida* on trunks with different bark surfaces. (**c**) Adult female of *Selenops cocheleti* (Selenopidae) on Myrtaceae. (**d**) Expected and observed frequencies of *Selenops cocheleti* on trees with exfoliating and not exfoliating barks in 2014 (Photographs: Yuri Fanchini Messas)

many shades of brown, gray, and green. A remarkable morphological feature that distinguishes the two species is the long, longitudinally striped abdomen of *E. sagana*, while *E. taquara* has a subtriangular, slightly lengthened abdomen (Fig. 7.5). This study was conducted in the same environment using a similar methodology to that employed for *E. perfida* (see Messas et al. 2014), making the results comparable.

To verify the spatial distribution of these two species, the researchers visually searched for spiders within plots inside the forest and on the forest edge. Both species were found living exclusively associated with shrub and herbaceous plants on the forest edge. Both *E. taquara* and *E. sagana* have cryptic coloration (from a human point of view) and rest on dry vegetation structures. To show that their distribution was not random, the frequencies of green (live) and dry (dead) vegetation was estimated for the plots on the edge of the forest, and posterior comparison was made with the frequency of sites (green or dry) effectively used by spiders to rest. Again, both species were more abundant in similar environments, with more than 90% of individuals occupying dry vegetation structures.

If both species occur in such similar environments, which factors determine the spatial segregation between *E. taquara* and *E. sagana*? The researchers tested the hypothesis that the altitude and the type of vegetation help determine the distribution of these two species. In fact, *E. taquara* occurred more frequently in the intermediate regions (1000 m above sea level), while most *E. sagana* individuals were found at lower elevations (750–850 m) of Serra Japi. The authors argued that these differences may be related to biotic (e.g., architecture of vegetation, availability of prey, and the presence of natural enemies) and abiotic factors (e.g., temperature, humidity, and solar radiation) (Turnbull 1973; Brown 1981; Janetos 1986; Lubin et al. 1991; Marshall and Rypstra 1999).

In Ecuador, Purcell and Avilés (2007) observed that the altitude can also affect the distribution of some species of *Anelosimus* (Theridiidae), mainly in response to biotic factors, such as the prey size and predator pressure. To verify whether spiders show specificity for certain plant species, the researchers estimated the diversity and frequency of plant species in plots on the forest edge (in the altitudes where each spider species shows greater abundance). The relative abundance of plant species (expected frequency) was compared with the relative abundance of plants that are effectively used as sites for web construction (observed frequency). *E. taquara* were found more frequently in plants belonging to the species *Conyza bonariensis* (Asteraceae), apparently avoiding web-building in *H. suaveolens* (Lamiaceae), which is preferably used by *E. sagana* (Fig. 7.5). Hesselberg and Triana (2010) also studied the specificity of *Eustala* for certain plant species, showing that the spiders *Eustala illicita* and *Eustala oblonga* are associated with the plants *Acacia collinsii* and *Acacia melanoceras* (Fabaceae) respectively, which present a complex plant–ant–spider interaction.

The arboreal araneid *E. perfida* (Messas et al. 2014) and both *E. taquara* and *E. sagana* (Souza et al. 2015) are sympatric spider species that are phylogenetically related. In all three species, the cryptic coloration matching the plant substrate seems to play a fundamental role in the history of these animals' lives. Studies

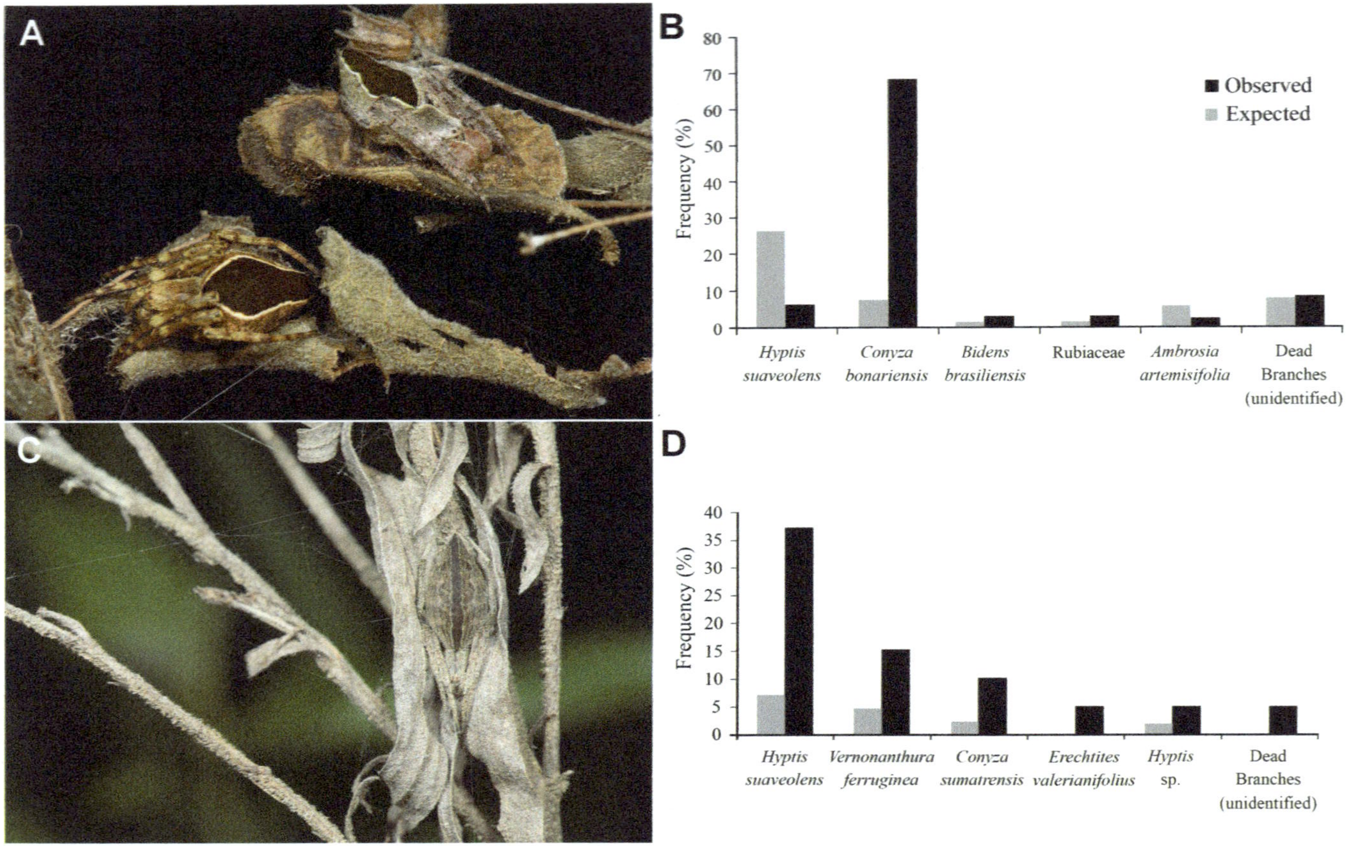

Fig. 7.5 (**a**) Male and female of *Eustala taquara* (Araneidae). (**b**) Expected and observed frequencies of *E. taquara* on different plant species. (**c**) Adult female of *Eustala sagana* (Araneidae). (**d**) Differences in frequency (expected and observed) of *E. sagana* on plants (Photographs: Yuri Fanchini Messas)

suggest that the camouflage and the color polymorphism may be a result of the selective pressure exerted by visually oriented predators such as birds and hymenopteran parasitoids. Souza et al. (2015) observed that *E. taquara* and *E. sagana* rest on specific plant structures, the former preferring dry capitula and the latter dry stems, which are similar in shape to the long body of the spider (Souza et al. unpublished data). Thus, it is likely that the spiders are choosing specific microhabitats that may promote the effectiveness of their cryptic coloration.

Phenological Synchrony and Lags in Plant–Spider Relationships

An interesting question in ecology is how populations of plants and spiders interact to maintain specific phenological associations. For example, do spiders have the same effects on plants throughout the year? Studies have shown that populations of spiders associated with vegetation often suffer directly from climatic factors or indirectly from changes in the availability of foraging sites or prey. However, the populations of these predators do not always show synchronized responses to biotic or abiotic variables.

Arango et al. (2000) studied the relationship between the spider *P. viridans* and the plant *Cnidoscolus aconitifolius* (Euphorbiaceae), which attracts floral visitors, including flies, bees, and wasps, in Mexico. There was a clear lag time between events such as the onset of rains, the flowering of the plant, the arrival of floral visitors, and an increase in the spider population. In May, the rains began and the plants flowered. In July, floral visitors increased, and in August, the spiders increased in number. A similar phenological pattern was observed in the system featuring the spider *M. argenteus* and the plant *T. adenantha*, which attracts herbivores and floral visitors, in Serra do Japi, Jundiaí (SP). Temporal lag analysis (with up to a 3-month delay) detected a 1-month delay between the start of rains and the flowering period of *T. adenantha*. An increase in the arthropod population (potential prey for *M. argenteus*) on the plant occurred in synchrony with the increase in the number of reproductive branches. The population of *M. argenteus* increased 2 months after the numerical response of arthropods (Romero 2001; Romero and Vasconcellos-Neto 2003).

These results indicate that climatic factors such as rainfall primarily shape the phenological pattern of plants. In response to increased rainfall, plants produce more reproductive branches. These branches, which are used as foraging sites by spiders, provide food resources in the form of several species of herbivores and pollinators (Arango et al. 2000; Romero 2001; Romero and Vasconcellos-Neto 2003, 2004a). If these resources are scarce at a particular time of the year, such as the dry season, the insects that directly depend on them will be scarce too. Consequently, the availability of prey and foraging sites for the spiders also decreases, reducing their populations. These results indicate that the systems studied by Arango et al. (2000) and Romero and Vasconcellos-Neto (2003, 2004a) are strongly influenced by bottom-up effects, when changes in the lower levels of the food chain, such as the producers, affect the levels above (Romero 2007). These studies reveal the importance of interactions between biotic and abiotic forces in determining the community structure of arthropods on plants.

P. rubrolineata and *P. flava* (Oxyopidae) are two species associated with *T. adenantha*, and population sizes and age structures of spiders are related to climatic variables, plant phenology, and abundance of prey, which may or may not result in synchrony and time lags in this system of tri-trophic interactions (Villanueva-Bonilla et al. in preparation).

Studies of the phenology of *T. adenantha* (Romero and Vasconcellos-Neto 2005d) and the natural history of *M. argenteus* reveal the lifecycle adjustments (phenogram) made by this spider species to climatic conditions, plant phenology, and prey availability (Romero and Vasconcellos-Neto 2003, 2004a). The age structure of the spider population throughout the year expresses the interactions of the spiders' lifecycle with biotic and abiotic conditions. During colder and drier periods of the year, the juvenile and subadult instars have longer durations. The longer development time for these phases may result from low availability of prey.

Negative Effects of the Presence of Spiders on Plants

Although spiders frequently occur on plants, their role as predators and the cascade effect of their presence on herbivores and plants have not been fully explored. Their effects may be positive or negative for the plant. In some cases, spiders prey on herbivores, favoring the plant's fitness (Fig. 7.6). These mutualistic relationships will be discussed later.

In other cases, spiders consume or interfere with pollinators, reducing the plant's fitness. Usually the negative effects of spiders occur through trait-mediated indirect interactions (TMIIs), defined by Abrams et al. (1996) as effects transmitted through changes in traits (e.g., behavioral, morphological, and life history) of affected species.

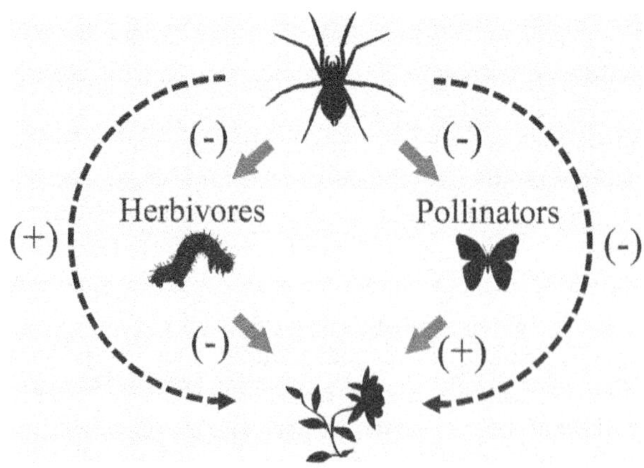

Fig. 7.6 Indirect effects (positives and negatives) of spiders on plants

The most studied TMIIs induced by spiders involve deterrence of pollinating insects from plants containing spiders; the insects can detect flowers that contain sit-and-wait spider predators and avoid them. Spiders can also negatively affect digestive mutualistic interactions between insects and plants. Furthermore, spiders are known to repel predators of phytophagous insects and/or consume insects (e.g., ants) that protect the plant against predators, facilitating the presence of herbivores on the plant. Some of these interactions will be discussed individually in the sections that follow.

Negative Effects of Spiders on Plant–Insect Digestive Mutualism

Although spiders frequently contribute to plant nutrition by producing feces, they can also negatively affect digestive mutualism. The heteropteran predator *Pameridea roridulae* (Miridae), for example, lives exclusively associated with the carnivorous plant *Roridula gorgonias* (Roridulaceae) in South Africa, and can contribute up to 70% of the total N used by its host plants (Ellis and Midgley 1996). However, in some regions, *R. gorgonias* is also inhabited by the spider *Synaema marlothi* (Thomisidae), which often decreases the density of the mutualist *P. roridulae* on the plant. Spiders do not defecate directly on the plant so they do not contribute N to it. Anderson and Midgley (2002) showed that in the presence of spiders, plants had low density of Heteroptera and were less enriched with nitrogen (^{15}N isotope).

Negative Effects of Spiders on Ant–Plant Mutualism

Gastreich (1999) showed that the spider *Dipoena banksii* (Theridiidae) exerts TMIIs in a mutualistic association between the ant *Pheidole bicornis* and the plant *Piper obliquum* (Piperaceae) in Costa Rica. This ant–plant interaction follows the general pattern of this type of association, wherein the plant provides food for the ant and the ants protect the plant against insect herbivores, reducing folivory and consequently increasing the fitness of the plant. The theridiid *D. banksii* constructs its web in the base of new leaves of *Piper* plants, and preys almost exclusively on *P. bicornis*. The presence of the web helps the spider to capture ants but at the same time allows the ants to detect and avoid the spiders, making it possible to study indirect interactions mediated by behavioral modification.

If TMIIs actually exist in this ant–plant interaction mediated by *D. banksii*, plants containing spiders would be expected to exhibit lower density of ants (whose standard foraging patterns would be altered by the presence of spiders). Consequently, these plants would exhibit an increased rate of folivory compared with plants lacking spiders. To test these hypotheses, Gastreich (1999) compared folivory rates among plants with and without spiders in the field. Subsequently, she investigated the effect of *D. banksii* on *P. bicornis*' behavior by removal experiments. She compared the numbers of ants patrolling leaves with spiders and on these same leaves after removal of the spider and, finally, after removal of the web. The removal of spiders and webs increased folivory and decreased the number of ants on plants that had contained spiders or webs, supporting the hypothesis that there is a TMII between *D. banksii* and *Piper* plants.

Negative Effects of Spiders on Plant–Pollinator Mutualism

Spiders that live on flowers can interfere in the dynamics of plant communities when they mediate the balance between pollination (e.g., by preying on or repelling pollinator insects) and herbivory by insects. Louda (1982b) was the first to investigate the negative effects of spiders on mutualistic relationships between plants and pollinators. Louda reported that the spider *P. viridans* (Oxyopidae), which lives in *Haplopappus venetus* (Asteraceae) in California (United States), was responsible for a significant reduction in the number of pollinated flowers and in the average fertility rate of the flowering branches.

Spiders can have strong negative effects on pollinator behavior and plant fitness (Dukas 2001; Dukas and Morse 2003; Suttle 2003; Heiling and Herberstein 2004). For example, Gonçalves-Souza et al. (2008) showed that the presence of artificial spiders designed to mimic species of the Thomisidae family interferes in the visitation behaviors of several species of pollinators of the plant *Rubus rosifolius* (Rosaceae), especially Hymenoptera (bees). Plants containing spider models produced 42% fewer seeds and the biomass of their fruits was reduced by approximately 50%.

Dukas and Morse (2003) showed that in Maine (United States), the bumblebee *Bombus ternarius* (Hymenoptera) visited *Asclepias syriaca* (Apocynaceae) less frequently when the plant contained Thomisidae spiders. The honeybee *Apis mellifera* showed a similar trend in behavior, although it was not significant. This decrease in visitation rate by pollinators can be explained both by direct effects (e.g., predation of pollinators by spiders) and indirect effects (e.g., avoidance of plants containing spiders by pollinators). Robertson and Maguire (2005) also showed a reduction in insect visitation of flowers of the plant *Lepidium papilliferum* L. (Brassicaceae), which housed the crab spider *Misumena vatia*. Flower visitors increased significantly after the spiders were removed.

In a later experiment, Dukas and Morse (2005) tested whether plants with crab spiders had fewer bee visitors than the plants without spiders, and verified whether the pollinia removal rate (indicating male fitness) and seed production rate (indicating female fitness) were lower in plants containing spiders than in those without spiders. In contrast with their earlier findings (Dukas and Morse 2003), the researchers found that the presence of spiders had no effect on the visitation of two species of bumblebee, *B. ternarius* and *Bombus vagans*, but *A. mellifera* visited significantly fewer plants containing spiders. This difference may be related to the higher rate of honeybee predation by spiders (as the honeybees are smaller and easier to capture) compared with bumblebee predation by spiders. Male and female plant fitness were not affected by the presence of *M. vatia* spiders in *A. syriaca*. Dukas and Morse (2005) hypothesized that the lack of an effect on plant fitness may be due to the low predation rate by spiders; in addition, spiders consume their prey slowly and other insects have a lengthy opportunity to visit and pollinate the plant while this is occurring.

These studies showed strong evidence that the presence of sit-and-wait spiders on flowers can negatively affect plant–pollinator mutualism. However, the components of this system and the mechanisms that affect it require clarification. Can the top-down effects of these spiders cascade to affect plant fitness, or are effects on fitness derived from TMIIs or density-mediated indirect interactions (DMIIs)? If the latter is true, there are probably adaptations related to traits of predators (e.g., foraging mode and morphology) and/or visual components that enable pollinators to recognize and avoid predation.

Gonçalves-Souza et al. (2008) tested these hypotheses through a series of experiments in the Atlantic Forest of southeastern Brazil. Artificial spiders were placed on flowers of the plant *R. rosifolius* (Rosaceae) (see details in Gonçalves-Souza et al. 2008), and randomized block experiments were conducted to test the effects of predator presence on pollinators and the power of the TMIIs over components of plant fitness (e.g., individual seed set and fruit biomass). The results showed that, in fact, some floral-visiting insects (e.g., Hymenoptera) can use visual cues to evaluate and avoid flowers containing objects that are similar to spiders or that mimic different morphological traits of spiders (e.g., abdomen and front legs). Thus, morphological traits, but not coloration, are responsible for the avoidance shown by insects. In addition, plants containing artificial spiders showed a considerable reduction in fitness, producing only about half of the individual seed set and fruit biomass. These findings showed that a reduction in the plant fitness is due to TMII related to the presence of spiders on flowers. Subsequently, Brechbühl et al. (2010) reported that different types of pollinators react differently to the presence of spiders (only solitary bees and syrphid flies avoided plants with spiders) and that these effects may also differ between plant species. They hypothesized that top-down effects of predators on plants via pollination depend on the degree of specialization of pollinators and the strength of their tendency to avoid spiders.

Gonçalves-Souza et al. (2008) used conspicuous spider models, and the coloration of crab spiders apparently had no effect on TMII mediated by the spiders. However, some species of Thomisidae, such as *M. vatia*, have similar coloration to the flowers of their hosts. Through two complementary studies, Ings and Chittka (2008, 2009) showed that this cryptic coloration can increase the TMII of bees in this system. In their 2009 study, these researchers exposed bees to predation risk experiences by placing cryptic robotic crab spiders on yellow flowers. After being exposed to the spiders, the bees were released and avoided yellow flowers even if they lacked spiders. Thus, it was demonstrated that when spider cryptic coloration causes avoidance by bees, it can negatively affect the reproductive success of plants containing cryptic spiders.

Some species of Thomisidae spiders, which appear cryptic from the human point of view, reflect ultraviolet wavelengths of light and thus attract their prey (Heiling et al. 2003, 2005a, b, Herberstein et al. 2009, see detailed review in Théry et al. 2011, see also Welti et al. 2016). Therefore, it is necessary to evaluate the function of the spiders' coloration from the point of view of their prey. Llandres and Rodríguez-Gironés (2011) conducted a study in Queensland, Australia that studied the response of *A. mellifera* to the presence of *T. spectabilis* spiders (which have chromatic white and yellow polymorphism) in inflorescences of the plant *Bidens alba* (Asteraceae). The authors used spectrophotometry to collect the data of reflectance from the spiders and inflorescences to determine how they are perceived by *A. mellifera*. Subsequently, they conducted a series of experiments to determine which traits of spiders (e.g., size, cryptic coloration, UV reflectance, and movement) result in higher rates of avoidance by bees. Unlike the results reported by Ings and Chittka (2009), the cryptic coloration did not play a strong role in avoidance behavior. However, spider size, movement, and UV reflectance did affect TMII by the spiders.

Arango et al. (2012) studied the system composed by the plant *Cnidoscolus multilobus* (Euphorbiaceae), its floral visitors, and the predatory spider *P. viridans* (Oxyopidae). The researchers evaluated the effect of spider presence on the plant on seed production during the whole year and showed that spiders may indirectly reduce the fitness (i.e., number of seeds) of plants, especially in months with few floral visitors.

However, according to Ribas and Raizer (2013), when spiders are rare on the plant and/or pollinators are very abundant, these predators have very small effects on the fitness of the plant, either low or neutral, and their presence does not significantly affect the production of seeds. Through two meta-analyses, Romero et al. (2011) and Romero and Koricheva (2011) synthesized the available literature regarding the risk effects of predation on the behavior of pollinators and the cascade effects of spiders on the fitness of plants, respectively. Romero et al. (2011) showed that different methods of foraging by spiders (e.g., sit-and-wait predation vs. active hunting) both caused avoidance behavior in pollinators. Furthermore, the effect of repelling pollinators was stronger in pollinators of smaller size. Romero and Koricheva (2011) reported that even though some studies show that spiders negatively affect pollination and the quantity of plant seeds, this has little effect on the global fitness of the plant. It is important to note that spiders can cause simultaneous positive and negative effects on plant fitness, and these effects are complementary and not mutually exclusive. Recent advances in our knowledge of these interactions have opened new perspectives for understanding the mechanisms of co-evolution in plant–pollinator–predator tri-trophic systems.

Negative Effects of Spider–Floral Herbivore Interactions on Plants

The plant–pollinator system can be affected by other aspects of the trophic chain, such as herbivore–predator interactions. These interactions can have direct effects (e.g., plant damage) or indirect effects (e.g., interruptions in pollination). Herbivory, specifically florivory, may influence plant breeding and plant population growth (Louda 1983; Marquis 1984). Florivory can directly reduce plant fitness by destroying reproductive tissues such as petals and sepals, which attract pollinators (Cardel and Koptur 2010; Botto-Mahan et al. 2011). The damage to these tissues can change the appearance of flowers and inflorescences, preventing pollinator visits (Møller and Sorci 1998). Predators on flowers can also cause indirect effects by reducing pollinator visits and time spent pollinating flowers (Romero et al. 2011), affecting the fitness of the plant. However, hardly any studies have evaluated the combined risk effect of floral herbivory and predation on the behavior of pollinators and the reproductive success of plants.

Antiqueira and Romero (2016) manipulated the floral symmetry and the presence of predators (artificial Thomisidae spiders) on flowers on the shrub *R. rosifolius* (Rosaceae) to evaluate the effect of these factors and the additive or interactive effects on the visitation of pollinators and the reproductive success of the plant. Their study randomly assigned flowers on 112 plants to the following groups: addition of artificial spider, manipulation of flower to produce asymmetry and addition of spider, asymmetry without spider, and control (no treatment). The artificial spiders simulated a thomisid that usually occurs in the flowers of *R. rosifolius*. Both asymmetry and the presence of a predator reduced the number of visits from pollinators (mostly Hymenoptera). The effects were additive, rather than interacting. Interestingly, the risk effect of predation was 62% greater than the effect of flower asymmetry on the avoidance behavior of pollinators. In addition, only the risk of predation significantly decreased the biomass of the fruits (by 33%) and the number of seeds (by 28%). It appears that although the asymmetry caused by herbivory can alter the quality of resources, this effect does not carry the same evolutionary pressure as do interactions between predators and prey.

Positive Effects of the Presence of Spiders on Plants

Several studies have reported the positive effects of spiders on plants (Louda 1982b; Carter and Rypstra 1995; Ruhren and Handel 1999; Whitney 2004; Romero and Vasconcellos-Neto 2011) due to their predation on herbivores. Spiders can affect herbivory even if they do not consume herbivores directly, which can have important implications for biological control programs. Signs of their presence, such as draglines, feces, or chemotactile cues, can alter the foraging behavior of insect herbivores and thereby reduce the damage to plants. Several studies have tested this hypothesis experimentally. Pest insects of soybean leaves reduced their foraging activity in the presence of spiders or spider cues (e.g., silk draglines and feces) under laboratory conditions (Hlivkro and Rypstra 2003). In another study, Rypstra and Buddle (2012) treated entire plants in the field with silkworm or spider silk, and compared the amount of herbivory they experienced. Herbivory was lowest in plants that received spider silk treatments, intermediate in plants treated with silkworm silk, and highest in control plants (which received no treatment). These results suggest that silk might be a mechanism for trait-mediated impacts of spiders and might be used in integrated pest management programs.

Bucher et al. (2015) also performed a field experiment to determine the extent of spiders' effects that are distinct from herbivore consumption, by enclosing *Urtica dioica* plants and removing all arthropods from them, then repeatedly placing *Pisaura mirabilis* spiders on them so that they could deposit cues. Control plants were enclosed in the same way, but did not have spiders. After cue deposition, the enclosures were removed to allow arthropods to colonize the plants and feed on them. The presence of chemotactile spider cues reduced leaf damage by 50% and also led to changes in the arthropod community. Smaller spiders avoided plants with spider cues. In contrast, the aphid-tending ant *Myrmica rubra* showed higher recruitment of workers on cue-bearing plants, possibly because the presence of more ant workers could protect aphids.

Work by Romero and Koricheva (2011) also supported the prediction that the strength and direction of terrestrial trophic cascades are strongly influenced by the relative effects of carnivores on pollinators vs herbivores, predator hunting mode, carnivore habitat domain and taxonomy, and presence and type of plant attractors. The net positive effect of carnivores on plant fitness suggests that carnivore effects on herbivores were stronger than their effects on pollinators.

Multitrophic Interactions and Mutualism

Although spiders are often involved in complex food webs or in direct or indirect interactions with other arthropods and plants (review in Romero and Vasconcellos-Neto 2007a), few studies have shown evidence of mutualism between plants and spiders (Louda 1982b; Ruhren and Handel 1999; Whitney 2004; Romero and Vasconcellos-Neto 2004a; Romero et al. 2008a; Morais-Filho and Romero 2010). Spider–plant mutualistic interactions fall into two categories: defensive/protective, in which spiders increase the fitness of plants by removing phytophagous insects, and digestive, in which spiders contribute to the nutrition of their host plants.

Protective Mutualism and Glandular Trichomes

Protective mutualism occurs when a symbiont reduces the negative effects of another symbiont or of a natural enemy (e.g., a predator) in a common host (Golubski and Abrams 2011). According to Krimmel and Pearse (2012), plants that produce sticky substances are common and often entrap and kill small insects, which can increase predator densities and potentially affect the plants' indirect defenses. The common tarweed (*Madia elegans*, Asteraceae) is an annual flowering plant that produces abundant glandular trichomes. Common predators on tarweed include the assassin bug *Pselliopus spinicollis*, the two stilt bugs *Hoplinus echinatus* and *Jalysus wickhami*, the green lynx spider *Peucetia* sp., and the crab spider *Mecaphesa schlingeri*. The researchers manipulated the abundance of insects' carrion entrapped on tarweed plants under natural field conditions, and found that carrion augmentation increased the abundance of a set of predators, decreased herbivory, and increased plant fitness. The carrion of entrapped insects may function broadly as food provided by the plant for predators.

Mutualism between spiders and plants with glandular trichomes was first investigated by Louda (1982b), who studied the interaction between *P. viridans* and the plant *H. venetus* (Asteraceae). The presence of the spider on the plant inflorescences reduced the number of fertilized ovules, indicating that its occurrence can harm the plant by disrupting plant–pollinator interactions. However, the presence of the spider also reduced the number of dry fruits (achenes) damaged by endophagous insects of the capitula, compared with inflorescences that did not have spiders. According to Louda (1982b), the benefits to the plant outweighed the costs. However, Romero and Koricheva (2011) used a meta-analysis metrics (log response ratio) which concluded that the positive and negative effects were similar in magnitude.

To test the effects of the spiders *P. flava* and *P. rubrolineata* on *T. adenantha,* a plant with glandular trichomes, Romero et al. (2008a) developed field experiments that compared plants with and without spiders. The plants without spiders showed a higher abundance of insects that are harmful to plants, such as leafhoppers, Lepidoptera larvae, and endophagous insects that feed on seeds. Plants with spiders experienced less damage from most of these insects and from leafminers. Moreover, seed damage by Geometridae sp. (Lepidoptera) larvae, a sessile insect, was 16 times higher in plants that lacked the presence of *Peucetia*. The most common species of endophagous insects were *Melanagromyza* spp. (Diptera, Agromyzidae) and *Trupanea* sp. (Diptera, Tephritidae). Spiders decreased the damage caused by *Trupanea* but did not affect *Melanagromyza*. Romero et al. (2008a) attributed these results to the different behavior of these two flies. Whereas *Trupanea* adult females remain on the plant for a long time to lay eggs (~30 min) and travel relatively long distances among the leaves to perform oviposition (18.8 cm on average), *Melanagromyza* females laid their eggs faster (~16 min) and moved much less on the plant (2.9 cm on average). Therefore, it is likely that *Trupanea* is more vulnerable to *Peucetia* spiders than *Melanagromyza*. In contrast with the data obtained by Louda (1982b), the two species of *Peucetia* studied by Romero et al. (2008a) tended ($p = 0.067$) to decrease the fitness of the plant *T. adenantha* via the effect on pollinators (Fig. 7.7).

Another species of spider (*M. argenteus*, Thomisidae) lives on the same plant, and also captures insect herbivores and floral visitors. Romero and Vasconcellos-Neto (2004a) tested whether these crab spiders increase or decrease plant fitness, and obtained results similar to those obtained by Romero et al. (2008a). For example, while the presence of *Misumenops* decreased seed damage caused by Geometridae, *Trupanea* and *Melanagromyza* were not affected. Herbivore vulnerability, as discussed above, may be valid here as well. Therefore, the spiders *P. flava*, *P. rubrolineata*, and *M. argenteus* all affected communities of phytophagous insects similarly. Furthermore, the trophic cascade of these predators affecting plant fitness was similar.

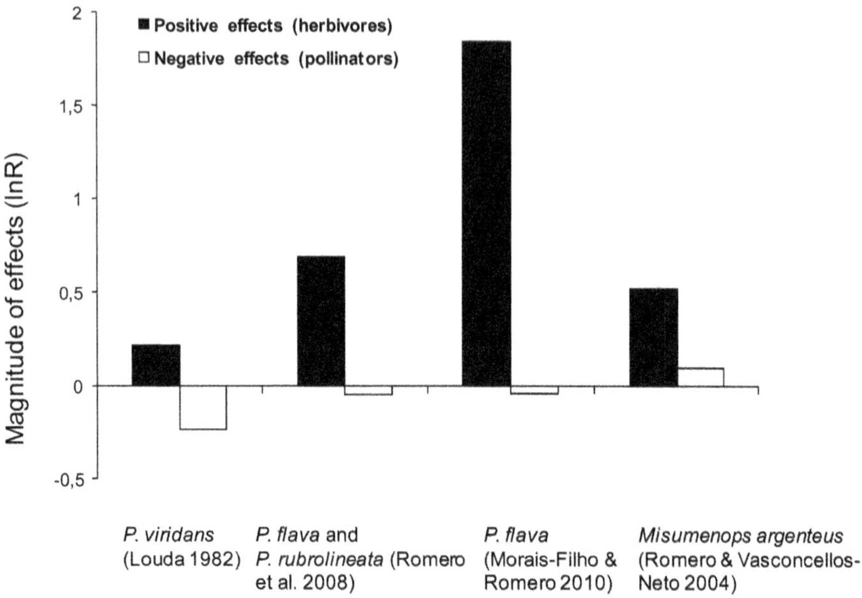

Fig. 7.7 Magnitude of effects of spiders on herbivores and pollinators on plants (from Romero and Vasconcellos-Neto 2011)

Interestingly, in the study that evaluated the effects of *M. argenteus* via pollinators, no decrease in the number of fertilized ovules was found in the capitula of the plants that were not previously damaged by phytophages. Yet fertilized ovules on damaged capitula were more frequent in plants with spiders. How could this happen? When the spiders forage on the capitula buds, they capture the endophagous insects that usually cause damage to the eggs. Therefore, the presence of spiders reduces the floral damage, resulting in capitula with more flowers, which are more attractive to flower visitors. In the absence of spiders, plants had many damaged capitula that were less attractive to pollinators. Although the spiders feed on floral visitors, capitula with flowers are widely spaced and spiders cannot forage on all of them at the same time (Romero and Vasconcellos-Neto 2003). Spiders may remain on one of the capitula, leaving the others free from predators. Thus, the spiders exert a dual beneficial effect on the plants in this system: they reduce herbivores in the capitula attacked by endophages and help attract pollinators.

The spider *P. flava* also occurs on *R. dichotoma* (Melastomataceae), a plant with glandular trichomes, in the northwest of São Paulo state (Brazil). Through field experiments, Morais-Filho and Romero (2010) showed a decline of the abundance of several guilds of insects in plants with spiders. To test whether spiders decrease leaf damage, the authors compared plants with and without spiders during different seasons. They found that the spiders do not affect leaf herbivory rates during the rainy season. In contrast, their presence reduced leaf damage by herbivores by 74% during the post-rain period. In this system, the role of spiders as bodyguards was temporally conditional. During the rainy season, the plants invest in growth by producing a huge amount of vegetative biomass, a phenomenon triggered by bottom-up forces (i.e., by the presence of rain and addition of nutrients in the system). This vegetative productivity supports a great quantity of herbivores, which exceed the capacity of the leaves, affecting the top-down effects of spiders on leaf herbivory. In contrast, during the post-rain period the plants do not grow, investing instead in reproductive tissues (inflorescences). This allows the spiders and herbivores to remain exposed on the leaves for a longer time, possibly strengthening the top-down effects of spiders in this system (Morais-Filho and Romero 2010). The presence of *P. flava* decreased the number of damaged flower buds, increasing plant fitness via herbivory. However, plants with and without spiders produced a similar number of seeds per fruit, indicating that spiders do not negatively affect the plants' fitness via pollinator inhibition.

It is intriguing that the effect of *Peucetia* spiders is strongly negative in some plants (Louda 1982b), less negative in other systems (Romero et al. 2008a), and exclusively positive in others (Morais-Filho and Romero 2010). The system studied by Louda (1982a, b) attracts many small pollinators that are appropriate prey for spiders. In addition, the capitula of *H. venetus* are very close to each other, forming a flat platform where the spiders forage. This type of architecture favors the capture of pollinators by spiders. In contrast, in the system studied by Romero et al. (2008a), the *T. adenantha* capitula are more spaced and spiders could not forage on all of them, although several *Peucetia* spiders can group the capitula and unite them with silk threads to forage. The plants studied by Morais-Filho and Romero (2010) present bigger flowers in great quantities and are spaced apart, attracting larger pollinators (*Bombus* spp., Hesperiidae butterflies) whose capture is difficult for the spiders. Therefore, apparently the architecture of flowers or inflorescences and the type of floral visitors affect the direction and intensity of trophic cascades via pollinators.

All the studies cited in this section involve sticky plants with glandular trichomes that entrap and kill small insects. These trichomes may provide an important pathway in the evolution of relationships between the *Peucetia* genus and these plants, helping to develop protective mutualism between spiders and plants.

Digestive Mutualism

Plants are exposed to selective pressure from insect herbivores, and have developed several defense mechanisms: intrinsic (chemical or mechanical), and extrinsic (including the protection of predators and parasitoids) (e.g., Lawton and McNeil 1979; Price et al. 1980; Crawley 1989; Fritz and Simms 1992; Coley and Barone 1996; Marquis and Whelan 1996; Lucas et al. 2000; Del-Claro and Torezan-Silingardi 2011; Marquis 2011). Mechanical defenses include the hardness of leaves, the presence of thorns, and uncinate and glandular trichomes that can trap insects (Levin 1973; Johnson 1975; Fernandes 1994; Fordyce and Agrawal 2001; Medeiros et al. 2004; Medeiros and Boligon 2007; Cardoso 2008).

A limiting factor for plants is the availability of nutrients, especially in poor soils. Although some plants can trap insects on the surfaces of leaves and stems, they are not necessarily able to absorb the nutrients from their decomposition (Anderson et al. 2012). Carnivory in plants seems to be an efficient way to obtain nutrients, particularly during adverse environmental conditions (Adamec 1997). This strategy arose independently in more than 600 species of plants and at least six different subclasses of angiosperms around the world (Albert et al. 1992; Ellison and Gotelli 2001). According to Givnish et al. (1984), carnivorous plants are defined as plants that are able to absorb nutrients from dead animals next to their surfaces and that possess morphological, physiological, or behavioral features whose primary effect is attraction, capture, or digestion of prey. Givnish et al. note that "plants capable of absorbing nutrients from dead animals, but which lack active means of prey attraction and prey digestion, and possess neither motile traps nor passive structures such as one-way passages whose primary result is immobilization of animals near plant surfaces must be considered saprophytes and not carnivorous plants." Chase et al. (2009) expanded this definition to include the ability to absorb the products of decomposition from organic matter by any tissue. According to this definition, it does not matter whether the decomposition is performed by the individual or whether the plant relies on species-specific mutualism to perform the decomposition (e.g., putrefactive bacteria).

If plants have some way to absorb the feces of the animals living in association with them, they can benefit nutritionally. These additional nutrients may allow plants to store energy reserves and grow more (e.g., Romero et al. 2006). According to Anderson and Midgley (2003), digestive mutualism, in which animals provide nutrients for plants, may represent a step toward the evolution of carnivory in plants. Mutualism involving the provisioning of plants with nutrients by animals (i.e., digestive mutualism) was documented in ant–plant systems, Heteroptera predators (*Pameridea* spp., Miridae) and plants, amphibians and bromeliads (Romero et al. 2010), and carnivorous plants (*Roridula* spp.) (Anderson and Midgley 2002, 2003). Only recently has this kind of mutualism been demonstrated in spider–plant associations.

Romero et al. (2006) were the first to show that spiders contribute to the nutrition of Bromeliaceae. These plants' leaves contain structures that are specialized to absorb water and nutrients (especially nitrogen). The researchers used isotopic techniques (stable isotope ^{15}N) to verify that the spider *P. chapoda* nourishes the bromeliad *B. balansae* with its feces and prey carrion. To enrich the feces of spiders with ^{15}N isotope, the authors first enriched yeast with a salt (ammonium sulfate) that had previously been enriched with nitrogen isotope. Then they mixed the yeast in a culture medium to feed *D. melanogaster*, which became enriched after consuming the yeast. After spiders consumed the flies, the spiders produced enriched feces. The feces and enriched flies were placed in the center of the rosette of *B. balansae* plants, whose leaves were then analyzed isotopically. The results showed that 15% of the total nitrogen of plants was derived from the spiders, and only 3% of the plants' nitrogen came from the flies. In another experiment, the authors kept plants with and without spiders for over a year, and showed that those containing spiders grew 15% more than did plants without spiders. As these bromeliads live in regions where the soil is very poor (e.g., *cerrado* vegetation), an association with spiders can allow the plants to grow faster.

The intensity of the digestive mutualism in this system can vary depending on the density of spiders in different areas. The isotopic nitrogen ^{15}N is 2–4‰ more positive at each trophic level. Therefore, plants inhabited by many spiders can absorb more nitrogen from these predators and present higher values of ^{15}N. In addition, since *P. chapoda* prefers bromeliads living in open areas, presumably such bromeliads derive more nitrogen from spiders than do bromeliads in the interior of the forest. These hypotheses were confirmed by Romero et al. (2008b), using forest fragments with varying numbers of spiders on bromeliads. There was a positive relationship between the density of spiders and the isotopic values of bromeliads. In addition, the nitrogen derived from animals was much higher in the bromeliads in open areas than in bromeliads in the forest interior. However, the bromeliads in the forest presented total nitrogen concentrations ($^{14}N + ^{15}N$) similar to those of plants from open areas. The forest plants may have absorbed nutrients from dry leaves that fell from the trees (plant litter), which have lower isotopic values than those from spider feces (Romero et al. 2008b). These results suggest that bromeliads in the forest derive more of their nutrition from leaf litter, while those in open areas derive more from spider feces.

In another study, Gonçalves et al. (2011) evaluated the role of feces from the spider *P. chapoda* and from *D. melanogaster* flies on the nutrition and growth of the host bromeliads *B. balansae*, *Ananas comosus* (pineapple), and *A. distichantha*, as well as the seasonal variation in the importance of this digestive mutualism. The researchers performed isotopic and physiological analyses using the isotope ^{15}N. Spiders contributed from 0.6% (dry season) to 2.7% (wet season) of the total nitrogen in *B. balansae*, 2.4% (dry) to 4.1% (wet) of the total in *A. comosus*, and 3.8% (dry) to 5% (wet) of the total in *A. distichantha*. Flies did not contribute to the nutrition of these bromeliads. Chlorophyll and carotenoid concentrations did not differ among treatments. Plants that received feces had higher soluble protein concentrations and showed leaf growth only during the wet season. These results indicate that the mutualism between spiders and bromeliads is seasonally restricted. Interspecific variation in nutrient uptake occurred, probably related to the performance of each species and to photosynthetic pathways. Whereas *B. balansae* seems to use nitrogen for growth, *A. distichantha* apparently stores nitrogen to balance out stressful nutritional conditions.

In southeastern Brazil (Serra do Cipó - Minas Gerais), *Alpaida quadrilorata* (Araneidae) inhabits almost exclusively *Paepalanthus bromelioides* (Eriocaulaceae), a plant with rosette-shaped leaves that has similar architecture to bromeliads. The spiders build their webs above the central tank of the plant and, when disturbed, weave a guide wire and dive into the liquid accumulated inside the rosette, using this strategy as defensive behavior against their predators (Figueira and Vasconcellos-Neto 1991; Vasconcellos-Neto et al. unpublished data). Few studies have demonstrated that specific associations of spiders with plants provide the spiders with protection from predators. *P. bromelioides* leaves are associated with multiple partners, such as spiders and termites, and the plant is considered a protocarnivorous species (Figueira et al. 1994). Nishi et al. (2013) used analysis of ^{15}N to show that the isotopic signature of *P. bromelioides* is similar to that of carnivorous plants, and is higher than that of the non-carnivorous plants in the study area. They showed that the presence of spiders on the rosettes of *P. bromelioides* resulted in overall nitrogen contributions of 26.5% (a top-down flux) due to spider feces and prey carrion. Although nitrogen flux was not detected from termites to plants via decomposition of labeled cardboard, the data on ^{15}N in natural nitrogen abundance indicated that 67% of nitrogen from *P. bromelioides* is derived from termites (a bottom-up flux). Bacteria did not affect nutrient cycling or nitrogen uptake from spider feces and prey carrion. The results suggest that the nitrogen used by *P. bromelioides* derives from associated predators and termites, despite differences in the rate of nitrogen cycling, which was higher in nitrogen derived from predators (leaves) than in nitrogen derived from termites (roots). This is the first study that demonstrates partitioning effects from multiple partners in a digestion-based mutualistic system. Although most of its nitrogen is absorbed through the roots (via termites), *P. bromelioides* has all the attributes of a carnivorous plant in the context of digestive mutualism. All these studies reinforce the beneficial role played by spiders in digestive mutualism.

Concluding Remarks

Spider families that actively hunt on vegetation were long thought to be wandering through the plant rather than specifically associated with it. Specific associations and adaptations, and examples of mutualism involving spiders and plants, were not known. Researchers reported that spiders which hunt by ambush (e.g., Thomisidae) chose specific substrates according to the optimal foraging theory (i.e., prey availability).

Recent evidence, however, suggests that the physical structure of the habitat may be a more important factor for spider communities, and that microhabitat selection is mostly influenced by plant architecture per se, not by microclimatic factors or prey availability. Although many spider families live on vegetation, it remains true that very few specific spider–plant associations are known, and it is not known which plant traits attract and facilitate spider populations.

Here, we have reported specific associations between spiders and certain plant species or plants that share similar traits, such as glandular trichomes (Vasconcellos-Neto et al. 2007), spines (Hormiga and Scharff 2014), rosette shape (Romero 2006), tree bark (Messas et al. 2014), and dry structures (Souza et al. 2015). Specific plant traits, such as rosette shape and presence of glandular trichomes, can mediate spider–plant mutualism in which spiders contribute to plant nutrition and growth or protect plants against herbivores. Very little is known about how associations between spiders and plants evolve toward mutualism. Most of the associations reported are occasional, and to achieve a better understanding of their evolution, it is necessary to investigate them while considering spatio-temporal variations.

As the order Araneae presents great diversity on vegetation and a variety of behaviors and lifestyles, we believe that many other specific associations and examples of spider–plant mutualism are waiting to be reported.

References

Abrams P, Menge BA, Mittelbach GG, Spiller D, Yodzis P (1996) The role of indirect effects in food webs. In: Polis GA, Winemiller KO (eds) Food webs: integration of patterns and dynamics. Chapman and Hall, New York

Adamec L (1997) Mineral nutrition of carnivorous plants: a review. Bot Rev 63:273–299

Albert VA, Williams SE, Chase MW (1992) Carnivorous plants: phylogeny and structural evolution. Science 257:1491–1495

Anderson B, Midgley JJ (2002) It takes two to tango but three is a tangle: mutualists and cheaters on the carnivorous plant Roridula. Oecologia 132:369–373

Anderson B, Midgley JJ (2003) Digestive mutualism, an alternate pathway in plant carnivory. Oikos 102:221–224

Anderson B, Kawakita A, Tayasu I (2012) Sticky plant captures prey for symbiotic bug: is this digestive mutualism? Plant Biol 14:888–893

Antiqueira PAP, Romero GQ (2016) Floral asymmetry and predation risk modify pollinator behavior, but only predation risk decreases plant fitness. Oecologia 181:475–485

Arango AM, Rico-Gray V, Parra-Tabla V (2000) Population structure, seasonality, and habitat use by the green lynx spider *Peucetia viridans* (Oxiopidae) inhabiting *Cnidoscolus aconitifolius* (Euphorbiaceae). J Arachnol 28:185–194

Arango AM, Portillo JL, Parra-Tabla V, Salazar LTH, Mávil JEM, Rico-Gray V (2012) Effect of the spider *Peucetia viridans* (Oxyopidae) on floral visitors and seed set of *Cnidoscolus multilobus* (Euphorbiaceae). Acta Bot Mex 100:1–14

Araújo VA, Melo SK, Araújo APA, Gomes MLM, Carneiro MAA (2007) Relationship between invertebrate fauna and bromeliad size. Braz J Biol 67:611–617

Askenmo C, von Bromssen A, Ekman J, Jansson C (1977) Impact of some wintering birds on spider abundance in spruce. Oikos 28:90–94

Balogh J, Loksa J (1956) Untersuchungen über die Zoozönose des Luzernefeldes. Acta Zool Acad Sci Hung 2:17–114

Barth FG, Seyfarth EA (1979) *Cupiennius salei* Keyserling (Araneae) in the highlands of central Guatemala. J Arachnol 7:255–263

Barth FG, Seyfarth EA, Bleckmann H, Schüch W (1988a) Spiders of the genus *Cupiennius* Simon 1891 (Araneae, Ctenidae). I. Range distribution, dwelling plants, and climatic characteristics of the habitats. Oecologia 77:187–193

Barth FG, Bleckmann H, Bohnenberger J, Seyfarth EA (1988b) Spiders of the genus *Cupiennius* Simon 1891 (Araneae, Ctenidae). II. On the vibratory environment of a wandering spider. Oecologia 77:194–201

Baurecht D, Barth FG (1992) Vibratory communication in spiders. I. Representation of male courtship signals by female vibration receptor. J Comp Physiol 171:231–243

Bell G (1985) On the function of flowers. P Roy Soc Lond B Bio 224:223–265

Bell AD (1991) Plant form: an illustrated guide to flowering plant morphology. Oxford University Press, New York

Bell AD, Roberts D, Smith A (1979) Branching patterns: the simulation of the plant architecture. J Theor Biol 81:351–375

Benzing DH (2000) Bromeliaceae: profile of an adaptive radiation. Cambridge University Press, Cambridge

Botto-Mahan C, Ramírez PA, Ossa CG, Medel R, Ojeda-Camacho M, González AV (2011) Floral herbivory affects female reproductive success and pollinator visitation in the perennial herb Alstroemerialigtu (Alstroemeriaceae). Int J Plant Sci 172:1130–1136

Brechbühl R, Kropf C, Bacher S (2010) Impact of flower-dwelling crab spiders on plant–pollinator mutualisms. Basic Appl Ecol 11:76–82

Brescovit AD (1993) *Thaloe* e *Bromelina*, novos gêneros de aranhas neotropicais da família Anyphaenidae (Arachnida, Araneae). Rev Bras Entomol 37:693–703

Brescovit AD, do Oliveira MEES (1994) *Rhoicinus urucu*, uma espécie nova de Rhoicininae para a Região Amazônica (Araneae, Trechaleidae). Biociencias 2:63–69

Brose U (2003) Bottom-up control of carabid beetle communities in early successional wetlands: mediated by vegetation structure or plant diversity? Oecologia 135:407–413

Brown KM (1981) Foraging ecology and niche partitioning in orb-weaving spiders. Oecologia 50:380–385

Brown JH (2014) Why are there so many species in the tropics? J Biogeogr 41:8–22

Bucher R, Menzel F, Entling MH (2015) Risk of spider predation alters food web structure and reduces local herbivory in the field. Oecologia 178:571–577

Cardel YJ, Koptur S (2010) Effects of florivory on the pollination of flowers: an experimental field study with a perennial plant. Int J Plant Sci 171:283–292

Cardoso MZ (2008) Herbivore handling of a plant's trichome: the case of *Heliconius charithonia* (L.) (Lepidoptera: Nymphalidae) and *Passiflora lobata* (Killip) Hutch. (Passifloraceae). Neotrop Entomol 37:247–252

Cardoso P, Pekár S, Jocqué R, Coddington JA (2011) Global patterns of guild composition and functional diversity of spiders. PLoS One 6(6):e21710

Carrel JE, Burgess HK, Shoemaker DM (2000) A test of pollen feeding by a linyphiid spider. J Arachnol 28:243–244

Carter PE, Rypstra AL (1995) Top-down effects in soybean agroecosystems: spider density affects herbivore damage. Oikos 72:433–439

Chase MW, Christenhusz MJM, Sanders D, Fay MF (2009) Murderous plants: Victorian Gothic, Darwin and modern insights into vegetable carnivory. Bot J Linn Soc 161:329–356

Coddington JA, Levi HW (1991) Systematics and evolution of spiders (Araneae). Ann Rev Ecol Syst 22:565–592

Cohen D, Shmida A (1993) The evolution of flower display and reward. Evol Biol 27:197–243

Coley PD, Barone JA (1996) Herbivory and plant defenses in tropical forests. Ann Rev Ecol Syst 27:305–335

Corcuera P, Valverde PL, Jiménez-Salinas E, Vite F, López-Ortega G, Pérez-Hernández MA (2010) Distribution of Peucetia viridans (Araneae: Oxyopidae) on Croton ciliatoglandulifer. Environ Entomol 39:320–327

Cotgrave P, Hill MJ, Middleton JAG (1993) The relationship between body size and population size in tank bromeliad fauna. Biol J Linn Soc 49:367–380

Cramer MJ, Willig MR (2002) Habitat heterogeneity, habitat associations, and rodent species diversity in a sand–shinnery-oak landscape. J Mammal 83:743–753

Crawley MJ (1989) Insect herbivores and plant population dynamics. Annu Rev Entomol 34:531–564

Dafni A, Lehrer M, Kevan PG (1997) Spatial flower parameters and insect spatial vision. Biol Rev 72:239–282

Dalingwater JE (1987) Chelicerate cuticle structure. In: Nentwig W (ed) Ecophysiology of spiders. Springer, Berlin

de Piccoli GCO (2011) História natural da aranha *Corinna* sp. nov. (Corinnidae): interações com bromélias e comportamento de submersão de fitotelmatas. Master thesis, Universidade Estadual Paulista, Instituto de Biociências, Letras e Ciências Exatas. São José do Rio Preto – SP

Del-Claro K, Torezan-Silingardi HM (2011) Ecologia das interações plantas–animais: uma abordagem ecológico-evolutiva. Technical Books editora, Rio de Janeiro

Dennis P, Young MR, Gordon IJ (1998) Distribution and abundance of small insects and arachnids in relation to structural heterogeneity of grazed, indigenous grasslands. Ecol Entomol 23:253–264

Denno RF, Finke DL, Langellotto GA (2005) Direct and indirect effects of vegetation structure and habitat complexity on predator–prey and predator–predator interactions. In: Barbosa P, Castellanos I (eds) Ecology of predator-prey interactions. Oxford University Press, Oxford

Dias SC, Brescovit AD (2003) Notes on the behavior of *Pachistopelma rufonigrum* Pocock (Araneae, Theraphosidae, Aviculariinae). Rev Bras Zool 20:13–17

Dias SC, Brescovit AD (2004) Microhabitat selection and co-occurrence of *Pachistopelma rufonigrum* Pocock (Araneae, Theraphosidae) and *Nothroctenus fuxico* sp. nov. (Araneae, Ctenidae) in tank bromeliads from Serra de Itabaiana, Sergipe, Brazil. Rev Bras Zool 21:789–796

Dias SC, Brescovit AD, Santos LT, Couto ECG (2000) Aranhas em bromélias de duas restingas do estado de Sergipe, Brasil. Biol Geral Exp 1:22–24

Dias SC, Carvalho LS, Bonaldo AB, Brescovit AD (2010) Refining the establishment of guilds in Neotropical spiders (Arachnida: Araneae). J Nat Hist 44:219–239

Diniz S (2011) Influência da complexidade arquitetural de ramos vegetativos na diversidade e abundância de aranhas e outros artrópodes. Master thesis, Universidade Estadual de Campinas

Draney ML, Hegnet JA, Johnson AL, Porter BC, Justmann CK, Forsythe PS (2014) Microhabitat distribution of *Drapetisca alteranda*, a tree trunk specialist sheet web weaver (Araneae: Linyphiidae). J Arachnol 42:195–198

Duffey SS (1986) Plant glandular trichomes: their partial role in defence against insects. In: Juniper B, Southwood R (eds) Insects and the plant surface. Edward Arnold, London

Dukas R (2001) Effects of perceived danger on flower choice by bees. Ecol Lett 4:327–333

Dukas R, Morse DH (2003) Crab spiders affect patch visitation by bees. Oikos 101:157–163

Dukas R, Morse DH (2005) Crab spiders show mixed effects on flower-visiting bees and no effect on plant fitness components. Ecoscience 12:244–247

Eggs B, Sanders D (2013) Herbivory in spiders: the importance of pollen for orb-weavers. PLoS One 8(11):e82637

Ellis AG, Midgley JJ (1996) A new plant–animal mutualism involving a plant with sticky leaves and a resident hemipteran. Oecologia 106:478–481

Ellison AM, Gotelli NJ (2001) Evolutionary ecology of the carnivorous plants. Trends Ecol Evol 16:623–629

Evans TA (1997) Distribution of social crab spiders in eucalypt forests. Aust J Ecol 22:107–111

Fernandes GW (1994) Plant mechanical defenses against insect herbivory. Rev Bras Entomol 32:421–433

Figueira JEC, Vasconcellos-Neto J (1991) *Paepalanthus*, cupins e aranhas. Ciência Hoje 13:20–26

Figueira JEC, Vasconcellos-Neto J, Jolivet P (1994) Une nouvelle plante protocarnivore *Paepalanthus bromelioides* Silv. (Eriocaulaceae) du Brésil. Rev Écol 49:3–9

Foelix RF (2011) Biology of the spiders, 3rd edn. Oxford University Press, Inc, New York

Fordyce JA, Agrawal AG (2001) The role of plant trichomes and caterpillar group size on growth and defence of the pipevine swallowtail *Battus philenor*. J Anim Ecol 70:997–1005

Frank JH, Sreenivasan S, Benshoff PJ, Deyrup MA, Edwards GB, Halbert SE, Hamon AB, Lowman MD, Mockford EL, Scheffrahn RH, Steck GJ, Thomas MC, Walker TJ, Welbourn WC (2004) Invertebrate animals extracted from native *Tillandsia* (Bromeliales: Bromeliaceae) in Sarasota County, Florida. Fla Entomol 87:176–185

Fritz RS, Simms EL (1992) Plant resistance to herbivores and pathogens: ecology, evolution, and genetics. The University of Chicago Press, Chicago

Gastreich KR (1999) Trait-mediated indirect effects of a theridiid spider on an ant-plant mutualism. Ecology 80:1066–1070

Givnish TJ, Burkhardt EL, Happel RE, Weintraub JD (1984) Carnivory in the bromeliad *Brocchinia reducta*, with a cost/benefit model for the general restriction of carnivorous plants to sunny, moist, nutrient-poor habitats. Am Nat 124:479–497

Golubski AJ, Abrams PA (2011) Modifying modifiers: what happens when interspecific interactions interact? J Anim Ecol 80:1097–1108

Gonçalves AZ, Mercier H, Mazzafera P, Romero GQ (2011) Spider-fed bromeliads: seasonal and interspecific variation in plant performance. Ann Bot London 107:1047–1055

Gonçalves-Souza T, Omena PM, Souza JC, Romero GQ (2008) Trait-mediated effects on flowers: artificial spiders deceive pollinators and decrease plant fitness. Ecology 89:2407–2413

Gonçalves-Souza T, Brescovit AD, Rossa-Feres DC, Romero GQ (2010) Bromeliads as biodiversity amplifiers and habitat segregation of spider communities in a Neotropical rainforest. J Arachnol 38:270–279

Gonçalves-Souza T, Almeida-Neto M, Romero GQ (2011) Bromeliad architectural complexity and vertical distribution predict spider abundance and richness. Austral Ecol 36:476–484

Greco CF, Kevan PG (1994) Contrasting patch choosing by anthophilous ambush predators: vegetation and floral cues for decisions by a crab spider (*Misumena vatia*) and males and females of an ambush bug (*Phymata americana*). Can J Zool 72:1583–1588

Greenstone MH (1984) Determinants of web spider species diversity: vegetation structural diversity vs. prey availability. Oecologia 62:299–304

Gunnarsson B (1983) Winter mortality of spruce-living spiders: effect of spider interactions and bird predation. Oikos 40:226–233

Gunnarsson B (1988) Spruce-living spiders and forest decline; the importance of needle-loss. Biol Conserv 43:309–319

Gunnarsson B (1990) Vegetation structure and the abundance and size distribution of spruce-living spiders. J Anim Ecol 59:743–752

Gunnarsson B (1992) Fractal dimension of plant and body size distribution in spiders. Funct Ecol 6:636–641

Gunnarsson B (1996) Bird predation and vegetation structure affecting spruce-living arthropods in a temperate forest. J Anim Ecol 65:389–397

Halaj J, Ross DW, Moldenke AR (1998) Habitat structure and prey availability as predictors of the abundance and community organization of spiders in western Oregon forest canopies. J Arachnol 26:203–220

Halaj J, Ross DW, Moldenke AR (2000) Importance of habitat structure of the arthropod food-web in Douglas-fir canopies. Oikos 90:139–152

Harmer AMT (2009) Elongated orb-webs of Australian ladder-web spiders (Araneidae: *Telaprocera*) and the significance of orb-web elongation. J Ethol 27:453–460

Harmer AMT, Herberstein ME (2009) Taking it to extremes: what drives extreme web elongation in Australian ladder web spiders (Araneidae: *Telaprocera maudae*)? Anim Behav 78:499–504

Harmer AMT, Herberstein ME (2010) Functional diversity of ladder-webs: moth specialization or optimal area use? J Arachnol 38:119–122

Hatley CL, MacMahon JA (1980) Spider community organization: seasonal variation and the role of vegetation architecture. Environ Entomol 9:632–639

Hawkins CP, MacMahon JA (1989) Guilds: the multiple meanings of a concept. Annu Rev Entomol 34:423–451

Heiling AM, Herberstein ME (2004) Predator–prey coevolution: Australian native bees avoid their spider predators. P Roy Soc Lond B Bio 271:S196–S198

Heiling AM, Herberstein ME, Chittka L (2003) Crab spiders manipulate flower signals. Nature 421:334

Heiling AM, Cheng K, Herberstein ME (2004) Exploitation of floral signals by crab spiders (*Thomisus spectabilis*, Thomisidae). Behav Ecol 15:321–326

Heiling AM, Chittka L, Cheng K, Herberstein ME (2005a) Colouration in crab spiders: substrate choice and prey attraction. J Exp Biol 208:1785–1792

Heiling AM, Cheng K, Chittka L, Goeth A, Herberstein ME (2005b) The role of UV in crab spider signals: effects on perception by prey and predators. J Exp Biol 208:3925–3931

Heiling AM, Cheng K, Herberstein ME (2006) Picking the right spot: crab spiders position themselves on flowers to maximize prey attraction. Behaviour 143:957–968

Herberstein ME (1998) Implications of microhabitat selection on prey capture for the web spider Neriene radiata (Walckenaer) (Araneae: Linyphiidae). In: Selden PA (ed) P European Coll Arachnol, vol 17. Burnham Beeches, Bucks, UK, pp 197–202

Herberstein ME, Heiling AM, Cheng K (2009) Evidence for UV-based sensory exploitation in Australian but not European crab spiders. Evol Ecol 23:621–634

Hesselberg T, Triana E (2010) The web of the acacia orb-spider *Eustala illicita* (Araneae: Araneidae) with notes on its natural history. J Arachnol 38:21–26
Hlivkro JT, Rypstra AL (2003) Spiders reduce herbivory: nonlethal effects of spiders on the consumption of soybean leaves by beetle pests. Ann Entomol Soc Am 96:914–919
Höfer H, Brescovit AD (2001) Species and guild structure of a Neotropical spider assemblage (Araneae) from Reserva Ducke, Amazonas, Brazil. Andrias 15:99–119
Hormiga G, Scharff N (2014) The strange case of *Laetesia raveni* n. sp., a green linyphiid spider from Eastern Australia with a preference for thorny plants (Araneae, Linyphiidae). Zootaxa 3811:83–94
Horvath R, Lengyel S, Szinetar C, Jakab L (2005) The effect of prey availability on spider assemblages on European black pine (*Pinus nigra*) bark: spatial patterns and guild structure. Can J Zool 83:324–335
Ings TC, Chittka L (2008) Speed-accuracy tradeoffs and false alarms in bee responses to cryptic predators. Curr Biol 18:1520–1524
Ings TC, Chittka L (2009) Predator crypsis enhances behaviourally mediated indirect effects on plants by altering bumblebee foraging preferences. P Roy Soc Lond B Bio 276:2031–2036
Jackson RR, Pollard SD, Nelson XJ, Edwards GB, Barrion AT (2001) Jumping spiders (Araneae: Salticidae) that feed on nectar. J Zool 255:25–29
Janetos AC (1986) Web-site selection: are we asking the right questions? In: Shear WA (ed) Spiders: webs, behavior, and evolution. Stanford University Press, Stanford, pp 9–48
Jiménez-Salinas E, Corcuera P (2008) Inflorescences and plant selection by the green lynx spider *Peucetia viridans* (Hentz) in a dry forest of western Mexico. Rev Iber Aracnol 15:63–66
Johnson HB (1975) Plant pubescence: an ecological perspective. Bot Rev 41:233–258
Krell FT, Krämer F (1998) Chemical attraction of crab spiders (Araneae, Thomisidae) to a flower fragrance component. J Arachnol 26:117–119
Krimmel BA, Pearse IS (2012) Sticky plant traps insects to enhance indirect defence. Ecol Lett 2012:1–6. https://doi.org/10.1111/ele.12032
Küppers M (1989) Ecological significance of above-ground architectural pattern in woody plants: a question of cost–benefit relationships. Trends Ecol Evol 4:375–379
Langellotto GA, Denno RF (2004) Responses of invertebrate natural enemies to complex-structured habitats: a meta-analytical synthesis. Oecologia 139:1–10
Lawton JH (1983) Plant architecture and the diversity of phytophagous insects. Annu Rev Entomol 28:23–39
Lawton JH, McNeil S (1979) Between the devil and the deep blue sea: on the problem of being a herbivore. In: Anderson RM, Turner BD, Taylor LR (eds) Population dynamics. Blackwell, Oxford, pp 223–244
Levin DA (1973) The role of thichomes in plant defense. Q Rev Biol 48:3–15
Llandres AL, Rodríguez-Gironés MA (2011) Spider movement, UV reflectance and size, but not spider crypsis, affect the response of honeybees to Australian crab spiders. PLoS One 6(2):e17136
Louda SM (1982a) Limitation of the recruitment of the shrub *Haplopappus squarrosus* (Asteraceae) by flower- and seed-feeding insects. J Ecol 70:43–53
Louda SM (1982b) Inflorescence spiders: a cost/benefit analysis for the host plant, *Haplopappus venetus* Blake (Asteraceae). Oecologia 55:185–191
Louda SM (1983) Seed predation and seedling mortality in the recruitment of a shrub, Haplopappusvenetus (Asteraceae), along a climatic gradient. Ecology 64:511–521
Lubin Y, Kotzman M, Ellner S (1991) Ontogenetic and seasonal changes in webs and websites of a desert widow spider. J Arachnol 19:40–48
Lucas PW, Turner IM, Dominy NJ, Yamashita N (2000) Mechanical defences to herbivory. Ann Bot 86:913–920
Macarthur RH, Macarthur JW (1961) On bird species diversity. Ecology 42:594–598
Maddison WP (1996) *Pelegrina* Franganillo and other jumping spiders formerly placed in the genus *Metaphidippus* (Araneae: Salticidae). Bull Mus Comp Zool 154:215–368
Marquis RJ (1984) Leaf herbivores decrease fitness of a tropical plant. Science 226:537–539

Marquis RJ (2011) Uma abordagem geral das defesas das plantas contra a ação dos herbívoros. In: Del-Claro K, Torezan-Silingardi HM (eds) Ecologia das interações plantas–animais: uma abordagem ecológico-evolutiva. Technical Books editora, Rio de Janeiro

Marquis RJ, Whelan C (1996) Plant morphology and recruitment of the third trophic level: subtle and little recognized defenses? Oikos 75:330–334

Marshall S, Rypstra A (1999) Spider competition in structurally simple ecosystems. J Arachnol 27:343–351

McCoy ED, Bell SS (1991) Habitat structure: the evolution and diversifications of a complex topic. In: Bell SS, McCoy ED, Mushinsky HR (eds) Habitat structure: the physical arrangement of objects in space. Chapman and Hall, New York

Medeiros L, Boligon DS (2007) Adaptations of two specialist herbivores to movement on the hairy leaf surface of their host, *Solanum guaraniticum* Hassl (Solanaceae). Rev Bras Entomol 51:210–216

Medeiros L, Boligon DS, Moreira GRP (2004) Morphological and behavioral adaptations to movement on different leaf surfaces: studies with cassidine larvae. In: Jolivet PH, Santiago-Blay JA, Schmidt M (eds) New contributions to the biology of *Chrysomelidae*. SPB Academic Publishers, Kugler Publications, Amsterdam

Meehan CJ, Olson EJ, Reudink MW, Kyser TK, Curry RL (2009) Herbivory in a spider through exploitation of an ant–plant mutualism. Curr Biol 19:892–893

Messas YF, Souza HS, Gonzaga MO, Vasconcellos-Neto J (2014) Spatial distribution and substrate selection by the orb-weaver spider *Eustala perfida* Mello-Leitão, 1947 (Araneae: Araneidae). J Nat Hist 48:2645–2660

Mestre LAM, Aranha JMR, Esper MLP (2001) Macroinvertebrate fauna associated to the bromeliad *Vriesea inflata* of the Atlantic Forest (Paraná State, Southern Brazil). Braz Arch Biol Technol 44:89–94

Michel AK, Winter S (2009) Tree microhabitat structures as indicators of biodiversity in Douglas-fir forests of different stand ages and management histories in the Pacific Northwest, U.S.A. For Ecol Manag 257:1453–1464

Mohsin M, Sulehria AQK, Yousuf I, Ejaz M, Yousuf MJ, Hussain A (2010) Comparison of spider guilds found in various oilseed crops of Pakistan. Biologia (Pakistan) 56:69–76

Møller AP, Sorci G (1998) Insect preference for symmetrical artificial flowers. Oecologia 114:37–42

Morais-Filho JC, Romero GQ (2008) Microhabitat use by *Peucetia flava* (Oxyopidae) on the glandular plant *Rhyncanthera dichotoma* (Melastomataceae). J Arachnol 36:374–378

Morais-Filho JC, Romero GQ (2009) Natural history of *Peucetia flava* (Araneae, Oxyopidae): seasonal density fluctuation, phenology and sex ratio on the glandular plant *Rhyncanthera dichotoma* (Melastomataceae). J Nat Hist 43:701–711

Morais-Filho JC, Romero GQ (2010) Plant glandular trichomes mediate protective mutualism in a spider–plant system. Ecol Entomol 35:485–494

Morse DH (1981) Prey capture by the crab spider *Misumena vatia* (Clerck) (Thomisidae) on three common native flowers. Am Midl Nat 105:358–367

Morse DH (1990) Leaf choices of nest-building crab spiders (*Misumena vatia*). Behav Ecol Sociobiol 27:265–267

Morse DH (1999) Choice of hunting site as a consequence of experience in late-instar crab spiders. Oecologia 120:252–257

Morse RD, Lawton JH, Dodson MM, Williamson MH (1985) Fractal dimension of vegetation and the distribution of arthropod body lengths. Nature 314:731–733

Nahas L, Gonzaga MO, Del-Claro K (2012) Emergent impacts of ant and spiders interactions: herbivory reduction in a tropical savanna tree. Biotropica 44:498–505

Nentwig W, Cutler B, Heimer S (1993) Spiders of Panama: biogeography, investigation, phenology, check list, key and bibliography of tropical spiders fauna. Sandihill Crane Press, Gainesville

Nicolai V (1986) The bark of trees: thermal properties, microclimate and fauna. Oecologia 69:148–160

Nicolai V (1989) Thermal properties and fauna on the bark of trees in two different African ecosystems. Oecologia 80:421–430

Nishi AH, Vasconcellos-Neto J, Romero GQ (2013) The role of multiple partners in a digestive mutualism with a protocarnivorous plant. Ann Bot 111:143–150

Nyffeler M (1982) Field studies on the ecological role of spiders as insect predators in agroecosystems (abandoned grassland, meadows, and cereal fields). PhD thesis, Swiss Federal Institute of Technology, Zurich

Nyffeler M (2016) Phytophagy in jumping spiders: the vegetarian side of a group of insectivorous predators. Peckhamia 137.1:1–17

Nyffeler M, Olson EJ, Symondson WO (2016) Plant-eating by spiders. J Arachnol 44:15–27

Omena PM, Romero GQ (2008) Fine-scale microhabitat selection in a bromeliad-dwelling jumping spider (Salticidae). Biol J Linn Soc 94:653–662

Omena PM, Romero GQ (2010) Using visual cues of microhabitat traits to find home: the case study of a bromeliad-living jumping spider (Salticidae). Behav Ecol 21:690–695

Pianka ER (1966) Convexity, desert lizards, and spatial heterogeneity. Ecology 47:1055–1059

Pollard SD, Beck MW, Dodson GN (1995) Why do male crab spiders drink nectar? Anim Behav 49:1443–1448

Post WM, Riechert SE (1977) Initial investigation into the structure of spider communities. J Anim Ecol 46:729–749

Preisser EL, Orrock JL, Schmitz OJ (2007) Predator hunting mode and habitat domain alter nonconsumptive effects in predator–prey interactions. Ecology 88:2744–2751

Price PW, Bouton CE, Gross P, Mcpheron BA, Thompson JN, Weis AE (1980) Interactions among three trophic levels: influence of plants on interactions between insect herbivores and natural enemies. Annu Rev Ecol Syst 11:41–65

Prinzing AJ (2001) Use of shifting microclimatic mosaics by arthropods on exposed tree trunks. Ann Entomol Soc Am 94:210–218

Prinzing AJ (2005) Corticolous arthropods under climatic fluctuations: Compensation is more important than migration. Ecography 28:17–28

Purcell J, Avilés L (2007) Smaller colonies and more solitary living mark higher altitude populations of a social spider. J Anim Ecol 76:590–597

Rembold K, Fischer E, Striffler BF, Barthlott W (2012) Crab spider association with the Malagasy pitcher plant *Nepenthes madagascariensis*. Afr J Ecol 51:188–191

Ribas ACA, Raizer J (2013) Spiders do not affect fruit set in *Byrsonima intermedia* (Malpighiaceae). JNR EEB 10:1–5

Ribas ACA, Brescovit AD, Raizer J (2011) The composition of spider assemblages varies along reproductive and architectural gradients in shrubs Byrsonima intermedia (Malpighiaceae). J Arachnol 39:537–540

Riechert SE, Gillespie RG (1986) Habitat choice and utilization in web-building spiders. In: Shear W (ed) Spiders: webs, behavior and evolution. Stanford University Press, Stanford

Riechert SE, Lockley T (1984) Spiders as biological control agents. Annu Rev Entomol 29:299–320

Riechert SE, Tracy CR (1975) Thermal balance and prey availability: bases for a model relating web-site characteristics to spider reproductive success. Ecology 56:265–284

Robertson IC, Maguire DK (2005) Crab spiders deter insect visitations to slickspot peppergrass flowers. Oikos 109:577–582

Rodrigues BV, Bonaldo AB (2014) Taxonomic revision of the species group rubripes of Corinna Koch, 1842 (Araneae; Corinnidae). Zootaxa 17(3815):451–493

Romero GQ (2001) Estudo experimental da associação de *Runcinioides argenteus* (Araneae, Thomisidae) em *Trichogoniopsis adenantha* (DC) (Asteraceae). Tese de Mestrado. Universidade Estadual de Campinas, Campinas

Romero GQ (2006) Geographic range, habitats and host plants of bromeliad-living jumping spider (Salticidae). Biotropica 38:522–530

Romero GQ (2007) Papel das aranhas como agentes de controle biológico em agroecossistemas. In: Gonzaga MO, Santos AJ, Japyassú HF (eds) Ecologia e comportamento de aranhas. Interciência, Rio de Janeiro

Romero GQ, Koricheva J (2011) Contrasting cascade effects of carnivores on plant fitness: a meta-analysis. J Anim Ecol 80:696–704

Romero GQ, Vasconcellos-Neto J (2003) Natural history of *Misumenops argenteus* (Thomisidae): seasonality and diet on *Trichogoniopsis adenantha* (Asteraceae). J Arachnol 31:297–304
Romero GQ, Vasconcellos-Neto J (2004a) Beneficial effects of flower-dwelling predators on their host plant. Ecology 85:446–457
Romero GQ, Vasconcellos-Neto J (2004b) Foraging by the flower-dwelling spider, *Misumenops argenteus* (Thomisidae), at high prey density sites. J Nat Hist 38:1287–1296
Romero GQ, Vasconcellos-Neto J (2004c) Spatial distribution patterns of jumping spiders associated with terrestrial bromeliads. Biotropica 36:596–601
Romero GQ, Vasconcellos-Neto J (2005a) The effects of plant structure on the spatial and microspatial distribution of a bromeliad-living jumping spider (Salticidae). J Anim Ecol 74:12–21
Romero GQ, Vasconcellos-Neto J (2005b) Population dynamics, age structure and sex ratio of the bromeliad-dwelling jumping spider, *Psecas chapoda* (Salticidae). J Nat Hist 39:153–163
Romero GQ, Vasconcellos-Neto J (2005c) Spatial distribution and microhabitat preference of *Psecas chapoda* (Peckham and Peckham) (Araneae, Salticidae). J Arachnol 33:124–134
Romero GQ, Vasconcellos-Neto J (2005d) Flowering phenology, seed set and arthropod guilds in Trichogoniopsis adenantha (Asteraceae) in south-east Brazil. Rev Bras Bot 28:171–178
Romero GQ, Vasconcellos-Neto J (2007a) Aranhas sobre plantas: dos comportamentos de forrageamento às associações específicas. In: Gonzaga MO, Santos AJ, Japyassú HF (eds) Ecologia e comportamento de aranhas, 1st edn. Interciência, Rio de Janeiro
Romero GQ, Vasconcellos-Neto J (2007b) Interações bióticas entre plantas, herbívoros e aranhas. In: Gonzaga MO, Santos AJ, Japyassú HF (eds) Ecologia e comportamento de aranhas, 1st edn. Interciência, Rio de Janeiro
Romero GQ, Vasconcellos-Neto J (2011) Interações entre aranhas e plantas: associações específicas e mutualismos. In: Del-Claro K, Torezan-Silingardi HM (eds) Ecologia de Interações. Technical Books, Rio de Janeiro
Romero GQ, Mazzafera P, Vasconcellos-Neto J, Trivelin PCO (2006) Bromeliad-living spiders improve host plant nutrition and growth. Ecology 87:803–808
Romero GQ, Santos AJ, Wienskoski EH, Vasconcellos-Neto J (2007) Association of two *Coryphasia* species (Araneae, Salticidae) with tank-bromeliads in southeastern Brazil: habitats and patterns of host plant use. J Arachnol 35:181–192
Romero GQ, Souza JC, Vasconcellos-Neto J (2008a) Anti-herbivore protection by mutualistic spiders and the role of plant glandular trichomes. Ecology 89:3105–3115
Romero GQ, Vasconcellos-Neto J, Trivelin PO (2008b) Spatial variation in the strength of mutualism between a jumping spider and a terrestrial bromeliad: evidence from the stable isotope 15N. Acta Oecol 33:380–386
Romero GQ, Nomura F, Gonçalves AZ, Dias NYN, Mercier H, Conforto EC, Rossa-Feres DC (2010) Nitrogen fluxes from treefrogs to tank epiphytic bromeliads: an isotopic and physiological approach. Oecologia 162:941–949
Romero GQ, Antiqueira PA, Koricheva J (2011) A meta-analysis of predation risk effects on pollinator behaviour. PLoS One 6(6):e20689
Root RB (1967) The exploitation pattern of the blue-grey gnatcatcher. Ecol Monogr 37:317–350
Root RB (1973) Organization of a plant–arthropod association in simple and diverse habitats: the fauna of collards (*Brassica oleracea*). Ecol Monogr 43:95–124
Rossa-Feres DC, Romero GQ, Gonçalves-de-Freitas E, Feres RJF (2000) Reproductive behavior and seasonal occurrence of *Psecas viridipurpureus* (Salticidae, Araneae). Rev Bras Biol 60:221–228
Ruhren S, Handel SN (1999) Jumping spiders (Salticidae) enhance the seed production of a plant with extrafloral nectaries. Oecologia 119:227–230
Rypstra AL, Buddle CM (2012) Spider silk reduces insect herbivory. Biol Lett 9(1). https://doi.org/10.1098/rsbl.2012.0948
Sanders D (2007) Ants and spiders in grassland food webs: top-down control and Intraguild interactions. Dissertação de Doutorado em Ciência Naturais e Matemáticas. Universidade Georg August, Göttingen

Sanders D, Nickel H, Ghützner T, Platner C (2008) Habitat structure mediates top-down effects of spiders and ants on herbivores. Basic Appl Ecol 9:152–160
Santos AJ, Romero GQ (2004) A new bromeliad-dwelling jumping spider (Araneae: Salticidae) from Brazil. J Arachnol 32:188–190
Santos RL, Almeida MG, Nunes JV (2002) Notes on the association of *Pachistopelma rufonigrum* Pocock 1901 (Theraphosidae) with phytotelm bromeliads in eastern Rio Grande do Norte State, NE-Brazil. J Bromeliad Soc 52:122–124
Scheuring I (1991) The fractal nature of vegetation and the species–area relation. Theor Popul Biol 39:170–177
Schimper AFW (1903) Plant-geography upon a physiological basis. Clarendon Press, Oxford
Schmalhofer VR (2001) Tritrophic interactions in a pollination system: impacts of species composition and size of flower patches on the hunting success of a flower-dwelling spider. Oecologia 129:292–303
Schoonhoven LM, Jermy T, van Loon JJA (1998) Insect-plant biology. Chapman and Hall, London
Schulze ED (1989) Air pollution and forest decline in a spruce *(Piceaabies)* forest. Science 244:776–783
Shear WA (1986) Spiders: webs, behavior, and evolution. Stanford University Press, Stanford
Simberloff D, Dayan T (1991) The guild concept and the structure of ecological communities. Annu Rev Ecol Syst 22:115–143
Smith RB, Mommsen TP (1984) Pollen feeding in an orb-weaving spider. Science 226(4680):1330–1332
Souza ALT (1999) Influência da arquitetura de ramos vegetativos e inflorescências na distribuição de aranhas em plantas. PhD thesis, Universidade Estadual de Campinas, Campinas
Souza ALT (2007) Influência da estrutura do habitat na abundância e diversidade de aranhas. In: Gonzaga MO, Santos AJ, Japyassú HF (eds) Ecologia e comportamento de aranhas. Editora Interciência, Rio de Janeiro
Souza ALT, Martins RP (2005) Foliage density of branches and distribution of plant-dwelling spiders. Biotropica 37:416–420
Souza ALT, Módena EDS (2004) Distribution of spiders on different types of inflorescences in the Brazilian pantanal. J Arachnol 32:345–348
Souza HS, Messas YF, Gonzaga MO, Vasconcellos-Neto J (2015) Substrate selection and spatial segregation by two congeneric species of *Eustala* (Araneae: Araneidae) in southeastern Brazil. J Arachnol 43:59–66
Stefani V, Pires TL, Torezan-Silingardi HM, Del-Claro K (2015) Beneficial effects of ants and spiders on the reproductive value of *Eriotheca gracilipes* (Malvaceae) in a Tropical Savanna. PLoS One 10(7):e0131843
Suetsugu K, Hayamizu M, Koike N (2014) *Clubiona* spider (Araneae: Clubionidae) visiting flowers of nectariferous orchid *Neottianthe cucullata*. Entomol Sci 17:262–264
Sugihara G, May RM (1990) Applications of fractals in ecology. Trends Ecol Evol 5:79–86
Sugiura S, Yamazaki K (2006) Consequences of scavenging behaviour in a plant bug associated with a glandular plant. Biol J Linn Soc 88:593–602
Sundberg I, Gunnarsson B (1994) Spider abundance in relation to needle density in spruce. J Arachnol 22:190–194
Suter RB (1999) An aerial lottery: the physics of ballooning in a chaotic atmosphere. J Arachnol 27:281–293
Suttle KB (2003) Pollinators as mediators of top-down effects on plants. Ecol Lett 6:688–694
Szinetár C, Horváth R (2005) A review of spiders on tree trunks in Europe (Araneae). Acta Zool Bulgar 1:221–257
Taylor RM (2004) Plant nectar contributes to the survival, activity, growth, and fecundity of the nectar-feeding wandering spider *Cheiracanthiuminclusum*(Hentz) (Araneae: Miturgidae). PhD dissertation, Ohio State University, Columbus
Taylor RM, Bradley RA (2009) Plant nectar increases survival, molting, and foraging in two foliage wandering spiders. J Arachnol 37:232–237

Taylor RM, Foster WA (1996) Spider nectarivory. Am Entomol 42:82–86

Tedore C, Johnsen S (2015) Immunological dependence of plant-dwelling animals on the medicinal properties of their plant substrates: a preliminary test of a novel evolutionary hypothesis. Arthropod Plant Interact 9:437–446

Tedore C, Johnsen S (2016) Disentangling the visual cues used by a jumping spider to locate its microhabitat. J Exp Biol. https://doi.org/10.1242/jeb.129122

Tews J, Brose U, Grimm V, Tielbörger K, Wichmann MC, Schwager M, Jeltsch F (2004) Animal species diversity driven by habitat heterogeneity/diversity: the importance of keystone structures. J Biogeogr 31:79–92

Théry M, Insausti TC, Defrize J, Casas J (2011) The multiple disguises of spiders. In: Stevens M, Merilaita S (eds) Animal camouflage: mechanisms and function. Cambridge University Press, Cambridge

Turnbull AL (1973) Ecology of the true spiders (Araneomorphae). Annu Rev Entomol 18:305–348

Uetz GW (1977) Coexistence in a guild of wandering spiders. J Anim Ecol 46:531–541

Uetz GW (1991) Habitat structure and spider foraging. In: Bell SS, ED MC, Mushinsky HR (eds) Habitat structure: the physical arrangement of objects in space. Chapman and Hall, New York

Uetz GW, Halaj J, Cady AB (1999) Guild structure of spiders in major crops. J Arachnol 27:270–280

Vasconcellos-Neto J, Romero GQ, Santos AJ, Dippenaar-Schoeman AS (2007) Associations of spiders of the genus *Peucetia* (Oxyopidae) with plants bearing glandular hairs. Biotropica 39:221–226

Villanueva-Bonilla GA (2015) Dinâmica populacional e seleção de habitat por *Selenops cocheleti* Simon, 1880 (Araneae: Sselenopidae) na Serra do Japi, São Paulo, Brasil. Master thesis, Universidade Estadual de Campinas

Vogelei A, Greissl R (1989) Survival strategies of the crab spider *Thomisusonustus* Walckenaer 1806 (Chelicerata, Arachnida, Thomisidae). Oecologia 80:513–515

Waldorf ES (1976) Spider size, microhabitat selection, and use of food. Am Midl Nat 96:76–87

Welti EAR, Putnam S, Joern A (2016) Crab spiders (Thomisidae) attract insect flower-visitors without UV signalling. Ecol Entomol 41:611–617

Whitney KD (2004) Experimental evidence that both parties benefit in a facultative plant–spider mutualism. Ecology 85:1642–1650

Wise DH (1993) Spiders in ecological webs. Cambridge University Press, Cambridge

Woodcock BA, Potts SG, Westbury DB, Ramsay AJ, Lambert M, Harris SJ, Brown VK (2007) The importance of sward architectural complexity is structuring predatory and phytophagous invertebrate assemblages. Ecol Entomol 32:302–311

World Spider Catalog (2017) World Spider Catalog. Natural History Museum Bern, version 18.0. http://wsc.nmbe.ch. Accessed 8 Apr 2017

Wunderlich J (1982) Mitteleuropäische Spinnen (Araneae) der Baumrinde. Z Angew Entomol 94:9–21

Young OP, Edwards GB (1990) Spiders in United States field crops and their potential effect on crop pests. J Arachnol 18:1–27

Young OP, Lockley TC (1989) Spiders of Spanish moss in the Delta of Mississippi. J Arachnol 17:143–148

Zanatta MF (2013) História natural, seleção de folhas e locais para nidificação e efeito do cuidado materno em *Aysha piassaguera* Brescovit, 1992 (Araneae: Anyphanidae) na Serra do Japi, Jundiaí-SP, Brasil. Master thesis, Universidade Estadual de Campinas

Zanatta MF, Romero GQ, Vasconcellos-Neto J (2016) Effect of maternal care on egg survival in *Aysha piassaguera* (Araneae: Anyphaenidae). Insect Soc 63:439–445

Chapter 8
Spiders as Plant Partners: Complementing Ant Services to Plants with Extrafloral Nectaries

Kleber Del-Claro, Vanessa Stefani, Larissa Nahas, and Helena Maura Torezan-Silingardi

Abstract In terrestrial communities, multitrophic interactions comprise a minimum of three trophic levels that interact among each other: plants, herbivores and their natural enemies. The top-down forces exerted by invertebrate predators on herbivores, and their cascading effect on plants, are very important to community structuring. Among major invertebrate predators, ants exert a strong impact on the density and spatial distribution of leaf and floral herbivores, which is reflected in the reproductive capacity of the plants. This important effect has only recently also been attributed to spiders. Studies of trophic interactions involving spiders and their impacts on the vegetation have increased considerably in the last few years. This is to be expected, considering that spiders are present in almost all terrestrial environments and occur in higher abundance in vegetation-rich areas. Regarded as excellent predators, spiders also use plants as foraging substrates, exploring differences in the plant architecture and in prey-capture strategies. Furthermore, spiders commonly prey on insect herbivores, which can result in a great decrease in herbivory rates, benefiting the host plants. In this chapter we will explore the hypothesis that spiders increase the reproductive value of plants with extrafloral nectaries, complementing the services provided by ants.

Plant–animal associations are ancient interactions of species responsible for the great diversity of relationships that we find today. This was possible because these

K. Del-Claro (✉) • V. Stefani • H.M. Torezan-Silingardi
Universidade Federal de Uberlândia, Instituto de Biologia, Laboratório de Ecologia Comportamental e de Interações, Uberlândia, MG, Brazil
e-mail: delclaro@ufu.br

L. Nahas
Universidade Federal de Uberlândia, Instituto de Biologia, Laboratório de Ecologia Comportamental e de Interações, Uberlândia, MG, Brazil

Universidade Federal de Uberlândia, Escola de Educação Básica, Uberlândia, MG, Brazil

C. Viera, M.O. Gonzaga (eds.), *Behaviour and Ecology of Spiders*,
https://doi.org/10.1007/978-3-319-65717-2_8

associations shaped terrestrial biodiversity through creation, extinction, and coevolution over evolutionary time (Thompson 2013; Del-Claro et al. 2016). The outcomes of these interactions depend on physical and biotic (internal) and environmental (external) factors (Del-Claro and Marquis 2015). Thus, evolutionary changes gradually influence the direction and magnitude of interactions (Del-Claro and Torezan-Silingardi 2012; Thompson 2013).

The pressure that herbivores exert on plant ontogeny and fitness leads plants to develop numerous defensive strategies (Strong et al. 1984). These defences may be constitutive, such as the presence of spines and chemicals, or temporary, such as those that are only induced upon the perception of attack, in order to ensure optimal resource allocation (Karban and Baldwin 1997; Campbell and Kessler 2013). From a consumer-resource perspective, plant defences against floral herbivores, such as sequential flowering, may represent an important defensive developmental strategy (Marquis and Lill 2010; Vilela et al. 2014). Biotic defences have a mutualistic character, whereby resources (from plants) are exchanged for services (from animals) and are mediated by interests, costs, and benefits. Sometimes the costs of biotic defences are high, because resources are the plant's own tissues when consumed by herbivores (Price 2002), whereas at other times the benefits of biotic defences are high, and plants receive protection against herbivores from third-partner species, such as ants, spiders, and wasps (Del-Claro et al. 2016). The most common resource that plants offer to attract and reward these predators is nectar (Nahas et al. 2016).

Nectar is an aqueous solution that is very rich in carbohydrates (mainly sucrose and/or fructose) and contains low concentrations of lipids, enzymes, amino acids, phenols, alkaloids, and volatile organic compounds (Koptur 1994; González-Teuber and Heil 2009). In contrast to floral nectar, which is clearly associated with beneficial plant–animal interactions (i.e., pollination; Faegri and Van der Pijl (1976), Torezan-Silingardi (2012)), extrafloral nectar (EFN) is secreted by both vegetative and reproductive plant parts (e.g., spike, pedicel, bud, calyx, leaves, shoots, petioles, bracts, and stems) and has no direct influence on pollination. After more than 100 years of studies that were initiated by Belt (1874), the majority of reviews explain that EFN is a valuable resource to ants (Byk and Del-Claro 2011), and that its production generally benefits plants through indirect reductions in herbivory (Oliveira and Freitas 2004; Rosumek et al. 2009; Zhang et al. 2015; Del-Claro et al. 2016). However, we now know that both spiders and wasps feed on EFN (Figs. 8.1 and 8.2) and are effective plant guards (Ruhren and Handel 1999; Cuautle and Rico-Gray 2003; Torezan-Silingardi 2011; Nahas et al. 2012, 2016; Alves-Silva et al. 2013; Stefani et al. 2015).

Spiders Feeding on Nectar

Until recently, the end of the twentieth century, almost all papers and books described spiders as obligate carnivores. Indeed, studies have shown that spiders basically prey on insects, other spiders, and small vertebrates, including fish.

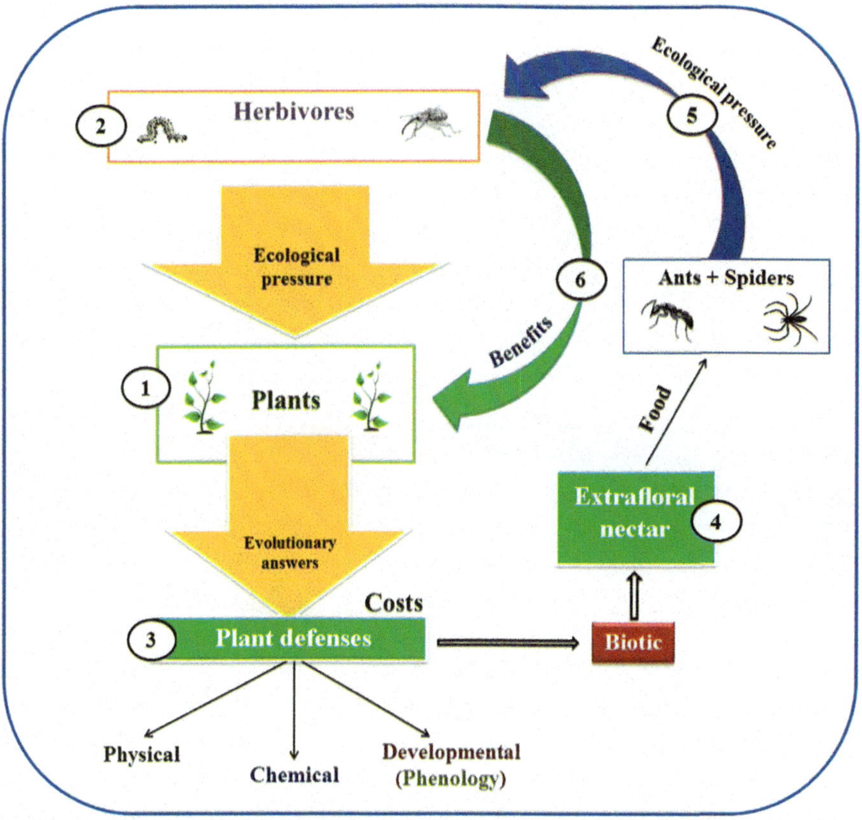

Fig. 8.1 The ecological system in different interacting trophic levels: plants–herbivores–predators. As producers, plants (*1*) incur constant pressure from herbivores (*2*), which are consumers at the base of the ecological trophic chain. In response, plants have evolved a wide range of defences against herbivores (*3*), including spines (mechanical), alkaloids (chemical), and the ability to resprout during periods when insects are less abundant (developmental). However, plants have also developed biotic defences, such as the production of nectar by extrafloral nectaries (*4*), which attract nectar-feeding predators, mainly ants and spiders. Because such predators also require protein, they often attack herbivores (*5*), thereby providing benefits to host plants (*6*)

However, spiders occasionally use plant food to supplement their insect diet. In a review, Nyffeler et al. (2016) tracked >60 spider species representing ten families that have been observed feeding on plant materials from over 20 plant families. These spiders feed on a wide diversity of plant-derived products including floral nectar, extrafloral nectar, stigmatic exudate, plant sap, honeydew, seeds, Beltian bodies, Müllerian bodies and pollen (originating from very different plant types such as coniferous and deciduous trees, herbaceous plants, and ferns). Furthermore, spiders have been shown to consume fungal spores in laboratory trials (Nyffeler et al. 2016). Reports of spiders feeding on plant products in nature first appeared in recent decades. Smith and Mommsen (1984) reported that the webs of immature

Fig. 8.2 Consumers of extrafloral nectar (EFN). (**a**) Ants, such as *Camponotus crassus*, which is shown feeding on the EFN of *Ouratea spectabilis* (Ochnaceae), and (**b**) *Neoponera villosa*, which is shown feeding on the EFN of *Chamaecrista brevipes* (Fabaceae), are recognized as the main protective biotic agents in ant–plant interactions (**c**). However, recent studies have also shown that EFN-feeding spiders, such as salticids, shown here feeding on the EFN of *Banisteriopsis malifolia* (Malpighiaceae), and (**d**) the oxyopid *Oxyopes macroscelides*, which is shown feeding on the EFN of *O. spectabilis* (Ochnaceae), may similarly provide benefits to host plants by preying on herbivores (**e**, **f**)

ecribellate orb-weavers are pollen collectors, and that the pollen grains adhering to the sticky threads are unintentionally ingested along with the old silk material when the spiderlings are recycling their webs. Pollard et al. (1995) as well as Taylor and Foster (1996) document cases of spider nectarivory from North America and Central America. Since those early reports, there have been many more discoveries of spiders feeding on plant materials (see Nyffeler et al. 2016). For example, Jackson et al. (2001) observed 31 salticid species using their chelicerae to make contact with floral structures (i.e., nectaries) in natural environments. In response to this observation, Jackson et al. (2001) designed an experiment that involved 90 salticid species to test whether the spiders were searching for sugar or just water. Briefly, the spiders were offered two smithereens of filter paper in a Petri dish, one saturated with 30% sucrose solution and the other with distilled water, and given 10 min to choose between the smithereens. Interestingly, all the species exhibited a clear preference for the sucrose solution.

Using the cold anthrone test, Nahas et al. (2016) investigated the presence and concentration of fructose in the bodies of wandering (Anyphaenidae, Oxyopidae, Pisauridae, Salticidae, and Thomisidae) and orb-weaving (Araneidae and Theridiidae) spiders that were collected from extrafloral nectary-bearing plants in a Neotropical savanna. The authors collected and tested adults and subadults of 301 spiders (39 species from seven families) for fructose ingestion, and found that 88.04% of the spiders tested positive. The fructose contents of the spiders ranged from 0.054 to 44.532 μg/μl, and exceeded 2 μg/μl for 175 (58.14%) of the individuals. Such values are often considered the standard limit, above which the presence of fructose can be determined confidently, even without the use of a spectrophotometer (Chen et al. 2010; Taylor and Pfannenstiel 2008). The highest fructose concentrations reported by Nahas et al. (2016) were obtained from an adult *Araneus venatrix* (44.532 μg/μl) that was collected at night on *Qualea grandiflora* (Vochysiaceae — tree) and an adult *Tmarus* species (32 μg/μl) that was collected in the morning on *Heteropterys pteropetala* (Malpighiaceae — shrub). The highest percentage of fructose-positive individuals was observed for members of the Pisauridae (100%), Salticidae (92.96%), and Thomisidae (90.76%), whereas the lowest percentage was observed for members of the Anyphaenidae (76%). The percentages of fructose-positive wandering and orb-weaving spiders were similar (Fig. 8.3).

Other authors have also used the cold anthrone test to demonstrate nectar feeding by spiders, including members of the Salticidae (e.g., Kuja et al. 2012; Chen et al. 2010), Oxyopidae, Thomisidae, Pisauridae, Lycosidae, Tetragnatidae, Araneidae, Agelenidae (Chen et al. 2010), and Miturgidae and Anyphaenidae (Taylor and Pfannenstiel 2008). As an exceptionally rich source of sugar that often contains significant quantities of amino acids and other nutrients, nectar may be an especially rewarding addition to the diet of predatory arthropods (Sanders 2013; Jackson et al. 2001). Byk and Del-Claro (2011) performed a long-term study to investigate whether EFN had a positive effect on the fitness of ant colonies by quantifying the growth rate and survival of colonies and the final weight of individuals. The results provided clear evidence that EFN can significantly improve the survivorship,

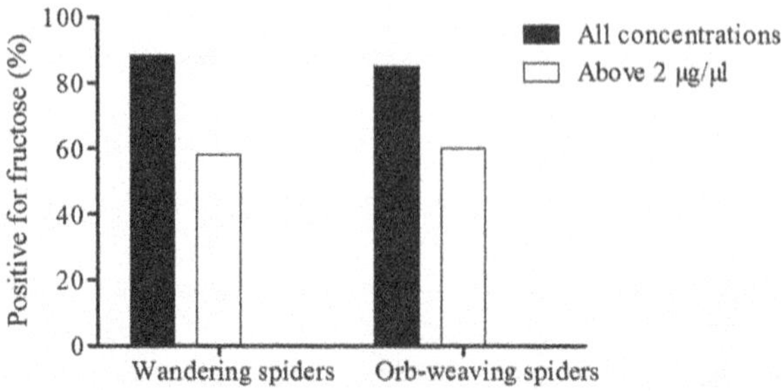

Fig. 8.3 The percentage of fructose-positive wandering and orb-weaving spiders that tested positive for the ingestion of fructose. *Black bars* indicate the percentage of positive spiders including all fructose concentrations and *white bars* indicate the percentage of positive spiders that present fructose concentration above 2 µg/µl (Nahas et al. 2016)

growth, and reproduction of ants. In fact, an EFN-rich diet (at least 30 cal per day) resulted in five times more individuals per colony, greater body weights, and greater egg production. Meanwhile, Patt and Pfannenstiel (2008) demonstrated that juvenile spiders can recognize and remember particular chemical stimuli associated with nectar. Pfannenstiel (2015) also demonstrated that the consumption of honeydew from white flies (*Aleurothrixus floccosus*, Homeptera: Aleyrodidae) increased the survivorship of nocturnal spiders by 73.5% (*Apollophanes punctipes*, Philodromidae), 266.7% (*Cesonia bilineata*, Gnaphosidae), 352.6% (*Dictyna* sp., Dictynidae), 130.9% (*Thiodina sylvana* - Salticidae), and 1102.5% (*Hibana futilis*, Anyphaenidae); and Taylor and Pfannenstiel (2009) demonstrated that feeding on the floral nectar of *Cheiracanthium inclusum* (Miturgidae) can improve spider fitness.

Evolution and Benefits of Spider–Plant Interactions

Vollrath and Selden (2007) suggested that spiders have existed in the Paleozoic era (400–360 mya). At the same time, plants began to abundantly occupy the terrestrial environment, offering new ecological niches to a wide range of terrestrial arthropods, including the first insects (Bernays 1998). During the Devonian, the Earth was inhabited by mosses (bryophytes), ferns (pteridophytes), and conifers (gymnosperms), which were able to sustain a great diversity of herbivorous insects that fed on their pollen, leaves, branches, roots, resin, and phloem liquids (Torezan-Silingardi 2012), and thereby provided a fertile environment for the recently arrived spiders.

Ants, which are currently among the most abundant arthropods that inhabit and forage on plants, arose soon afterwards to share the designation of 'herbivores' greatest predators' with spiders (Floren et al. 2002). Moreau et al. (2006) reported that the most ancient ant fossil is dated at between 168 and 140 mya, co-occurring with the diversification of Angiosperms, and proposed two possible mechanisms for ant diversification that can also be applied to arboreal spiders: a) angiosperm forests were more heterogeneous and diverse than previously existing ones, and offered several new niches for ants to utilize, and b) angiosperm forests were exploited by diversifying herbivorous insects (Torezan-Silingardi 2012) and became an important source of supplemental protein. In addition, trophobiont herbivores (i.e., membracids and aphids) may also have attracted ants by offering honeydew, a carbohydrate-, fat-, and amino acid-rich solution that ants collect and maintain using specialized behaviour (Del-Claro et al. 2013; Moreira and Del-Claro 2005; Blüthgen et al. 2004; Hölldobler and Wilson 1990) and that could have contributed to the evolution of ant sociality (Moreau et al. 2006).

Extrafloral nectaries may have evolved relatively recently in the evolution of Angiosperms, arising around the Eocene (60–50 mya; Marazzi and Sanderson 2010). Weber and Keeler (2013) reported that extrafloral nectaries occur in 3941 species of vascular plants, representing 745 genera in 108 families. More specifically, extrafloral nectaries have been identified in four fern families (39 species seven genera), monocots (260 species from 82 genera), including some true grasses (22 species from five genera), various dioscorea (71 species) and many orchids (77 species in 45 genera). However, extrafloral nectaries are most common in eudicots (3642 species, representing 654 genera in 89 families, mainly the Fabaceae, Passifloraceae, and Malvaceae). Until the 1970s, the function of extrafloral nectaries was controversial, yet scientist agreed that the organs were unrelated to pollination (Rico-Gray and Oliveira 2007; Koptur 2005). Now, extrafloral nectaries are recognized as important structures that are related to mutualistic associations with predatory arthropods, such as ants, spiders, and wasps, that feed on EFN and, in turn, ward off or consume potential herbivores (Stefani et al. 2015; Koptur et al. 2015; Katayama and Suzuki 2011) and confer increased performance.

Recent research has shown that both spiders and ants that feed on extrafloral nectaries reduce herbivore abundance and increase fruit set. For example, Ruhren and Handel (1999) observed a positive correlation between the feeding of salticid spiders on the extrafloral nectaries and the production of fruits and seeds by *Chamaecrista nictitans* (Fabaceae). Similarly, Whitney (2004) reported higher seed production by *Acacia ligulata* (Fabaceae) that contained nests of the subsocial spider *Phryganoporus candidus* (Desidae), and even though the study failed to identify the plant's EFN as an important reward to this spider species, *A. ligulata* was found to host a greater number of spider colonies per plant and nest volume than other host plants that lacked extrafloral nectaries.

In the Brazilian tropical savanna, Nahas et al. (2012) demonstrated that ants and spiders have complementary impacts on the reduction of herbivory of trees, such as *Qualea multiflora* (Vochysiaceae), with extrafloral nectaries, and that the influences of ants and spiders on the diversity and abundance of herbivores are more significant when both predator types are present (Fig. 8.4). At the same site, Stefani et al. (2015) found a significantly higher number of fruits per bud, seeds per fruit, and seed viability in specimens of the tree *Eriotheca gracilipes* (Malvaceae) that hosted both spiders and ants. These data, along with direct observations of spiders feeding on EFN (e.g., Ruhren and Handel 1999; Taylor and Pfannenstiel 2008), suggest that

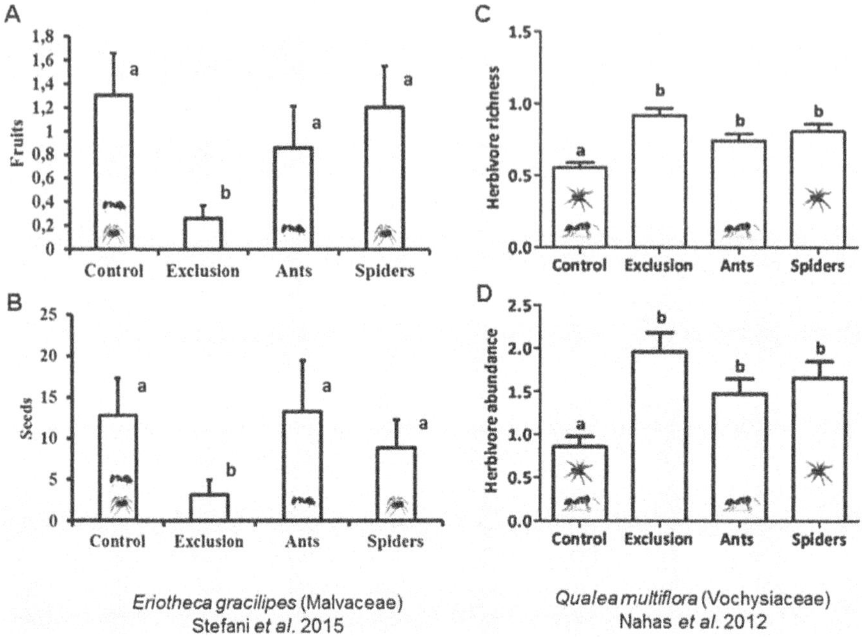

Fig. 8.4 Outcomes of predators on plants. Spiders and ants can provide complementary services to plants with extrafloral nectaries, and although each predator can provide benefits to host plants independently, the co-occurrence of the two predator types provides the maximum benefits, in terms of fruit per bud (**a**) and number of viable seeds (**b**). production, as demonstrated in *Eriotheca gracillipes* (Malvaceae) by Stefani et al. (2015), as well as in the reduction of herbivore diversity (**c**) and abundance (**d**) as demonstrated in *Qualea multiflora* (Vochysiaceae) by Nahas et al. (2012). In the group '*Control*'—with both predators, '*Exclusion*'—without predators, '*Ant*'—with spiders removed, and '*Spiders*'—with ants removed. Means +1SE are presented. *Different letters over bars* point to statistical significant difference

extrafloral nectaries mediate mutualistic spider–plant interactions, in the same way that they mediate ant–plant interactions.

According to Stefani et al. (2015), the complementarity of ecological services provided by different predators, such as ants and spiders, which possess different abilities and cognitive and predatory capacities, may result in a direct benefit to the host plant, and similar patterns have been reported for wasps that visit extrafloral nectaries (e.g. Torres-Hernández et al. 2000; Torezan-Silingardi 2011; Alves-Silva et al. 2013). Nevertheless, generalist predators, such as spiders and ants, that use the same foraging substrates may compete for prey and, as a result, attack one another (Stefani et al. 2015). Stefani et al. (2015) observed that the species richness of spiders was significantly greater in the absence of ants, although the opposite was not true, possibly due to the composition of distinct ant and spider species and, consequently, to the different types of interactions among them. Similarly, Nahas et al. (2012) also found that spider abundance and richness were significantly lower on plants that were visited by ants, and that ant abundance and richness were unaffected by the presence or absence of spiders. Therefore, both spiders and ants are active at the third trophic level, and may directly affect the diversity, abundance, and behaviour of host plant herbivores and indirectly affect the fitness of the plants they visit.

Extrafloral Nectary-Bearing Plant Phenology Influences Evolution of Spider—Plant Interactions

For the evolution of interactions between ants and extrafloral nectary-bearing plants, three mechanisms have recently been reported that probably also apply to the evolution of spider–plant interactions.

Vilela et al. (2014) hypothesized that the sequential flowering of related plants could result in the development of a shared herbivore guild that could, subsequently, be quite harmful to the related plant species and make associations with ants critical for reproductive success. Indeed, the authors confirmed the hypothesis in four co-occurring members of the Malpighiaceae that exhibit sequential flowering, possibly in order to sustain pollinator populations. Vilela et al. (2014) demonstrated that the sequential development (resprouting and flowering) of these species also provided an uninterrupted food supply and, over the course of the season, allowed shared herbivores to move from one plant species to the next. Interestingly, ants exhibited the same pattern, following sequential EFN production among the plants. However, it remains unclear whether spiders may be inserted in this type of circumstances, whether seasonal changes in plant ontogeny and phenotypic plasticity directly affect the abundance and structure of arboreal spider communities that feed on EFN, or whether spiders are capable of moving between plant species to enhance their survivorship.

Second, since sugar sources are vital to ant nutrition and colony survivorship (Byk and Del-Claro 2011), it is expected that ants will forage on flowers to obtain nectar (Santos et al. 2014; Rico-Gray 1993). However, floral visitation by ants might be detrimental to plant reproduction, especially because flower-visiting ants might deter, expel, or prey upon pollinators and, moreover, reduce the amount of nectar available for effective pollinators (Assunção et al. 2014). Similarly, Taylor and Pfannenstiel (2009) demonstrated that the fitness of spiders was improved by feeding on the floral nectar of *Cheiracanthium inclusum* (Miturgidae). Thus, it is also expected that spiders will forage on flowers, regardless of whether plants possess extrafloral nectaries, and could also reduce visitation by effective pollinators, as observed for ants (e.g., Ness 2006; Ness et al. 2009; Assunção et al. 2014).

Finally, recent studies have used tools derived from graph theory to investigate the organization of ecological interactions in different ecosystems around the world. In this new perspective, several studies have used a network approach to describe the structure of interactions between ants and plants with extrafloral nectaries

(Bascompte 2009; Dáttilo et al. 2013, 2014a, b, Lange and Del-Claro 2014). However, because no current publications have used a network approach to investigate spider–plant interactions, we suggest that this is a promising target for future research.

In particular, we suggest that seasonally dry tropical forests are important ecosystems for investigating the uncertainties of spider–plant interactions, because such environments are one of the world's main ecosystems in which plant phenology is strongly influenced by climatic seasonality (Del-Claro et al. 2016).

Acknowledgements To two anonymous referees that made valuable improvements, and thanks to CNPq, FINEP, and FAPEMIG for financial support.

References

Alves-Silva E, Barônio GJ, Torezan-Silingardi HM, Del-Claro K (2013) Foraging behavior of *Brachygastra lecheguana* (hymenoptera: Vespidae) on *Banisteriopsis malifolia* (Malpighiaceae): extrafloral nectar consumption and herbivore predation in a tending ant system. Entomol Sci 16:162–169

Assunção MA, Torezan-Silingardi HM, Del-Claro K (2014) Do ant visitors to extrafloral nectaries of plants repel pollinators and cause an indirect cost of mutualism? Flora 2:244–249

Bascompte J (2009) Disentangling the web of life. Science 325:416–419

Belt T (1874) The naturalist in Nicaragua. Dent, London

Bernays EA (1998) The value of being a resource specialist: behavioral support for a neural hypothesis. Am Nat 151:451–464

Blüthgen N, Stork NE, Fiedler K (2004) Bottom-up control and cooccurrence in complex communities: honeydew and nectar determine a rainforest ant mosaic. Oikos 106:344–358

Byk J, Del-Claro K (2011) Ant–plant interaction in the Neotropical savanna: direct beneficial effects of extrafloral nectar on ant colony fitness. Popul Ecol 53:327–332

Campbell SA, Kessler A (2013) Plant mating system transitions drive the macroevolution of defense strategies. Proc Nat Acad Sci USA 110:3973–3978

Chen X, Chen Y, Wu L, Peng Y, Chen J, Liu F (2010) A survey of nectar feeding by spiders in three different habitats. B Insectol 63:203–208

Cuautle M, Rico-Gray V (2003) The effect of wasps and ants on the reproductive success of the extrafloral nectaried plant *Turnera ulmifolia* (Turneraceae). Funct Ecol 17:417–423

Dáttilo W, Guimarães PR, Izzo T (2013) Spatial structure of ant–plant mutualistic networks. Oikos 122:1643–1648

Dáttilo W, Díaz-Castelazo C, Rico-Gray V (2014a) Ant dominance hierarchy determines the nested pattern in ant–plant networks. Biol J Linn Soc 113:405–414

Dáttilo W, Fagundes R, Gurka CAQ, Silva MSA, Vieira MCL, Izzo TJ, Díaz-Castelazo C, Del-Claro K, Rico-Gray V (2014b) Individual based ant–plant networks: diurnal-nocturnal structure and species-area relationship. PLoS One 9(6):e99838

Del-Claro K, Marquis RJ (2015) Ant species identity has a greater effect than fire on the outcome of an ant protection system in Brazilian Cerrado. Biotropica 47:459–467

Del-Claro K, Stefani V, Lange D, Vilela AA, Nahas L, Velasques M, Torezan-Silingardi HM (2013) The importance of natural history studies for a better comprehension of animal–plant interactions networks. Biosci J 29:439–448

Del-Claro K, Rico-Gray V, Torezan-Silingardi HM, Alves-Silva E, Fagundes R, Lange D, Dáttilo W, Vilela AA, Aguirre A, Rodriguez-Morales D (2016) Loss and gains in ant–plant interactions mediated by extrafloral nectar: fidelity, cheats, and lies. Insect Soc 63:207–221

Faegri K, Van der Pijl L (1976) The principles of pollination ecology. Pergamon Press, Oxford
Floren A, Biun A, Linsenmair KE (2002) Arboreal ants as key predators in tropical lowland rainforest trees. Oecologia 131:137–144
González-Teuber M, Heil M (2009) Nectar chemistry is tailored for both attraction of mutualists and protection from exploiters. Plant Signal Behav 4:809–813
Hölldobler B, Wilson EO (1990) The ants. Springer, Berlin
Jackson RR, Pollard SD, Nelson XJ, Edwards GB, Barrion AT (2001) Jumping spiders (Araneae: Salticidae) that feed on nectar. J Zool 255:25–29
Karban R, Baldwin IT (1997) Induced responses to herbivory. The University of Chicago Press, Chicago
Katayama N, Suzuki N (2011) Anti-herbivory defense of two *Vicia* species with and without extrafloral nectaries. Plant Ecol 212:743–752
Koptur S (1994) Floral and extrafloral nectars of Neotropical *Inga* trees: a comparison of their constituents and composition. Biotropica 26:276–284
Koptur S (2005) Nectar as fuel for plant protectors. In: Wäckers FL, van-Rijn PCJ, Bruin J (eds) Plant-provided food for carnivorous insects. Cambridge University Press, Cambridge, pp 75–108
Koptur S, Jones IM, Pena JE (2015) The influence of host plant extrafloral nectaries on multitrophic interactions: an experimental investigation. PLoS One 10(9):e0138157
Kuja JO, Jackson RR, Sune GO, Karanja RNH, Lagat ZO, Carvell GE (2012) Nectar meals of a mosquito-specialist spider. Psyche:898721. https://doi.org/10.1155/2012/898721
Lange D, Del-Claro K (2014) Ant–plant interaction in a tropical savanna: may the network structure vary over time and influence on the outcomes of associations? PLoS One 9(8):e105574
Marazzi B, Sanderson M (2010) Large-scale patterns of diversification in the widespread legume genus *Senna* and the evolutionary role of extrafloral nectaries. Evolution 64:3570–3592
Marquis RJ, Lill JT (2010) Impact of plant architecture versus leaf quality on attack by leaf-tying caterpillars on five oak species. Oecologia 163:203–213
Moreau CS, Bell CD, Vila R, Archibald SB, Pierce NE (2006) Phylogeny of the ants: diversification in the age of angiosperms. Science 312:101–104
Moreira VSS, Del-Claro K (2005) The outcomes of an ant–treehopper association on *Solanum lycocarpum* St. Hil: increased membracid fecundity and reduced damage by chewing herbivores. Neotrop Entomol 34:881–887
Nahas L, Gonzaga MO, Del-Claro K (2012) Emergent impacts of ant and spider interactions: herbivory reduction in a tropical savanna tree. Biotropica 44:498–505
Nahas L, Gonzaga MO, Del-Claro K (2016) Wandering and web spiders feeding on the nectar from extrafloral nectaries in Neotropical savanna. J Zool 301:125–132
Ness JH (2006) A mutualism's indirect costs: the most aggressive plant bodyguards also deter pollinators. Oikos 113:506–514
Ness JH, Morris WF, Bronstein JL (2009) For ant-protected plants, the best defense is a hungry offense. Ecology 90:2823–2831
Nyffeler M, Olson EJ, Symondson WOC (2016) Plant-eating by spiders. J Arachnol 44:15–27
Oliveira PS, Freitas AVL (2004) Ant plant herbivore interactions in the Neotropical cerrado savanna. Naturwissenschaften 91:557–570
Patt JM, Pfannenstiel RS (2008) Odor-based recognition of nectar in cursorial spiders. Entomol Exp Appl 127:64–71
Pfannenstiel RS (2015) Extended survival of spiders (Aranaeae) feeding on whitefly (Homoptera: Aleyrodidae) honeydew. J Entomol Sci 50:110–118
Pollard SD, Beck MW, Dodson GN (1995) Why do male crab spiders drink nectar? Anim Behav 49:1443–1448
Price PW (2002) Species interactions and the evolution of biodiversity. In: Herrera CM, Pellmyr O (eds) Plant–animal interactions: an evolutionary approach. Blackwell Science, Oxford, pp 3–25
Rico-Gray V (1993) Use of plant-derived food resources by ants in the dry tropical lowlands of coastal Veracruz, Mexico. Biotropica 25:301–315

Rico-Gray V, Oliveira PS (2007) The ecology and evolution of ant–plant interactions. The University of Chicago Press, Chicago
Rosumek FB, Silveira FAO, Neves FS, Barbosa NP, Diniz L, Oki Y, Pezzini F, Fernandez WG, Cornelissen T (2009) Ants on plants: a meta-analysis of the role of ants as plant biotic defenses. Oecologia 160:537–549
Ruhren S, Handel SN (1999) Jumping spiders (Salticidae) enhance the seed production of a plant with extrafloral nectaries. Oecologia 119:227–230
Sanders D (2013) Herbivory in spiders. In: Nentwig W (ed) Spider ecophysiology. Springer, Bern, pp 385–391
Santos GMM, Dáttilo W, Presley SJ (2014) The seasonal dynamic of ant–flower networks in a semi-arid tropical environment. Ecol Entomol 39:674–683
Smith RB, Mommsen TP (1984) Pollen feeding in an orb-weaving spider. Science 226:1330–1333
Stefani V, Pires TL, Torezan-Silingardi HM, Del-Claro K (2015) Beneficial effects of ants and spiders on the reproductive value of *Eriotheca gracilipes* (Malvaceae) in a tropical savanna. PLoS One 10(7):e0131843
Strong DR, Lawton JH, Southwood R (1984) Insects on plants: community patterns and mechanisms. Harvard University Press, Cambridge, Massachuttes
Taylor RM, Foster WA (1996) Spider nectarivory. Am Entomol 42:82–86
Taylor RM, Pfannenstiel RS (2008) Nectar feeding by wandering spiders on cotton plants. Environ Entomol 37:996–1002
Taylor RM, Pfannenstiel RS (2009) How dietary plant nectar affects the survival, growth, and fecundity of a cursorial spider *Cheiracanthium inclusum* (Araneae: Miturgidae). Environ Entomol 38:1379–1386
Thompson JN (2013) Relentless evolution. University of Chicago Press, Chicago
Torezan-Silingardi HM (2011) Predatory behavior of *Pachodynerus brevithorax* (hymenoptera: Vespidae, Eumeninae) on endophytic herbivore beetles in the Brazilian tropical savanna. Sociobiology 57:181–189
Torezan-Silingardi HM (2012) Flores e animais, uma introdução à história natural da polinizacão. In: Del-Claro K, Torezan-Silingardi HM (eds) Ecologia das interações plantas-animais: uma abordagem ecológico-evolutiva. Technical Books, Rio de Janeiro, pp 111–142
Torres-Hernández L, Rico-Gray V, Castillo-Guevara C, Vergara JA (2000) Effect of nectar-foraging ants and wasps on the reproductive fitness of *Turnera ulmiflora*, Turneraceae in a costal sand dune in Mexico. Acta Zool Mex 81:13–21
Vilela AA, Torezan-Silingardi HM, Del-Claro K (2014) Conditional outcomes in ant–plant–herbivore interactions influenced by sequential flowering. Flora 209:359–366
Vollrath F, Selden P (2007) The role of behavior in the evolution of spiders, silks, and webs. Annu Rev Ecol Evol Syst 38:819–846
Weber MG, Keeler KH (2013) The phylogenetic distribution of extrafloral nectaries in plants. Ann Bot 111:1251–1261
Whitney KD (2004) Experimental evidence that both parties benefit in a facultative plant-spider mutualism. Ecology 85:1642–1650. doi:10.1890/03-0282
Zhang S, Zhang Y, Keming MA (2015) The equal effectiveness of different defensive strategies. Sci Rep 5:13049. https://doi.org/10.1038/srep13049

Chapter 9
Foraging Strategies of Cursorial and Ambush Spiders

Rodrigo H. Willemart and Mariángeles Lacava

Abstract Food consumption in animals is a complex task with multiple steps. Choosing an adequate foraging site is the very first one, and involves not only the presence of prey and predators but also abiotic conditions. Because spiders are usually cannibalistic, conspecifics fall within these two categories in addition to being competitors. Specifically for ambush and cursorial spiders, the type of substrate is also very relevant because spiders often rely on substrate-borne vibrations to find their prey, and distinct substrates propagate vibrations differently. At this point or after contacting the prey, spiders have to decide whether or not to attempt capture. Such a decision involves profitability, prey defenses, and the physiological state of the spider. To capture prey, ambush and cursorial spiders may rely on web sheets, adhesive setae on the tips of the legs, glue-spitting, and venom directly injected from the fangs of the chelicerae. The actual mode of ingestion also varies among species. For almost every step from picking a foraging place to prey consumption, multiple sensory modalities may be used, such as vision, contact chemoreception, olfaction, detection of substrate-borne vibrations, and air displacement. Adequately choosing where to forage, properly detecting, choosing, capturing, and handling prey may have important fitness implications. In this chapter, we summarize the knowledge on these topics with regard to Neotropical cursorial and ambush spiders, detecting gaps and areas better covered within the topics above. Finally, we attempt to suggest promising model species to investigate these different steps of foraging in these animals.

R.H. Willemart (✉)
Universidade de São Paulo, USP Leste, Rua Arlindo Béttio,
1000 - Ermelino Matarazzo, 03828-000 Sao Paulo, SP, Brazil
e-mail: willemart@usp.br

M. Lacava
Centro Universitario de Rivera, Universidad de la República, Rivera, Uruguay

Laboratorio Ecología del Comportamiento, Instituto de Investigaciones Biológicas Clemente Estable, Montevideo, Uruguay

C. Viera, M.O. Gonzaga (eds.), *Behaviour and Ecology of Spiders*,
https://doi.org/10.1007/978-3-319-65717-2_9

Spiders are a diverse group of predators, and their evolution has been accompanied by the evolution of their main prey, insects (Vollrath and Selden 2007). Strategies for searching out and capturing prey vary greatly among spiders, which may prey upon flying, running, walking, jumping, and even aquatic prey. Cardoso et al. (2011) classified how spiders exploit resources in different guilds or functional groups. Within spiders, some species use webs to attract and capture prey ("sensing web", "sheet web", "space web"), but others do not use webs and rely mainly on their legs and chelicerae to immobilize prey. Many of these spider families fall within the guilds of "cursorial hunters", "ambush hunters" and "other hunters" (Cardoso et al. 2011). Among spiders, the guild of "ground hunters" consists of 19 Neotropical families, including Lycosidae, Oonopidae, Corinnidae, and Paratropididae, among others. The guild of "ambush hunters" comprises six families, five of which occur in Neotropical regions: Deinopidae, Thomisidae, Microstigmatidae, Sicariidae, and Selenopidae. In the "other hunters" guild, the authors included the Neotropical families Clubionidae, Senoculidae, and Ctenidae, among others. Spiders with "uncommon" predatory habits, such as the species of Scytodidae, which spits on prey to capture them, are also in this group (Cardoso et al. 2011).

In this chapter we describe through a discussion of diet how non-web-building spiders manage to acquire food; the choice of an adequate site to forage; and detecting, capturing, and handling prey, emphasizing Neotropical species. Because in many cases there are no examples of Neotropical species, we often refer to studies conducted in the Northern Hemisphere. We aim to provide complementary information to great previous reviews of spider foraging such as the book chapters by Riechert and Luczak (1982), Romero and Vasconcellos-Neto (2007), and Nelson and Jackson (2011), and the classic book *Biology of Spiders* by Foelix (2011).

A Brief Introduction to Spider Diet

Spider diet certainly varies greatly (Fig. 9.1), ranging from polyphagous species to oligophagous and monophagous ones (Nyffeler 1999; Nelson and Jackson 2011). Detailed studies of diet breadth are sometimes done in the laboratory because the percentage of feeding spiders in the field varies from 0.4% to 8% (references in Nentwig 1986). Some spiders may feed on nectar (Salticidae: Jackson et al. 2001; Carvell et al. 2015) and some may scavenge (Sicariidae: Sandidge 2003; Cramer 2008; Vetter 2011). Spiders that feed predominantly on ants, or are specialized in woodlice, in other spiders, or in blood-fed mosquitoes, are dealt with in Chap. 10.

Polyphagous species do not just eat anything. Several factors are known to influence spider diet. The ontogeny of the spider may influence prey taxa, diversity, and size in the diet (Bartos 2011). In addition, cannibalism is common (Rypstra and Samu 2005), more often involving juveniles (Wise 2006), and more likely to occur in hungry individuals (Samu et al. 1999; Mayntz and Toft 2006; Wise 2006), which may change their locomotor activity compared with well-fed spiders (Walker et al. 1999). Finally, sexual cannibalism has also been observed in several species (e.g., Schwartz et al. 2014; Toft and Albo 2016).

Fig. 9.1 Prey items captured by some Neotropical spider species. (**a**) Salticid feeding on a moth (by O. Pulgarín). (**b**) *Peucetia sp.* (Oxyopidae) feeding on a vespid wasp (by O. Pulgarín). (**c**) Lycosid feeding on an asylid fly. (**d**) *Ctenus* sp. feeding on a conspecific. (**e**) *Ctenus* sp. feeding on a gryllid: the spider can capture a prey without releasing the previous catch (by L. F. Garcia). (**f**) *Ctenus sp.* feeding on a manaosbiid harvestmen (by L.F. Garcia)

Prey-to-predator size ratio matters, and spiders' prey may vary a lot in size (Erickson and Morse 1997). They tend to attack prey smaller than themselves, often less than two-thirds their size (Henschel 1994). However, prey twice the size of the spider can also be subdued (Salticidae: Bartos 2004; Thomisidae: Guseinov 2006).

Spiders may also select prey according to nutrient composition and, which is maybe more impressive, selectively ingest protein or lipids from a prey item according to its own nutritional status (Mayntz et al. 2005). Eating selectively may be very important, and feeding indirectly on nectar (by eating prey that fed on nectar), for example, may increase survival, growth, and fecundity in *Cheiracanthium inclusum* (Taylor and Pfannenstiel 2009). In *Schizocosa*, spiders with a high-quality diet mature faster, are larger, and have better body condition indexes than spiders fed on a low-quality diet (Hebets et al. 2008). Long-term diet may affect some traits associated with mating success also in *Schizocosa stridulans* (Rosenthal and Hebets 2015). In *Paratrechalea ornata*, males in good condition may have higher mating success than males in poor condition (Albo et al. 2014). Finally, past diet influences future prey choice (Schmidt et al. 2012). Foraging in spiders is therefore a complex matter, and these animals are by no way generalists that just eat the first animal they find. There is a high interspecific variation, and prey choice is really important.

Diet in Neotropical Spiders

Ctenidae is the most studied Neotropical family of cursorial spiders when it comes to foraging and behavior as a whole. They are medium to large spiders that wander or sit and wait on trees and on leaf litter depending on the species, with males often wandering more than females (Schmitt et al. 1990; Schuster et al. 1994; Salvestrini and Gasnier 2001; Gasnier et al. 2002). *Ctenus*, *Cupiennius* and *Enoploctenus* leave their retreat at night, preying upon cockroaches, crickets, earwigs, flies, grasshoppers, moths, termites, and *Ctenus* spiders (Barth and Seyfarth 1979; Hofer et al. Höfer et al. 1994; Willemart and Kaneto 2004). *Enoploctenus cyclothorax* was found to usually reject the armored harvestmen *Mischonyx cuspidatus* (Willemart and Kaneto 2004), and Willemart and Pellegatti-Franco (2006) found that almost 80% of the spiders rejected this harvestman even after having been starving for more than 2 months. In contrast, crickets offered simultaneously for control spiders were all eaten within 13 h. It was later found that the cause of rejection is probably the thick exoskeleton of armored harvestmen (Laniatores) as a whole (see below).

Few Neotropical cursorial species have had their diet studied in detail. Nentwig (1986) studied the diet of seven species of cursorial spiders in the laboratory. He offered a variety of prey. Relatively softly chitinised insects (Diptera, Lepidoptera, Homoptera, Miridae, Ensifera) had acceptance rates of more than 50% by spiders of the families Lycosidae, Thomisidae, Salticidae, and Linyphiidae. Arthropods with a thick cuticle (e.g., Coleoptera) and arthropods that are aggressive and/or are chemically defended (e.g., Formicoidea, many Heteroptera, Myriapoda) were mainly refused. The large *Cupiennius* was an exception, accepting sometimes chemically defended and armored prey. Nentwig (1986) has shown that *Cupiennius* (Ctenidae) is at the polyphagous end of the spectrum, and *Misumena* (Thomisidae) was the most specialized studied species, with *Pisaura* (Pisauridae), *Evarcha* (Salticidae), *Xysticus* (Thomisidae), *Pardosa* (Lycosidae), and *Tibellus* (Philodromidae) in between, in this order. Nentwig (1986) suggests that the degree of polyphagy is influenced by the spider habitat and the availability of prey. For example, whereas *Cupiennius* has a large array of prey items available in the forest, *Tibellus* in meadows and *Misumena* on flowers have a less diverse array of prey available.

Thomisids (popularly known as crab spiders) are commonly found on flowers where they wait for prey. Romero and Vasconcellos-Neto (2003) found that the crab spider *Misumenops argenteus* feeds on a variety of insects that visit the flowers it forages on (Asteraceae). The authors have observed 76 spiders feeding in the field, and prey consisted of herbivores (43.5%), pollinators (8%), parasitoids (12%), and predators (23%). The rest were other arthropods. Although *M. argenteus* is polyphagous, it attacked mostly prey that stayed longer on the plant (such as prey that got stuck within the trichomes of the leaves) or wingless insects (Romero and Vasconcellos-Neto 2003).

The spitting spider *Scytodes longipes* ("other hunters" guild) is a polyphagous species that exhibit preferences based on prey morphology. This synanthropic species with a body length of 10 mm in captivity studies show a great preference for Ensifera and other spiders as a prey. They refuse hard-chitinised prey such as beetles, and dangerous prey such as bees and wasps and prey larger than 24 mm (Nentwig 1985). Differently from the spitting spider, the diet of the recluse spider *Loxosceles* includes hard-bodied prey such as beetles and isopods, and dangerous prey such as ants (Fischer et al. 2006; García et al. 2016). Other species known to include ants in their diet are the sand-dwelling spiders *Allocosa alticeps* and *A. brasiliensis* (Lycosidae) (Aisenberg et al. 2009). The remaining prey items consumed are insects, mainly beetles and dipterans, the latter being caught mainly during their nuptial swarms. The authors suggest that these *Allocosa* are actually opportunistic feeders (Aisenberg et al. 2009).

Lycosids may also possibly prey upon vertebrates. Though vertebrates are often considered as spider predators and not as their prey, spiders from several families, including Ctenidae, Pisauridae, Trechaleidae, Lycosidae, Sparassidae, and mygalomorphs were observed consuming this prey category. Theraphosids, for example, were observed preying upon several species of anurans and caecilians (Menin et al. 2005), snakes (Borges et al. 2016), and lizards (Vieira et al. 2012). Arboreal species of the genus *Avicularia* have been observed preying on small mammals such as bats (Nyffeler and Knörnschild 2013). Because *Ctenus* and *Ancylometes* are very abundant on the forest ground in Amazonia, Menin et al. (2005) have suggested that predation on vertebrates by spiders is ecologically important. Finally, there are several records of large ctenid and trechaleid spiders preying on various species of fish and anurans (Höfer and Brescovit 2000; Zina and Gonzaga 2006; Nyffeler and Pusey 2014).

Choice of Foraging Site

When food resources become scarce, spiders are at risk of starving if they stay at the same locality for a long time. This is often a determining factor triggering displacement to another site (Wagner and Wise 1997). In two North-American species of the genus *Pardosa* and *Hogna*, experiments showed that well-fed individuals have a lower mobility when compared with starved individuals (Walker et al. 1998). Once the spider arrives at a new locality, it must evaluate several factors before deciding to stay there, including the presence of prey and predators. The decrease in foraging activities in the presence of predators is a common behavior in several animals including spiders. Eiben and Persons (2007) evaluated the effect of vibrations, chemical and visual cues of the predatory North American lycosid spider *Pardosa milvina* on the activity of another lycosid *Rabidosa rabida*. The predator cues, mainly chemical ones, inhibited the activity of *R. rabida*. *Pardosa* in turn, are preyed upon by the larger *Hogna*. *Pardosa* discriminates the quantity of predatory cues and how old the cues are, behaving accordingly (Persons and Rypstra 2001; Barnes et al. 2002; Rypstra et al. 2007). Predator diet also matters: cues from the larger spider *Hogna* fed the smaller *Pardosa* elicited a significantly greater reduction in activity in *Pardosa* than *Hogna* fed crickets (Persons et al. 2001). Moreover, *Pardosa* have been shown to avoid pitfall traps with *Hogna* but did not avoid empty traps or traps with crickets, indicating the important role of olfaction (Schonewolf et al. 2006).

The presence of prey or prey cues when choosing a foraging site also matters. The spider *Pardosa ramulosa* is more frequently found next to pools of water with aquatic insects such as dipteran larvae (*Aedes* and *Ephydra*) and the heteropteran *Trichocorixa*, common prey for these spiders (Greenstone 1983). The European wolf spider *Pardosa saltans* chooses foraging sites containing silk and feces left by smaller spiders and conspecifics, both potential prey (Wetter et al. 2012). The North-American wolf spider *Schizocosa ocreata* remains longer when exploring sites with cricket cues (Persons and Uetz 1996). Chemicals may sometimes repel spiders: individuals of *Pisaura mirabilis* move away from different extracts of flowers that contain b-caryophyllene and nerolidol. However, the thomisid *Misumena vatia* that usually forages on flowers is not repelled by these chemicals (Junker et al. 2011).

The substrate to wait for prey is also considered by spiders. The thomisid *Misumenops argenteus* prefers flowers of *Trichogoniopsis adenantha* that match the spider colors (Heiling et al. 2005). Romero (2001) showed that this spider occurs most frequently on this plant when compared to other plants in the same environment. These flowers also offer a wide variety of prey to the spider by attracting different phytophagous insects such as heteropterans, orthopterans, aphids, and lepidopterans, though also parasitoids such as Braconidae and Pteromalidae (Romero and Vasconcellos-Neto 2003). Nevertheless, because some pollinator insects avoid flowers with crab spiders (Heiling et al. 2003; Dukas and Morse 2005; Gonçalves-Souza et al. 2008; Romero et al. 2011; Llandres and Rodríguez-Gironés 2011), and assuming that predators may also do it, an alternative hypothesis is that the flower color choice is actually an antipredatory behavior.

Specifically when ambush prey, the type of substrate is also relevant because cursorial spiders often rely on substrate-borne vibrations to find their prey, and distinct substrates propagate vibrations differently. Some Neotropical ctenids, such as *Cupiennius salei*, prefer to forage on leaves of banana plants, which are good conductors of vibrations produced by prey (Barth et al. 1988; Barth 2002). Finally, habitat complexity may negatively affect prey capture in *Pardosa milvina*, but it offers protection against the large spider *Hogna helluo* (Rypstra et al. 2007). Therefore, a good foraging site is definitely not a randomly chosen one but depends on the physiological conditions of the spider, presence of prey and predators or their cues, adequate substrate, and habitat complexity.

Prey Detection

Spider legs, in addition to being used in locomotion, prey capture, mating behavior, and web building, possess various receptors of external stimuli (Foelix 2011). One of the main receptors is the trichobothrium, a structure located on the tarsi, metatarsi, and tibia of pedipalps and legs (Barth 1982). A trichobothrium is usually a long and thin hair that emerges from a socket with sensory cells connected to the base of the hair (Reissland and Görner 1985). About 900 trichobothria can be found on the legs and pedipalps of the ctenid *Cupiennius salei* (Barth and Holler 1999). Because groups of trichobothria may be arranged in different spatial combinations and because the length of the hair shafts are variable, they respond to a wide range of wavelength frequencies, including vibrations produced by some flying insects such as flies that tend to be very turbulent and easily detectable (Barth 2002). Other important sensors for detecting prey are the metatarsal lyriform organs, which are slit sensilla located on the distal region of the metatarsus. The slit sensilla are areas with thinner cuticle in the exoskeleton that are very susceptible to deformations (Young et al. 2014), which are readily detected and transmitted to the nervous system (Barth et al. 1993).

The Neotropical spider *Cupennius salei* has been used as a study model of receptors and sensory organs (Barth 1985, 2002, Patil et al. 2006b; McConney et al. 2009; Young et al. 2014), providing most of the information we know about trichobothria and metatarsal lyriform organs. The intensity of the vibrations produced by the movements of the prey varies with the distance and therefore indicates how far the prey is from the spider (Hill 2009). In addition, spiders only respond to some of a range of vibrational cues, and can distinguish between vibrations produced by conspecifics of the opposite sex from other sources such as abiotic factors (wind) or potential prey (Barth 2002; Hill 2009).

Olfactory receptors are also present in spiders, being located on distal parts of the legs and pedipalps, in a structure called the tarsal organ, which encloses six or seven innervated sensilla with a pore at the tip (Foelix and Chu-Wang 1973). During prey detection, olfactory receptors have been shown to play a role. The spider *Cupiennius salei* (Hostettler and Nentwig 2006) is able to distinguish between cricket gel models with and without cricket smell, preferring those with scent, suggesting these spiders use olfactory cues for prey detection. Another example is *Falconina gracilis* (Coriniidae), which is able to detect the pheromones produced by its prey, the ant *Acromyrmex landolti fractixcornis* (Fowler 1981).

Visual cues can also be used for prey choice and capture behavior. Spiders generally possess eight eyes that are arranged in pairs and named according to their positions: anterior median, anterior lateral, posterior median and posterior lateral eyes. The anterior median eyes can receive and transmit more complex stimuli than other eyes (Barth 2002), having sometimes extraordinary spatial resolution. That is the case with spiders in the family Salticidae, which have anterior median eyes much larger than the other eyes. The African jumping spider *Evarcha culicivora* is able to visually distinguish *Anopheles* mosquitoes which recently fed on vertebrate blood from *Anopheles* which did not feed on blood, and to distinguish *Anopheles* from *Culex* by their posture when resting (Jackson et al. 2005; Nelson and Jackson 2006, 2012). With regard to prey capture, *Corythalia albicincta* jumps on the prey from a greater distance (~2.5 times its body length) when dealing with fast-moving crickets or adult flies compared to fly larvae (~1 time its body length) (Aguilar-Argüello and García-Chávez 2015). These jumping spiders also behaved differently when dealing with different prey: the latency to jump on larvae is higher, maybe because they do not move as much as the other two insects used as prey items. These behaviors are probably possible because of salticid's great vision. Other spiders that rely on vision for prey capture are the ogre-faced spiders (genus *Deinopis*). These are commonly seen at night in Neotropical forests holding their nets waiting for prey. Getty and Coyle (1996) have described the prey-capture behavior of these amazing spiders. They may either strike forward to capture wandering prey or backwards towards mechanical stimuli (vocalizations or vibrating tuning forks, aerial strikes). *Deinopis* has recently been found to also rely heavily on its posterior median gigantic eyes to capture prey (Stafstrom and Hebets 2016). Interestingly, another nocturnal species, *Cupienius salei*, which is known to heavily rely on mechanical stimuli to detect prey and does not have particularly large eyes, may also use its eyes to detect movement and attack (Fenk et al. 2010). In lycosids, visual information may also influence patch residence time (Persons and Uetz 1997).

There is therefore evidence of spiders using an array of sensory stimuli to detect prey, including air- and substrate-borne vibrations, volatiles, and visual stimuli. We do not know much about the interaction and relative importance of these sensory modalities in Neotropical spiders for detecting prey. However, studies on salticids, lycosids, and ctenids mentioned above have shown that these animals can definitely use more than one sensory modality when foraging.

Prey Capture with Emphasis on Neotropical Spiders

Most predators try not to be detected by their prey, and camouflage may therefore be useful: the light-colored salticid *Yllenus arenarius* attacked from closer distances, approached prey faster, and was more successful when camouflaged in light substrates than in darker ones (Bartos et al. 2013). However, other species do the opposite and actually expose themselves to attract prey, sometimes exploiting the sensory system of their prey: crab spiders hunting honeybees on flowers may attract prey with UV reflectance on their bodies (Llandres and Rodríguez-Gironés 2011). The white patches on the forelegs of *Dolomedes raptor* also attract prey: dummies with patches attracted more grasshopper prey than dummies without the patches, and grasshoppers were more attracted to spiders when their white patches were present (Tso et al. 2016). After prey attraction, spiders need to efficiently attack and hold them. Holding prey requires adhesive setae on the ventral region of the legs in some species, with the predator jumping on the prey with the legs forming a basket that involves the prey (Lycosidae: Rovner 1980). In contrast, *Scytodes* immobilize prey by spitting a mixture of glue, silk, and venom from the chelicerae (Suter and Stratton 2005). Horizontal web sheets may also be used in cursorial species such as the recluse spider in the genus *Loxosceles* (Cramer 2015). Behaviors displayed when capturing prey may vary depending on the prey and ontogenetically in the speed, direction of approach, and prey manipulation among others (Bartos 2007, 2008; Pekár and Lubin 2009; Pekár and Haddad 2011; Bartos and Szczepko 2012).

Animals have to save energy whenever possible if food availability is unpredictable. *Cupiennius salei* can control the amount of the costly venom they inject when subduing prey according to the size of the prey and the intensity and duration of struggling movement (Malli et al. 1998, 1999). Prey items that are easier to capture such as stick insects and crickets receive less venom than blowflies and beetles, which are harder to subdue (Wigger et al. 2002). Moreover, individuals of *C. salei* are aware of the amount of venom available in their glands: when experimentally venom-depleted, they orient towards prey less often and display a decrease in their attack rates (Hostettler and Nentwig 2006). Moreover, they choose prey accordingly. Prey more sensitive to their venom are preferred over prey less sensitive when their glands are experimentally emptied (Wullschleger and Nentwig 2002). What is even more amazing is that they can make such choices using olfaction only (Hostettler and Nentwig 2006). This ability to control venom use according to each prey makes sense in the diverse Neotropical forests where these spiders live, where a wide array of prey are available for this polyphagous species. If there is a relationship between stenophagy and euryphagy and venom control, we could predict that specialized spiders such as some dysderids (woodlouse eaters) or salticids (spider eaters) would vary less the amount of venom injected in their prey.

Dealing with Dangerous Prey

Biting may require extra care when spiders are dealing with well-defended prey such as arthropods that are chemically defended. The chemical defenses used by harvestmen (Arachnida, Opiliones), if experimentally applied to a palatable prey, may repel trechaleid and ctenid spiders (Machado et al. 2005), which sometimes drop the prey and rub their mouthparts against the substrate after contacting the defensive droplet (but see Souza and Willemart 2011: *Enoploctenus cyclothorax* did not release prey after contacting the secretion of *Discocyrtus invalidus*). The predation strategy of recluse spiders (*Loxosceles*) is more efficient in avoiding defensive secretions because the spider bites only the legs of these harvestmen, therefore decreasing the chances of coming into contact with their defensive secretions (Segovia et al. 2015a).

Some armored harvestmen in the suborder Laniatores have dangerous sharp spines on legs IV that can pierce a spider abdomen (Segovia et al. 2015b). Such pinching with legs IV startles large ctenids: they may move away from the prey even when the spines do not touch the spider (Dias et al. 2014). *Loxosceles*, however, carefully approach such armored harvestmen, avoiding proximity with legs IV, spiny pedipalps, and chelicerae of these prey (Segovia et al. 2015a). In a spider-spider interaction, the spitting spider *Scytodes globula* occasionally invades webs of other spiders such as *Metaltella simony* to prey upon them. However, the host spider may react defensively and make the spitting spider spit defensively, aborting the attack (Escalante et al. 2015).

Whereas ants are often avoided by some spiders, others include ants in their diet. That is the case of *Loxosceles*, polyphagous animals that efficiently subdue these dangerous prey (García et al. 2016). However, some spider species actually specialize in ants. The aphantochilid *Aphantochilus rogersi* in late instars preferentially attacks ants from behind, probably avoiding their dangerous mandibles (Oliveira and Sazima 1984; Castanho and Oliveira 1997). Younger specimens use a different strategy because probably they can only seize the ant's petiole tightly if they approach the ant from the front (Castanho and Oliveira 1997). While spiders are eating, patrolling ants may pass nearby and occasionally approach the spider, which raises the dead ant and shows it to the approaching ant, as if it were an ant carrying a dead nest mate (Oliveira and Sazima 1984). The morphological similarity between *A. rogersi* and their model ants of the tribe Cephalotini, however, is thought to help them avoiding predation, since visually guided predators often avoid ants (Oliveira and Sazima 1984).

Dealing with Prey Armor

Some arthropods with a hard exoskeleton are sometimes rejected by spiders after these touch the prey (Nentwig 1985; Eisner et al. 2004; Carvalho et al. 2012; Souza and Willemart 2013; Dias and Willemart 2013). Because laniatorid harvestmen are well protected within a rigid armor, spiders that bite in random areas of the prey have low success (successful captures: *Enoploctenus cyclothorax* 2/40, *Ctenus ornatus*: 5/34) (Souza and Willemart 2011; Dias and Willemart 2013). The spitting spider *Scytodes globula* has been shown to never overcome prey defenses (0/33) (Carvalho et al. 2012). In contrast, the recluse spider (*Loxosceles gaucho*) only bites the non-sclerotized distal parts of the legs and the soft articulations. It successfully killed and ate armored harvestmen in 31/38 opportunities (Segovia et al. 2015a). The scorpion *Bothriurus bonariensis* stings at the mouth, another soft part of a harvestman's body. Its success was 35/58 (Albin and Toscano-Gadea 2015). The spider *Ctenus fasciatus* is known to feed on Goniosomatinae harvestmen in caves (Gnaspini 1996), contrasting with studied ctenids that are often unsuccessful when attacking armored harvestmen (Souza and Willemart 2011; Dias and Willemart 2013). Prey capture has not been studied in *Ctenus fasciatus*, but its strategy is possibly similar to that of other ctenids: jumping on the prey, sometimes manipulating it a bit and biting in random areas. However, because *C. fasciatus* are very large, heavy-bodied, and have big chelicerae, they possibly overpower the armor by simply breaking it with their fangs, which *E. cyclothorax* rarely manages to do and *C. ornatus* only occasionally does.

Some spider species included in the families Lycosidae, Trechaleidae, and Ctenidae quickly jump on the prey and bite it. However, spiders such as *Scytodes* and *Loxosceles* (Ades and Ramires 2002; Segovia et al. 2015a; García et al. 2016) carefully approach and tap prey before spitting venom or biting. We know nothing about what information these spiders are accessing when tapping prey. Spiders do have contact chemoreceptors on their legs, but are they accessing prey chemicals in order to decide whether or not to continue attacking? If they are, we could be able to experimentally fool these spiders by applying chemicals of other animals such as mates or predators. Is contact required to identify prey or only to search for vulnerable areas to bite? Why do some spiders tap prey, while others do not and simply jump on the prey? With regard to the adhesive hairs needed to hold prey after jumping on it (Rovner 1980), we know autotomy brings costs to foraging spiders by reducing capture rate in complex environments (Wrinn and Uetz 2008). Since many legs are used for holding prey (Rovner 1980), is it possible that autotomy affects capture success (Amaya et al. 2001)? Do ctenids or lycosids have more adhesive hairs than sicariids or scytodids? These and several other questions remain still unknown.

Handedness in Spiders?

Laterality is common in animals, both invertebrates and vertebrates, including humans (see references in Benelli et al. 2015). Gorillas may be right-handed for most of their frequent intraspecific gestures (Prieur et al. 2016), left-handed humans may be favored for example by their surprise effects when fighting the majority of right-handed fighters (Pollet et al. 2013), and octopuses may use one eye more than the other to look at stimuli using monocular vision (Byrne et al. 2004). Cesar Ades, one of the researchers responsible for the spread of ethological studies in Brazil, with his former student Eduardo Ramires, provided evidence that spiders may also have behavioral lateralization (Ades and Ramires 2002). The authors first detected that *Scytodes globula* collected in the field were missing left legs I and II more often than right legs I and II. Knowing that these spiders usually attack after touching the prey, the authors hypothesized that they use left legs more often than their right counterparts to tap prey. They brought *Loxosceles* to the laboratory to use them as prey and, indeed, when *Loxosceles* managed to bite a leg of *Scytodes*, the latter autotomized the leg caught. In one case where the spider did not autotomize the leg, it died. Their results showed that *Scytodes* as a whole used their left legs I and II more often than the right legs I and II to probe prey. Though the authors could not provide a clear explanation as to why lateralization occurs, the results were clear and the first example of lateralization in arachnids. As suggested by the authors, it would be nice to know whether there is an individual consistency in handedness.

Should We Expect Differences in Foraging in the Neotropics Compared to Temperate Regions?

Before we start discussing this question, we have to bear in mind that the Neotropical region encompasses very distinct ecosystems, with differences in biotic and abiotic factors. We are grouping Patagonia, where temperatures may reach −15 Celsius in winter, with Manaus in the Amazon rainforest where temperatures are rarely under 20 °C and often above 30 °C. However, the Neotropical region that is actually between or close to the tropics is fairly different from temperate regions as a whole, for example. We can therefore attempt to extract potentially useful information in an attempt to understand different selective pressures that may influence foraging.

Moya-Laraño (2010) proposed that higher temperatures and water availability (which is often the case in the tropics) could permit higher rates of mobility in organisms, maybe leading to higher rates of encounter among individuals. This could lead to the prediction that spiders in Neotropical forests would wander more than spiders in temperate forests. However, spiders are prey to a diverse array of predators among invertebrates (including other spiders) and vertebrates (amphibians, reptiles, mammals), which are both more diverse in the tropics than in temperate regions. Moreover, there is evidence at least in invertebrates and fishes that predation pressure increases toward the tropics (Schemske et al. 2009). This would lead to the opposite prediction; that spiders should wander less in the tropics to avoid predation. Of course, other factors should also be considered: for example, higher temperatures per se may increase the locomotor activity of wandering spiders (Ford 1978), which may bias comparisons between temperate and tropical forests.

Several spiders in temperate regions overwinter as immature or adults in areas where temperature may be well below 0 °C. They deal with such conditions by choosing appropriate microhabitats, increasing resistance to cold or reducing their metabolic rate (Foelix 2011). Some long-lived spiders in temperate regions, such as *Dolomedes* (Pisauridae), may take 2–3 years to become adults in northern regions of their distribution (Jones et al. 2001). Individuals of *Pardosa lugubris* (Lycosidae) in Scotland or Netherlands may need 2 years to become adults in nature (Jones et al. 2001). That means they overwinter as immature, dealing with very low temperatures in winter. Do low temperatures influence foraging? In two species of crab spiders (Thomisidae) from New Jersey (USA), there is no relationship between the number of prey captured in the field and temperature, controlled per prey type (Schmalhofer and Casey 1999). With laboratory experiments, Schmalhofer and Casey (1999) also asked whether temperature affected the hunting performance (HP) of these spiders on flies. They measured HP in the laboratory according to the equation: $HP = ab/c$, where “a” indicates whether or not a spider made a kill (yes = 1, no = −1), “b” is the number of strikes made by a spider, and “c” indicates the number of opportunities the spider had to strike at prey. It is important to state that such a formula does not directly evaluate physiological effects of temperature but only the final outcome, that is, what the spider actually ingests. The authors have shown that temperature did not influence HP, which they attribute to muscle physiological adaptations for burst activity and/or the use of venom. Hunting performance being unaffected is important not only because of daily variations but also because of seasonal variations, allowing these crab spiders to maintain feeding rates also in the reproductive season, when temperature declines for one of the species studied. Foraging success is correlated with reproductive success also in crab spiders (Morse and Stephens 1996). We can ask if foraging thomisids in the Neotropical region, where the spiders do not have to deal with such low temperatures, are also unaffected by temperature. Or is it an adaptation to areas with greater temperature variation? Similar tests could be done comparing pisaurids and lycosids from temperate and Neotropical regions. We do not know much about this topic in Neotropical species, but in some warm desert web-building species, low temperatures may affect latency to attack, duration of prey-capture sequences, web mass, and time allocated to feeding in spiders in the families Eresidae, Theridiidae, and Agelenidae (Riechert and Tracy 1975; Lubin and Henschel 1990; Henschel et al. 1992; Turner et al. 1993). Spiders may also thermoregulate behaviorally (Humphreys 1978, 1987), which can also influence metabolism and therefore foraging, but this is widely unexplored in spiders. While most studies on the influences of temperature on behavior and physiology in arthropods have been conducted in insects (Chown and Nicolson 2004), the wide distribution of some spiders certainly contributes to studies comparing latitudinal effects of temperature on foraging behavior.

Concluding Remarks

Hunting spiders are widely distributed and present a high diversity in the Neotropics, also comprising a wide diversity of feeding strategies. When compared to the temperate regions, knowledge about feeding behavior on Neotropical spiders is scarce, since there are few studied species and in some cases the same species is used as a model for many studies, such as *C. salei*. Additionally, research on Neotropical hunting spiders has focused mainly on sexual and reproductive behaviors, while studies about feeding behavior are still lacking. Neotropical hunting spiders are good models for the study of predatory behavior because they include a wide variety of prey from small or dangerous insects such as ants, up to vertebrates such as bats. They also play an important role as predators of different ecosystems such as crops (see Chap. 11), and are suitable models for the study of evolutionary and ecological questions, and also of applied problems related to trophic ecology. All these characteristics call for the need to intensify studies of the predatory behavior of these spiders.

References

Ades C, Ramires EN (2002) Asymmetry of leg use during prey handling in the spider *Scytodes globula* (Scytodidae). J Insect Behav 15:563–570

Aguilar-Argüello SO, García-Chávez JH (2015) Importance of hunger and prey type on predatory behavior stages in *Corythalia albicincta* (Araneae: Salticidae). J Arachnol 43:143–151

Aisenberg A, González M, Laborda A, Postiglioni R, Simó M (2009) Reversed cannibalism, foraging, and surface activities of *Allocosa alticeps* and *Allocosa brasiliensis*: two wolf spiders from coastal sand dunes. J Arachnol 37:135–138

Albín A, Toscano-Gadea CA (2015) Predation among armored arachnids: *Bothriurus bonariensis* (scorpions, Bothriuridae) versus four species of harvestmen (harvestmen, Gonyleptidae). Behav Process 121:1–7

Albo MJ, Melo-González V, Carballo M, Baldenegro F, Trillo MC, Costa FG (2014) Evolution of worthless gifts is favoured by male condition and prey access in spiders. Anim Behav 92:25–31

Amaya CC, Klawinski PD, Formanowicz DR (2001) The effects of leg autotomy on running speed and foraging ability in two species of wolf spider, (Lycosidae). Am Midl Nat 145:201–205

Barnes MC, Persons MH, Rypstra AL (2002) The effect of predator chemical cue age on chemically-mediated antipredator behavior in the wolf spider *Pardosa milvina* (Araneae: Lycosidae). J Insect Behav 15:269–281

Barth FG (1982) Vibratory communication in a spider. In: Barth FG (ed) Joint Symp Neurobiology and Strategies of Adaptation. Universitatsdruck, Frankfurt am Main, pp 1–9

Barth FG (1985) Neuroethology of the spider vibration sense. In: Barth FG (ed) Neurobiology of arachnids. Springer, Berlin/Heidelberg/New York/Tokyo, pp 203–229

Barth FG (2002) A spider ´s world: senses and behavior. Springer, Berlin

Barth FG, Holler A (1999) Dynamics of arthropod filiform hairs. V. The response of spider trichobothria to natural stimuli. Phil Trans R Soc Lond B 354:183–192

Barth FG, Seyfarth EA (1979) *Cupiennius salei* keys. (Araneae) in the highlands of central Guatemala. J Arachnol 7:255–263

Barth FG, Bleckmann H, Bohnenberger J, Seyfarth EA (1988) Spiders of the genus *Cupiennius* Simon 1891 (Araneae, Ctenidae). II On the vibratory environment of a wandering spider Oecologia 77:194–201

Barth FG, Wast U, Humphrey JAC, Devarakonda R (1993) Dynamics of arthropod filiform hairs. II. Mechanical properties of spider trichobothria (*Cupiennius salei*). Phil Trans R Soc Lond B 340:445–461

Bartos M (2004) The prey of *Yllenus arenarius* (Araneae, Salticidae). Bull Br Arachnol Soc 13:83–85

Bartos M (2007) Hunting prey with different escape potentials—alternative predatory tactics in a dune dwelling salticid. J Arachnol 35:499–508

Bartos M (2008) Alternative predatory tactics in a juvenile jumping spider. J Arachnol 36:300–305

Bartos M (2011) Partial dietary separation between coexisting cohorts of *Yllenus arenarius* (Araneae: Salticidae). J Arachnol 39:230–235

Bartos M, Szczepko K (2012) Development of prey-specific predatory behavior in a jumping spider (Araneae: Salticidae). J Arachnol 40:228–233

Bartos M, Szczepko K, Stanska M (2013) Predatory response to changes in camouflage in a sexually dimorphic jumping spider. J Arachnol 41:381–386

Benelli G, Romano D, Messing RH, Canale A (2015) Population-level lateralized aggressive and courtship displays make better fighters not lovers: evidence from a fly. Behav Process 115:163–168

Borges LM, Da Rosa CM, Franzoi Dri G, Bertani R (2016) Predation of the snake *Erythrolamprus almadensis* (Wagler,1824) by the tarantula *Grammostola quirogai* Montes De Oca,D'Elía and Pérez-Miles, 2016. Herpetology Notes 9:321–322

Byrne RA, Kuba MJ, Meisel DV (2004) Lateralized eye use in *Octopus vulgaris* shows antisymmetrical distribution. Anim Behav 68:1107–1114

Cardoso P, Pekár S, Jocqué R, Coddington JA, Catterall C (2011) Global patterns of guild composition and functional diversity of spiders. PLoS One 6:e21710. https://doi.org/10.1371/journal.pone.0021710

Carvalho LA, Souza ES, Willemart RH (2012) Behavioral analysis of the interaction between the spitting spider *Scytodes globula* (Araneae: Scytodidae) and the harvestman *Discocyrtus invalidus* (Opiliones: Gonyleptidae). J Arachnol 40:332–337

Carvell GE, Kuja JO, Jackson RR (2015) Rapid nectar-meal effects on a predator's capacity to kill mosquitoes. R. Soc. open sci 2:140426. https://doi.org/10.1098/rsos.140426

Castanho LM, Oliveira PS (1997) Biology and behaviour of the Neotropical ant-mimicking spider *Aphantochilus rogersi* (Araneae: Aphantochilidae): nesting, maternal care and ontogeny of ant-hunting techniques. J Zool 242:643–650

Chown SL, Nicolson SW (2004) Insect physiological ecology mechanisms and patterns. Oxford University Press, Oxford

Cramer KL (2008) Are brown recluse spiders, *Loxosceles reclusa* (Araneae, Sicariidae) scavengers? The influence of predator satiation, prey size, and prey quality. J Arachnol 36:140–144

Cramer KL (2015) Activity patterns of a synanthropic population of the brown recluse spider, *Loxosceles reclusa* (Araneae: Sicariidae), with observations on feeding and mating. J Arachnol 43:67–71

Dias BC, Souza ES, Hara MR, Willemart RH (2014) Intense leg tapping behavior by the harvestman Mischonyx cuspidatus (Gonyleptidae): An undescribed defensive behavior in Opiliones? J Arachnol 42:123–125

Dias BC, Willemart RH (2013) The effectiveness of post-contact defenses in a prey with no pre-contact detection. Zoology 116:168–174

Dukas R, Morse DH (2005) Crab spiders show mixed effects on flower-visiting bees and no effect on plant fitness components. Écoscience 12:244–247. https://doi.org/10.2980/i1195-6860-12-2-244.1

Eiben B, Persons M (2007) The effect of prior exposure to predator cues on chemically-mediated defensive behavior and survival in the wolf spider *Rabidosa rabida* (Araneae : Lycosidae). Behaviour 144:889–906

Eisner T, Rossini C, González A, Eisner M (2004) Chemical defense of an opilionid (*Acanthopachylus aculeatus*). J Exp Biol 207:1313–1321

Erickson KS, Morse DH (1997) Predator size and the suitability of a common prey. Oecologia 109:608–614

Escalante I, Aisenberg A, Costa FG (2015) Risky behaviors by the host could favor araneophagy of the spitting spider *Scytodes globula* on the hacklemesh weaver *Metaltella simoni*. J Ethol 33:125–136

Fenk LM, Hoinkes T, Schmid A (2010) Vision as a third sensory modality to elicit attack behavior in a nocturnal spider. J Comp Physiol A 196:957–961. https://doi.org/10.1007/s00359-010-0575-8

Fischer ML, Vasconcellos-Neto J, dos Santos Neto LG (2006) The prey and predators of *Loxosceles intermedia* Mello-Leitao 1934 (Araneae, Sicariidae). J Arachnol 34:485–488

Foelix RF (2011) Biology of spiders, 3rd edn. Oxford University Press, New York

Foelix RF, Chu-Wang IW (1973) The morphology of spider sensilla II. Chemoreceptors. Tissue Cell 5(3):461–478. https://doi.org/10.1016/S0040-8166(73)80038-2

Ford MJ (1978) Locomotory activity and the predation strategy of the of the wolf-spider *Pardosa amentata* (Clerck) (Lycosidae). Anim Behav 26:31–35

Fowler HG (1981) Behavior of two myrmecophiles of Paraguayan leaf-cutting ants. Revista Chilena Entomología. Retrieved from http://www.insectachile.cl/rchen/pdfs/VOL._11_(1981)/Fowler_1981.pdf

García LF, Franco V, Robledo-Ospina LE, Viera C, Lacava M, Willemart RH (2016) The predation strategy of the recluse spider *Loxosceles rufipes* (Lucas, 1834) against four prey species. J Insect Behav. https://doi.org/10.1007/s10905-016-9578-9

Gasnier TR, Azevedo CS, Torres-Sanchez MP, Hubert H (2002) Adult size of eight hunting spider species in central Amazonia: temporal variations and sexual dimorphisms. J Arachnol 30:146–154

Getty RM, Coyle FA (1996) Observations on prey capture and anti-predator behaviors of ogre-faced spiders (*Deinopis*) in southern Costa Rica (Araneae, Deinopidae). J Arachnol 24:93–100
Gnaspini P (1996) Population ecology of *Goniosoma spelaeum*, a cavernicolous harvestman from southeastern Brazil (Arachnida: Opiliones: Gonyleptidae). J Zool 239:417–435
Gonçalves-Souza T, Llandres AL, Omena PM, Souza JC, Romero GQ (2008) Trait-mediated effects on flowers: artificial spiders deceive pollinators and decrease plant fitness. Ecology 89:2407–2413. https://doi.org/10.1890/07-1881.1
Greenstone MH (1983) Site-specificity and site tenacity in a wolf spider : a serological dietary analysis. Oecologia 56:79–83
Guseinov EFO (2006) The prey of a lithophilous crab spider *Xysticus loeffleri* (Araneae, Thomisidae). J Arachnol 34:37–45
Hebets EA, Wesson J, Shamble PS (2008) Diet influences mate choice selectivity in adult female wolf spiders. Anim Behav 76:355–363
Heiling AM, Herberstein ME, Chittka L (2003) Crab spiders manipulate flower signals. Nature 421:334
Heiling A, Chittka L, Cheng K, Herberstein M (2005) Colouration in crab spiders: substrate choice and prey attraction. J Exp Biol 208:1785–1792
Henschel JR (1994) Diet and foraging behaviour of huntsman spiders in the Namib dunes (Araneae: Heteropodidae). J Zool 34:239–251
Henschel JR, Ward D, Lubin YD (1992) The importance of thermal factors for nest-site selection, web construction and behaviour of *Stegodyphus lineatus* (Araneae: Eresidae) in the Negev Desert. J Therm Biol 17:97–106
Hill P (2009) How do animals use substrate-borne vibrations as an information source? Naturwissenschaften 96:1355–1371
Höfer H, Brescovit A (2000) A revision of the Neotropical spider genus *Ancylometes* Bertkau (Araneae: Pisauridae). Insect Syst Evol 31:323–360
Höfer H, Brescovit AD, Gasnier T (1994) The wandering spiders of the genus *Ctenu*s (Ctenidae: Araneae) of Reserva Ducke, a rainforest reserve in central Amazonia. Andrias 13:81–98
Hostettler S, Nentwig W (2006) Olfactory information saves venom during prey-capture of the hunting spider *Cupiennius salei* (Araneae: Ctenidae). Funct Ecol 20:369–375
Humphreys WF (1978) The thermal biology of *Geolycosa godeffroyi* and other burrow inhabiting Lycosidae (Araneae) in Australia. Oecologia 31:319–347
Humphreys WF (1987) The thermal biology of the wolf spider *Lycosa tarantula* (Araneae: Lycosidae) in northern Greece. Bull Br Arachnol Soc 7:117–122
I-Min T, Zhang S, Tan WL, Peng P, Blamires SJ (2016) Prey-luring coloration of a nocturnal semi-aquatic predator. Ethology 122:671–681
Jackson RR, Pollard SD, Nelson XJ, Edwards GB, Barrion AT (2001) Jumping spiders (Araneae: Salticidae) that feed on nectar. J Zool 255:25–29. https://doi.org/10.1017/S095283690100108X
Jackson RR, Nelson HJ, Sune GO (2005) A spider that feeds indirectly on vertebrate blood by choosing female mosquitoes as prey. PNAS 102:15155–15160
Jones D, Ledoux JC, Emerit M (2001) Guide des araignées et opilions d'Europe. Delachaux and Niestle, Geneve
Junker R, Bretscher S, Dötterl S, Blüthgen N (2011) Phytochemical cues affect hunting-site choices of a nursery web spider (*Pisaura mirabilis*) but not a crab spider (*Misumena vatia*). J Arachnol 39:113–117
Llandres AL, Rodríguez-Gironés MA (2011) Spider movement, UV reflectance and size, but not spider crypsis, affect the response of honeybees to australian crab spiders. PLoS ONE 6:e17136
Lubin YD, Henschel JR (1990) Foraging at the thermal limit: burrowing spiders (*Seothyra*, Eresidae) in the Namib Desert dunes. Oecol 84:461–467
Machado G, Carrera PC, Pomini AM, Marsaioli AJ (2005) Chemical defense in harvestmen (Arachnida, Opiliones): do benzoquinone secretions deter invertebrate and vertebrate predators? J Chem Ecol 31:2519–2539
Malli H, Imboden H, Kuhn-Nentwig L (1998) Quantifying the venom dose of the spider *Cupiennius salei* using monoclonal antibodies. Toxicon 36:1959–1969

Malli H, Kuhn-Nentwig L, Imboden H, Nentwig W (1999) Effects of size, motility and paralysation time of prey on the quantity of venom injected by the hunting spider *Cupiennius salei*. J Exp Biol 202:2083–2089

Mayntz D, Toft S (2006) Nutritional value of cannibalism and the role of starvation and nutrient imbalance for cannibalistic tendencies in a generalist predator. J Anim Ecol 75:288–297

Mayntz D, Raubenheimer D, Salomon M, Toft S, Simpson SJ (2005) Nutrient-specific foraging in invertebrate predators. Science 307:111–113

McConney ME, Schaber CF, Julian MD (2009) Surface force spectroscopic point load measurements and viscoelastic modelling of the micromechanical properties of air flow sensitive hairs of a spider (*Cupiennius salei*). J R Soc Interface 6:81–94

Menin M, Rodrigues DJ, Azevedo CS (2005) Predation on amphibians by spiders (Arachnida, Araneae) in the Neotropical region. Phyllomedusa 4:39–47

Morse DH, Stephens EG (1996) The consequences of adult foraging success on the components of lifetime fitness in a semelparous, sit and wait predator. Evol Ecol 10:361–373

Moya-Laraño J (2010) Can temperature and water availability contribute to the maintenance of latitudinal diversity by increasing the rate of biotic interactions? Open Ecol J 3:1–13

Nelson XJ, Jackson RR (2006) A predator from east Africa that chooses malaria vectors as preferred prey. PLOS ONE Issue 1:e132

Nelson XJ, Jackson RR (2011) Flexibility in the foraging strategies of spiders. In: Herberstein ME (ed) Spider behaviour: flexibility and versatility. Cambridge University Press, Cambridge, MA, pp 31–56

Nelson XJ, Jackson RR (2012) The discerning predator: decision rules underlying prey classification by a mosquito-eating jumping spider. J Exp Biol 215:2255–2261

Nentwig W (1985) Feeding ecology of the tropical spitting spider Scytodes Longipes (Araneae, Scytodidae). Oecologia 65:284–288

Nentwig W (1986) Non-webbuilding spiders: prey specialists or generalists? Oecologia 69:571–576

Nyffeler M (1999) Prey selection of spiders in the field. J Arachnol 27:317–324

Nyffeler M, Knörnschild M (2013) Bat predation by spiders. PLoS One 8:e58120

Nyffeler M, Pusey B (2014) Fish predation by semi-aquatic spiders: a global pattern. PLoS One 9:e99459

Oliveira PS, Sazima I (1984) The adaptive bases of ant-mimicry in a Neotropical aphantochilid spider (Araneae: Aphantochilidae). Biol J Linn Soc 22:145–155

Patil B, Prabhu S, Rajashekhar KP (2006) Lyriform slit sense organs on the pedipalps and spinnerets of spiders. J Biosci 31:75–84

Pekár S, Haddad C (2011) Trophic strategy of ant-eating *Mexcala elegans* (Araneae: Salticidae): looking for evidence of evolution of prey-specialization. J Arachnol 39:133–138

Pekár S, Lubin Y (2009) Prey and predatory behavior of two zodariid species (Araneae, Zodariidae). J Arachnol 37:118–121

Persons M, Rypstra A (2001) Wolf spiders show graded antipredator behavior in the presence of chemical cues from different sized predators. J Chem Ecol 27:2493–2504

Persons M, Uetz G (1996) The influence of sensory information on patch residence time in wolf spiders (Araneae: Lycosidae). Anim Behav 51:1285–1293

Persons M, Uetz G (1997) The effect of prey movement on attack behavior and patch residence decision rules of wolf spiders (Araneae: Lycosidae). J Insect Behav 10:737–752

Persons MH, Walker SE, Rypstra AL, Marshall SD (2001) Wolf spider predator avoidance tactics and survival in the presence of diet-associated predator cues (Araneae: Lycosidae). Anim Behav 61:43–51

Pollet T, Stulp G, Groothuis T (2013) Born to win? Testing the fighting hypothesis in realistic fights: left-handedness in the ultimate fighting championship. Anim Behav 86:39–84

Prieur J, Pika S, Barbu S, Blois-Heulin C (2016) Gorillas are right-handed for their most frequent intraspecific gestures. Anim Behav 118:165–170

Reissland A, Görner P (1985) Trichobothria. In: Neurobiology of arachnids. Springer, Berlin pp 138–161

Riechert SE, Luczak J (1982) Spider foraging: behavioral responses to prey. In: Witt PN, Rovner JS (eds) Spider communication: mechanisms of ecological significance. Princeton University Press, Princeton, pp 353–385
Riechert SE, Tracy CR (1975) Thermal balance and prey availability: bases for a model relating web-site characteristics and spider reproductive success. Ecology 56:265–284
Romero GQ (2001) Experimental study of the associatio between *Runcinioides argenteus* (Araneae, Thomisidae) and *Trichogoniopsis adenantha* (DC) AsteraceaeDissertation. State University of Campinas, São Paulo
Romero GQ, Vasconcellos-Neto J (2003) Natural history of *Misumenops argenteus* (Thomisidae): seasonality and diet on *Trichogoniopsis adenantha* (Asteraceae). J Arachnol 31:297–304
Romero GQ, Vasconcellos-Neto J (2007) Aranhas sobre plantas:dos comportamentos de forrageamento às associações específicas. In: Gonzaga MO, Santos AJ, Japyassú HF (eds) Ecologia e comportamento de aranhas. Interciência, Rio de Janeiro
Romero GQ, Antiqueira PAP, Koricheva J (2011) A meta-analysis of predation risk effects on pollinator behaviour. PLoS One 6:e20689. https://doi.org/10.1371/journal.pone.0020689
Rosenthal MF, Hebets EA (2015) Temporal patterns of nutrition dependence in secondary sexual traits and their varying impacts on male mating success. Anim Behav 103:75–82
Rovner JS (1980) Morphological and ethological adaptations for prey capture in wolf spiders (Araneae, Lycosidae). J Arachnol 8:201–215
Rypstra AL, Samu F (2005) Size dependent intraguild predation and cannibalism in coexisting wolf spiders (Araneae, Lycosidae). J Arachnol 33:390–397
Rypstra AL, Schmidt JM, Reif BD (2007) Tradeoffs involved in site selection and foraging in a wolf spider: effects of substrate structure and predation risk. Oikos 116:853–863
Salvestrini FMD, Gasnier TR (2001) Differences in the activity of juveniles, females and males of two hunting spiders of the genus *Ctenus* (Araneae, Ctenidae): active males or inactive females? J Arachnol 29:276–278
Samu F, Toft S, Kiss B (1999) Factors influencing cannibalism in the wolf spider *Pardosa agrestis* (Araneae, Lycosidae). Behav Ecol Sociobiol 45:349–354
Sandidge JS (2003) Scavenging by brown recluse spiders. Nature 426:30
Schemske DW, Mittelbach GG, Cornell HV, Sobel JM, Roy K (2009) Is there a latitudinal gradient in the importance of biotic interactions? Annu Rev Ecol Evol Syst 40:245–269
Schmalhofer VR, Casey TM (1999) Crab spider hunting performance is temperature insensitive. Ecol Entomol 24:345–353
Schmidt JM, Sebastian P, Wilder SM, Rypstra AL (2012) The nutritional content of prey affects the foraging of a generalist arthropod predator. PLoS One 7:e49223. https://doi.org/10.1371/journal.pone.0049223
Schmitt A, Schuster M, Barth FG (1990) Daily locomotor activity patterns in three species of *Cupiennius* (Araneae, Ctenidae): the males are the wandering spiders. J Arachnol 18:249–255
Schonewolf K, Bell R, Rypstra A, Persons M (2006) Field evidence of an airborne enemy-avoidance kairomone in wolf spiders. J Chem Ecol 32:1565–1576
Schuster M, Baurecht D, Mitter E, Schmitt A, Barth FG (1994) Field observations on the population structure of three ctenid spiders (*Cupiennius*, Araneae, Ctenidae). J Arachnol 22:32–38
Schwartz SK, Wagner WE Jr, Hebets EA (2014) Obligate male death and sexual cannibalism in dark fishing spiders. Anim Behav 93:151–156
Segovia JMG, Del-Claro K, Willemart RH (2015a) Delicate fangs, smart killing: the predation strategy of the recluse spider. Anim Behav 101:169–177
Segovia JMG, Del-Claro K, Willemart RH (2015b) Defences of a Neotropical harvestman against different levels of threat by the recluse spider. Behaviour 152:757–773
Souza EDS, Willemart RH (2011) Harvest-ironman: heavy armature, and not its defensive secretions, protects a harvestman against a spider. Anim Behav 81:127–133
Stafstrom JA, Hebets EA (2016) Nocturnal foraging enhanced by enlarged secondary eyes in a net-casting spider. Biol Lett 12:20160152. https://doi.org/10.1098/rsbl.2016.0152
Suter RB, Stratton GE (2005) *Scytodes* vs. *Schizocosa*: Predatory techniques and their morphological correlates. J Arachnol 33:7–15

Taylor RM, Pfannenstiel RS (2009) How dietary plant nectar affects the survival, growth, and fecundity of a cursorial spider *Cheiracanthium inclusum* (Araneae: Miturgidae). Environ Entomol 38:1379–1386

Toft S, Albo MJ (2016) The shield effect: nuptial gifts protect males against pre-copulatory sexual cannibalism. Biol Lett 12:20151082. https://doi.org/10.1098/rsbl.2015.1082

Tso I-M, Zhang S, Tan W-L, Peng P, Blamires SJ, Herberstein M (2016) Prey luring coloration of a nocturnal semi-aquatic predator. Ethology 122(8):671–681

Turner JS, Henschel JR, Lubin YD (1993) Thermal constraints on prey-capture behaviour of a burrowing spider in a hot environment. Behav Ecol Sociobiol 33:35–43

Vetter RS (2011) Scavenging by spiders (Araneae) and its relationship to pest management of the brown recluse spider. J Econ Entomol 104:986–989

Vieira WLS, Gonçalves MBR, Nóbrega RP (2012) Predation on *Tropidurus hispidus* (Squamata: Tropiduridae) by *Lasiodora klugi* (Aranea: Theraphosidae) in the semiarid caatinga region of northeastern Brazil. Biota Neotrop 12:1–3

Vollrath F, Selden P (2007) The role of behavior in the evolution of spiders, silks, and webs. Annu Rev Ecol Evol Syst 38:819–846

Wagner JD, Wise DH (1997) Influence of prey availability and conspecifics on patch quality for a cannibalistic forager: laboratory experiments with the wolf spider *Schizocosa*. Oecologia 109:474–482

Walker SE, Marshall SD, Rypstra AL, Taylor DH (1999) The effects of hunger on locomotory behaviour in two species of wolf spider (Araneae, Lycosidae). Anim Behav 58:515–520

Wetter MB, Wernisch B, Tof S (2012) Tests for attraction to prey and predator avoidance by chemical cues in spiders of the beech forest floor. Arachnologische Mitt 43:49–54

Wigger E, Kuhn-Nentwig L, Nentwig W (2002) The venom optimisation hypothesis: a spider injects large venom quantities only into difficult prey types. Toxicon 40:49–52

Willemart RH, Kaneto GE (2004) The use of retreats, mobility, and notes on the natural history of the Neotropical hunting spider *Enoploctenus cyclothorax* (Araneae, Ctenidae). Bull Br Arachnol Soc 13:53–59

Willemart RH, Pellegatti-Franco F (2006) The spider *Enoploctenus cyclothorax* (Araneae: Ctenidae) avoids preying on the harvestman *Mischonyx cuspidatus* (Opiliones: Gonyleptidae). J Arachnol 32:649–652

Wise DH (2006) Cannibalism: food limitation, intraspecific competition, and the regulation of spider populations. Annu Rev Entomol 51:441–465

Wrinn KM, Uetz GW (2008) Effects of autotomy and regeneration on detection and capture of prey in a generalist predator. Behav Ecol 19:1282–1288

Wullschleger B, Nentwig W (2002) Influence of venom availability on a spider's prey-choice behaviour. Funct Ecol 16:802–807

Young SL, Chyasnavichyus M, Erko M, Barth FG, Fratzl P, Zlotnikov I, Politi Y, Tsukruk VV (2014) A spider's biological vibration filter: micromechanical characteristics of a biomaterial surface. Acta Biomater 10:4832–4842

Zina J, Gonzaga MO (2006) Aplastodiscus arildae (green tree frog) predation. Herpetol Rev 37:440

Chapter 10
Trophic Niches and Trophic Adaptations of Prey-Specialized Spiders from the Neotropics: A Guide

Stano Pekár, Luis Fernando García, and Carmen Viera

Abstract Spiders are the most diversified group of terrestrial predators. They employ a wide variety of feeding strategies, and exploit several prey types, from invertebrates up to small vertebrates. Many studies on the trophic ecology of spiders have focused on generalist and euryphagous species. Thus, our knowledge of prey specialist (and stenophagous) species is very limited despite the high number of endemic species occurring in the Neotropics, many of which are most probably specialized. In this chapter, we provide a guide on how to study the trophic niches of spiders in order to encourage other researchers to investigate prey-specialized species. At the beginning, we define the term trophic niche and identify its dimensions (prey type, size, and availability). We critically outline methodological approaches on how to study it. A narrow trophic niche is paralleled by the evolution of specific cognitive, behavioural, metabolic, morphological, and venomic adaptations used in prey capture. We provide an overview of these adaptations and focus on approaches to reveal them. On the basis of an extensive bibliographic review, we summarize the current state-of-the-art with respect to knowledge on the trophic ecology of Neotropical spiders, with particular emphasis on specialists. Finally, we provide recommendations for future research.

S. Pekár (✉)
Department of Botany and Zoology, Faculty of Science, Masaryk University, Kotlářská 2, 611 37 Brno, Czech Republic
e-mail: pekar@sci.muni.cz

L.F. García
Centro Universitario de la Región Este, sede Treinta y Tres, Universidad de la República, Treinta y Tres, Uruguay

Laboratorio Ecología del Comportamiento, Instituto de Investigaciones Biológicas Clemente Estable, Montevideo, Uruguay

C. Viera
Entomología, Biología Animal, Facultad de Ciencias, Universidad de la República, Montevideo, Uruguay

Laboratorio Ecología del Comportamiento, Instituto de Investigaciones Biológicas Clemente Estable, Montevideo, Uruguay

C. Viera, M.O. Gonzaga (eds.), *Behaviour and Ecology of Spiders*, https://doi.org/10.1007/978-3-319-65717-2_10

Spiders are considered the most diversified group of terrestrial predators (Coddington and Levi 1991). They are true predators (sensu Begon et al. 2006), as they kill prey before consuming it, and feed on a number of individuals over their life cycle. The diets of spiders have frequently been studied; however, because of the huge species richness of spiders, to date we have data on the diets of only about 1.5% of species (Pekár et al. 2012). These data show that most spider species are euryphagous, that is, they capture and consume a wide variety of prey types. Stenophagous species, i.e., species that capture and consume a restricted array of prey, are much less common, representing up to 30% of species. Of these, only about one-third are expected to be specialized, that is, feeding only on a few prey types and possessing specialized trophic adaptations.

Comparative analysis of spider diets has also revealed that stenophagy is much more common in the tropics than in temperate or subtropical zones (Pekár et al. 2012), most likely because of more intensive interspecific competition in the tropics. The Neotropics primarily include the tropic zone, but also the subtropical and temperate zones of Central and South America (Morrone 2014). Thus, it is expected that the species richness of, in particular, specialized species should be high there. However, in general, very little is known about the trophic ecology or adaptations of spiders in this region.

Here, we aim to fill this gap. With particular reference to background theory and suitable methodological approaches, we provide a guide on how to study the trophic niches of spiders. We begin by defining what a trophic niche is and proposing approaches that have been used to study it. Then, we outline various trophic adaptations that specialists have evolved and present an overview of approaches used to reveal them. We also summarize the current state-of-the-art with respect to research into the trophic niches of Neotropical spiders, with particular emphasis on specialists. Finally, we provide suggestions for future research in this area.

Trophic Categories

We use the definition of trophic categories proposed by Pekár and Toft (2015), which was constructed on the basis of ecological and evolutionary contexts. The ecological context stands for the ecological conditions to which the spider species is exposed. The evolutionary context stands for traits which are the result of evolutionary adaptation. On the basis of these contexts, four categories are distinguished: euryphagous generalists, stenophagous generalists, euryphagous specialists, and stenophagous specialists.

The category *euryphagous generalists* includes the majority of spider species, that is, those with a wide diet breadth (i.e., composed of a variety of prey species) and possessing generalised adaptations. It is the generalization of these adaptations that allows the exploitation of diverse prey species. The category *stenophagous generalists* includes local populations or individuals of a population of a euryphagous generalist species which have a narrow trophic niche as a result of ecological circumstances or individual specialization. These species possess generalized adaptations. The species may have narrow diets dominated by a certain taxon in one habitat (Líznarová et al. 2013), or in a certain season or at a certain stage of development (Lesar and Unzicker 1978). The category *stenophagous specialists* includes the most extreme examples of species (not only populations or individuals, as is the case of stenophagous generalists) with a very narrow prey range and a dependence on the focal prey for growth and reproduction. They should possess specialized and stereotyped adaptations that may limit the possibilities of exploiting alternative prey. The category *euryphagous specialists* includes species with a wider diet breadth (several prey taxa) and more flexible specialized adaptations compared with stenophagous specialists. They show a clear preference for a focal prey taxon, but may have the capacity to grow and reproduce on prey outside the focal group. The diet breadth may be maintained by moderately generalized adaptations.

To classify a spider into any of these categories, it is essential to have information on the breadth of its trophic niche and the specificity of its adaptations. Thus, it is necessary to perform field observations to determine the realized trophic niche. If the niche is wide (i.e., it includes different prey types), it is very likely that the species is a euryphagous generalist. However, if it is narrow, then experiments should be performed to identify the fundamental trophic niche and the specificity of its trophic adaptations. For example, *Zodarion germanicum* (C. L. Koch) was first reported to catch only *Formica* ants in nature (Wiehle 1953). Then, in the laboratory, it accepted several ant species in addition to *Formica*, but did not accept other arthropods (Pekár 2004). Later, it was found to possess metabolic adaptations to formicine ants (Pekár et al. 2008), recognizing only their trail pheromone (Cárdenas et al. 2012). Taken together, the evidence available for this species shows that it is a stenophagous specialist.

Trophic Niche

The acquisition of food forms a central part of a spider's life-history. The trophic niche of a species, a part of the total species niche space, is an attribute formed over its evolutionary history and determined by diverse factors, including interactions with other species (mutualism, predation, etc.) (Soberon and Peterson 2005). As well as environmental conditions and food resources, the trophic niche is formed by interactions between the traits of predators and those of their prey. The trophic niche may differ among populations of a species and during the evolutionary history of a species; however, for simplicity, we will assume it to be constant. For a wide trophic niche, we will use the term euryphagous, and for a narrow niche, we will use the term stenophagous. These are only the extreme ends of a continuum (Nelson and Jackson 2011). To obtain more precise estimates we need to use some metrics (see below).

To describe the trophic niche of a spider species we use the concept proposed by Hutchinson (1957), who defined niche as a multidimensional hypervolume with

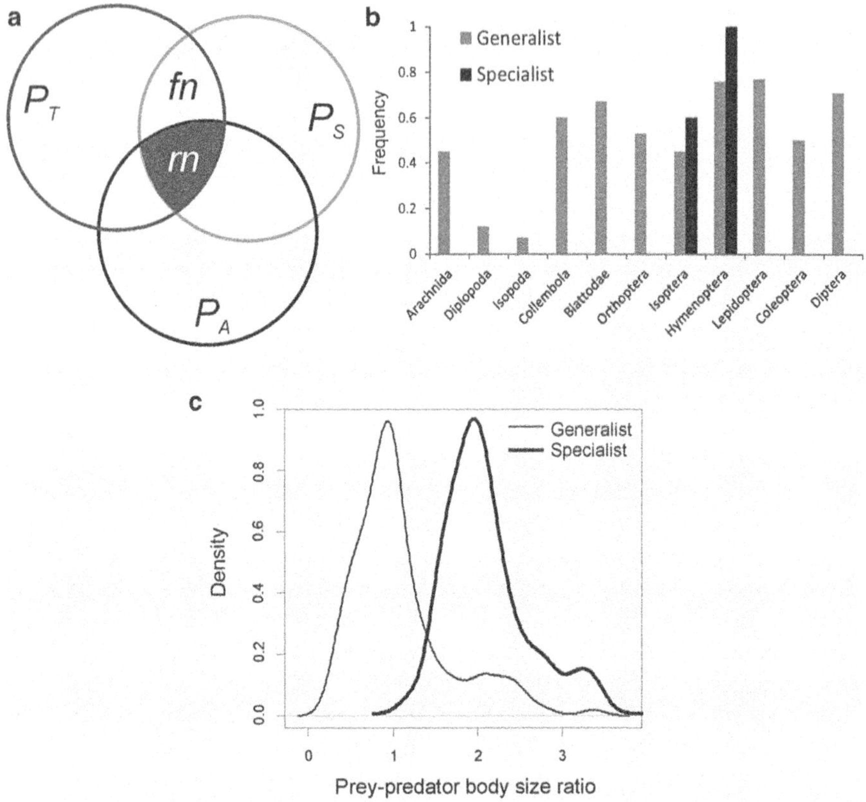

Fig. 10.1 Comparison of the trophic niche space of specialists and generalists. (**a**) Venn diagram of three niche dimensions, prey type (P_T), prey size (P_S), and prey availability (P_A), yielding fundamental (*fn*) and realized (*rn*) trophic niches. (**b**) Comparison of relative capture frequencies of 11 prey types (orders) captured by a specialist (*Zodarion*) and a generalist (*Selamia*). (**c**) Density plot of relative prey size (estimated as the ratio of the total prey body length to the length of spider prosoma) captured by a specialist (*Zodarion*) and a generalist (*Selamia*)

permissive conditions and requisite resources as its axes. He distinguished two types of niches: fundamental and realized. The latter takes into account interactions with other species in the community. To visualize them, we use a Venn diagram (Fig. 10.1a). The particular dimensions of the trophic hyperspace are: prey type (P_T), prey size (P_S), and prey availability (P_A). These dimensions are not independent (orthogonal), as some prey types (e.g., Collembola) are inherently tiny, while others (e.g., Odonata) are large. The resulting overlap of prey type and prey size defines the fundamental trophic niche (*fn*): $fn = P_T \cap P_S$. The realized trophic niche (*rn*) is then defined as the space in which these two dimensions overlap with prey availability: $rn = P_T \cap P_S \cap P_A$ (Fig. 10.1a).

Fundamental Trophic Niche

As mentioned above, the dimensions of fundamental trophic niche include prey size and prey type. A few other dimensions, namely prey quality and prey ontogenetic stage, can be distinguished. Prey quality characterizes defensive chemicals as well as the macro- and micro-nutritional compositions of prey. Recently, it has been found that spiders even aim to attain a nutritional target, which is a combination of amounts of macronutrients (Wilder 2011), by adjusting their feeding preferences. The target seems to be achieved differently in specialists and generalists; while the former achieve it consuming a single type of prey (Pekár et al. 2010), the latter needs a variety of prey (Mayntz et al. 2005). Prey quality is nested within prey type; therefore, this dimension is not distinguished. The ontogenetic stage of prey is important, in particular, for prey species with holometabolous development. The particular developmental stages (eggs, larvae, pupas, imagoes) can be classified as different prey types, as they may differ in defensive traits and nutritional composition.

Similarly, a few variables inherent to the spider predator, namely ontogenetic stage and sex, can be distinguished as other dimensions. This is because the trophic niche of, for example, adult males is often different from that of adult females and juveniles: males often cease hunting or switch to easy food, such as scavenging (Cangialosi 1990), kleptoparasitism (Cangialosi 1990, 1997), and plant-derived food (Pollard et al. 1995). Rather than distinguishing separate niche dimensions, it is more convenient to estimate trophic niches separately for sex/stage.

Let us look closely at the particular dimensions. Prey type is a categorical dimension. Types of prey can be classified in several ways. Spiders may classify prey in a different way to humans, but due to phylogeny and the resulting similarity in the traits of related species, the taxonomic classification of prey seems to be the most effective. However, even taxonomic classification has several levels. Ideally, the observer should identify every prey into species. As this is often not possible, particularly in the tropics, where many species have not been described, classification on a higher taxonomic level, e.g., family or order, might be sufficient. For certain prey types, the classification should be at a lower level than order. For example, among Hymenoptera, some wasps and bees are often accepted as prey by spiders, whereas ants are consistently rejected by some species (von Drees 1952; Edwards and Jackson 1994; Jackson and Van Olphen 1991; Korenko et al. 2014). Therefore, within this order, prey should be classified into families (or even into subfamilies if only ants are accepted). The range and spacing of categories should be constant in order to allow comparison of niche metrics among studies.

Prey size is a continuous dimension of relative prey/predator body size (e.g., body length). Consequently, both ontogenetic changes in the size of prey (Hemimetabola) and that of predators can be adjusted. In spiders, it is more reliable to measure the size (length or width) of the prosoma, as the size of the abdomen changes with satiation (Anderson 1974). On a wide scale, there is unimodal response along this dimension, i.e., too small prey is ignored, and too large prey is rejected both on account of their low profitability. Generally, there is a positive relationship between prey and spider size (e.g., Bartos 2011). But in solitary generalists, the majority of spider prey is smaller than the spider (e.g., Bartos 2011), whereas prey of spider specialists is larger (Pekár et al. 2014).

The fundamental niche can only be accurately measured experimentally by studying the response of a species to the mentioned trophic dimensions. It cannot be inferred from field data, for example from the absence/presence of prey, because prey availability (see below) strongly affects the response. An efficient means of studying the fundamental trophic niche is the acceptance experiment (e.g., Korenko et al. 2014; Petráková et al. 2015)

The procedure of acceptance experiments is based on observing the predator's willingness or refusal (i.e., a binary scoring of the response) to catch and consume a certain prey. The most effective design for these experiments is a randomised block design in which a reasonable number of spider specimens (~30) is used. In this design, each spider is offered every type of prey in a randomised order. Spiders are confined singly in a container that provides sufficient space, e.g., to build a retreat or a web, and at the same time allows the hunting sequence to be performed. Typically, the area of the container is several times larger than the body size of the spider. The size of the container should not be too large, as this will reduce the chance of prey capture and increase the latency to catch. Some species may require a specific substrate to initiate hunting; thus, the container should imitate their natural microhabitat. However, the micro-environment should not be too complex, in order to prevent the prey from hiding and to allow interactions to be observed.

Spiders must be satiated prior to experiments and subjected to experimental trials after a fixed number of days without prey to control for their hunger. Starving in order to initiate prey capture should be moderate, as lengthy starvation affects prey handling (Framenau et al. 2000) and attack rates (Rossi et al. 2006). Following capture of an individual prey, the prey must be removed after approximately 1 h to standardize the consumption time of differently sized prey. If the prey is not attacked within a short time (e.g., 10 min) following prey recognition, it is replaced with a different one. The procedure is repeated until one prey is accepted.

A set of potential prey types/stages is selected. The potential prey includes those which are sympatric (same microhabitat) and synchronic (same season) with the studied species. As the diversity of such prey is usually very high, use of all types is unrealistic. Therefore, 10–20 prey types are selected to represent the main and distinct taxonomic groups.

When conducting experiments on the prey size dimension, spider individuals are offered a variety of prey sizes so that the prey/predator size ratio varies considerably across the range of 0.2–4 (the ratio between the total body length of the prey and the length of the spider prosoma). Since the size of the prey can be a limiting factor, using different instars of the studied spiders might compensate for the size difference between the spiders and their prey. Alternatively, a mixture of prey which includes closely related species of different sizes might be used. In such a case, it is important to select appropriate prey because the cost of prey capture could vary widely (Japyassú and Viera 2002).

A useful metric for estimating the trophic niche breadth from acceptance experiments is the standardized version of the Simpson–Yule index (Southwood and Henderson 2000). Confidence intervals of the index value can be estimated using a formula derived by delta method or bootstrap (Smith 1982).

In specialists, the fundamental niche breadth of the prey type is narrow, i.e., the value of the Simpson–Yule index is between 0 and 0.02, while that of generalists is wide, i.e., larger than 0.02 (Fig. 10.1b). The fundamental niche breadth of specialists with respect to prey size is similar to that of generalists. The values of the ratio are, however, shifted to values larger than 1 in specialists (Fig. 10.1c).

Realized Trophic Niche

The realized trophic niche space is obviously smaller than the fundamental space, as it is limited by yet another variable, prey availability. Prey availability imposes physical limits on prey exploitation, as it is determined by the spatial and temporal co-occurrence of predator and prey, inter- and intraspecific competition among predators, mutualism, and the prey population dynamic, etc. (Begon et al. 2006).

To determine the realized trophic niche of a species, analysis of the natural diets of different populations across the species' geographic distribution during all seasonal activity is needed. This is often hard to obtain, and unrealistic for species with a large distribution range. Thus, to minimize the bias, at least populations occurring in different microhabitats should be surveyed. Otherwise, one can erroneously observe too narrow a breadth.

Natural diets can be investigated using a number of methods with varying specificity, reliability, labour input, and expense, etc. We will focus here only on methods that can be used to estimate diet breadth. A priori knowledge of the life-history, microhabitat preference, and circadian and seasonal hunting activity of a spider species is essential to maximise efficiency and minimize bias. Such information can be deduced from previous continuous sampling or from the literature on habitat preference and phenology. It is recommended to perform observations during the maturation period of the species under study, because juvenile spiders might not be reliably identified to species level. The spectrum of available prey can vary over seasons so that the diets of generalists change accordingly (Salomon 2011), while those of specialists are rather constant (Haddad et al. 2016).

The oldest and still most frequently used method to study the realized trophic niche is direct in-situ observation. This can be performed either by an observer or by means of video surveillance (Sunderland 1988). The latter option is now widely available, as digital video cameras (e.g., security cameras) are small and cheap, are capable of recording video in time-lapse, and can possess a motion detection mode (i.e., photo-traps); in addition, they benefit from long battery endurance. However, to observe several individuals, a large number of cameras is needed. Very recently, pictures of spiders with prey have been made available via numerous Internet servers. Such pictures often provide evidence of prey capture on unusual prey, and this type of information can eventually be used to evaluate the realised trophic niche of the respective spider species (see Nyffeler and Knörnschild 2013; Nyffeler and Pusey 2014).

Typically, an observer searches for the spider of interest, and focuses on individuals that captured a prey. The probability of finding a spider with a prey is rather low, on average 7%; thus, it is necessary to find more than 500 spider individuals to obtain sufficient data. This amounts to dozens of hours (Greenstone 1999) to complete a full study. Captured prey might be identified directly without disturbance to the spider, or, most frequently, must be removed for identification. Then, in the laboratory, the spider and the prey are identified to the lowest taxonomic unit possible; their size is measured, and their sex and ontogenetic stage are determined. Direct observations can only be used to survey spiders that hunt in exposed situations (e.g., not in crevices), and are most effective for sedentary species, such as web-building (Nentwig 1985; Smithers 2005) or ambush (Huseynov 2014) spiders. Direct observations are particularly effective for specialised species, as they often hunt aggregated prey (e.g., ants) in close proximity to prey, e.g., near nests and/or on foraging trails.

Another frequent method of studying the realized trophic niche is the analysis of prey carcasses, i.e., the survey of prey captured by sedentary species that leave the carcasses of prey nearby (e.g., Fischer et al. 2006; Moreno-Mendoza et al. 2012; Pompozzi et al. 2013; García et al. 2014). This particularly applies to web-building species, which store prey carcasses in their web (Salomon 2011), or burrowing species, which store prey carcasses inside or outside their burrows (Miller et al. 2012). Typically, the entire web together with the spider is collected (Salomon 2011; Líznarová et al. 2013) and then the content is analysed in the laboratory to record prey identity, ontogenetic stage, and size. Carcasses are often found at different stages of decomposition depending on the sclerotization of the prey and its damage during feeding, allowing variable levels of identification.

Modern methods of natural prey analysis are based on post-mortem gut content analysis. Spiders are collected by traditional sampling methods, such as the use of live pitfall traps (Wirta et al. 2015), sweeping, beating, or vacuum sampling. It is important to prevent contamination of the target spider individuals (i.e., to avoid damaged individuals). The spider needs to be placed in a reagent (pure ethanol, RNAlater) or stored at −20 °C as early as possible to reduce the microbial breakdown of consumed fluids. Remnants of DNA can be detected for either a few or many days following digestion: for example, at 5 °C the detection limit lasted several days (Petráková et al. 2016), whereas at 20 °C, it only lasted for a day (Chapman et al. 2013). Similarly to direct observations, an observer must have a priori information on hunting conditions in order to avoid biased estimates. These methods are mostly suited to species with cryptic feeding, i.e., those which consume extremely small prey (e.g., mites), and hunt in inaccessible spaces (e.g., in litter) and at night, etc.

As spiders possess extraoral digestion and ingest only liquefied tissue, the gut content is without any cuticle particles; thus, serological methods, electrophoresis, molecular methods, or stable isotope analysis must be applied. Molecular techniques have recently become the most common means of studying predator–prey interactions, because of their rapid development and decreasing costs (Symondson 2002); thus, we will deal only with DNA-based assays here. Molecular methods can be applied on the gut, faeces, and silk of spiders (Xu et al. 2015). The latter two options are non-invasive; however, the quantity of prey DNA is limited, and thus the analysis of gut contents is preferred. As the predator's DNA is far more abundant than prey DNA in the body of the spider, it is recommended to dissect the alimentary system and use only its contents for analysis. DNA is extracted using kits for tissue, food, or faeces sampling.

To detect multiple preys, either simple PCR, multiplex PCR, or genome sequencing can be used (e.g., Wirta et al. 2015). These methods target a certain DNA region sufficiently specific to identify the prey taxon. Most frequently, cytochrome c oxidase subunit I (COI) is used because it is sufficiently specific and is available for many organisms in databases (GeneBank, BOLD). The specificity is achieved by a fragment about 300 bp long, which amplifies fragments still too variable to identify prey taxa at species level (King et al. 2008). It is also necessary to extract DNA from all potential prey to make sure that the COI of prey is known.

Simple and multiplex PCR require the development of a number of taxa-specific primers with a varying level of specificity (e.g., an order). A number of such primers are currently available: Collembolla (Kuusk and Agustí 2008), Platygastridae, Aphididae (Chapman et al. 2010), Diptera (Sint et al. 2012). The design of primers specific enough to target an array of potential prey requires thorough cross-testing for the desired specificity to avoid false-positive results. Multiplex PCR further requires conditions to be optimised during PCR so that all primer pairs work at the same annealing temperature.

Genome sequencing through, for example, next-generation sequencing (NGS) makes possible the simultaneous amplification, and subsequent identification of thousands of prey DNA sequences (Hebert et al. 2003). NGS also requires the use of specific or universal primers. Specific primers must be designed to amplify the DNA of prey but not that of the predator, which is present in a much greater amount (Petráková et al. 2015). For this purpose, a combination of specific primers might be used, or predator-blocking primers must be applied (Piñol et al. 2014). To assign prey sequences to individual predators, MID identifiers (10-bp tags) are added to the forward and reverse primers (Zaidi et al. 1999). Following enrichment PCR, one- or bi-directional sequencing is performed on high-throughput sequencing machines (e.g., Ion Torrent PGM, Ion Proton, Illumina, Roche 454, GS-FLX). As NGS produces millions of sequences, bioinformatics skills are necessary to process the wealth of data (Pompanon et al. 2011). The sequences can be processed using, for example, the Galaxy platform (Giardine et al. 2005) and assigned to molecular operational taxonomic units, which are later compared to existing records in databases. A similarity above 99% allows a unit to be classified even to species level.

Recently, stable isotope analysis has begun to be used for the assessment of trophic interactions in spiders (e.g., McNabb et al. 2001; Mestre et al. 2013). Stable isotopes (^{13}C and ^{15}N) occur naturally in the environment; and because they are rare, hazard-safe, and do not decay, they are useful tracers of trophic interactions for longer periods (Hood-Nowotny and Knols 2007). This method is based on the fact that stable isotope ratios ($\delta^{15}N = {}^{15}N/{}^{14}N$, $\delta^{13}C = {}^{13}C/{}^{12}C$) in the proteins of the predator reflect those ratios of the prey. Isotopic signatures are, however, only indirect evidence of prey consumption. Furthermore, this method only provides resolution at gross trophic level; thus, the estimate of diet breadth is problematic (Bearhop et al. 2004).

As well as data on natural prey, it is essential to survey available prey to estimate the realised trophic niche breadth (Moreno-Mendoza et al. 2012; Liznarová and Pekár 2015). A number of standard sampling methods are available: trapping by means of pitfall traps (ground hunters), sticky traps (crawlers, fliers), interception window traps (fliers), hand collection, sweeping, beating, and baiting, etc. (Chapman et al. 2013; García et al. 2014). As available prey will be very variable, it is necessary to focus on the prey which would be potentially consumed by the spider. For example, the analysis should exclude extremely large, extremely small, or dangerous prey which would be ignored or avoided by the spider.

To compare the mentioned methods, the biases, advantages, and disadvantages of each need to be considered. The majority of the mentioned methods provide short-term prey analysis, while stable isotope analysis and carcass analysis allow for long-term assessment. Potential bias in indirect observations may include the presence of an observer next to the hunting spider (e.g., flying insects may avoid the area), while carcass analysis may produce overestimates because not all prey captured in a web are consumed (Maupin and Riechert 2001; Blackledge 2011). Additionally, carcass analysis will not provide data on whether or how the prey was captured/consumed. Although widely used, molecular methods are far more expensive and labour-intensive than other methods. These also introduce several biases, as they cannot distinguish predation from scavenging, cannibalism, or secondary predation (the detection of prey that was previously eaten by the prey of the spider) (Putnam 1967). These methods cannot be used to identify prey stage, size, or sex, and may produce false positive results (due to contamination). The drawbacks of molecular analyses can be minimized by combining such approaches with other methods such as acceptance experiments to, for example, exclude cases of secondary predation. Direct observations have some advantages over other methods, as they provide data on whether the prey was consumed or only captured, and how it was captured. Therefore, direct observation seems the most informative, precise and cost-effective method, which should be used whenever possible.

All methods, except for carcass analysis, only produce qualitative data (at the level of individual predator). Carcass analysis produces multiple prey items per individual. These quantitative data are, however, not independent, and thus cannot be used directly to estimate proportions of prey types but must be transformed into qualitative data. Quantification in molecular methods is problematic; therefore, it is not possible to estimate how many individuals of a certain prey were eaten. This is because DNA is gradually degraded in a live spider and its amount depends on a number of variables: the size of DNA fragments, ingestion time, predator hunger, the amount of prey tissue consumed, and the rate of digestion, which is temperature-dependent (Piñol et al. 2014). These variables are not, unfortunately, under control in spiders collected in the wild.

The realized trophic niche breadth can be measured by indices that take into account data on prey availability, such as the Smith index (Smith 1982). In specialists, the realised trophic niche should be narrow with respect to prey type, while in generalists, it should be wide. The selectivity for certain prey can be estimated by, for example, the selection index (Manly et al. 2002), which will indicate positive selection (preference), no selection (indifference), or negative selection (avoidance). Inferences on index values can be obtained by means of the chi-square test. Specialists should have a marked preference for a certain prey type, while generalists should exhibit positive selection towards a few prey types.

Trophic Adaptations

The exclusive exploitation of a certain resource should lead to increased capture and consumption efficiency, reflected in the evolution of specialised adaptations. We will use the term generalist for species possessing generalised adaptations that allow the exploitation of a variety of prey types, and the term specialist for species which are constrained to exploit only a small number of prey types. Adaptations observed in spiders can be classified into five major categories: morphological, cognitive, behavioural, metabolic, and venomic.

The methods of studying levels of adaptations differ among types of adaptations. Morphological adaptations are often represented by unusual shapes of trophic structures, such as forelegs, spinnerets, palps, and chelicerae (including fangs). The role of an adaptation can be demonstrated directly via experiment (and manipulation) or indirectly by comparative analysis. For example, different shapes of chelicera in *Dysdera* spiders were found to be differently effective in the prey capture of woodlice (Řezáč et al. 2008). Comparative analysis is an efficient tool if the studied traits and phylogeny are known among the compared species. In such an analysis, the presence or shape of a certain morphological trait is related to the presence of a functional trait by means of phylogenetic comparative methods (Liu et al. 2016). For example, in theridiid spiders, the presence of several traits, namely the presence of web, gumfoot lines, modified hairs on tarsi, and long cheliceral fangs, has been positively related to specialisation on ants (Fig. 10.2a).

Cognitive adaptations include sensory tuning to focal prey by means of visual, chemical, or mechanical cues that lead to the formation of selective attention. As specialists often prey on their focal prey from the first instar, their selective attention is assumed to be innate (Pekár and Cárdenas 2015). Approaches to the study of cognitive adaptations include preference experiments, in which the response of the spider to cues of different origin is investigated. Preference experiments can be performed by means of an olfactory device (olfactometer), in which chemicals are tested (Cárdenas et al. 2012), or other experimental devices used to study, for example, visual cues (Nelson and Jackson 2011). Both qualitative (e.g., frequency of

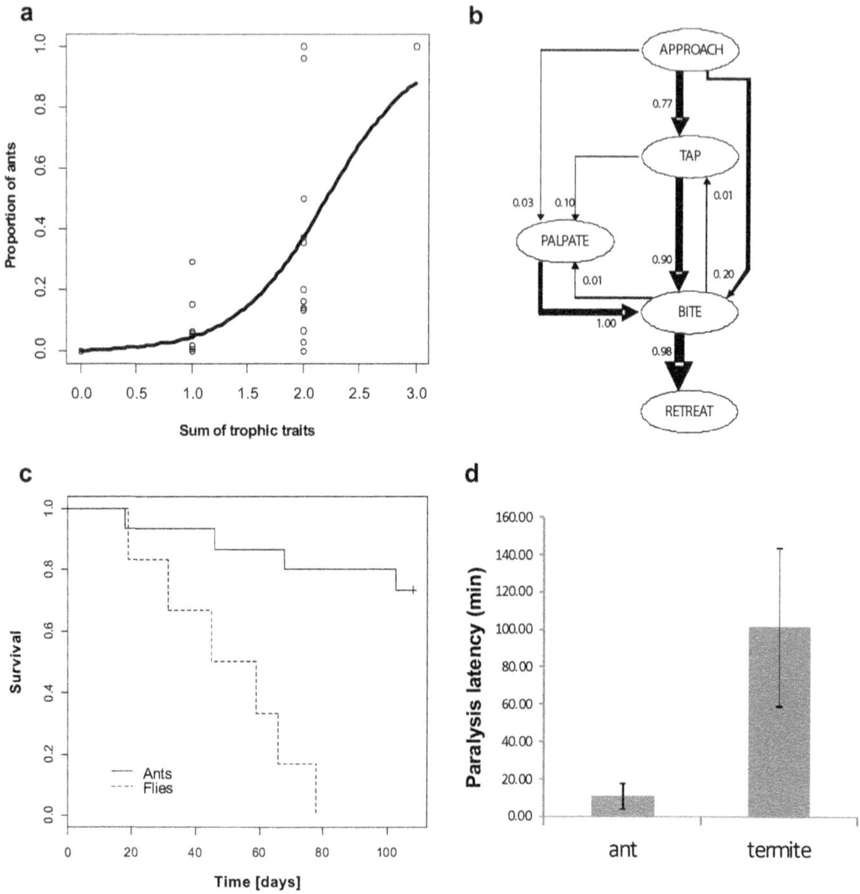

Fig. 10.2 Examples of various adaptations found in specialists. (**a**) Relationship between myrmecophagy (expressed as the proportion of ants in the diet) and the sum of trophic traits (web, gumfoot lines, modified hairs on tarsi, long cheliceral fangs) in theridiid spiders. (**b**) Kinematic diagram of the capture behaviour of *Zodarion* spiders when catching ants. (**c**) Survival of juvenile *Zodarion* spiders over 4 months reared on two diet types, pure ants and pure *Drosophila* flies. (**d**) Comparison of the paralysis latency (measured as the time between bite and immobilisation) of *Zodarion* spiders for two prey types, ants and termites

selection) and quantitative (e.g., latency to selection) data are recorded and compared among cue types. Available data show that specialist species are attracted to only a few cues, whereas generalists are either not attracted to a particular cue or attracted to a general cue. For example, myrmecophagous *Zodarion* spiders possess specific olfactory attention to chemical cues produced by the Dufour gland of some Formicinae ant species only (Cárdenas et al. 2012).

Behavioural adaptations can be studied by means of detailed observations of prey capture. Spiders are offered a variety of prey types, in a similar set-up to that for acceptance experiments (see above), and their behaviours during prey capture sequences for different prey types are identified in order to form ethograms. As the prey capture of specialists is often very fast, the use of a high-speed camera might be necessary. Then, transition probabilities between behaviours are estimated to form a transition matrix (Lehner 1999). The ethogram and the transition probabilities of behaviour occurrence can then be used to construct kinematic diagrams. The complexity of the kinematic diagram can be evaluated by information theory statistics, such as the Shannon entropy index or the stereotypy index (Lehner 1999). The capture behaviour of specialists is often very stereotyped; thus, they might not be able to catch alternative prey. The complexity of their capture behaviour is lower than that of generalists (Fig. 10.2b). For example, the capture sequence of myrmecophagous *Zodarion* was highly stereotyped towards the few prey types it accepted (Pekár 2004).

Metabolic adaptations can be studied by means of performance experiments or by the nutritional analysis of predator body composition (Toft et al. 2010). Performance experiments aim to reveal the effect of prey on the fitness of a predator (e.g., Toft 1999). The spiders are fed with a certain prey type (often constant throughout the experiment) and kept in optimal conditions (temperature, light duration, humidity). Ideally, such experiments should begin with spider hatchlings and end with their natural mortality in the adult stage so that age-specific survival, age-specific fecundity and fertility, and the timing of moulting are recorded. If the experiment is expected to end before the production of offspring, body mass is recorded so that fecundity can be estimated from the average body mass using a general regression model of spider mass and egg clutch (Marshall and Gittleman 1994). From the obtained parameters, predator fitness, measured as the contribution to the next generation, can be estimated. Fitness is estimated by calculating the dominant eigenvalue (λ_1) of the Leslie transition matrix for an age-structured population (alternatively, the Lefkovitch matrix for ontogenetic stages can be used) (Caswell 2001). The components of the Leslie matrix are survival and fecundity. The size of the transition matrix is derived from arbitrary time intervals (e.g., weeks or months). The dominant eigenvalue λ_1 of the transition matrix gives the population's asymptotic growth rate: the larger the value, the higher the fitness and the faster the population increase. An estimate of 95% confidence intervals of $\lambda_1$can be made by means of bootstrap (Caswell 2001). It has repeatedly been found that specialists have a higher fitness when feeding on their preferred prey (Fig. 10.2c), while generalists perform better when feeding on a mixture or nutrient-enriched prey. For example, *Zodarion* spiders attained their highest fitness on a pure ant diet composed of their preferred ant species (Pekár et al. 2008).

Venomic adaptations can be studied by means of the analysis of venom composition and the effectiveness of venom with respect to prey immobilization. Pure venom can be extracted from the spider using electro-stimulation and collected into micro-capillaries or by dissection of the venom glands (Garb 2014). To analyse venom, samples are subjected to SDS-PAGE for the separation of proteins (>10 kDa) and MALDI-TOF (Palagi et al. 2013) for the profiling of peptides and small proteins (<10 kDa). Profiling does not provide an exact number of peptides or proteins, because similarly-sized proteins will be merged in a single band/peak.

However, the diversity of composition can be roughly estimated using, for example, the Shannon entropy index. Specialists are expected to exhibit a lower diversity of compounds than generalists, these compounds being especially effective on the focal prey. The efficacy of venom can be investigated by measuring the latency to paralysis in different prey types. Venom can be applied in a range of doses (of constant concentration) to different prey by injection, so that the LD50 can be estimated and used for comparison (Wullschleger and Nentwig 2002). The venom of specialists should have a lower LD50 for focal prey than for alternative prey. Such venom is more effective (i.e., latency is shorter) on focal prey than on other kinds (Fig. 10.2d), whereas in generalists it is the other way around. For example, in *Zodarion* spiders, the paralysis latency was much shorter for focal than for other ant species (Pekár et al. 2005).

Specialization in Neotropical Species

So far, the trophic niches of only 72 spider species (see list in the Appendix) occurring in the Neotropics have been investigated (excluding cosmopolitan species). The majority of these species (64.3%, $N = 72$) are web-building spiders belonging to two families, Araneidae and Theridiidae (51.4% altogether) (Fig. 10.3a). The most frequent prey captured by these spiders was Diptera, followed by Hymenoptera and Hemiptera (Fig. 10.3b). For 90% ($N = 72$) of species, the realised trophic niche was studied. The number of prey analysed ranged from ten to thousands of individuals (Fig. 10.3c). Thus, the majority of the species appear to be euryphagous generalists, with only 12.9% being specialists (Fig. 10.3d). Some species have been found to be stenophagous generalists; for example, *Latrodectus mirabilis* (Pompozzi et al. 2013) and *Oecobius concinnus* both preyed mostly on ants in the field (García et al. 2014).

Pekár and Toft (2015) found that, globally, spiders have specialized on six prey types: ants, termites, crustaceans, spiders, moths, and dipterans. There is also laboratory evidence that an oonopid species could be specialised on Collembola (Korenko et al. 2014). Altogether, some evidence is available for the existence of stenophagous and euryphagous specialists from the Neotropics (Table 10.1). However, the majority of these species have so far been inadequately studied; thus, it may turn out in the future that they are not, in fact, specialists.

Neotropical spider species from seven families are myrmecophagous (48%, $N = 23$). Anecdotal evidence of realized trophic niches is available for several species, namely *Attacobius*, *Falconina*, *Eilica*, *Gallianoella*, *Bucranium*, *Strophius*, and *Tmarus* (Table 10.1). Acceptance experiments were performed with *Hentzia palmarum* (Cutler 1980). This salticid captured various ant species, but also *Drosophila* flies. Ants were captured as frequently as flies in choice experiments. Similar results were obtained for another salticid, *Anasaitis canosa* (Edwards et al. 1974). In the laboratory, it captured a number of ant species, though with a different level of efficiency (Fig. 10.4a). Ants succumbed to the venom within a few minutes.

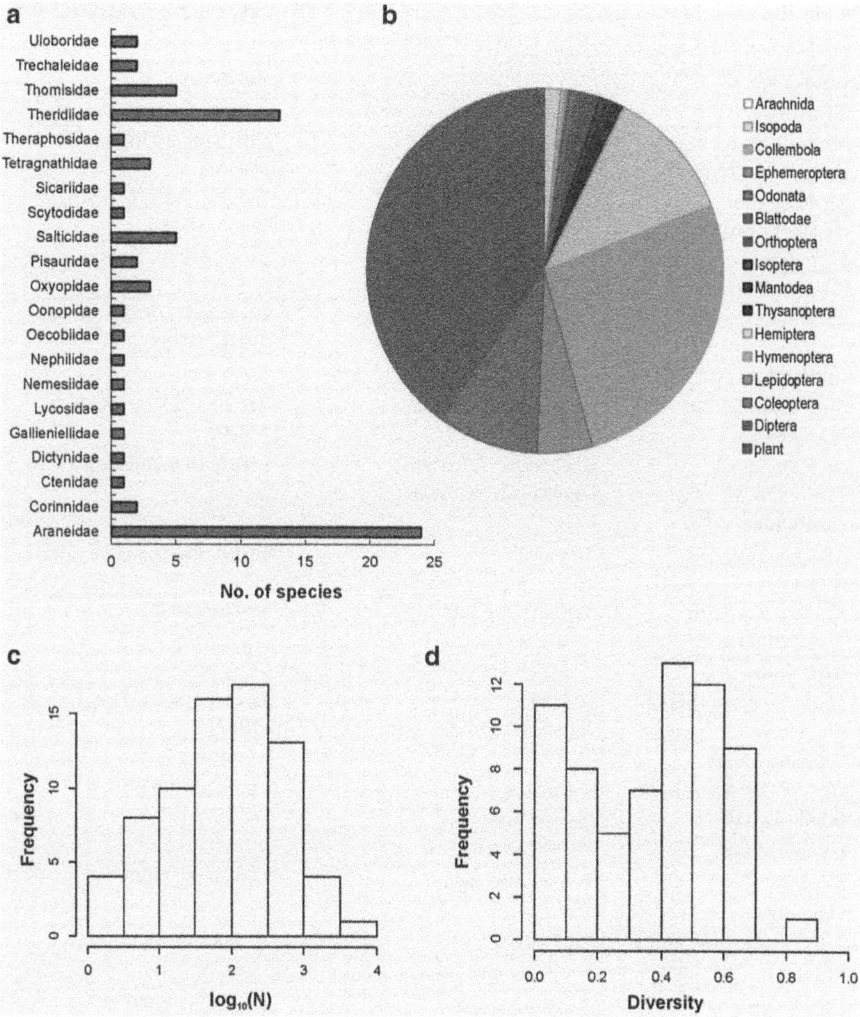

Fig. 10.3 Summary of the trophic niche measures of 72 Neotropic species. (**a**) Barplot of the number of spider species per 21 families investigated so far. (**b**) Pie chart showing the proportions of different types of prey recorded in the diet of 72 spider species. (**c**) Histogram of the number of individual prey recorded in the diet of 72 spider species. Notice that the numbers are $\log_{10}$ transformed. (**d**) Histogram of standardised indices of the niche breadth of prey type. Species with values <0.2 are considered stenophagous

These spiders caught ants using conditional strategies: frontally or from the rear. Spiders were even capable of capturing ants larger than themselves. Preference trials conducted with this species also suggest specialisation on ants. The thomisid *Aphantochilus rogersi* is not only an ant-mimic but a highly specialized predator of *Cephalotes* ants, as evidenced by numerous observations in the field. All spider instars seem to feed on *Cephalotes* ants even when the prey is larger (Castanho and

Table 10.1 List of prey-specialised spiders (stenophagous and euryphagous specialists) of the Neotropics and the prey taxon they seem to be specialised on

Family/species	Prey	References
Araneidae		
Mastophora cornigera [juv., male] (Hentz)	Nematocera	Eberhard (1980), Stowe (1986)
M. dizzydeani [female] Eberhard	Lepidoptera	Eberhard (1977, 1980), Stowe (1986, 1988)
Scoloderus cordatus (Taczanowski)	Lepidoptera	Stowe (1978, 1986)
Taczanowskia sp.	Pyralidae	Eberhard (1981a), Stowe (1986, 1988)
Caponidae		
Nops guanabacoae MacLeay	*Rhopalurus* scorpion	Teruel and Sánchez-Ruiz (2000)
Nops sp.	Oecobiidae, Lycosidae, Dipluridae, Sicariidae	García (unpublished)
Corinnidae		
Attacobius attarum (Roewer)	*Atta* ants	Erthal and Tonhasca (2001)
Falconina gracilis (Keyserling)	*Acromyrmex* ants	Fowler (1984)
Dictynidae		
Mallos gregalis (Simon)	Muscidae Tabanidae	Diguet (1909), Tietjen et al. (1987)
Gallieniellidae		
Galianoella leucostigma (Mello-Leitão)	Formicinae, Myrmicinae	Goloboff (2000)
Gnaphosidae		
Eilica sp.	*Acromyrmex* ants	Goloboff (2000)
Oonopidae		
Triaeris stenapsis Simon	Collembola	Korenko et al. (2014)
Mimetidae		
Gelanor sp.	*Leucauge* spiders	Gonzaga (2007)
Mimetus sp.	*Leucauge* spiders	Romero-Ortiz and Flórez-Daza (2014)
Palpimanidae		
Ohiothops birabeni Mello-Leitão	Lycosidae, Theridiidae	García (unpublished).
Salticidae		
Anasaitis canosa (Walckenaer)	Formicidae	Edwards et al. (1974), Jackson and Van Olphen (1991)
Hentzia palmarum (Hentz)	*Myrmica* ants	Cutler (1980)
Theridiidae		
Chrosiothes portalensis Levi	Termites	Pérez de la Cruz et al. (2007)

(continued)

Table 10.1 (continued)

Family/species	Prey	References
C. tonala (Levi)	*Tenuirostritermes termites*	Eberhard (1991)
Janula sp.	*Nasutitermes termites*	Marshall et al. (2015)
Phoroncidia studo Levi	Sciaridae	Eberhard (1981b)
Thomisidae		
Aphantochilus rogersi O. P.-Cambridge	*Cephalotes* ants	Oliveira and Sazima (1984), Castanho and Oliveira (1997)
Bucranium taurifrons O. P.-Cambridge	*Cephalotes* ants	Bristowe (1939), Oliveira (1987)
Strophius nigricans Keyserling	*Camponotus* ants	Oliveira and Sazima (1985)
Tmarus stoltzmanni Keyserling	Formicinae Dolichoderinae	Lubin (1983)
Zodariidae		
Leprolochus birabeni Mello-Leitão	*Acromyrmex* ants	Jocqué (1988), G. Pompozzi (pers. comm.)

Oliveira 1997). The spiders show an ontogenetic hunting shift: early instars use frontal attacks, while later instars use rear attacks. A zodariid, *Leprolochus birabeni*, is a stenophagous specialized predator of ants (Fig. 10.4b). It is a nocturnal predator, foraging on ant trails and/or at nest entrances. Investigations of its trophic niche showed that this species accepted only ants and, in the field, it hunted *Acromyrmex lobicornis* Emery in particular (G. Pompozzi, pers. com.). We expect that myrmecophagy will be found, in particular, in other theridiid, salticid, gnaphosid, corinnid, and zodariid species.

Approximately 13% of species, all from Theridiidae, are termitophagous. The rich carcass analysis of *Chrosiothes portalensis* webs revealed specialisation on termites (Pérez de la Cruz et al. 2007). Eberhard (1991) reported specialized predatory behaviour in *Chrosiothes* sp., which hunted termites by descending from a horizontal thread. Similar field observations on *Janula* sp. (Fig. 10.4c) indicate specialization on only one termite species, as the spiders were only attracted to the nests of *Nasutitermes* (Marshall et al. 2015). Termitophagy is not especially common but might be found in other theridiid species. As there is high termite diversity in the Neotropics, many more termitophagous species, e.g., among salticids, are expected to be found.

Dipterophagous species were reported within three families. Anecdotal observations suggest the specialization of the theridiid *Phoroncidia* sp. on sciarid flies, but further evidence is needed to support this hypothesis. In another species, *Mallos gregalis*, carcass analysis of the web along with preference experiments provide quite strong support for dipterophagy. The webs of this social species were found to attract muscid and tabanid flies (Tietjen et al. 1987). In the case of male bolas spiders of the genus *Mastophora*, a morphological adaptation in the form of spiny forelegs and the production of an allomonal blend to attract psychodid flies support their strict specialization (Stowe 1986).

Fig. 10.4 Neotropic spider specialists. (**a**) *Anasaitis canosa* with an ant (Photo: D. Hill). (**b**) *Leprolochus birabeni* (Photo: O. Michálek). (**c**) *Janula* sp. with a termites (Photo: S. Marshall). (**d**) *Mastophora cornigera* with a swinging bolas (Photo: M. Coors). (**e**) *Nops* sp. with a lycosid spider (Photo: O. Michálek). (**f**) *Othiotops birabeni* with a theridiid spider (Photo: A. Laborda)

Several spider species from Araneidae are lepidopterphagous specialists. Evidence is particularly strong for bolas spiders of the genus *Mastophora* (Fig. 10.4d). Large juvenile and adult females of this genus were observed to catch only a few species of male moths by means of a bolas. This spider does not locate its prey but attracts the moth to itself by means of the production of an allomonal blend (Eberhard 1980). Its restricted prey range together with its specialised capture behaviour show that it is a stenophagous specialist. However, it remains to be shown whether this spider also possesses cognitive, metabolic, and venomic adaptations. For two other species from Araneidae, namely *Scoloderus* and *Taczanowskia*, only data on the realized trophic niche suggest trophic specialization (Stowe 1986, Eberhard 1981a).

Although araneophagy has been reported in some Neotropical species of the families Mimetidae (Romero-Ortiz and Flórez-Daza 2014), Oxyopidae (Gonzaga et al. 1998), and Scytodidae (Escalante et al. 2015), most records are anecdotal observations. For two species, the evidence is stronger. Although a recent record suggests that spiders of the genus *Nops* are able to prey on scorpions (Teruel and Sánchez-Ruiz 2000), detailed investigation suggests that the trophic niches of Colombian populations of *Nops* sp. (Fig. 10.4e) are restricted to spiders alone, though only those of a few families. This species also exhibits stereotyped capture behaviour and high venom efficiency, providing strong evidence that it is a stenophagous specialist (García, unpublished). The other case is the palpimanid species *Otiothops birabeni*, which has been observed to catch theridiid (Fig. 10.4f) and lycosid spiders, which may be even larger than itself. This species occurs on the ground, remains hidden under rocks during the day, and is active mainly at night, when it stalks both wandering and other spiders. Similarly to other palpimanid genera, it has a thick cuticle and forelegs empowered with dense scopulae (García, unpublished).

Conclusions

Despite the fact that the spider fauna of the Neotropics is very rich, knowledge of the trophic ecology of Neotropic spider species is very limited. Several of these species can potentially be recognized as specialists, yet strong evidence of narrow trophic niches and specialised adaptations that constrain the utilisation of alternative prey is lacking. Our experience of studying specialists in other geographical regions shows that the investigation of specialists is often difficult, yet feasible. In addition, researchers are rewarded with exciting results, such as cases of aggressive mimicry or the discovery of very specific adaptations. We believe that this paper could be useful as an initial framework for investigators interested in working with trophic specialization in spiders. Thus, overall, we encourage other researchers to study Neotropical trophic specialists.

Acknowledgments We thank L. Petráková and to anonymous reviewers for comments on the manuscript; also thanks to David Hill, Steven Marshall, Alvaro Laborda, and Ondřej Michálek, who kindly provided nice photos of spiders. SP was supported by the Czech Science Foundation (grant no. GA15-14762S). LFG was supported by Doctoral grant 8880 from the Uruguayan National Agency for Research and Innovation (ANII) and PEDECIBA (Postgraduare Programm).

Appendix

List of species from the Neotropics for which data on trophic niche were collected and used in the analysis. Only species for which more than ten prey records have been made are listed.

Araneidae: *Alpaida tuonabo* (Chamberlin & Ivie) [Shelly 1983], *A. variabilis* (Keyserling) [Flórez et al. 2004], *Argiope argentata* (Fabricius) [Nentwig 1985], *A. aurantia* Lucas [Bilsing 1920, Uetz et al. 1978, Nyffeler et al. 1987, Blackledge and Wenzel 1999], *A. savignyi* Levi [Nentwig 1985], *Aspidolasius branicki* (Taczanowski) [Calixto and Levi 2006], *Cyclosa caroli* (Hentz) [Ibarra-Núnez et al. 2001, Moreno-Mendoza et al. 2012], *C. turbinata* (Walckenaer) [Nyffeler et al. 1986, Nyffeler and Sterling 1994], *Cyrtophora moluccensis* (Doleschall) [Lubin 1974], *Eriophora edax* (Blackwall) [Ceballos et al. 2005], *Eriophora fuliginea* (C. L. Koch) [Nentwig 1985], *Gasteracantha cancriformis* (Linnaeus) [Gregory 1989, Ibarra-Núnez et al. 2001], *Gea heptagon* (Hentz) [Nyffeler et al. 1989, Nyffeler and Sterling 1994], *Mastophora cornigera* (Hentz) [Stowe 1986], *M. dizzydeani* Eberhard [Eberhard 1980], *Mecynogea lemniscata* (Walckenaer) [Wise and Barata 1983], *Micrathena gracilis* (Walckenaer) [Uetz and Biere 1980, Uetz and Hartsock 1987], *M. schreibersi* (Perty) [Shelly 1984], *Neoscona arabesca* (Walckenaer) [Bilsing 1920, Culin and Yeargan 1982], *Nephila clavipes* (Linnaeus) [Nentwig 1985], *Parawixia bistriata* (Rengger) [Fowler and Diehl 1978, Fowler and Gobbi 1988], *Scoloderus cordatus* (Taczanowski) [Stowe 1978], *Zygiella x-notata* (Clerck) [Nentwig 1983]; Corinnidae: *Attacobius attarum* (Roewer) [Erthal and Tonhasca 2001], *Falconina gracilis* (Keyserling) [Fowler 1984]; Ctenidae: *Cupiennius salei* (Keyserling) [Nentwig 1986, 1990]; Dictynidae: *Mallos gregalis* (Simon) [Tietjen et al. 1987]; Gallieniellidae: *Galianoella leucostigma* (Mello-Leitao) [Goloboff 2000]; Lycosidae: *Allocosa brasiliensis* (Petrunkevitch) [Aisenberg et al. 2009]; Nemesiidae: *Acanthogonatus francki* Karsch [Pinto and Saiz 1997]; Oecobiidae: *Oecobius concinnus* Simon [García et al. 2014]; Oonopidae: *Triaeris stenaspis* Simon [Korenko et al. 2014]; Oxyopidae: *Oxyopes salticus* Hentz [Nyffeler et al. 1992, Nyffeler and Sterling 1994] *Peucetia flava* Keyserling [Gonzaga et al. 1998], *P. viridans* (Hentz) [Randall 1982, Nyffeler et al. 1992, Arango et al. 2012]; Pisauridae: *Architis tenuis* Simon [Nentwig 1985], *Dolomedes triton* (Walckenaer) [Zimmermann and Spence 1989]; Salticidae: *Anasaitis canosa* (Walckenaer) [Edwards et al. 1974, Jackson and Van Olphen 1991], *Bagheera kiplingi* Peckham and Peckham [Meehan et al. 2009], *Cobanus mandibularis* (Peckham and Peckham) [Jackson 1989], *Pelegrina galathea* (Walckenaer) [Dean et al. 1987], *Zuniga magna* Peckham and Peckham [Oliveira 1988]; Scytodidae: *Scytodes longipes* Lucas [Nentwig 1985]; Sicariidae: *Loxosceles intermedia* Mello-Leitão [Fischer et al. 2006]; Tetragnathidae: *Leucage mariana* (Taczanowski) [Ibarra-Núnez et al. 2001], *L. venusta* (Walckenaer) [Bilsing 1920, Hénaut et al. 2006], *Homalometa nigritarsis* Simon [Edwards and Edwards 2000]; Theraphosidae: *Acanthoscurria atrox* Vellard [Lourenco 1978]; Theridiidae: *Anelosimus baeza* Agnarsson [Guevara and Avilés 2009], *A. eximius* (Keyserling) [Nentwig 1985,

Pasquet and Krafft 1992], *A. jucundus* (O. P.-Cambridge) [Nentwig and Christenson 1986], *Ariamnes attentuatus* O. P.-Cambridge [Eberhard 1979], *Helvibis longicauda* Keyserling [Gonzaga et al. 2006], *Chrosiothes portalensis* Levi [Pérez de la Cruz et al. 2007], *C. tonala* (Levi) [Eberhard 1991], *Chrysso cambridgei* (Petrunkevitch) [Moreno-Mendoza et al. 2012], *Latrodectus mirabilis* (Holmberg) [Pompozzi et al. 2013], *Parasteatoda tesselata* (Keyserling) [Ibarra-Núnez et al. 2001], *Romphaea projiciens* O. P.-Cambridge [Eberhard 1979], *Theridion australe* Banks [Nyffeler et al. 1988], *T. evexum* Keyserling [Barrantes and Weng 2007], *Tidarren haemorrhoidale* (Bertkau) [Nyffeler et al. 1988, Moreno-Mendoza et al. 2012]; Thomisidae: *Aphantochilus rogersi* O. P.-Cambridge [Castanho and Oliveira 1997], *Misumenops pallidus* (Keyserling) [Cheli et al. 2006, González et al. 2009], *Runcinioides argenteus* Mello-Leitão [Romero and Vasconocellos-Neto 2003], *Strophius nigricans* Keyserling [Oliveira and Sazima 1985], *Tmarus stolzmanni* Keyserling [Lubin 1983]; Trechaleidae: *Trechalea extensa* (O. P.-Cambridge) [Van Berkum 1982], *Trechaleoides biocellata* (Mello-Leitão) [Van Berkum 1982]; Uloboridae: *Miagrammopes intempus* Chickering [Lubin et al. 1978], *Uloborus trilineatus* Keyserling [Moreno-Mendoza et al. 2012].

References

Anderson JF (1974) Responses to starvation in the spiders *Lycosa lenta* Hentz and *Filistata hibernalis* (Hentz). Ecology 55:576–585

Bartos M (2011) Partial dietary separation between coexisting cohorts of *Yllenus arenarius* (Araneae: Salticidae). J Arachnol 39:230–235

Bearhop S, Adams CE, Waldron S, Fueller RA, MacLeod H (2004) Determining trophic niche width: a novel approach using stable isotope analysis. J Anim Ecol 73:1007–1012

Begon M, Townsend CR, Harper JL (2006) Ecology: from individuals to ecosystems, 4th edn. Oxford, Blackwell Publishing

Blackledge TA (2011) Prey capture in orb weaving spiders: are we using the best metric? J Arachnol 39:205–210

Bristowe WS (1939) The comity of spiders I. Ray Society, London

Cangialosi KR (1990) Life cycle and behaviour of the kleptoparasitic spider *Argyrodes ululans* (Araneae, Theridiidae). J Arachnol 18:347–358

Cangialosi KR (1997) Foraging versatility and the influence of host availability in *Argyrodes trigonum* (Araneae, Theridiidae). J Arachnol 25:182–193

Cárdenas M, Jiroš P, Pekár S (2012) Selective olfactory attention of a specialised predator to intraspecific chemical signals of its prey. Naturwissenschaften 99:1–9

Castanho LM, Oliveira PS (1997) Biology and behaviour of the Neotropical ant-mimicking spider *Aphantochilus rogersi* (Araneae: Aphantochilidae): nesting, maternal care and ontogeny of ant-hunting techniques. J Zool 242:643–650

Caswell H (2001) Matrix population models: construction, analysis and interpretation, 2nd edn. Massachusetts, Sinauer Associates

Chapman EG, Romero SA, Harwood JD (2010) Maximizing collection and minimizing risk: does vacuum suction sampling increase the likelihood for misinterpretation of food web connections? Mol Ecol Resour 10:1023–1033

Chapman EG, Schmidt JM, Welch KD, Harwood JD (2013) Molecular evidence for dietary selectivity and pest suppression potential in an epigeal spider community in winter wheat. Biol Control 65:72–86
Coddington JA, Levi HW (1991) Systematics and evolution of spiders (Araneae). Annu Rev Ecol Syst 22:565–592
Cutler B (1980) Ant predation by *Habrocestum pulex* Hentz (Araneae: Salticidae). Zool Anz 204:97–101
Diguet L (1909) Le mosquero nid d'araignée employé dans certaines régions du Mexique comme piége a mouches. Bull Soc Natl Acclim France 56:368–337
Drees von O (1952) Untersuchungen uber die angeborenen Verhaltensweisen bei Springspinnen (Salticidae). Z Tierpsychol 9:169–207
Eberhard WG (1977) Aggressive chemical mimicry by a bolas spider. Science 198:1173–1175
Eberhard WG (1980) The natural history and behavior of the bolas spider *Mastophora dizzydeani* sp. n. (Araneidae). Psyche 87:143–169
Eberhard WG (1981a) Notes on the natural history of *Tazanowskia* sp. (Araneae: Araneidae). Bull Br Arachnol Soc 5:175–176
Eberhard WG (1981b) The single line web of *Phoronocidia studo* Levi (Araneae, Theridiidae) — a prey attractant? J Arachnol 9:229–232
Eberhard WG (1991) *Chrosiothes tonala* (Araneae, Theridiidae): a web-building spider specializing on termites. Psyche 98(1):7–19
Edwards GB, Jackson RR (1994) The role of experience in the development of predatory behaviour in *Phidippus regius*, a jumping spider (Araneae, Salticidae) from Florida. N Z J Zool 21:269–277
Edwards GB, Carroll JF, Whitcomb WH (1974) *Stoidis aurata* (Araneae: Salticidae) a spider predator of ants. Fla Entomol 57:337–345
Erthal M, Tonhasca A (2001) *Attacobius attarum* spiders (Corinnidae): myrmecophilous predators of immature forms of the leaf-cutting ant *Atta sexdens* (Formicidae). Biotropica 33:374–376
Escalante I, Aisenberg A, Costa FG (2015) Risky behaviors by the host could favor araneophagy of the spitting spider *Scytodes globula* on the hacklemesh weaver *Metaltella simoni*. J Ethol 33:125–136
Fischer ML, Vasconocellos-Neto J, dos Santos Neto LG (2006) The prey and predators of *Loxosceles intermedia* Mello-Leitao 1934 (Araneae, Sicariidae). J Arachnol 34:485–488
Fowler HG (1984) Note on a clubionid spider associated with attine ants. J Arachnol 12:117
Framenau VW, Finley LA, Allan K, Love M, Shirley D, Elgar MA (2000) Multiple feeding in wolf spiders: the effect of starvation on handling time, ingestion rate, and intercatch intervals in *Lycosa lapidosa* (Araneae: Lycosidae). Aust J Zool 48:59–65
Garb JE (2014) Extraction of venom and venom gland microdissections from spiders for proteomic and transcriptomic analyses. J Vis Exp 93:e51618
García LF, Lacava M, Viera C (2014) Diet composition and prey selectivity by the spider *Oecobius concinnus* (Araneae: Oecobiidae) from Colombia. J Arachnol 42:199–201
Giardine B, Riemer C, Hardison RC, Burhans R, Elnitski L, Shah P, Zhang Y, Blankenberg D, Albert I, Taylor J, Miller W, Kent WJ, Nekrutenko A (2005) Galaxy: a platform for interactive large-scale genome analysis. Genome Res 15:1451–1455
Goloboff PA (2000) The family Gallieniellidae (Araneae, Gnaphosidae) in the Americas. J Arachnol 28:1–6
Gonzaga MO (2007) Araneofagia e cleptoparasitismo. In: Gonzaga MO, Santos AJ, Japyassú HF (eds) Ecologia e Comportamento de Aranhas. Interciência, Rio de Janeiro, pp 239–255
Gonzaga MO, dos Santos AJ, Dutra GF (1998) Web invasion and araneophagy in *Peucetia tranquillini* (Araneae, Oxyopidae). J Arachnol 26:249–250
Greenstone MH (1999) Spider predation: how and why we study it. J Arachnol 27:333–342
Haddad CR, Brabec M, Pekár S, Fourie R (2016) Seasonal population dynamics of a specialized termite-eating spider (Araneae: Ammoxenidae) and its prey (Isoptera: Hodotermitidae). Pedobiologia 59:105–110

Hebert PDN, Cywinska A, Ball SL, de Waard JR (2003) Biological identifications through DNA barcodes. Proc Biol Sci 270:313–321

Hood-Nowotny R, Knols BGJ (2007) Stable isotope methods in biological and ecological studies of arthropods. Entomol Exp Appl 124:3–16

Huseynov EFO (2014) Natural prey of the crab spider *Xysticus marmoratus* (Araneae: Thomisidae) on *Eryngium* plants. J Arachnol 42:130–132

Hutchinson GE (1957) Concluding remarks. Cold Spring Harb Symp Quant Biol 22:415–427

Jackson RR, van Olphen A (1991) Prey-capture techniques and prey preferences of *Corythalia canosa* and *Pystira orbiculata*, ant-eating jumping spiders (Araneae: Salticidae). J Zool 223:577–591

Jocqué R (1988) An updating of the genus *Leprolochus* (Araneae: Zodariidae). Stud Neotropical Fauna Environ 23:77–87

Japyassú HF, Viera C (2002) Predatory plasticity in *Nephilengys cruentata* (Araneae: Tetragnathidae): relevance for phylogeny reconstruction. Behaviour 139(4):529–544

King RA, Read DS, Traugott M, Symondson WOC (2008) Molecular analysis of predation: a review of best practice for DNA-based approaches. Mol Ecol 17:947–963

Korenko S, Hamouzová K, Pekár S (2014) Trophic niche and predatory behaviour of the goblin spider Triaeris stenaspis (Oonopidae): springtail specialist? J Arachnol 42:74–78

Kuusk AK, Agustí N (2008) Group-specific primers for DNA-based detection of springtails (Hexapoda: Collembola) within predator gut contents. Mol Ecol Resour 8:678–681

Lehner PN (1999) Handbook of ethological methods, 2nd edn. Cambridge, Cambridge University Press

Lesar DE, Unzicker JD (1978) Life history, habits and prey preferences of *Tetragnatha laboriosa* (Araneae: Tetreagnathidae). Environm Entomol 7(6):879–884

Liu J, May-Collado LJ, Pekár S, Agnarsson I (2016) A revised and dated phylogeny of cobweb spiders (Araneae, Araneoidea, Theridiidae): a predatory cretaceous lineage diversifying in the era of the ants (hymenoptera, Formicidae). Mol Phylogenet Evol 94:658–675

Liznarová E, Pekár S (2015) Trophic niche of *Oecobius maculatus* (Araneae: Oecobiidae): evidence based on natural diet, prey capture success, and prey handling. J Arachnol 43:88–193

Líznarová E, Sentenská L, García LF, Pekár S, Viera C (2013) Local trophic specialisation in a cosmopolitan spider (Araneae). Zoology 116:20–26

Lubin YD (1983) An ant-eating crab spider from the Galapagos. Not Galapagos 37:18–19

Manly BFJ, McDonald LL, Thomas DL, McDonald TL, Erickson WP (2002) Resource selection by animals: statistical design and analysis for field studies, 2nd edn. New York, Kluwer Academic Publishers

Marshall SD, Gittleman JL (1994) Clutch size in spiders: is more better? Funct Ecol 8:118–124

Marshall SA, Borkent A, Agnarsson I, Otis GW, Fraser L, d'Entremont D (2015) New on a Neotropical termite-hunting theridiid spider: opportunistic nest raiding, prey storage, and ceratopogonid kleptoparasites. J Arachnol 43:419–421

Maupin JL, Riechert SE (2001) Superfluous killing in spiders: a consequence of adaptation to food-limited environments? Behav Ecol 12(5):569–576

Mayntz D, Raubenheimer D, Salomon M, Toft S, Simpson SJ (2005) Nutrient-specific foraging in invertebrate predators. Science 307:111–113

McNabb DM, Halaj J, Wise DH (2001) Inferring trophic positions of generalist predators and their linkage to the detrital food web in agroecosystems: a stable isotope analysis. Pedobiologia 45:289–297

Mestre L, Piňol J, Barrientos JA, Espalader X, Brewitt K, Werner C, Planter C (2013) Trophic structure of the spider community of a Mediterranean citrus grove: a stable isotope analysis. Basic Appl Ecol 14:413–422

Miller J, Griswold C, Scharff N, Řezáč M, Szuts T, Marhabaie M (2012) The velvet spiders: an atlas of the Eresidae (Arachnida, Araneae). ZooKeys 195:1–144

Moreno-Mendoza SD, Nunez-Ibarra G, Chame-Vazquez ER, Valle-Mora FJ (2012) Gama de presas capturadas por cuatro especies de aranas tejedoras (Arachnida: Araneae) en un agroecosistema de cacao en Chipas, Mexico. Trop Subtrop Agroecosyst 15:457–469
Morrone JJ (2014) Biogeographical regionalisation of the Neotropical region. Zootaxa 3782:1–110
Nelson XJ, Jackson RR (2011) Flexibility in the foraging strategies of spiders. In: Herberstein ME (ed) Spider behaviour: flexibility and versatility. Cambridge University Press, Cambridge, pp 31–56
Nentwig W (1985) Prey analysis of four species of tropical orb-weaving spiders (Araneae: Araneidae) and a comparison with araneids of the temperate zone. Oecologia 66:580–594
Nyffeler M, Knörnschild M (2013) Bat predation by spiders. PLoS One 8:e58120
Nyffeler M, Pusey BJ (2014) Fish predation by semi-aquatic spiders: a global pattern. PLoS One 9:e99459
Oliveira PS (1987) Ant-mimicry in some spiders from Brazil. Bull Soc Zool France 111:297–311
Oliveira PS, Sazima I (1984) The adaptative bases of ant-mimicry in a Neotropical aphantochilid spider (Aphantochilidae). Biol J Linn Soc 22:145–155
Oliveira PS, Sazima I (1985) Ant-hunting behaviour in spiders with emphasis on *Strophius nigricans* (Thomisidae). Bull Br Arachnol Soc 6:309–312
Palagi A, Koh JMS, Leblanc M, Wilson D, Dutertre S, King GF, Nicholson GM, Escoubas P (2013) Unravelling the complex venom landscapes of lethal Australian funnel-web spiders (Hexathelidae: Atracinae) using LC–MALDI–TOF mass spectrometry. J Proteome 80:292–310
Pekár S (2004) Predatory behavior of two European ant-eating spiders (Araneae, Zodariidae). J Arachnol 32:31–41
Pekár S, Cárdenas M (2015) Innate prey preference overridden by familiarisation with detrimental prey in a specialised myrmecophagous predator. Sci Nat 102(1–2):1257
Pekár S, Toft S (2015) Trophic specialisation in a predatory group: the case of prey-specialised spiders (Araneae). Biol Rev 90:744–761
Pekár S, Král J, Lubin YD (2005) Natural history and karyotype of some ant-eating zodariid spiders (Araneae, Zodariidae) from Israel. J Arachnol 33:50–62
Pekár S, Toft S, Hrušková M, Mayntz D (2008) Dietary and prey-capture adaptations by which *Zodarion germanicum*, an ant-eating spider (Araneae: Zodariidae), specialises on the Formicinae. Naturwissenschaften 95:233–239
Pekár S, Mayntz D, Ribeiro T, Herberstein ME (2010) Specialist ant-eating spiders selectively feed on different body parts to balance nutrient intake. Anim Behav 79:1301–1306
Pekár S, Coddington JA, Blackledge T (2012) Evolution of stenophagy in spiders (Araneae): evidence based on the comparative analysis of spider diets. Evolution 66:776–806
Pekár S, Šedo O, Líznarová E, Korenko S, Zdráhal Z (2014) David and the Goliath: potent venom of an ant-eating spider (Araneae) enables capture of a giant prey. Naturwissenschaften 101:533–540
Pérez de la Cruz MS, Sanchez Soto M, Ortiz García CF, Zapata Mata R, Pérez de la Cruz A (2007) Diversity of insects captured by weaver spiders (Arachnida: Araneae) in the cocoa agroecosystem in Tabasco, Mexico. Neotrop Entomol 36:90–101
Petráková L, Líznarová E, Sentenská L, Haddad CR, Pekár S, Symondson WOC (2015) Discovery of a monophagous true predator, a specialist termite-eating spider (Araneae: Ammoxenidae). Sci Rep 5:14013
Piñol J, San Andrés V, Clare EL, Mir G, Symondson WOC (2014) A pragmatic approach to the analysis of diets of generalist predators: the use of next-generation sequencing with no blocking probes. Mol Ecol Resour 14(1):18–26
Pollard SD, Beck MW, Dodson GN (1995) Why do male crab spiders drink nectar? Anim Behav 49:1443–1448
Pompanon F, Deagle BE, Symondson WOC, Brown DS, Jarman SN, Taberlet P (2011) Who is eating what: diet assessment using next generation sequencing. Mol Ecol 21:1931–1950

Pompozzi G, Ferretti N, Schwerdt L, Copperi S, Ferrero A, Simó M (2013) The diet of the black widow spider *Latrodectus mirabilis* (Theridiidae) in two cereal crops of central Argentina. Iheringia – Ser Zool 103:388–392

Putnam WL (1967) Prevalence of spiders and their importance as predators in Ontario peach orchards. Can Entomol 99:160–170

Řezáč M, Pekár S, Lubin Y (2008) How oniscophagous spiders overcome woodlouse armour. J Zool 275:64–71

Romero-Ortiz C, Flórez-Daza E (2014) Un caso de araneofagia de *Mimetus* sp. (Araneae, Mimetidae), sobre *Leucauge* sp. (Araneae, Tetragnathidae) en el nororiente de Colombia. Rev Iberica Aracnol 24:102–103

Rossi MN, Reigada C, Godoy WAC (2006) The effect of hunger level on predation dynamics in the spider *Nesticodes rufipes*: a functional response study. Ecol Res 21:617–623

Salomon M (2011) The natural diet of a polyphagous predator, *Latrodectus hesperus* (Araneae: Theridiidae), over one year. J Arachnol 39:154–160

Sint D, Raso L, Traugott M (2012) Advances in multiplex PCR: balancing primer efficiencies and improving detection success. Methods Ecol Evol 3:898–905

Smith EP (1982) Niche breadth, resource availability, and inference. Ecology 63:1675–1681

Smithers P (2005) The diet of the cave spider *Meta menardi* (Latreille 1804) (Araneae, Tetragnathidae). J Arachnol 33:243–246

Soberon J, Peterson AT (2005) Interpretation of models of fundamental ecological niches and species' distributional areas. Biodiv Inform 2:1–10

Southwood TRE, Henderson PA (2000) Ecological methods. Blackwell Science, Oxford

Stowe MK (1978) Observations of two nocturnal orbweavers that build specialized webs: *Scoloderus cordatus* and *Wixia ectypa*. J Arachnol 6:141–146

Stowe MK (1986) Prey specialization in the Araneidae. In: Shear WA (ed) Spiders, webs, behavior, and evolution. Stanford University Press, Stanford, pp 101–131

Stowe MK (1988) Chemical mimicry. In: Spencer KC (ed) Chemical mediation of coevolution. Academic Press, San Diego, pp 513–580

Sunderland KD (1988) Quantitative methods for detecting invertebrate predation occurring in the field. Ann Appl Biol 112:201–224

Symondson WOC (2002) Molecular identification of prey in predator diets. Mol Ecol 11:627–641

Teruel R, Sánchez-Ruiz A (2000) Nota sobre la depredación de un escorpión (Scorpiones: Buthidae) por una araña (Araneae: Caponiidae). Biodivers Cuba Orient 4:82–83

Tietjen WJ, Ayyagari LR, Uetz GW (1987) Symbiosis between social spiders and yeast: the role in prey attraction. Psyche 94:151–115

Toft S (1999) Prey choice and spider fitness. J Arachnol 27:301–307

Toft S, Li D, Mayntz D (2010) A specialized araneophagic predator's short-term nutrient utilization depends on the macronutrient content of prey rather than on prey taxonomic affiliation. Physiol Entomol 35:317–327

Wiehle H (1953) Spinnentiere oder Arachnoidea (Araneae) IX: Orthognatha — Cribellatae — Haplogynae — Entelegynae (Pholcidae, Zodariidae, Oxyopidae, Mimetidae, Nesticidae). In: Dahl F (ed) Die Tierwelt Deutschlands und der angrenzenden Meeresteile nach ihren Merkmalen und nach ihrer Lebensweise 42. Teil. Gustav Fischer, Jena

Wilder SM (2011) Spider nutrition: an integrative perspective. Adv Insect Phys 40:87–136

Wirta HK, Vesterinen EJ, Hambäck PA, Weingartner E, Rasmussen C, Reneerkens J, Schmidt NM, Gilg O, Roslin T (2015) Exposing the structure of an Arctic food web. Ecol Evol 5:3842–3856

Wullschleger B, Nentwig W (2002) Influence of venom availability on a spider's prey-choice behaviour. Funct Ecol 16:802–807

Xu CC, Yen IJ, Bowman D, Turner CR (2015) Spider web DNA: a new spin on noninvasive genetics of predator and prey. PLoS One 10:e0142503

Zaidi RH, Jaal Z, Hawkes NJ, Hemingway J, Symondson WOC (1999) Can multiple-copy sequences of prey DNA be detected amongst the gut contents of invertebrate predators? Mol Ecol 8:2081–2088

Additional References

Aisenberg A, González M, Laborda A, Postiglioni R, Simó M (2009) Reversed cannibalism, foraging, and surface activities of *Allocosa alticeps* and *Allocosa brasiliensis*: two wolf spiders from coastal sand dunes. J Arachnol 37:135–138

Arango AM, López-Portillo J, Parra-Tabla V, Hernández-Salazar LT, Morales-Mávil JE, Rico-Gray V (2012) Effect of the spider *Peucetia viridans* (Oxyopidae) on floral visitors and seed set of *Cnidoscolus multilobus* (Euphorbiaceae). Acta Bot Mex 100:1–14

Barrantes G, Weng JL (2007) Natural history, courtship, feeding behaviour and parasites of *Theridion evexum* (Araneae: Theridiidae). Bull Br Arachnol Soc 14:61–65

Bilsing WS (1920) Quantitative studies in the food of spiders. Ohio J Sci 20:215–260

Blackledge T, Wenzel J (1999) Do stabilimenta in orb webs attract prey or defend spiders? Behav Ecol 10:372–376

Calixto A, Levi HW (2006) Notes on the natural history of *Aspidolasius brancki* (Araneae: Araneidae) at Tinigua National Park, Colombia, with revision of the genus. Bull Br Arachnol Soc 13:314–230

Ceballos L, Hénaut Y, Legal L (2005) Foraging strategies of *Eriophora edax* (Araneae, Araneidae): a nocturnal orb-weaving spider. J Arachnol 33:509–515

Cheli G, Armendano A, González A (2006) Preferencia alimentaria de arañas *Misumenops pallidus* (Araneae: Thomisidae) sobre potenciales insectos presa de cultivos de alfalfa. Rev Biol Trop 54:505–513

Culin JD, Yeargan KV (1982) Feeding behavior and prey of *Neoscona arabesca* (Araneae: Araneidae) and *Tetragnatha laboriosa* (Araneae: Tetragnathidae) in soybean fields. Entomophaga 27:417–424

Dean DA, Sterling WL, Nyffeler M, Breene RG (1987) Foraging by selected spider predators on the cotton fleahopper and other prey. Southwest Entomol 12:263–270

Eberhard WG (1979) *Argyrodes attenuatus* (Theridiidae): a web that is not a snare. Psyche 86:407–413

Edwards R, Edwards A (2000) Observations on the webs and life history of *Homalometa nigritarsis* (Araneae: Tetragnathidae): a spider that lays its eggs in rows. Psyche 103:37–48

Flórez E, Pinzon J, Sabogal A, Barreto N (2004) Selección de presas y composicion de la dieta de la araña *Alpaida variabilis* (Araneae: Araneidae) en pastizales de la sabana de Bogota, Colombia. RIA 9:241–248

Fowler HG, Diehl J (1978) Biology of Paraguayan colonial orb-weaver *Eriophora bistriata* (Rengger) (Araneae, Araneidae). Bull Br Arachnol Soc 4:241–250

Fowler HG, Gobbi N (1988) Cooperative prey capture by an orb-web spider. Naturwissenschaften 75:208–209

Gonzaga MO, Leiner NO, Santos AJ (2006) On the sticky cobwebs of two theridiid spiders (Araneae: Theridiidae). J Nat Hist 40:293–306

González A, Liljesthrom G, Minervino E, Castro D, Gonzalez S, Armendano A (2009) Predation by *Misumenops pallidus* (Araneae: Thomisidae) on insect pests of soybean cultures in Buenos Aires Province, Argentina. J Arachnol 37:282–286

Gregory BM (1989) Field observations of *Gasteracantha cancriformis* (Araneae, Araneidae) in a Florida mangrove stand. J Arachnol 17:119–120

Guevara J, Avilés L (2009) Elevational changes in the composition of insects and other terrestrial arthropods at tropical latitudes: a comparison of multiple sampling methods and social spider diets. Insect Conserv Divers 2:142–152

Hénaut Y, García-Ballinas JA, Alauzet C (2006) Variations in web construction in Leucage venusta (Araneae, Tetragnathidae). J Arachnol 34:234–224

Ibarra-Núnez G, García JA, López JAL, Lachaud JP (2001) Prey analysis in the diet of some Ponerine ants (Hymenoptera: Formicidae) and web-building spiders (Araneae) in coffee plantations in Chiapas, Mexico. Sociobiology 37:723–755

Jackson RR (1989) The biology of *Cobanus mandibularis*, a jumping spider (Araneae: Salticidae) from Costa Rica: intraspecific interactions, predatory behaviour, and silk utilisation. N Z J Zool 16:383–392

Lourenco WR (1978) Notas sobre a biologia de *Acanthoscurria atrox* Vellard, 1924 (Araneae, Theraphosidae). Rev Bras Biol 38:161–164

Lubin YD (1974) Adaptive advantages and the evolution of colony formation in Cyrthophora (Araneae: Araneidae). Zool J Linn Soc 54:321–339

Lubin YD, Eberhard WG, Montgomery GG (1978) Webs of *Miagrammopes* (Araneae: Uloboridae) in the Neotropics. Psyche 85:1–23

Meehan CJ, Olson EJ, Reudink MW, Kyser TK, Curry RL (2009) Herbivory in a spider through exploitation of an ant–plant mutualism. Curr Biol 19:R892–R893

Nentwig W (1983) The prey of web-building spiders compared with feeding experiments (Araneae: Araneidae, Linyphiidae, Pholcidae, Agelenidae). Oecologia 56:132–139

Nentwig W (1986) Non-web-building spiders: prey specialists or generalists? Oecologia 69:571–576

Nentwig W (1990) Stick insects (Phasmida) as prey of spiders. Size, palatability and defence mechanisms in feeding tests. Oecologia 82:446–450

Nentwig W, Christenson TC (1986) Natural history of the non-solitary sheetweaving spider Anelosimus jucundus (Araneae: Theridiidae). Zool J Linn Soc 87:27–35

Nyffeler M, Sterling WL (1994) Comparison of the feeding niche of polyphagous insectivores (Araneae) in a Texas cotton plantation: estimates of niche breadth and overlap. Environ Entomol 23:1294–1303

Nyffeler M, Dean DA, Sterling WL (1986) Feeding habits of the spiders Cyclosa turbinata (Walckenaer) and Lycosa rabida Walckenaer. Southwest Entomol 11:195–201

Nyffeler M, Dean DA, Sterling WL (1987) Feeding ecology of the orb-weaving spider *Argiope aurantia* (Araneae, Araneidae) in a cotton agroecosystem. Entomophaga 32:367–376

Nyffeler M, Dean DA, Sterling WL (1988) Prey records of the web-building spiders *Dictyna segregata* (Dictynidae), *Theridion australe* (Theridiidae), *Tidarren haemorrhoidale* (Theridiidae), and *Frontinella pyramitela* (Linyphiidae) in a cotton agroecosystem. Southwest Nat 33:215–218

Nyffeler M, Dean DA, Sterling WL (1989) Prey selection and predatory importance of orb-weaving spiders (Araneae: Araneidae, Uloboridae) in Texas cotton. Environ Entomol 18:373–380

Nyffeler M, Dean DA, Sterling WL (1992) Diets, feeding specialization, and predatory role of two lynx spiders, *Oxyopes salticus* and *Peucetia viridans* (Araneae: Oxyopidae), in a Texas cotton agroecosystem. Environ Entomol 21:1457–1465

Oliveira PS (1988) Ant-mimicry in some Brazilian salticid and clubionid spiders (Araneae: Salticidae, Clubionidae). Biol J Linn Soc 33:1–15

Pasquet A, Krafft B (1992) Cooperation and prey capture efficiency in a social spider, *Anelosimus eximius* (Araneae, Theridiidae). Ethology 90:121–133

Petráková L, Michalko R, Loverre P, Sentenská L, Korenko S, Pekár S (2016) Intraguild predation among spiders and their effect on the pear psylla during winter. Agric Ecosyst Environ 233:67–74

Pinto C, Saiz F (1997) Uso del recurso trofico por parte de *Acanthogonatus franckii* Karsch, 1880 (Araneae: Nemesiidae) en el bosque esclerofilo del Parque Nacional "La Campana", Chile central. Rev Chil Entomol 24:45–59

Randall JB (1982) Prey records of the green lynx spider, *Peucetia viridans* (Hentz) (Araneae, Oxyopidae). J Arachnol 10:19–22

Romero GQ, Vasconocellos-Neto J (2003) Natural history of *Misumenops argenteus* (Thomisidae): seasonality and diet on *Trichogoniopsis adenantha* (Asteraceae). J Arachnol 31:297–304

Shelly TE (1983) Prey selection by the Neotropical spider, *Alpaida tuonado*, with notes on web-site tenacity. Psyche 90:123–133

Shelly T (1984) Prey selection by the Neotropical spider *Micrathena schreibersi* with notes on web-site tenacity. Proc Entomol Soc Wash 86:493–502

Uetz GW, Biere JM (1980) Prey of *Micrathena gracilis* (Walckenaer) (Araneae: Araneidae) in comparison with artificial webs and other trapping devices. Bull Br Arachnol Soc 5:101–107

Uetz GW, Hartsock SP (1987) Prey selection in an orb-weaving spider: *Micrathena gracilis* (Araneae, Araneidae). Psyche 94:103–116

Uetz GW, Johnson AD, Schemske DW (1978) Web placement, web structure, and prey capture in orb-weaving spiders. Bull Br Arachnol Soc 4:141–148

Van Berkum FH (1982) Natural history of a tropical, shrimp-eating spider (Pisauridae). J Arachnol 10:117–121

Wise D, Barata JL (1983) Prey of two syntopic spiders with different web structures. J Arachnol 11:271–281

Zimmermann M, Spence JR (1989) Prey use of the fishing spider *Dolomedes triton* (Pisauridae, Araneae): an important predator of the neuston community. Oecologia 80:187–194

Chapter 11
Spiders Associated with Agroecosystems: Roles and Perspectives

Marco Antonio Benamú, Mariángeles Lacava, Luis Fernando García, Martín Santana, and Carmen Viera

Abstract Spiders are considered one of the most abundant groups of predators in different ecosystems related to human production activities. Crops represent one of the most important economical sources for the Neotropical region. In consequence, agroecosystems are in continuous expansion and with them the associated spiders. In this chapter, we will present current knowledge about different studies concerning the role of spiders in agroecosystems, with emphasis on studies in the Neotropical region. Subsequently, we will focus on the diversity of predatory strategies in spiders, and how this trait allows them to capture different insect-pests, presenting also evidence with regard to the importance of spiders in controlling pest populations. Also, we will show how the effects of different human activities in agroecosystems may affect the predatory potential and reproductive traits in neotropical web-building and wandering spiders, focusing on the negative sub-lethal effects caused by pesticides used in crops.

M.A. Benamú (✉) • M. Lacava
Centro Universitario de Rivera, Universidad de la República, Rivera, Uruguay

Laboratorio Ecología del Comportamiento, Instituto de Investigaciones Biológicas Clemente Estable, Montevideo, Uruguay
e-mail: marcobenamu@gmail.com

L.F. García
Laboratorio Ecología del Comportamiento, Instituto de Investigaciones Biológicas Clemente Estable, Montevideo, Uruguay

Centro Universitario de la Región Este, sede Treinta y Tres, Universidad de la República, Treinta y Tres, Uruguay

M. Santana • C. Viera
Entomología, Biología Animal, Facultad de Ciencias, Universidad de la República, Montevideo, Uruguay

Laboratorio Ecología del Comportamiento, Instituto de Investigaciones Biológicas Clemente Estable, Montevideo, Uruguay

C. Viera, M.O. Gonzaga (eds.), *Behaviour and Ecology of Spiders*,
https://doi.org/10.1007/978-3-319-65717-2_11

Intensification of agriculture in recent decades has modified the heterogeneity of landscapes, affecting biological groups through loss or reduction of their habitats and undoubtedly becoming a threat to the conservation of species (Di Castri and Younés 1996; Anderson and Danielson 1997; Atauri and De Lucio 2001; Hunter and Gibbs 2006).

In Latin America, an extensive monoculture style of agriculture is predominant, *involving* the intensive use of agrochemicals such as fertilizers, fungicides, and insecticides. But traditional agriculture practices have also been proven to have a negative impact on biodiversity as a result of the reduction of original habitat by practices such as deforestation (Altieri and Letourneau 1982; Altieri 1995; Flint and Roberts 1988), or through the intensity of agricultural practice, where susceptible species are widely affected (Pickett and White 1985). Reduction in biodiversity diminishes ecological interactions, altering processes such as predation or competition, encouraging as a consequence the development of phytophagous species and leading them to become plagues.

Several studies show that spiders are a dominant group of generalist predators in several agroecosystems (Wise 1993; Foelix 2010; Benamú et al. 2011), and their habits increase their potential as the main predator of certain plagues (Mansour et al. 1980; Ibarra-Nuñez 1990; Minervino 1996; Medina 1994; Maloney et al. 2003; Pearce et al. 2004; Cheli et al. 2006; Benamú 2010; Samiayyan 2014).

Spiders have a generalist diet, including not only adult insects, but also eggs and insect larvae (Lepidoptera and Coleoptera) (Whitcomb 1974; Riechert and Lockley 1984; Nyffeler et al. 1990; Young and Edwards 1990; Green 1996). Despite this, they can occasionally have a restricted diet when a prey is found in large densities, as happens with plagues in monocultures (Liljesthröm et al. 2002; Benamú 2010; Almada and Sosa 2011). Their permanent presence and abundance during all of the stages of the crop cycle allow them to play a role as effective natural enemies of phytophagous insects (Aguilar 1988; Benamú and Aguilar 2001; Benamú 2004, 2010) as a result of their capability to settle in different agroecosystems (Riechert and Lockley 1984; Marc et al. 1999; Nyffeler et al. 1994; Symondson et al. 2002; Maloney et al. 2003; Benamú 2010; Cantor 2014). In addition to being indicators of environmental quality in agricultural fields (Clausen 1986), they are the dominant component of the generalist predator ensemble (Sunderland 1999). Many studies identify spiders as a significant component of the complex of polyphagous predators in agroecosystems (Young and Edward 1990; Minervino 1996; Sunderland and Greenstone 1999; Hagen et al. 1999; Samu et al. 1999; Wardle et al. 1999; Halaj et al. 2000; Liljesthröm et al. 2002; Büchs 2003; Jeanneret et al. 2003; Benamú 2004; Perafán and Flórez 2004; Pearce et al. 2005; Beltramo et al. 2006; Marshall et al. 2006; Armendano 2008).

Spiders can play a complementary role in the time lag between the initial increase of the plague and the numeric response from other specific enemies (Nyffeler et al. 1994; Benamú 2001; Brown et al. 2003; Hoefler et al. 2006; Saavedra et al. 2007). One of the most important characteristics of neotropical agricultural landscapes are the frequent disturbances caused by human activities, such as the spatial distribution of the crops, reduction and fragmentation of habitats, application of insecticides, herbicides, fungicides, synthetic fertilizers, pruning, harvest, and so on. These activities alter habitats dramatically, and spider populations are under constant pressure, the product of the permanent expansion in agroecosystems, which without sustainable management can disturb native biodiversity in a significant way, greatly reducing spider populations (Topping 2002; Benamú 2004; Alcayaga et al. 2013).

Spiders have been cited as natural enemies of weevil populations in alfalfa and cereal crops (Harcourt et al. 1986; Lanteri et al. 1998), several plagues of Lepidoptera and Hemiptera in soybean (Minervino 1996; Liljeström et al. 2002; Beltramo et al. 2006; Molinari and Minervino 2006), and other crops, whether seasonal or fruit-bearing ones.

Spiders in Seasonal Crops

Soybean

According to Benamú (2010), in soybean crops (*Glicine max*), the soil habitat is more consistent in structure and the community of araneofauna is more stable, being the community of transient foliage for many spiders (Almada and Sosa 2011).

Sampling results in soybean carried out by many authors match in relation to the number of spider families found (Molinari 1987; Minervino 1996; Liljesthröm et al. 2002; Perafán and Flórez 2004; Beltramo et al. 2006; Benamú 2010; Almada and Sosa 2011), the main families being: Lycosidae, Linyphiidae, Salticidae, Oxyopidae, Gnaphosidae, Thomisidae, Theridiidae, Araneidae, Tetragnathidae, and Anyphaenidae. Species richness and abundance of these families are limited by the stratification and vegetation height (Benamú 2010; Almada and Sosa 2011). The herbaceous stratum represents the highest percentage of the occupied area, the most abundant guilds being web-builder spiders, ambush hunters, ground hunters, ambushers, and foliage runners. All of them can be found inside the soybean crop as well as in adjacent vegetation, throughout the whole cycle of the crop in Argentina (Molinari and Minervino 2006; Beltramo et al. 2006; Benamú 2010) (Fig. 11.1). These results coincide with those found by Perafán and Flórez (2004), in a soybean–corn crop system in Colombia. The presence of spontaneous vegetation in the margins of the cultivated area contributes to the colonization by spiders of aerial habits, as they encourage recolonization after the application of pesticides, acting as refuge zones. The family Linyphiidae is the most abundant, and is strongly influenced by the composition of the landscape (Liljesthröm et al. 2002; Perafán and Flórez 2004; Beltramo et al. 2006; Benamú 2010). As they have great dispersal ability and are probably the first predators reaching a new habitat, they should have an important role in the development of the community structure (Benamú 2004, 2010). The presence of several spider species belonging to different families, with different hunting strategies, phenology, and size, would significantly increase control over the populations of a plague, the maintenance of natural

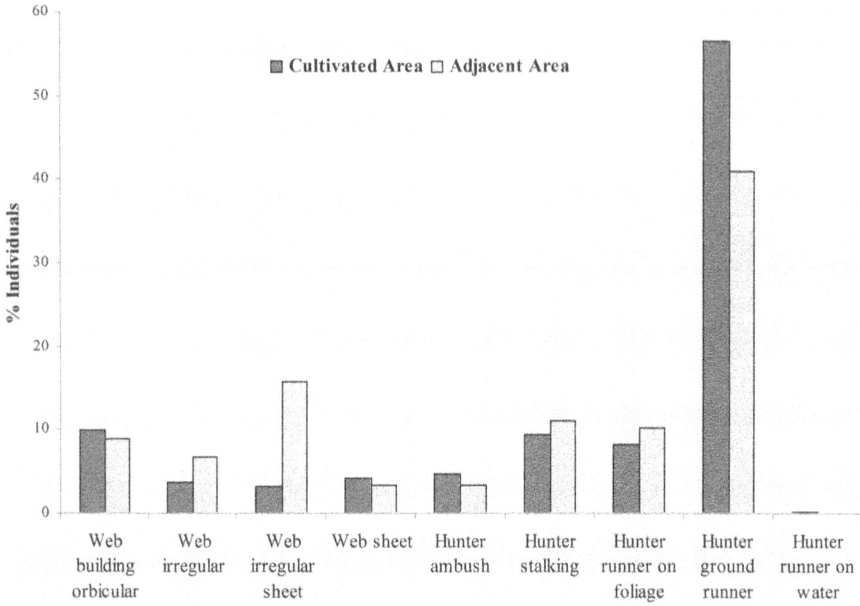

Fig. 11.1 Guilds of spiders in the second soybean seeding and vegetation adjacent area in Chivilcoy, Buenos Aires, Argentina (Benamú)

areas in biological control techniques by conservation of native predators being important (Liljesthröm et al. 2002; Beltramo et al. 2006; González et al. 2009; Benamú 2010; Almada and Sosa 2011).

Wheat and Alfalfa

Wheat crops (*Triticum aestivum)* and alfalfa (*Medicago sativa*) are usually affected by three groups of primary plagues (Lepidoptera defoliators, curculionid beetles, and aphids). Spider assemblages in these crops are important within the predator group, as they are present in the totality of the phenological development, with peaks of abundance in spring and summer. This kind of environment shows major complexity in vegetation structure, and the spider community shows a higher number, richness, and diversity of species. The guilds that predominate are ambushing hunting spiders (32.99%), stalkers (11.77%), ground hunters (10.84%) and cobweb builders (27.56%) (Armendano and González 2010). In both crops, Thomisidae is the most abundant family in the herbaceous stratum, Araneidae and Oxyopidae are abundant as well. Within the soil stratum, the most abundant families are: Lycosidae, Hahniidae, Linyphiidae and Tetragnathiidae (Yeargan and Dondale 1974; Armendano 2008; Armendano and González 2010; Armendano and González 2011a, b).

Predator–prey interaction in wheat and alfalfa crops occurs with dominant species, e.g., *Misumenops pallidus*, *Oxyopes salticus* and *Araneus* sp. feed on representative plague insects (aphids, weevils, and Lepidoptera larvae) (Armendano 2008). The dominant spider species prefer thinner cuticle preys (Lepidoptera larvae) rather than curculionids and aphids. Several studies noted that the defoliator species *Rachiplusia nu* (Noctuidae) was the most consumed and *Spodoptera frugiperda* (Noctuidae) was the least captured (Cheli et al. 2006, Armendano 2008; Armendano and González 2010).

In winter studies on wheat crops, Armendano and González (2011b) show that spiders are associated with habitats of vegetation adjacent to the margin of the crop, recording a greater number of spider species in such vegetation than in the crop itself, finding Thomisidae and Araneidae in herbaceous stratum, and Lycosidae in soil stratum. These results may be related to a sustained disturbance by harvesting and tillage among other field work. Moreover, the observed similarity between families from both communities of the margin could suggest that the process of colonization begins in adjacent areas.

Cotton

The first recorded study in cotton crops (*Gossypium hirsutum*) is from Peru, the product of 9 years of observations (1959–1968) of spider presence in cotton terminals (Aguilar 1968). Such observations differentiate two groups of spiders: frequent spiders in terminals and leaves, such as *Theridion calcynatum, Theridula gonygaster* (Theridiidae), *Leucauge* sp. (Tetragnathidae), *Oxyopes gracilipes, Peucetia* sp. (Oxyopidae), *Aysha* sp. (Anyphaenida), *Misumena amabilis, Misumenops variegatus* (Thomisidae), and frequent spiders inside bolls, flowers, and acorns, such as *Steatoda andina* (Theridiidae), *Gasteracantha cancriformis* (Araneidae), *Lycosa* sp. (Lycosidae), *Clubiona* sp. (Clubionidae), *Anyphaena* sp. (Anyphaenidae), *Metaphidippus* sp., *Phiale* sp. (Salticidae). The abundance of spiders is affected by the application of pesticides. Further studies on the population of spiders in cotton (Aguilar 1974) relate their presence to the development of plants, estimating that they can reach 30 individuals per bush on average, where over 50% of the population of spiders can consist of small individuals, mostly juveniles. It should be noted that the majority of spiders from the families Anyphaenidae, Thomisidae, Clubionidae, and Salticidae, and the families Araneidae, Tetragnathidae, Theridiidae, Oxyopidae, and Lycosidae, cannot increase their population in a cultivated field, due to management of the field itself (Aguilar 1975, 1976; Aguilar and Güerovich 1978). It has been found that the growth of the spider population within the cultivated area depended on the high number of juveniles found, favoring the recovery of the population at the beginning of each season (Aguilar 1978, 1979).

Generally, spiders are present during the whole development season of the crop, with peaks of abundance between flowering and maturity of the capsules, in agreement with spider guilds present in vegetable stratum and in soil (Aguilar 1968, 1974, 1975, 1976; Gómez and Flórez 2005; Almada et al. 2011).

Comparative studies of spider assemblages in conventional cotton crops and transgenic (Bt) crops in Colombia (Gómez and Flórez 2005), and conventional cotton without chemical control in Argentina (Almada et al. 2011), showed that differences did not exist in spider diversity between them: composition, abundancy, richness, and diversity were similar between them.

Rice

The first studies in Neotropical rice crops come from Colombia (Medina 1994; Bastidas et al. 1994a), where the spiders present in the crop were identified and their population fluctuation recorded, the most frequent being *Tetragnatha* sp. (Tetragnathidae), *Phidippus clarus* (Salticidae), *Synaemops rubropunctatum* (Thomisidae), *Oxyopes salticus* (Oxyopidae), *Argiope argentata* (Araneidae), and *Pardosa* sp. (Lycosidae). In parallel, studies were held in areas with crops of rice in irrigation and rice fed by rain, during all the stages of development of the crop (Cuevas 1994), finding spiders from the period of soil preparation, in rice shoots and weeds, and in the beginning of the establishment of the crop. The colonizing species are: *Alpaida. veniliae* (Araneidae), *P. clarus*, *Paraphidippus* spp. (Salticidae), *Centromerus* spp. (Linyphiidae), and *Tetragnatha* spp. (Tetragnathidae), the greater number of spider species being recorded during the maturation stage.

It has been found that spiders present in the crop carry out a natural controlling function on phytophagous insects, such as *O. salticus* y *A. veniliae,* that were found consuming *Tagosodes* sp., *Hortensia similis, Hydrellia* sp. (Diptera Ephydridae) and many species of Agromyzidae. The species *Tetragnatha* sp. and *Leucauge argyra* (Tetragnathidae) were found consuming *H. similis* and *Rupela albinell* (Lepidoptera, Pyrallidae) (Cuevas 1994). *A. argentata and P. argyra* were found consuming Diptera (35.5%) and Homoptera (48.2%) (Bastidas et al. 1994a, b).

Studies on spider fauna during the different stages of rice crops in Brazil (Rodrigues et al. 2008) found a predominance of Araneidae, Anyphaenidae, Oxyopidae, and Tetragnathidae, the most abundant species being *A. veniliae, Tetragnatha nitens, Ashtabula* sp.1, and *Tetragnatha aff. jaculator*, which constituted 45% of the individual adult spiders collected. Rodrigues et al. (2009) sampled the different stages of rice crops and their margin areas (forest edge, pasture), finding the same araneofauna. Most adults were found on the edge of the forest (62 spp), with less abundance within the cultivated area (38 spp) and in the pasture (26 spp), with eight common morphospecies in all areas. Observed diversity was greater at the forest edge, which suggests it could be an important refuge for fauna that live in areas with high anthropogenic disturbance such as agroecosystem activity.

Abundance and richness increase after harvest, a constant colonization of the habitat being found, given the high number of juvenile spiders present throughout all stages of rice development, favored by the presence of adjacent vegetation.

The predominance of some functional groups, such as hunter–ambush spiders, followed by the cobweb builders, supports the reduction of the number of phytophagous insects. The disturbance caused by sowing and harvesting alters the structure of the environment, leading to a change in spider diversity in terms of species richness and composition (Medina 1994; Bastidas et al. 1994a; Cuevas 1994; Rodrigues et al. 2008, 2009).

Sorghum

Cultivation of sorghum (*Sorghum bicolor*) used in animal feed and in the production of fodder, of productive importance for some countries, has been one of the few crops in which the composition of entomophagous arthropods is taken into account. In the few studies carried out on entomophagous arthropods, spiders stand out due to the high number of species, according to Campos et al. (1999) in studies of different sorghum genotypes in Brazil. Among them, the predatory activity is highlighted of the weaver spider *Alpaida veniliae*, the nocturnal spider *Cheiracanthium inclusum*, and the ambushing hunter spider *Misumenops pallidus*, which were the most abundant. According to the results of this study, they were able to establish that the panicles from dry seeding could collect a greater number of entomophagous arthropods in relation to wet seeding.

Spiders in Fruit Trees

Apple Tree

Studies carried out in apple tree crops (*Red Delicious* variety) show that the physiognomy of the plant changes with the growth of the crop, affecting the physical separation of predators and plagues, altering the efficiency and relative preference of prey capture (Benamú 1999a, b, 2000, 2001; Benamú and Aguilar 2001). In three agricultural systems (ecological, integrated, and conventional) (Benamú and Aguilar 2001), some families of spiders (Theridiidae, Thomisidae, Oxyopidae, Araneidae, Salticidae, Lycosidae, Clubionidae) were noted for their abundant presence and their importance as possible biological controllers when consuming a variety of phytophagous plague insect (Table 11.1).

Most of the spiders collected belong to the ecological agricultural system (64.38% of the total collected), characterized by the lack of insecticide applications and greater soil cover (cultivation of adjacent legumes, weeds, and vegetation). In this type of agricultural system, a greater number of spiders are found on the same plant, during the phenological stage of maturation of the fruit (Fig. 11.2). It was

Table 11.1 Family frequency of spiders in apple orchard (*Malus domestica*) and prey they eat. Mala Valley, Santa Enriqueta (Lima-Perú), 1997–1998

Preys of spiders		Predatory behavior of spiders											
		Web building				Wandering hunter		Stalking hunter		Night hunter			
Order	Family	Araneidae	Linyphiidae	Tetragnathidae	Theridiidae	Oxyopidae	Salticidae	Anypahenidae	Thomisidae	Corinnidae	Gnaphosidae	Lycosidae	Miturgidae
Orthoptera	Grillidae									X		X	X
Homoptera	Aleyrodidae	X		X	X	X	X		X				
	Aphididae	X			X		X						
	Cicadellidae	X		X	X		X	X	X	X	X		
	Delphacidae	X		X	X	X	X	X			X		
Hemiptera	Lygaeidae				X			X				X	
	Neididae			X	X								
	Pyrrhocoridae				X	X	X	X					
Coleoptera	Bruchidae				X	X		X				X	
	Curculionidae							X				X	X
	Tenebrionidae				X			X				X	X
Lepidoptera	Noctuidae	X			X	X	X	X		X	X	X	X
	Olethreutidae	X	X	X	X	X	X		X	X	X		
Diptera	Drosophilidae	X		X					X				
	Dolichopodidae	X		X	X	X			X				
	Muscidae	X			X	X			X	X		X	
	Syrphidae				X	X							
Hymenoptera	Formicidae		X	X	X						X		
	Platygasteridae	X			X								
	Pompilidae												

Benamú and Aguilar (2001)

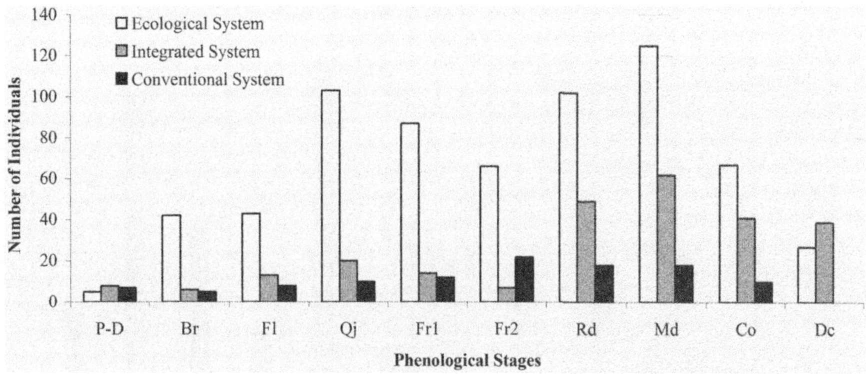

Fig. 11.2 Comparison of total spiders collected during the various phenological stages and agricultural activities of the apple (*Malus domestica*), Mala Valley, Santa Enriqueta (Lima, Peru) (Benamú). *P-D* pruning and defoliation, *Br* sprouting, *Fl* flowering, *Qj* setting (pollinated flower), *Fr1* formation of fruit-1, *Fr2* formation of fruit-2, *Rd*: coloring characteristic of the variety, *Md* maturation, *Co* harvesting, *Dc* rest of the crop

followed in abundance by the integrated agricultural system (25% of the total collected), while the conventional agricultural system was the least represented, with 10.62% of the total spiders collected.

These results show that in each agricultural system studied, the conditions were different, so the presence and behavior of the spiders was also modified. For example, the ecological agricultural system, which had a large cover of spontaneous vegetation (weeds) that grew along with the different stages of the crop, constituted new refuges for spiders on branches, leaves, and inflorescences. The composition of the araneofauna, in an ecological system (biodynamic) compared to a conventional (intensive) system, reveals that the ecological system has a greater density and diversity of spiders. In the conventional agricultural system, the excessive application of insecticides, fungicides, and herbicides, as well as the elimination of rest between cropping campaigns (Benamú 1999a), makes repopulation and colonization by spiders difficult. In this system, only dominant species will prevail (Benamú and Aguilar 2001).

Studies in another type of apple tree (Ana de Israel variety) with ecological management (Benamú 2001) revealed that during the first half of the growing period there was an increased abundance of spiders in the trees, decreasing in the second half. At the soil level, the development of different weeds could be observed, reaching a greater cover in the second half of the crop's development (Fig. 11.3). This finding coincided with the highest number of insects on the trees, during the first phenological stages of the crop, such as sprouting, flowering, setting (pollinated flower) and fruit formation. The final stages of fruit formation, maturation, and harvest coincided with the growth of weeds, allowing the possibility for insects to migrate to the weeds in a greater proportion, followed by the spiders due to a large supply of prey (Fig. 11.3). It was found that certain guilds of spiders were found in

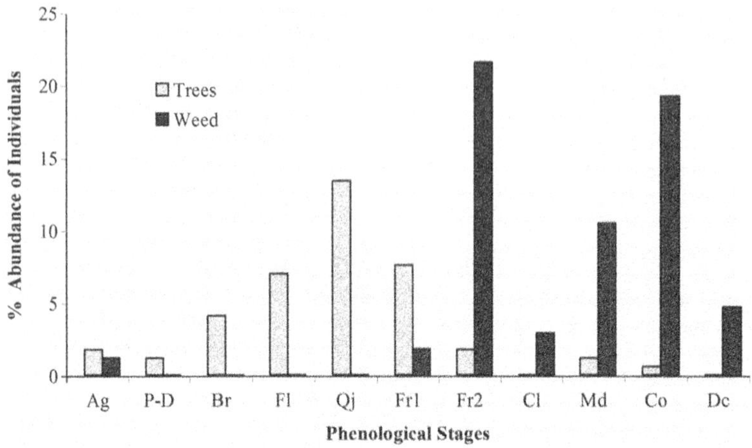

Fig. 11.3 Spiders registered in apple trees and weeds during crop phenological stages, Mala Valley Santa Enriqueta (Lima-Peru) (Benamú). *Ag* starving (suspension of irrigation), *P-D* pruning and defoliation, *Br* sprouting, *Fl* flowering, *Qj* setting (pollinated flower), *Fr1* formation of fruit-1, *Fr2* formation of fruit-2, *Cl* color characteristic of the variety, *Md* maturation, *Co* harvesting, *Dc* rest of the crop

the weeds, which were characterized by the type of foliage, inflorescences, and size (Benamú 2001).

Citrus

The first studies focused on spiders from American citrus were made by Muma in Florida (1973, 1975), and Mansour et al. (1982), presented a review about this topic. Jiménez and Tejas (1996) studied the seasonal variation of spiders in several citrus species (*Citrus aurantius, Citrus limetta, Citrus* spp.) from Baja California. These authors found that spring and autumn showed the highest spider abundance, and related the spider presence with the potential control of some pests.

In the Neotropical zone, some studies concerning tangerines from seedlings in Perú showed that spiders are effective generalist predators on this crop, and several species from the families Salticidae, Araneidae, Dictynidae, Gnaphosidae, Thomisidae, and Tetragnathidae were found feeding on a wide variety of pest insects (Benamú 1999b) (Table 11.2).

Studies concerning population fluctuation of insects and spiders associated with orange crops (*Citrus sinencis*) in Costa Rica (Elizondo 2002), comparing a crop with and without the use of agrochemicals, found a greater diversity of beneficial insects in the system without insecticides. Spiders were numerous with fluctuations in both systems, emphasizing their predatory action especially on *Phyllocnistis citrella*, highlighting the genera *Hibana velox*, *Clubiona* sp., *Thiodina* sp., *Carabella* sp. and *Phiale* sp. considered the most important predators of the leaf citrus miner.

Table 11.2 Insect prey of spiders present in crops of mandarins (*Citrus reticula*), (Lima-Perú), 1997

Spider family	Prey of spiders							
	Syrphidae	Ephydridae	Culicidae	Dolichopodidae	Delphacidae	Cicadellidae	Aphididae	Aleyrodidae
Anyphaenidae								X
Araneidae	X	X	X					
Clubionidae					X			X
Dictynidae			X				X	X
Gnaphosidae					X			
Salticidae	X	X		X		X	X	
Tetragnathidae	X			X				
Thomisidae	X					X	X	X

Benamú (1999b)

Similar results are reported in Uruguay on lemon crops. When insecticide-free crops were compared with fields under traditional management, a higher species richness and density was found in pesticide-free fields (62.08% total abundance) compared with traditional fields (37.92%). The vegetal architecture constituted by weeds and others plants may be a possible explanation for the spider abundance, as they create microenvironments inside the crops (Benamú 2004). According to this author, vegetation also promotes the migration and establishment of several spider guilds, and enhances its potential use as biological control agents. The wide diversity of potential prey such as phytophagous insects (Fig. 11.4) is also used as a possible explanation for the higher spider abundance. According to the same study, spider diversity in soil was higher when compared to other microenvironments.

The lower abundance of spiders found in fields under traditional management might be explained by the management of weeds and other plants, in addition to the continuous application of insecticides probably affecting the establishment of spiders and reducing the number of available prey, and acting as a cause of spider mortality..

In both fields, abundance and diversity of spiders varied according to the plant stratum; these parameters were higher in the medium stratum since it has the most complex vegetal structure (Fig. 11.5). After evaluating the temporal variation, it was observed that the highest spider abundance occurred during fruit development. Similar results were reported for orange crops *C. sinencis* in Argentina, where the complexity of orange trees provides refuge and food sources for insects, similar to natural systems.

We present here a review on the spider biodiversity in Neotropical citrus crops (Fig. 11.6). We recorded a total of 40 families in Neotropical citrus, Salticidae, Anyphaenidae, and Clubionidae being the most abundant. This spider diversity is also potentially useful for integrated pest management in citrus, since these are natural enemies for several pests. In the particular case of the hemipteran *Diaphorina citri* (Hemiptera, Psyllidae), a vector for the citrus disease HLB which has spread quickly across America causing huge financial losses, spiders constitute a promising biological control agent. In citrus fields in the USA for example, spiders constitute an effective predator complex against natural enemies of *D. citri* crops. We found at least four potential families (Clubionidae, Anyphaenidae, Salticidae, and Sicariidae) as natural enemies for nymphs, and Linyphiidae as adult predators. Further behavioral studies are evaluating the diet and capture efficiency of these families against pests on *D. citri* in order to incorporate spiders into IPM programs for citrus.

Coffee

Although coffee (*Coffea arabica*) is a representative crop in several countries from South and Central America, studies concerning spiders associated with coffee plants are scarce. A survey of orb-weaving spiders in Mexican coffee crops revealed that

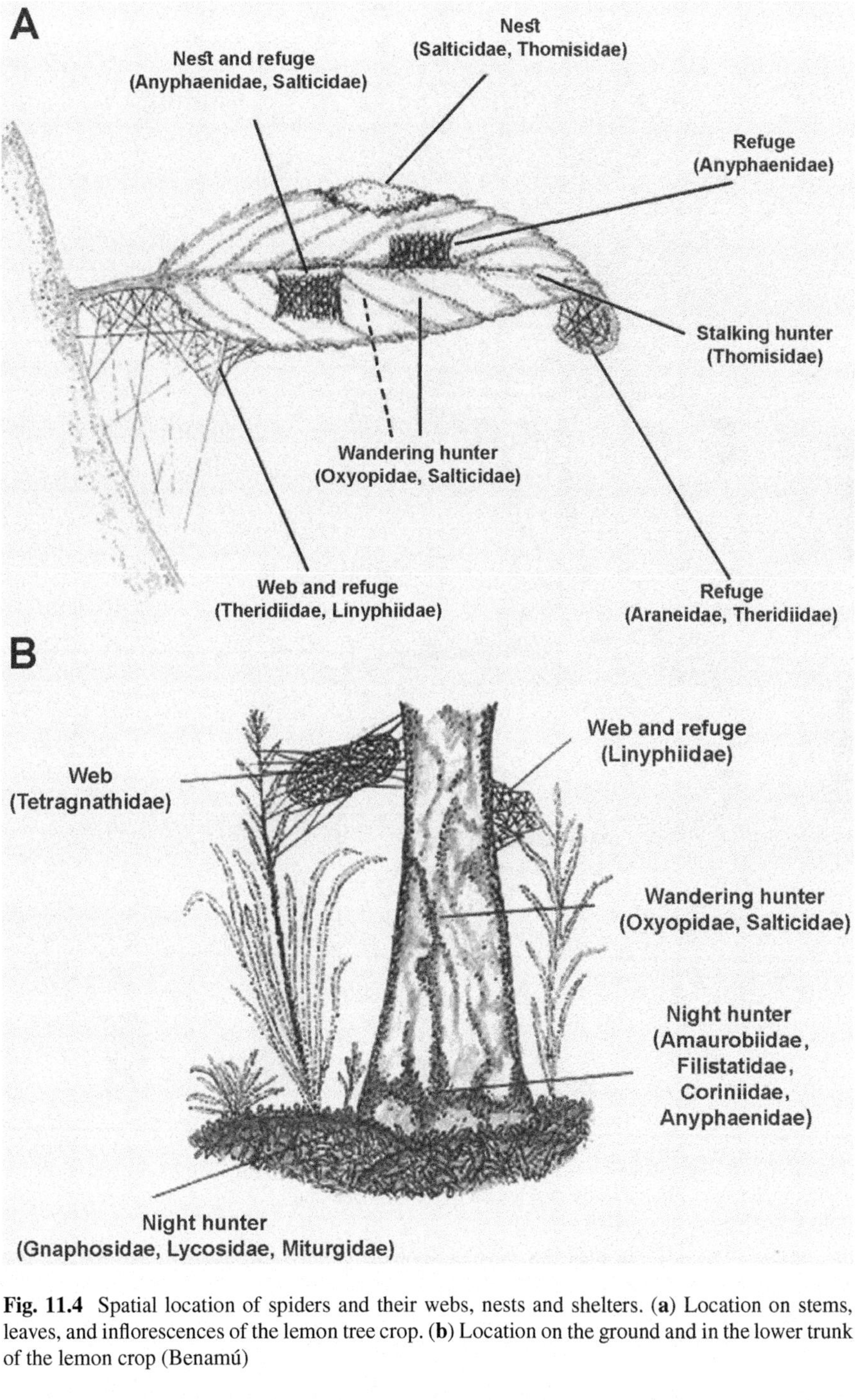

Fig. 11.4 Spatial location of spiders and their webs, nests and shelters. (**a**) Location on stems, leaves, and inflorescences of the lemon tree crop. (**b**) Location on the ground and in the lower trunk of the lemon crop (Benamú)

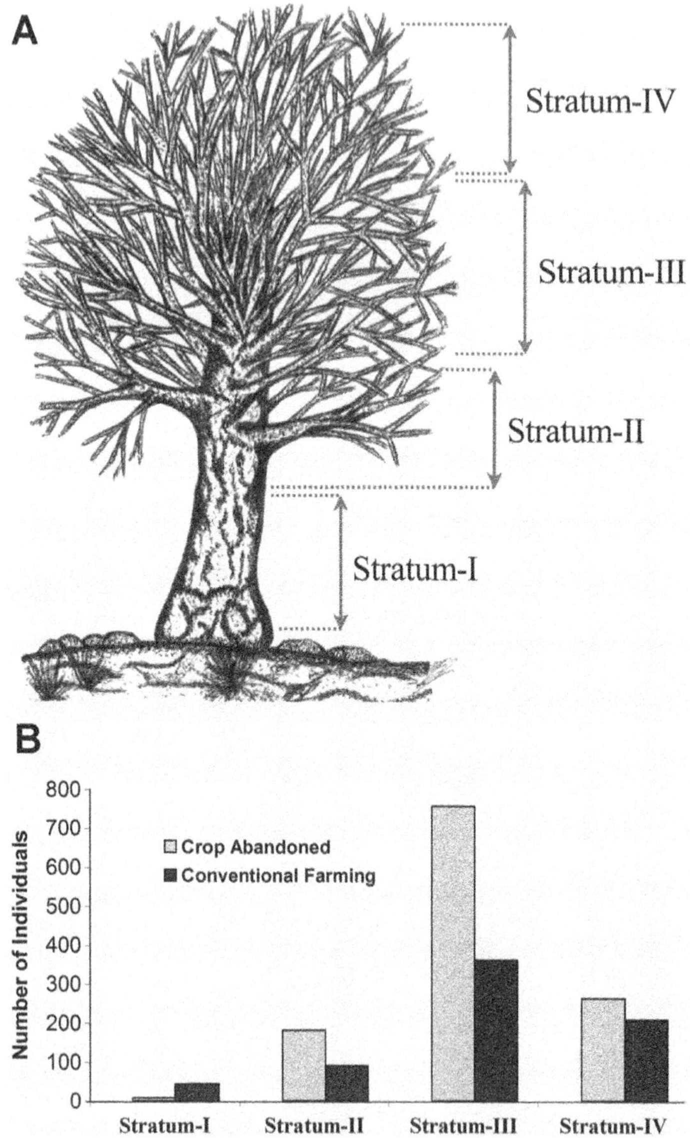

Fig. 11.5 (**a**) Representation of the strata examined in the lemon tree crops. (**b**) Abundance of spider in the level stratum in the two lemon tree crops (Benamú)

the family Theridiidae was the most numerous group, followed by Araneidae and Tetragnathidae (Ibarra-Nuñez and García-Ballinas 1998). Further studies compared the effect of crop management (organic vs traditional) on spider diversity (Pinkus et al. 2006), showing a seasonal effect on diversity only in the organically managed crop. Overall, alpha diversity did not exhibit significant differences with regard to

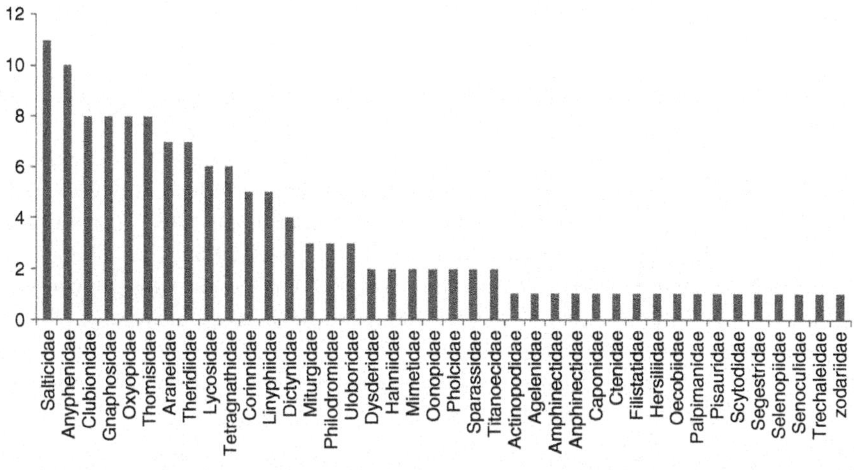

Fig. 11.6 Frequency of spider families recorded in citrus crops in the Neotropical region, obtained from a bibliographical search (Lacava)

crop management; nevertheless, beta diversity showed that the kind of management has a strong effect on species composition.

Similar studies performed by Leon and Cepeda (2009) in Colombian coffee crops, comparing the effect of traditional and organic crops on spider and beetle diversity, found that the latter group was more sensitive to disturbances, with strong differences in diversity depending on how crops were managed. Nevertheless, in the case of spiders a higher biomass was found in organic crops. Authors attribute the differences in biomass to the fact that organic crop increases functional diversity by having a higher biomass of different trophic groups such as herbivores and detritivores, which are a potential food source for predatory beetles and spiders, increasing as a consequence the biomass of the latter two groups.

Watermelon

Cunha et al. (2015) compared the spider fauna associated with watermelon crops (*Citrus lanatus*) under organic and traditional management as well as native vegetation from Colombia. These authors found seven spider families shared among all the management types, where lycosid spiders were the most abundant in agriculture systems, and corinnids were the most common in natural fields. When comparing the different guilds, these authors found that ground hunter spiders were the most abundant, followed by the irregular web builders.

Almond

A survey of spiders made in Chilean almond crops (*Prunus dulcis*) by Orellana et al. (2012), evaluated species present in the crop and surrounding areas. These authors found that the most abundant family was Linyphiidae, followed by Gnaphosidae, Dysderidae, and Anyphaenidae. The families Anyphaenidae, Linyphiidae, Dyctinidae, Gnaphosidae, and Salticidae represented 70% of total numbers. In this study, the almond crop was characterized by a highest abundance of wandering spiders, while web-building spiders were the most frequent in surrounding areas. These results might be explained by the higher plant diversity found in peripheral zones, which offer a higher structural complexity for web-building spiders. In contrast, disturbances due to agricultural practices allowed the establishment of other species like wandering spiders.

Cocoa

Studies by Lucio-Palacio and Ibarra-Nuñez (2015), about the diet composition from web-building spiders found in Mexican cocoa crops, evaluated 54 species of the families, Araneidae, Theridiidae, Tetragnathidae, Uloboridae, Pholcidae, Dyctinidae, and Linyphiidae. These authors found 1,749 insects consumed by spiders, belonging to ten orders and 93 families; 74% of total insects were grouped in the orders Coleoptera, Diptera, and Hemiptera. The family Araneidae consumed most of the recorded insect groups, except Isoptera, which was captured mainly by the family Theridiidae. The wide diversity of trapped insects suggests that web-building spiders could be more efficient against particular pests.

Lucio-Palacio and Ibarra-Nuñez (2015) compared the diversity of spiders found in crops under two types of management (organic vs modernized), and showed that the highest values for species richness were found in the dry season for traditional management and in the wet season for modernized management. Therefore, it is possible that the association between the season and management affects the use of spiders as natural enemies. It has been shown that agroecosystem structure, environmental factors such as temperature and humidity, and crop management, as well as spider migration, affect the population dynamics of spiders. (De la Cruz-Pérez et al. 2015).

Spiders as Potential Control Agents in the Neotropical Region

Cuevas (1994) showed that some abundant spiders found in rice crops like *Oxyopes salticus* feeds on potential pest insects such as *Tagosodes* sp., *Hortensia similis, Hydrellia* sp. (Diptera Ephydridae), and several agromyzid species. The same author also reports web-building spiders such as *Tetragnatha* sp. and *Pleisometa argyra* preying on *H. similis* and *Rupela albinell* (Lepidoptera, Pyrallidae). Similar studies in rice record a population control of *Tagosodes orizicolus* (Homoptera, Cercopidae) between 25% and 68.6%, at a density of four *Tetragnatha*/m^2 (Bastidas et al. 1994a). In this study, it was recorded in laboratory conditions that *Tetragnatha* sp. consumes 3.5 average adults of *Tagosode* per day, 2.5 first instar larvae of *Spodoptera frugiperda* per day, 1.4 second instar larvae per day, and 0.6 sixth instar larvae per day. *Argiope argentata* consumes 4.1 average adults of *Tagosodes* per day, *Phidippus clarus* consumes 1.8 nymphs of *Tagosodes* per day, 1.4 third instar nymphs, and 0.6 fourth instar nymphs of the species *Oebalus ornatus* per day.

Similar studies in grasslands have evaluated the diet composition, and found that *Alpaida variabilis* (Araneidae) feeds mainly on Homoptera (Cicadellidae) and Diptera (Sciaridae and Bibionidae), and that the diet is mainly influenced by prey abundance rather than a preference for a particular prey type. (Flórez et al. 2004).

Armendano and González (2011a) used the most frequent species on alfalfa, *Misumenops pallidus*, *Oxyopes salticus*, *Araneus* sp., and *Lycosa poliostoma*, and evaluated the consumption on pests. These authors found that, overall, spiders preferred lepidopterans (93.3%) instead of other pests such as aphids (25.33%) and curculionid beetles (11.67%).

Cheli et al. (2006) also studied the feeding preference of the spider *Misumenops pallidus* (Araneae, Thomisidae) on different alfalfa pests, and found a preference for soft-bodied prey without defensive glands such as *Rachiplusia nu* (Lepidoptera, Noctuidae); it can feed also on other prey such as *Horciasinus argentinus* and *Halticus spegazzinii* (Heteroptera, Miridae), beetles *Colaspis* sp., and *Diabrotica speciosa* (Coleoptera, Chrysomelidae). When evaluating the feeding preference of the same species under field conditions, González et al. (2009) found that *M. pallidus* showed a polyphagous diet, with a slight preference for *R. nu*.

Potential use of spiders as natural enemies has also been evaluated in fruit crops. Four species have been reported as natural predators for leafminers in lime trees: *Cheiracanthium inclusum* (Clubionidae), *Hibana velox* (Anyphaenidae), *Trachelas volutus* (Corinnidae), and *Hentzia palmarum* (Salticidae) (Amalin and Peña 1999). When evaluating the predatory behavior of these species, Amalin et al. (2001) found that they all have nocturnal habits and are able to detect and capture the leafminer larvae and pupae of *Phyllocnistis citrella* using the vibrations produced on the leaves.

Sublethal Effects of Agrochemicals on Neotropical Spiders

As well as being an important group of predators in agroecosystems, spiders have also been used as bioindicators of disturbances as a consequence of human activities. This is the case with agrochemicals such as acaricides, insecticides, fungicides, herbicides, which might have negative lethal and sublethal effects on spiders (Pekár, 2012), not only affecting their survival but also their feeding performance. Contamination on spiders might come from different sources; while some individuals might be exposed by contact, such as walking on contaminated surfaces, some others might feed on contaminated prey or receive the pesticide directly on their body. In general, use of agrochemicals can reduce the population density of spiders as well as promote emigration, and reduce the predation capability or reproductive rate (Benamú et al. 2010; Wrinn et al. 2012; Rodrigues et al. 2013). Many of these effects occur as several agrochemicals are stored in the body of the spider (Benamú et al. 2007, 2010, 2013).

Given the abundance of spiders in several agroecosystems, these can be used as a suitable model to evaluate the negative effects of several agrochemicals on natural enemies. Herein, we present a summary of the main studies about sublethal effects of two groups of agrochemicals, namely herbicides and insecticides.

Insecticides

Most studies about sublethal effects of insecticides on spiders in the Neotropical region have used web-building species as a model, and wide-spectrum as well as selective insecticides. Benamú et al. (2007) evaluated the effect of neurotoxic insecticides such as endosulfan and spinosad by contaminating by ingestion the web-building spider *Araneus pratensis,* and using different dosages. These authors found that spinosad produced a higher mortality than endosulfan. Both insecticides affected prey consumption and web-building behavior; similarly, fertility decreased as a consequence of insecticide side-effects, as contaminated spiders were not able to build eggsacs properly and produced deteriorated eggs.

A similar study was performed using the spider *Alpaida veniliae* in Argentina. This spider was selected as it is considered one of the main predators found in soybean crops (Benamú 2010; Benamú et al. 2013). Using this species, three insecticides commonly applied for pest control in soybean were used, namely spinosad, cypermethrin, and endosulfan. These authors found that the spinosad caused the most serious sublethal effects, which included reduction on prey consumption, web building, and fertility, even at lower dosages than those recommended for application. Cypermethrin produced the same negative effects although in lower proportions, while endosulfan affected web-building behavior and reproduction (Figs. 11.7, 11.8, 11.9 and 11.10).

By contaminating the spider *Parawixia audax* with neonicotinoids by ingestion, Benamú et al. (2017) found that physical properties as well as structure and amino-acid composition were affected in the contaminated spiders. Authors attribute negative effects on contaminated spiders as a consequence of metabolic stress caused by insecticides.

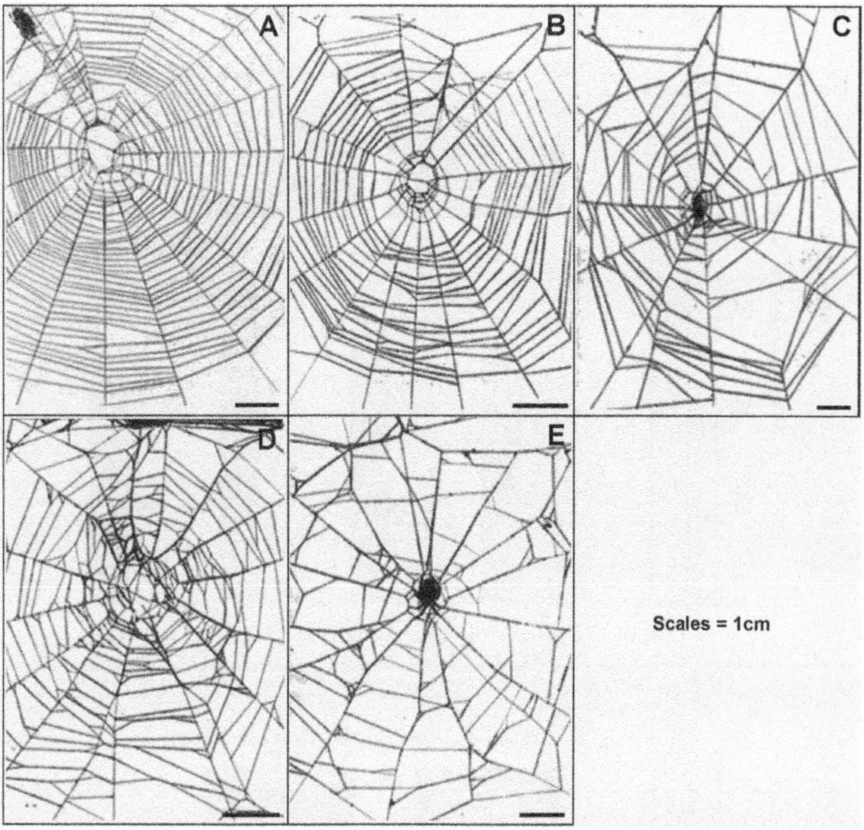

Fig. 11.7 Orbicular webs of *Alpaida veniliae* exposed to three neurotoxic insecticides. (**a**) Normal web (control). (**b**) Cypermethrin 8.25 mg a.i./l (25% of full field recommended concentration). (**c**) Cypermethrin18.75 mg a.i./l (75% of full field recommended concentration). (**d**) Endosulfan 25 mg a.i./l (23.8% of full field recommended concentration). (**e**) spinosad (2.5% of full field recommended concentration) (Benamú)

Herbicides

Although glyphosate has been considered an innocuous product for natural enemies, Benamú et al. (2010) showed that *A. veniliae* contaminated via ingestion with glyphosate (Fig. 11.8), showed similar patterns to those recorded for insecticides, affecting fecundity and fertility and postembryonic development (Fig. 11.11). Web-building behavior was also affected; spiders built webs without spiral threads, and with irregular structures (Fig. 11.12).

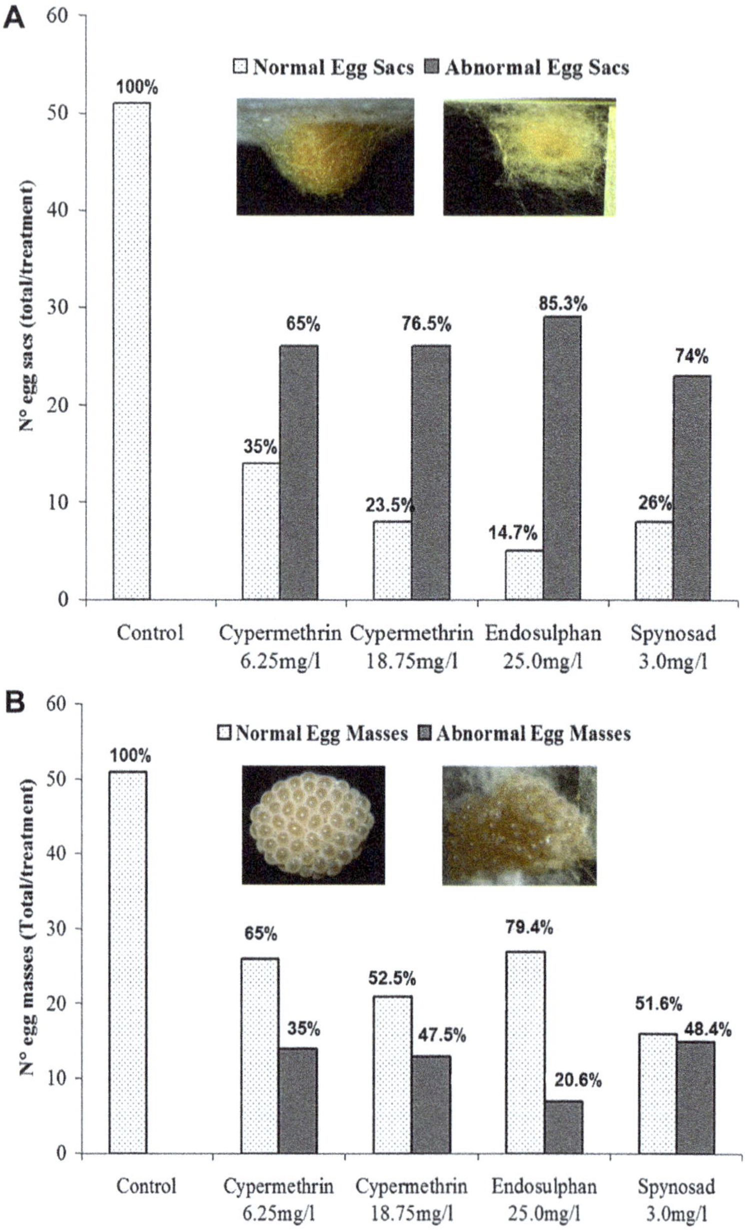

Fig. 11.8 Abnormalities in silk eggsacs and egg masses of *Alpaida veniliae* females treated by ingestion with three neurotoxic insecticides. (**a**) Effects on silk eggsacs. (**b**) Effects on egg masses. The percentage on each bar denotes the proportion of abnormalities versus normalities in the egg-sacs and egg masses respectively. The pictures show the abnormalities in the eggsacs and egg masses respectively (Benamú)

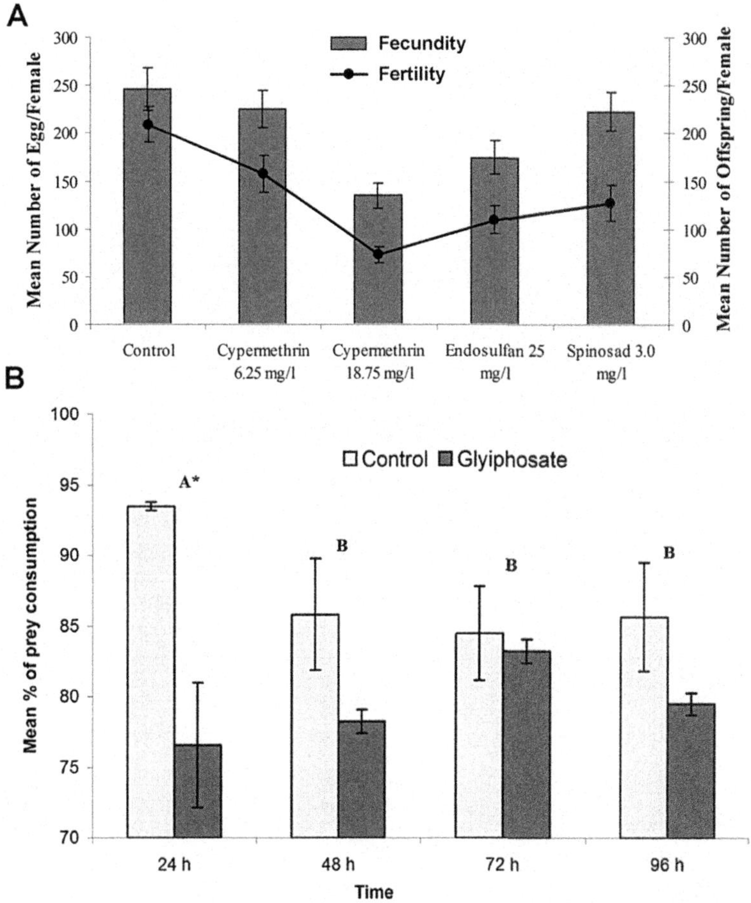

Fig. 11.9 Long-term effect of insecticides on fecundity and fertility of *Alpaida veniliae* females exposed by ingestion through chronic toxicity experiments. Bars with the same letter do not differ significantly (Benamú)

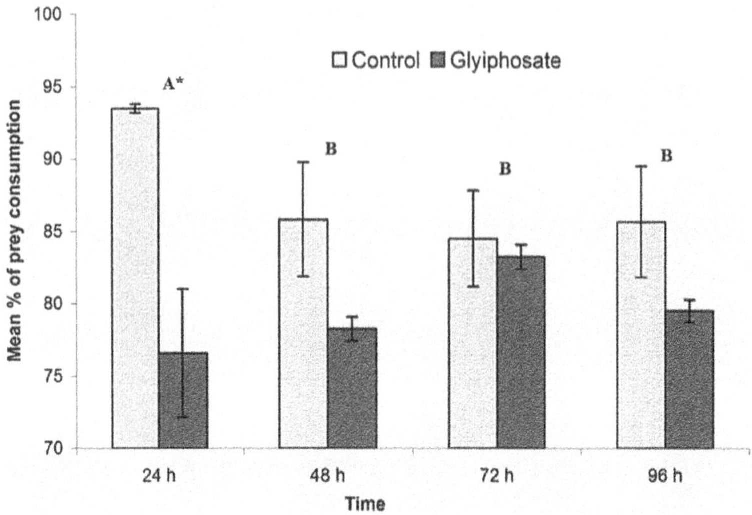

Fig. 11.10 Effect of glyphosate at 24, 48, 72, and 96 h after treated and control prey consumption. *Vertical lines* indicate standard errors. *Capital letters* indicate differences within control treatment. * Indicates significant differences between control and glyphosate treatments ($P = 0.007$) (Benamú)

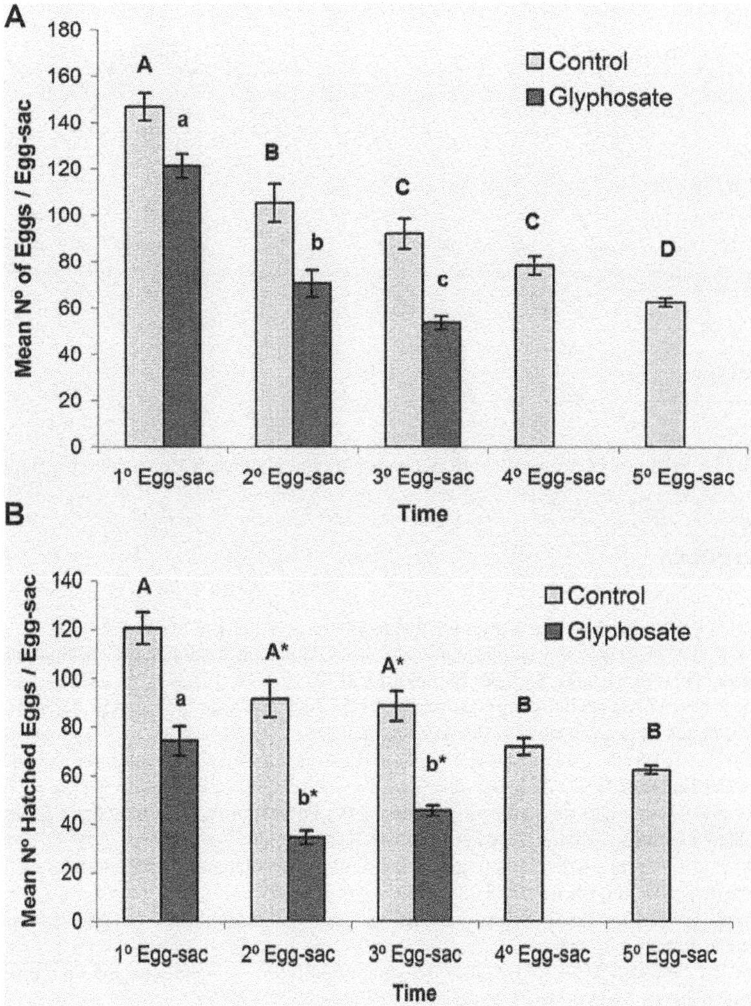

Fig. 11.11 Effect of glyphosate in *Alpaida venilae*, (**a**) on the mean number of eggs per eggsac, and (**b**) mean number of eggs hatched per eggsac. *Vertical lines* indicate standard errors. *Capital letters* indicate differences within control treatment, and *lowercase* within glyphosate treatment. * Indicates significant differences between control and glyphosate treatments ($P = 0.05$) (Benamú)

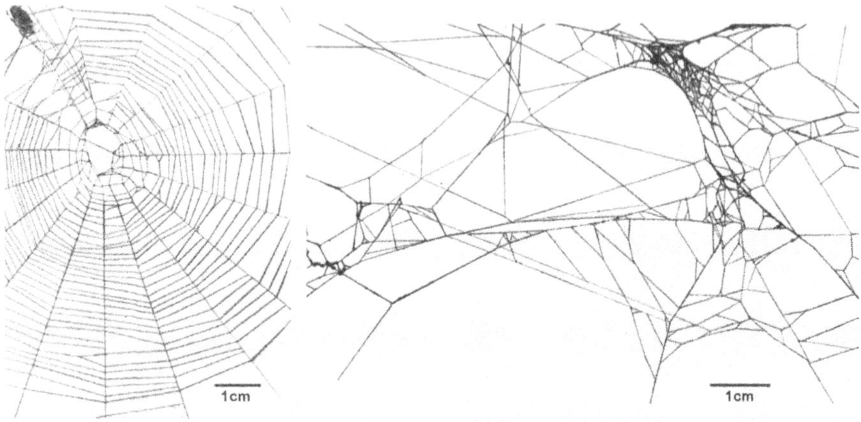

Fig. 11.12 Webs of *Alpaida veniliae* from control (**a**) and glyphosate treatments (**b**) (Benamú)

References

Aguilar P (1968) Araneidas frecuentes en el algodonero. Rev Per Entomol 11:92–95

Aguilar P (1974) Arañas del campo cultivado — I: Población de araneidos en algodonales de Cañete, Huaura y Rimac. Rev Per Entomol 17:21–27

Aguilar P (1975) Arañas del campo cultivado — II: fluctuación de las familias de araneidos en algodonales de la costa central. Rev Per Entomol 18:25–28

Aguilar P (1976) Arañas del campo cultivado — III: Araneidos en algodonales del Valle de Lurin. Rev Per Entomol 19:71–72

Aguilar P (1978) Arañas del campo cultivado — IV: Población de araneidos del algodonero en Chillon y Chancay–Huaral. Rev Per Entomol 21:39–41

Aguilar P (1979) Arañas del campo cultivado — VI: Observaciones en algodonales de la costa norte del Perú. Rev Per Entomol 22:71–73

Aguilar P (1988) Las arañas como controladores de plagas insectiles en la agricultura peruana. Rev Per Entomol 31:1–8

Aguilar P, Güerovich M (1978) Arañas del campo cultivado — V: Frecuencia de araneidos en algodonales de Mala, Asia y Chilca. Rev Per Entomol 21:42–45

Alcayaga O, Pizaro-Araya J, Alfaro F, Cepeda-Pizaro J (2013) Arañas (Arachnida, Araneae) asociadas a agroecosistemas en el Valle de Elqui (Región de Coquimbo, Chile). Rev Colomb Entomol 39:150–154

Almada M, Sosa M (2011) Relevamiento de arañas en cultivos de soja transgénica en el norte de Santa Fe. En Mercosoja 2011. Quinto congreso de la Soja del Mercosur / Primer Foro de la Soja Asia-Mercosur. Bolsa de Comercio de Rosario, Santa Fe

Almada M, Sosa M, González A (2011) Araneofauna (Arachnida: Araneae) en cultivos de algodón (*Gossypium hirsutum*) transgénicos y convencionales en el norte de Santa Fe, Argentina. Rev Biol Trop 60:611–623

Altieri MA (1995) Agroecology: the science of sustainable agriculture. Westview Press, Boulder

Altieri M, Letourneau D (1982) Vegetation management and biological control in agroecosystems. Crop Prot 1:405–430

Amalin D, Peña J (1999) Predatory spiders in lime orchards and their importance in the control of citrus leafminer, *Phyllocnistis citrella* (Lepidoptera: Gracillariidae). Proc Fla State Hort Soc 112:222–224

Amalin D, Reiskind J, Peña J, McSorley R (2001) Predatory behavior of three species of sac spiders attacking citrus leafminer. J Arachnol 29:72–81

Anderson G, Danielson B (1997) The effects of landscape composition and physiognomy on metapopulation size: the role of corridors. Landsc Ecol 12:261–271

Armendano A (2008) Estudio de la araneofauna presente en agroecosistemas de importancia económica (trigo y alfalfa). Dissertation, Universidad Nacional de La Plata, Facultad de Ciencias Naturales y Museo, La Plata, Argentina

Armendano A, González A (2010) Comunidad de arañas (Arachnida, Araneae) del cultivo de alfalfa (*Medicago sativa*) en Buenos Aires, Argentina. Rev Biol Trop 58:757–767

Armendano A, González A (2011a) Efecto de las arañas (Arachnida: Araneae) como depredadoras de insectos plaga en cultivos de alfalfa (*Medicago sativa*) (Fabaceae) en Argentina. Rev Biol Trop 59:1651–1662

Armendano A, González A (2011b) Spider fauna associated with wheat crops and adjacent habitats in Buenos Aires, Argentina. Rev Mex Biodivers 82:1176–1182

Atauri J, De Lucio J (2001) The role of landscape structure in species richness distribution of birds, amphibians, reptiles and lepidopterans in Mediterranean landscapes. Landsc Ecol 16:147–159

Bastidas H, Pantoja A, Fernández M (1994a) Consumo de presas por *Argiope argentata* (F.) (Araneae: Araneidae) Y *Plesiometa argyra* (Walkenaer) (Araneae: Tetragnathidae) en arroz irrigado en Colombia. Man Integ Plag 32:30–32

Bastidas H, Pantoja A, Murillo A, Zuluaga J, Duque M (1994b) Reconocimiento, fluctuación y pruebas de consumo de presas por arañas en cultivos de arroz, en el Valle del Cauca. Rev Colomb Entomol 20:149–160

Beltramo J, Bertolaccini I, González A (2006) Spiders of soybean crops in santa fe province, Argentina: influence of surrounding spontaneous vegetation on lot colonization. Braz J Biol 66:891–898

Benamú M (1999a) Araneofauna presente en cultivos de manzano (*Malus domestica*) variedad "delicia de Viscas", en el valle de Mala (Cañete, Lima). Dissertation, Universidad Ricardo Palma, Facultad de Ciencias Biológica, Lima, Perú

Benamú M (1999b) Estudio preliminar de la araneofauna presente en mandarina cultivada en Vitarte, Lima, Perú. Rev Per Entomol 41:154–157

Benamú M (2000) La influencia de dos sistemas agrícolas (ecológico y convencional) en la diversidad de arañas. In: Felipe-Morales C, Manrique A (eds) Agrobiodiversidad en la región Andina y Amazónica. XX, Lima, Perú, pp 306–319

Benamú M (2001) Observaciones sobre "crianza", liberación y distribución de arañas en un huerto de manzano del Valle de Mala, Lima. Perú Rev Per Entomol 42:211–217

Benamú M (2004) Estudio comparativo de la diversidad de arañas de un campo en abandono y un cultivo convencional de limonero (*Citrus limon* [L.] Burm.) en Rincón del Cerro, Montevideo, Uruguay. Dissertation, Universidad de la República, Facultad de Ciencias, PEDECIBA, Montevideo, Uruguay

Benamú M (2010) Composición y estructura de la comunidad de arañas en el sistema de cultivo de soja transgénica. Dissertation, Universidad Nacional de La Plata, Facultad de Ciencias Naturales y Museo, La Plata, Argentina

Benamú M, Aguilar P (2001) Araneofauna presente en huertos de manzano del Valle de Mala, Lima. Perú Rev Per Entomol 42:199–210

Benamú M, Schneider M, Pineda S, Sánchez N, González A (2007) Sublethal effects of two neurotoxican insecticides on *Araneus pratensis* (Araneae: Araneidae). Comm Appl Biol Sci 72:557–559

Benamú M, Schneider M, Sánchez N (2010) Effects of the herbicide glyphosate on biological attributes of *Alpaida veniliae* (Araneae, Araneidae), in laboratory. Chemosphere 78:871–876

Benamú M, Sánchez N, González A (2011) Postembryonic development and population parameters of *Alpaida veniliae* (Araneae, Araneidae) reared in the laboratory. J Nat Hist 45(25–26):1607–1617

Benamú M, Schneider M, González A, Sánchez N (2013) Short and long-term effects of three neurotoxic insecticides on biological and behavioural attributes of the orb-web spider *Alpaida veniliae* (Araneae, Araneidae): implications for IPM programs. Ecotoxicology 22:1155–1164

Benamú M, Lacava M, García LF et al (2017) Nanostructural and mechanical property changes to spider silk as a consequence of insecticide exposure. Chemosphere 181:241–249

Brown M, Schmeitt J, Abraham B (2003) Seasonal and diurnal dynamics of spiders (Araneae) in West Virginia orchards and the effect of orchard management on spider communities. Environ Entomol 32:830–839

Büchs W (2003) Biodiversity and agro-environmental indicators-general scopes and skills with special reference to the habitat level. Agric Ecosyst Environ 98:35–78

Campos A, Filho E, Lara F, Rinaldi I (1999) Composição da artropodofauna entomófoga associada a diferentes genótipos de sorgo granífero no Cerrado do sudeste do Brasil. An Soc Entomol Brasil 28:703–714

Cantor C (2014) Ecología funcional de la araneofauna de tejedoras (Arachnida: Araneae) en agroecosistemas cafeteros de Cundinamarca y su valor en la conservación. Dissertation, Thesis MSC, Universidad Nacional de Colombia Facultad de Ciencias, Departamento de Biología Bogotá, Colombia

Cheli G, Armendano A, González A (2006) Preferencia alimentaria de arañas *Misumenops pallidus* (Araneae: Thomisidae) sobre potenciales insectos presa de cultivos de alfalfa. Rev Biol Trop 54:505–513

Clausen I (1986) The use of spiders (Araneae) as ecological indicators. Bull Br Arachnol Soc 7:83–86

Cuevas A (1994) Las arañas: controladores naturales de insectos fitófagos en el cultivo de arroz en Norte de Santander, Colombia. Rev Colomb Entomol 20:179–186

Cunha J, Andrade E, Silva P, Barros R (2015) Araneofauna (Arachnida, Araneae) in conventional and organic crops of watermelon (*Citrullus lanatus*) in northeastern Brazil. Rev Colomb Entomo l41:68–75

De La Cruz-Pérez A, Pérez-De La Cruz M, Sánchez-Soto S, Torres-De La Cruz M (2015) Fluctuación poblacional de arañas (Araneae, Tetragnathidae, Pholcidae) en el agroecosistema cacao en Tabasco, México. Rev Colomb Entomol 41:132–138

Di Castri F, Younés T (1996) Biodiversity, science and development. CAB International/IUBS, Paris, Francia, pp 1–11

Flint M, Roberts P (1988) Using crop diversity to manage pest problems: some Californian examples. Am J Alternative Agr 3:164–167

Flórez E, Pinzón J, Sabogal A, Barreto N (2004) Selección de presas y composición de la dieta de la araña *Alpaida variabilis* (Araneae: Araneidae) en pastizales de la sabana de Bogotá, Colombia. Rev Ibérica Aracnol 9:241–248

Foelix R (2010) Biology of spiders, 3rd edn. Oxford University Press, New York/Oxford

Gómez L, Flórez E (2005) Estudio comparativo de las comunidades de arañas (Araneae) en cultivos de algodón convencional y transgénico en el departamento del Tolima, Colombia Acta Biológica Colombiana. Acta Biol Colombiana 10:93–94

González A, Liljesthröm G, Minervino E, Castro D, González S, Armendano A (2009) Predation by *Misumenops pallidus* (Araneae: Thomisidae) on insect pests of soybean cultures in Buenos Aires Province, Argentina. J Arachnol 37:282–286

Green J (1996) Sampling method and time determines composition of spider collections. J Arachnol 27:176–182

Hagen K, Mills N, Gordh G, McMurtry J (1999) Terrestrial arthropod predators of insect and mite pests. In: Bellows TS, Fisher TW (eds) Handbook of biological control: principles and applications of biological control. AP Academic Press, San Diego

Halaj J, Cady A, Uetz G (2000) Modular habitat refugia enhance generalist predators and lower plant damage in soybeans. Environ Entomol 29:383–393

Harcourt D, Aragón J, González R (1986) Plagas de la alfalfa. In: Bariggi C, Marble BL, Itria CD, Brun JM (eds) investigación, tecnología y producción de alfalfa, vol Tomo XXII. Colección Científica del INTA, Buenos Aires, 488 pp

Hoefler C, Chen A, Jakob E (2006) The potential of a jumping spider, *Phidippus clarus*, as a biocontrol agent. J Econ Entomol 99:432–436

Hunter M, Gibbs J (2006) Fundamentals of conservation biology (3rd Edition). John Wiley and Sons (ed), Oxford, Wiley

Ibarra-Nuñez G (1990) Los artrópodos asociados a cafetos en un cafetal mixto del Soconusco, Chiapas, México. I. Variedad y abundancia. Folia Entomol Mex 79:207–231

Ibarra-Nuñez G, García-Ballinas J (1998) Diversidad de tres familias de arañas tejedoras (Araneae, Araneidae, Tetragnathidae, Theridiidae) en cafetales del Soconusco, Chiapas, Mexico. Folia Entomol Mex 102:11–20

Jeanneret P, Schüpbach B, Luka H (2003) Quantifying the impact of landscape and habitat features on biodiversity in cultivated landscapes. Agric Ecosyst Environ 98:311–320

Jiménez M, Tejas A (1996) Variación temporal de la araneofauna en frutales de la región del Cabo, Baja California Sur México. Southwest Entomol 21:331–336

Lanteri A, Loiacono M, Díaz N, Marvaldi A (1998) Natural enemies of weevils (Coleoptera: Curculionidae) harmful to crops in Argentina. Rev Bras Ent 41:517–522

Leon T, Cepeda J (2009) Aproximación comparativa a la diversidad de arañas y coleópteros en cultivos de café bajo manejo convencional y orgánico. Rev Bras Agroecol 4:2063–2066

Liljesthröm G, Minervino E, Castro D, González A (2002) La Comunidad de Arañas del Cultivo de Soja en la Provincia de Buenos Aires, Argentina. Neotrop Entomol 31:197–210

Lucio-Palacio C, Ibarra-Núñez G (2015) Arañas arborícolas de cacaotales con diferente tipo de manejo en Chiapas, México. Rev Mex Biodivers 86:143–152

Maloney D, Drummond F, Alford R (2003) Spider predation in agroecosystems: can spiders effectively control pest populations?. Maine Agricultural and Forest Experiment Station. The University of Maine, Orono. Technical Bulletin 190:32 pp

Mansour F, Rosen D, Shulov A (1980) A survey of spider populations (Araneae) in sprayed and unsprayed apple orchards in Israel and their ability to feed on larvae of *Spodoptera littoralis* (Boisd.) Ecol Appl 1:189–197

Mansour F, Ross J, Edwards G, Whitcomb W, Richman D (1982) Spiders of Florida citrus groves. Fla Ent 65:514–522

Marc P, Canard A, Ysnel F (1999) Spiders (Araneae) useful for pest limitation and bioindication. Agric Ecosyst Environ 74:229–273

Marshall S, Walker S, Rypstra A (2006) Two ecologically-divergent generalists predators have different responses to landscape fragmentation. Oikos 114:241–248

Medina A (1994) Las arañas: controladores naturales de insectos en el cultivo de arroz en Norte de Santander. Rev. Colomb Entomol 20:179–186

Minervino E (1996) Estudio biológico y ecobiológico de arañas depredadoras de plagas de la soja. Dissertation, Universidad Nacional de La Plata, Facultad de Ciencias Naturales y Museo, La Plata, Buenos Aires, Argentina

Molinari A (1987) Conceptos y descripción de especies entomófagas asociadas a insectos plaga del cultivo de soja. Publicación Miscilánea Nro 19 E.E.A., Oliveros, INTA. 13 pp.

Molinari A, Minervino E (2006) Características generales e importancia de las arañas como agentes depredadores de insectos plagas. Centro Regional Santa Fe, estación experimental agropecuaria Oliveros, grupo protección vegetal — sección entomología

Muma M (1973) Comparison of ground surface spiders in four central Florida ecosystems. Fla Ent 56:173–196

Muma M (1975) Spiders in Florida citrus groves. Florida Ent 58(2):83–90

Nyffeler M, Breene R, Dean D, Sterling W (1990) Spiders as predators of arthropod eggs. J Appl Entomol 109:490–501

Nyffeler M, Sterling W, Dean D (1994) Insectivorous activities of spiders in United States field crops. J Appl Entomol 118:113–128

Orellana M, Ávila I, Estrada P (2012) Diversity of spiders in an almond *Prunus dulcis* (mill.) D.A. Webb orchard in the metropolitan region of Chile (Central Chile). Idesia 30:17–24

Pekár S (2012) Spiders (Araneae) in the pesticide world: an ecotoxicological review. Pest Manag Sci 68:1438–1446

Pearce S, Zalucki M, Hassan E (2005) Spider ballooning in soybean and non-crop areas of southeast Queensland. Agric Ecosyst Environ 105:273–281

Perafán C, Flórez E (2004) Composición y distribución espacio-temporal de las comunidades de arañas (*Arachnida: Araneae*) en el sistema de cultivo maíz-soya de la altillanura plana colombiana, municipio de Puerto López, Meta. Acta Biol Colomb 9:71–71

Pickett S, White P (1985) The ecology of natural disturbance and patch dynamics. Academic Press, San Diego, 472 pp

Pearce JL, Venier LA, Eccles G, Pedlar J, McKenney D (2004) Influence of habitat and microhabitat on epigeal spider (Araneae) assemblages in four stand types. Biodivers Conserv 13:1305–1334

Pinkus M, Ibarra-Núñez G, Parra-Tabla V, García-Ballinas J, Hénaut Y (2006) Spider diversity in coffee plantations with different management in southeast Mexico. J Arachnol 34:104–112

Riechert S, Lockley T (1984) Spiders as biological control agents. Annu Rev Entomol 29:299–320

Rodrigues E, Mendonça M, Ott R (2008) Fauna de aranhas (Arachnida, Araneae) em diferentes estágios do cultivo do arroz irrigado em Cachoeirinha, RS, Brasil. Iheringia 98:362–371

Rodrigues E, Mendonça M, Ott R (2009) Spider diversity in a rice agroecosystem and adjacent areas in southern Brazil. Rev Colomb Entomol 35:89–97

Rodrigues ENL, Mendonça M, Fritz LL, Heinrichs EA, Fiuza L (2013) Effect of the insecticide lambda-cyhalothrin on rice spider populations in southern Brazil. Fortschr Zool 30:615–622

Saavedra E, Flórez E, Fernández C (2007) Capacidad de depredación y comportamiento de *Alpaida veniliae* (Araneae: Araneidae) en el cultivo de arroz. Rev Colomb Entomol 33:74–76

Samu F, Toft S, Kiss B (1999) Factors influencing cannibalism in the wolf spider *Pardosa agrestis* (Araneae, Lycosidae). Behav Ecol Sociobiol 45:349–354

Samiayyan K (2014) Chapter 15 Spiders — the generalist super predators in agro-ecosystems. In: Abrol DP (ed) Integrated Pest management. Academic Press, San Diego, pp 283–310

Sunderland K (1999) Mechanisms underlying the effects of spiders on pest populations. J Arachnol 27:308–316

Sunderland K, Greenstone M (1999) Summary and future directions for research on spiders in agroecosystems. J Arachnol 27:397–400

Symondson W, Sunderland K, Greenstone M (2002) Can generalist predators be effective biocontrol agents? Annu Rev Entomol 47:561–594

Topping C (2002) Spider model description. ALMaSS (Animal, Landscape and Man Simulation System). National Environmental Research Institute, Department of Landscape Ecology. http://www.dmu.dk/NR/rdonlyres/08109388-D514-46AF-8474-70584B9963CF/73284/SpiderModelDescription.pdf

Wardle D, Nicholson K, Bonner K, Yeates G (1999) Effects of agricultural intensification on soil-associated arthropod population dynamics, community structure, diversity and temporal variability over a seven year period. Soil Biol Biochem 31:1691–1706

Whitcomb W (1974) Predacious insects, spiders and mites of Arkansas cotton fields. Ark Agric Exp Snt Bull 690:1–84

Wrinn KM, Evans SC, Rypstra AL (2012) Predator cues and an herbicide affect activity and emigration in an agrobiont wolf spider. Chemosphere 87:390–396

Wise D (1993) Spiders in ecological webs. Cambridge University Press, Cambridge, 328 pp

Yeargan K, Dondale C (1974) The spider fauna of alfalfa fields in northern California. Ann Entomol Soc Am 67:681–682

Young O, Edward G (1990) Spiders in United States field crops and their potential effect on crop pests. J Arachnol 18:1–27

Chapter 12
Sexual Selection in Neotropical Spiders: Examples from Selected Groups

Luiz Ernesto Costa-Schmidt, María José Albo, Fedra Bollatti, Franco Cargnelutti, Lucía Calbacho-Rosa, Sofía Copperi, Nelson Ferretti, Matías Izquierdo, and Alfredo V. Peretti

Abstract Spiders have long been noted as classic examples of sexual behavior among arachnids, including extreme sexual dimorphism in some groups, and behavioral adaptations to diverse mating patterns. In recent decades, studies on the biology of Neotropical spiders have offered novel information on processes related to reproductive biology, including sexual selection. The present chapter synthesizes the large amount of knowledge on sexual selection and associated subjects in spiders from the Neotropics. Some of the groups considered in this review are mygalomorphs, lycosids and related, orb-weaving species, tetragnathids, social species, pholcids, and oonopids, among others. Concepts, patterns, mechanisms, and prospects on different areas of sexual selection are shown in detail for all these groups. In particular, here we highlight selected examples of the different contexts in which male–female interactions occur, such as mate choice, sexual cannibalism, sperm competition, and cryptic female choice. We outline the potential evolutionary consequences according to those contexts, with a final selection of model groups for specific experimental and comparative investigations.

L.E. Costa-Schmidt
Programa de Pós-Graduação em Biologia, Universidade do Vale do Rio dos Sinos (UNISINOS), Av. Unisinos, 950 - CEP 93022-000, São Leopoldo, RS, Brazil

M.J. Albo
Laboratório de Etología, Ecología y Evolución, Instituto de Investigaciones Biológicas Clemente Estable, Montevideo, Uruguay

F. Bollatti • F. Cargnelutti • L. Calbacho-Rosa • M. Izquierdo • A.V. Peretti (✉)
Instituto de Diversidad y Ecología Animal, CONICET – Universidad Nacional de Córdoba, Vélez Sarsfield 299 (5000), Córdoba, Argentina
e-mail: aperettibec@gmail.com

S. Copperi
Instituto de Ciencias Biológicas y Biomédicas del Sur (INBIOSUR), Universidad Nacional del Sur-CONICET, San Juan 670, Bahia Blanca (8000), Argentina

N. Ferretti
Centro de Estudios Parasitológicos y de Vectores (CEPAVE) UNLP, CONICET, Boulevard 120s/n e/61 y 62, La Plata (1902), Argentina

C. Viera, M.O. Gonzaga (eds.), *Behaviour and Ecology of Spiders*,
https://doi.org/10.1007/978-3-319-65717-2_12

Among the foundations of evolutionary theory, sexual selection is undoubtedly characterized by dynamic development. Recent decades have been astonishing in terms of technological advances, allowing us to understand more about the subtleties of the intricate process called reproduction. This led us to theoretical advances, getting deep into the mechanisms behind the reproductive strategies, something hardly imaginable decades ago. The emergence of disciplines such as evolutionary biology and behavioral ecology has resulted in the establishment of a rigorous scientific discourse, sustained in careful observations obtained throughout experimental approaches in both laboratory and field conditions.

Most sexual selection studies are based on "biological models", usually a species taken as an object of observation. Such an approach allows us to evaluate how sexual selection interacts with the remaining life-history components of the species. Most of the success of spiders as model organisms for sexual selection studies may be associated with their diversity in terms of number of species and reproductive strategies. Nowadays more than 46,896 spider species are described (World Spider Catalog 2017), and it is reasonable to assume that the observed diversity of behavioral strategies is partly associated to the phylogenetic structure of these organisms. However, the efforts to understand the impact of sexual selection over the diversification of this megadiverse taxon remain embryonic, restricted to the scientific effort of a few research groups specialized in a small number of study models. Given its historical origin, a strong bias exists in terms of the number of spider species used as model organisms for sexual selection studies developed across the temperate zones, as most of our knowledge originated from Palearctic and Nearctic species. Such bias results in the description of interesting patterns involving large groups of species (e.g., Peretti 2014), but also in the recognition of exclusive reproductive behaviors with no correspondence to any of the remaining investigated species.

The Neotropical region is well known for its biological diversity, being considered a natural laboratory for field investigations of new models in sexual selection (Macedo and Machado 2014). The gradual inclusion of new Neotropical model species has had several consequences in our theoretical knowledge of sexual selection (Peretti 2014), but it has also had an impressive impact on the emergence of new research centers across the Neotropics. In addition to that, another particular aspect must be emphasized: several of these new models are the Neotropical equivalent of some Palearctic models (e.g., the Neotropical genus *Paratrechalea* vs the Palearctic *Pisaura mirabilis*).

The main goal of this chapter is to provide a general synthesis of the patterns and mechanisms of the sexual selection domain emerging from Neotropical spider models. It is not our intention to cover the entire subject in such a limited space, and for that reason we have selected some Neotropical models, optimizing most of the diversity of reproductive strategies studied so far. Another important criterion for selection was to cover most of the taxonomic range provided by the available studies by including both mygalomorphs and araneomorphs, and haplogyne and entelegyne species.

We start by covering the theoretical rationale behind sexual selection studies using Neotropical models, ranging from pre-copulatory to post-copulatory mechanisms. Later we present spiders' idiosyncrasies that have direct influence over the interpretation of the observed sexual selection patterns within the group. For the analyses of the models, we opted to organize our analysis by taxonomic groups, ranging from infraorder to species-specific description of the available discoveries. We finish this chapter by presenting some perspectives for the study of sexual selection using Neotropical models.

Sexual Selection in a Nutshell

Sexual selection is rooted in the fact that males and females have distinct interests in terms of how the reproductive process should be conducted (Eberhard 1996; Arnqvist and Rowe 2005), and it is the leading force behind the evolution of reproductive strategies in order to mediate such intersexual conflicts. Sexual selection is an evolutionary process that acts upon traits that confer an advantage for the competing individuals in terms of access to mates or gametes (Andersson 1994). This definition still holds as the foundation of the subject, even though recent theoretical reviews call attention to important changes in the way we consider sexual selection and its mechanisms (Birkhead 2010; Shuker 2014), leading to an improvement in the explanatory power of the theory [see also Roughgarden (2015) for an interesting synthesis of the evolution of sexual selection definition]. Most of such changes were addressed towards the premises behind the traditional definition of sexual selection. For example, the evolutionary interpretation of traits once explained exclusively by sexual selection is now better interpreted by considering the concerted action of natural and sexual selection processes, especially if the competition for mates has some influence over fecundity (Shuker 2014 p. 22).

One of the most important theoretical advances in sexual selection occurred during the 1970s with the inclusion of post-copulatory mechanisms into the repertoire by which sexual selection operates (Parker 1970; Thornhill 1983; Eberhard 1996; Simmons 2001). Before this, sexual selection was assumed to act only during pre-copulatory stages of the reproductive process by means of two broadly defined mechanisms: intrasexual competition for mates and intersexual mate choice (Darwin 1871). During the pre-copulatory stage, both sexes may compete for access to the other sex and/or choose the individual that better fits its preference (Davies et al. 2012, p. 201; Dugatkin 2014, p. 203), though male–male competition and female mate choice are the most common ways that sexual selection operates. With regard to the post-copulatory stage, intrasexual competition has over the years become strongly associated with sperm competition, while intersexual mate choice can include cryptic female choice and cryptic male choice as well (Eberhard 2015).

The descriptions of pre-copulatory and copulatory stages of the reproductive process have received most of the attention in sexual selection studies because of the possibility of direct observation, where well-designed studies allow us to make causal inferences between behavior and reproductive success. It is intrinsically difficult to link post-copulatory mechanisms to non-random fertilization success, especially because the entire process occurs inside the female reproductive tract for species with internal fertilization. Given this difficulty, the addition of the word "cryptic" for mate-choice mechanisms indirectly refers to the lack of a direct correlation between mating success and fertilization success.

The recognition of how sexual selection operates is crucial for the proposition of testable hypotheses (Gowaty 2015). The operational mechanisms mentioned above are most of the time sufficient to explain the diversity of reproductive strategies observed in nature, which evolve under the constraints imposed by the evolutionary history of the organisms and the ecological conditions they experience. Many of the explanations concerning reproductive strategies fall into the realm of mating systems, which traditionally deal with the classification of reproductive strategies in terms of the number of individuals involved in the mating and fertilization processes (Kokko et al. 2014, p. 42). In this sense, a strategy is an emerging pattern described for a population derived from individual decisions (actions), which under the scope of mating systems refers to how the selective pressures act upon individuals, resulting in predictable behavioral patterns.

Spiders: Idiosyncrasies Shaping Reproductive Strategies

Even considering recent definitions of sexual selection as a theoretical construct that could be easily regarded as a *broad-sense* natural selection (Carranza 2009; Shuker 2014, p. 24), spiders have some idiosyncrasies supporting the distinction between sexual and natural selection. Spiders reproduce by internal fertilization, and the entire reproductive process encompasses several stages, highly variable among species in terms of both complexity and sequential organization. Here we present a brief description of those major stages (for further compiled information, see Schneider and Andrade 2011). This segmented approach is obviously an attempt to understand this complex process, and it is important to keep in mind the synergetic action of these stages determining the emergence and maintenance of reproductive strategies.

Sperm Induction

The first stage of the reproductive process of a spider is sperm induction by the males, consisting of charging of palps with sperm (Costa 1998; Foelix 2011). Male genital organs are modified structures located on the tips of their palps, used to transfer sperm during copulation. Sperm induction starts with the construction of a tiny triangular sheet web by the male, used to support a drop of sperm released by the male. Next, the male contacts the tip of the palps in the sperm drop, which is adsorbed by capillarity (Costa 1998).

Little is known about which facts stimulate males to perform sperm induction, and most of the studies completely ignore this step or assume it as an innate behavior of mature males. Since the sperm induction occurs even before a mate search, variation in sperm quality is interpreted as an effect of the male physiological condition. This may reduce the repertoire of reproductive strategies related to sperm transfer, restricting it to the amount/rate of sperm transfer, but not to its quality.

Mate Search

Mate search is typically performed by males in spiders, with few examples of sex role reversal for this trait (Aisenberg et al. 2007). Given its importance, mate search may be a valuable aspect in terms of reproductive effort invested by males, acting as a trade-off against other life history activities (foraging, for example). Depending on the ecological and social context of the focus population, mate search may be associated with selection towards traits that increase searching efficiency, resulting in a specific class of intrasexual competition for mates, called scramble competition (Andersson 1994; Shuker 2014).

Another possible outcome of mate search is the emergence of extreme reproductive strategies such as genital mutilation or even copulatory suicide. Those classes of reproductive strategies are intimately associated with a low probability of survival during mate search and/or a low probability of finding a receptive mate. Under these circumstances, finding a partner may represent the single opportunity for achieving reproductive success, and any trait resulting in an increase of reproductive success will be selected in the population.

Courtship

Once they have found a partner, spiders resume the courtship process based on the exchange of mutual information by using multimodal sensory drivers (e.g., Elias et al. 2006, see also Partan and Marler 2005; Hebets et al. 2016). This is by far the most studied aspect of the reproductive process of spiders, though we still have little understanding of the causal relationship for most of its components with mating and/or fertilization success. Courtship complexity varies tremendously among species, with a strong phylogenetic influence in its structure, and it has traditionally been interpreted as a source for species recognition (Costa 1998; Ferretti et al. 2013). Recent theoretical advances highlight the role of courtship as the main stage for mate choice (mostly female mate choice), conferring a different meaning for species recognition as the outcome of an interaction outside the preference range of the mating pair (Ryan and Rand 1993).

Copulation and Copulatory Courtship

Copulation in spiders is as variable as the courtship process among species, and its function goes beyond the obvious sperm transfer. It involves the necessary insertion of the male's palp within the female genitalia, where additional reproductive strategies may take place. Some of those strategies are associated with the execution of specific genital movements inside the female reproductive tract, which may represent some sort of copulatory courtship inducing female acceptance by extending copulation duration (Eberhard 1994, 2004; Huber 1998a, 2005; Schäfer and Uhl 2002), or could even be used to remove the sperm of previous males in non-virgin females (Calbacho-Rosa et al. 2012).

The mating system of most spider species is based on polyandry, that is, the female copulates with several males before the fertilization of her eggs. In this sense, traits such as copulation duration are assumed to be directly correlated to the amount of sperm transfer, and can be selected for lasting beyond the necessary time required for fertilization. Extending the copulation duration is interpreted as a mechanism of sperm competition in polyandric species, although other mechanisms cannot be excluded from acting at the same time. For example, since most of the copulatory process occurs inside the female genital tract, the assumption of a constant rate of sperm transfer is weak, especially if males can extract information regarding the female reproductive status and adjust their behavior accordingly in an optimized way.

Other mechanisms such as placing mating plugs, genital mutilation, and even copulatory suicide, have been interpreted as emergent sexual selected strategies. From the perspective of males, such strategies have strong influence over the occurrence and/or effective fertilization success provided by polyandry. From the perspective of females, these strategies may be a source for mate choice, where those males able to impose their reproductive interest can be interpreted as the best-fitted mates.

Post-copulation and Fertilization

The post-copulatory stage for spiders represents a rich opportunity for sexual selection. This richness has a strong association with the internal morphology of the female reproductive tract through the expression of cryptic selective mechanisms under female control (Eberhard 1996; Peretti and Aisenberg 2015). The female genital morphology in spiders is of ultimate importance for the evolution of reproductive strategies for the control of the fertilization process. This is possible because the internal morphology of female spider reproductive tracts is composed of sperm-storage organs (spermathecae, receptaculum, depending on the taxonomic group) where the sperm remains viable for a long time. Since the deposition of sperm occurs almost directly inside the sperm-storage organs, especially in the spermathecae, we may assume a natural stratification of the sperm pool available to the female for fertilization, and such a pattern may lead to several predictions about the fertilization success of a mating solely based on the morphology of the spermathecae (Austad 1984). However, it is important to bear in mind the possible role of male strategies to overcome the predicted sperm priority pattern by mixing the sperm pool during the embolic insertions (Uhl 2002, see also Elias et al. 2011).

Another important mechanism of post-copulatory mate choice performed by females is the active removal of sperm transferred by a specific male, a process called sperm-dumping (e.g., Peretti and Eberhard 2010). The moment of its occurrence is variable among the species that perform such a strategy, and has as a consequence the almost complete exclusion of the selected sperm from the mating pool available for fertilization. Such strategy depends, of course, on several physiological and morphological conditions of the females. Sperm-dumping may be considered the extreme post-copulatory mate choice performed by females, given the complete removal of the sperm from the female body. Less extreme manipulations of the sperm pool, such as sperm assimilation for nutritional purposes, may occur as a result of female post-copulatory mate choice, but with unknown frequency.

Selected Spider Models

In this section, we present detailed information about the reproductive processes of selected spider models from the Neotropics. The idea is to provide a broad characterization of the possible sexual selection mechanisms acting upon such models, looking for patterns and peculiarities within and between groups. The choice of the models followed two basic criteria: the inclusion of a large range of taxonomic groups within spiders, and availability of experimental data for each case. We emphasize the incompleteness of our selection, justified by the limitations of space in this compilation.

Mygalomorphae

General Aspects of Sexual Behavior of Mygalomorphae Spiders

Mygalomorphae spiders comprise some of the largest and most fascinating spiders, such as tarantulas or trapdoor spiders. Although many species can reach large body sizes, are long-lived, and are usually captured in high quantities, the knowledge about features of their sexual behavior is scarce (see Ferretti et al. 2013 for a review). Among arachnids, spiders constitute a basal group, and the oldest fossil is dated from the Triassic (Selden and Gall 1992). Nowadays, 16 Mygalomorphae families are formally recognized and distributed worldwide, of which 15 occur in the Neotropics (World Spider Catalog 2017).

Most Mygalomorphae families show terrestrial habits and live in shelters surrounded or covered by silk, constructing burrows of different shapes on the ground or in crevices under stones or logs (Dippenaar-Schoeman 2002). The entrances of the shelters can be open or closed by a door formed with silk and detritus. Many species are nocturnal; thus, during the day they hide inside their refuges and at night they stalk for a potential prey at the entrance of their burrows. This situation changes during the reproductive period. Once individuals reach their sexual maturity, males leave their shelters in search for females (Baerg 1928; Costa and Pérez-Miles 2002). Although the reproductive periods vary according to the species, usually they mate once a year or twice in those species of large size, and mainly during warmer months, for example in spring or summer (Costa and Pérez-Miles 2002). Adult males have the last segment of the palp modified as a copulatory bulb that usually shows a simple shape. The palpal tarsus (named *cymbium*) has an extension that consists of a piriform bulb, which carries a spiraling duct that opens at the tip (embolus) through a small pore (Foelix 2011). This bulb acts as a reservoir for sperm until mating occurs. The males transfer the sperm during mating by introducing the embolus into the genital opening of the females. The genital opening is located ventrally at the abdomen, between the anterior pair of book lungs, and leads to a couple of receptacles or spermathecae of sac shape that connect directly with the external uterus, the place where fertilization takes place. Because of the genital shape consisting of a "*cul-de-sac*" blind sac, and assuming the absence of sperm-mixing, it is expected that the last sperm introduced will be the first to get out and to fertilize the eggs, and the spermatic preference could be towards the last male that mated (Austad 1984). This could be related to the reproductive strategy of most Mygalomorphae species, which consists in males trying to mate with as many females as possible, while females usually have numerous mating attempts during the same period. However, many aspects of sperm priority in those species showing this genital morphology are still unknown (Yoward 1996; Uhl 2002).

Males of Mygalomorphae are capable of detecting chemical signals and exchanging tactile signals with females, enhancing courtship behavior (Costa and Pérez-Miles 2002; Ferretti et al. 2013; Costa et al. 2015). In this way, the pheromones associated to the silk facilitate the sexual encounter and specific recognition. However, it was found recently that in two sympatric and synchronic tarantula species of the genera *Acanthoscurria* and *Eupalaestrus* from Uruguay, females confuse the specificity of courtship, even preferring non-specific males. The authors of this work concluded that these females could be choosing a more effusive courtship from males of the other species (Costa et al. 2013).

Male courtship behavior also includes vibratory and seismic stimuli transmitted through the substrate and, when individuals contact each other, tactile and chemotactile signals. Palpal drumming and leg tapping are two of the most spread behavioral units, whose function is to send acoustic and vibratory communication signals (Ferretti et al. 2013). Females of some species respond to male courtship by performing body vibrations and leg tapping, which indicate their sexual receptivity and willingness to mate. After contact, males engage many behavioral units such as spasmodic beats, leg tapping, and brushing over the female body, that help female relaxation and induction of the opening of fangs so males can raise them and insert their palpal organs (Costa and Pérez-Miles 2002; Ferretti et al. 2013).

The typical mating position of Mygalomorphae involves individuals facing each other, with the male clasping female fangs with the first leg pair. After the clasp, the male raises the female and begins a series of palpal insertion attempts (Jackson and Pollard 1990; Costa and Pérez-Miles 2002; Ferretti et al. 2013). It is remarkable that some species have spines or apophyses on the first pair of legs that facilitate the clasp and lock the female fangs. Although previous studies proposed a protection function of that behavior for males against females, today it could be interpreted as intrasexual communication, or even that females could be using the different shapes of spines and apophyses of males as a basis for mate choice (Jackson and Pollard 1990; Ferretti et al. 2013).

The courtship display *in copula* has been reported in a few species, but it is expected to be more widespread than current data suggest, which we will probably confirm when more detailed studies on sexual behavior of Mygalomorphae are achieved. This courtship involves vibrations or leg tapping over the female body at the same time that the male is inserting his palps, which suggests that females could be testing male copulatory ability (Costa and Pérez-Miles 2002; Ferretti et al. 2013). Females of many species remain in an immobile state or catalepsis during and after the mating. The number of palpal insertions is variable, alternated, and, depending on the species, involves a great number of palpal insertion failures, i.e., the male cannot insert the embolus and usually changes to the other palp (Pérez-Miles and Costa 1992; Ferretti et al. 2013). This could lead us to think that maybe some sexual selection mechanism, such as the cryptic female choice, could be operating inside the females (Eberhard 1996). Moreover, even when females are in a cataleptic state, they could provide mechanical challenges to male copulatory abilities, and in this way evaluate male performance and adjust their own behavior and/or physiology to maximize their fitness.

The real risk of sexual cannibalism during and after mating is a controversial aspect inside this group of spiders. Although the early studies on sexual behavior proposed a “slaughter” from females towards males (Bücherl 1952), nowadays it is unclear whether this is true (Jackson and Pollard 1990; Costa and Pérez-Miles 2002; Ferretti et al. 2013). However, it is known that some male features could have evolved to avoid sexual cannibalism risk, such as the tibial apophyses of males that clasp and lock the female fangs, or the retreated behavior after the mating. Arguably, many females are in a cataleptic state that could last for some minutes, giving the males the chance to escape to a safe distance. There are also some species that do not possess tibial apophyses; thus, they don’t clasp the female fangs, and mating takes place without aggression (Ferretti et al. 2013).

Current State of Knowledge of Neotropical Mygalomorphae Families

With regard to the different Mygalomorphae families from the Neotropics, the common characteristics of courtship and mating can vary (Table 12.1). It is important to note that the sexual behavior of the Neotropical species is only known from some representatives of Dipluridae, Mecicobothriidae, Microstigmatidae, Nemesiidae, and Theraphosidae (Ferretti et al. 2013).

Spiders belonging to Dipluridae are recognized by their ability to build large sheet webs constructed by their extremely long posterior lateral spinnerets. Nine genera of the 24 currently recognized are distributed in the Neotropics (World

Table 12.1 Courtship and mating behaviors performed by females (F) and males (M) of Neotropical Mygalomorphae

Family/genus	Courtship behavioral units							Mating behavioral units						References
	Body vibration	Leg tapping	Brush-ing	Palpal drumming	Scratch-ing	Spasmodic beats	Stretch-ing	Body jerks	Cata-lepsis	Biting fangs	Clasp	Leg beating	Palpal boxing	
Dipluridae														
Euagrus	M	F; M							F		M			(a)
Mecicobothriidae														
Mecicobothrium	M	M		M					F		M			(b) (c)
Microstigmatidae														
Xenonemesia		M					M		F		M	M	M	(d)
Nemesiidae														
Acanthogonatus		M	M		M	M	M	F	F		M		M	(e) (f)
Stenoterommata		M	M		M	M	M	F	F		M		M	(g)
Theraphosidae														
Acanthoscurria	F; M	F; M	M	M	M	M			F	M	M		M	(c) (h) (i) (j) (k) (l)
Avicularia	F; M	F; M	M	M	M	M			F	M	M		M	(c)
Brachypelma	F; M	F; M	M	M	M	M			F	M	M		M	(m) (n)
Catumiri	F; M	F; M	M	M	M	M			F	M	M		M	(h)
Eupalaestrus	F; M	F; M	M	M	M	M			F	M	M		M	(c) (h) (i) (j) (k) (l) (o)

Grammostola	F; M	F; M	M	M	M	M			F	M	M		M	(c) (h) (p) (q) (r) (s)
Homoeomma	F; M	F; M	M	M	M	M			F	M	M		M	(h)
Plesiopelma	F; M	F; M	M	M	M	M			F	M	M		M	(h) (t)
Sickius	F; M	F; M	M	M	M	M			F	M	M		M	(c) (u)
Vitalius	F; M	F; M	M	M	M	M			F	M	M		M	(v)

The table does not include occasional observational data
[a] Coyle (1986); [b] Costa and Pérez-Miles (1998); [c] Ferretti et al. (2013); [d] Ferretti et al. (2012); [e] Pérez-Miles and Capocasale (1982); [f] Ferretti et al. (2011); [g] Schwerdt and Copperi (2014); [h] Costa and Pérez-Miles (2002); [i] Costa et al. (2013); [j] Pérez-Miles et al. (2005); [k] Quirici and Costa (2005); [l] Quirici and Costa (2007); [m] Schmidt and Krause (1994); [n] Yañez et al. (1999); [o] Pérez-Miles et al. (2007); [p] Ferretti and Ferrero (2008); [q] Pérez-Miles and Costa (1992); [r] Postiglioni and Costa (2006); [s] Copperi et al. (2012); [t] Costa and Pérez-Miles (1992); [u] Bertani et al. (2008); [v] Ferretti et al. (2015); [w] Gerhardt (1929)

Spider Catalog 2017). However, information about the reproductive behavior of those genera is scarce, the single piece of information available being that published by Coyle (1986) for an unidentified species of the genus *Euagrus* Ausserer, 1875. Diplurid males begin courtship when they contact the female silk threads, and sexual signals involve body vibrations, and leg tapping with forelegs and palps against the sheet web. Females usually respond to the courtship by leg and palpal tapping. The mating position comprises clasping with male second leg pair towards the femora of female second leg. Females stay cataleptic during mating.

Spiders of the family Mecicobothriidae are represented only by two species in the Neotropics, both of the genus *Mecicobothrium* (World Spider Catalog 2017). They are small spiders that build sheet webs on small crevices under stones or logs. The only information about their sexual behavior comes from *M. thorelli* Holmberg, 1882, a species distributed in Argentina and Uruguay (Costa and Pérez-Miles 1998). Unlike most Mygalomorphae, this species shows its reproductive period during the coldest months of winter in the Southern Hemisphere. With regard to male sexual behavior, the primary sexual communication mechanism is through vibrations transmitted by the silk threads.

The contact between sexes in *M. thorelli* exhibits a unique feature due to the presence of cheliceral apophyses of males, which females clasp and bite during mating. The palpal insertions are complex, as the right bulb must be inserted into the right female spermathecal receptacle because of a spiral complementary configuration of both genitalia. As Ferretti et al. (2013) mentioned, *"(…) palpal insertions may cause intense genitalic stimulation, which could determine sexual selection by cryptic female choice"*. Finally, males remain close to females after mating (mate guarding) and repel other males when they approach, an unusual behavior for mygalomorphs.

Microstigmatidae is a small family in terms of diversity since it only has seven genera and 17 species worldwide, which 11 are distributed in the Neotropics (World Spider Catalog 2017). Given that they are small spiders and live between the leaf litters of rainforests, they are hard to find and collect. In fact, only some aspects of the reproductive behavior of *Xenonemesia platensis* Goloboff, 1988 are known (Ferretti et al. 2012). Males of this species initiate courtship once they make contact with the female body, and this could be related to the low use of silk in these spiders. Males do not have tibial apophyses, and clasp females between the palpal base and the chelicerae. Then, males raise females and begin a series of palpal insertion attempts. It is important to note that males court females when mating and during each palpal insertion they vigorously beat with the second pair of legs over the female bodies.

With regard to the Nemesiidae family, of the 45 known genera, 18 are distributed in the Neotropics (World Spider Catalog 2017). Although it comprises a diverse family with about 393 species worldwide, knowledge about sexual behavior is reported just for representatives of the genera *Acanthogonatus* and *Stenoterommata* (Pérez-Miles and Capocasale 1982; Ferretti et al. 2011, Schwerdt and Copperi 2014). Usually, male courtship behavior begins when he contacts the female silk threads. The courtship involves long-distance vibrational signals that need to convince the female to leave the silk tube. The mating takes place at the entrance of the silk tube or inside it, and females perform a huge flexion of the pedicel, reaching a carapace–abdomen angle of about 60–80°. In some cases, females protrude and expand the genital area during mating, but the possible function or relevance of this behavior is unknown. Moreover, it is notable that females display high frequencies of body somersaults by flexion and expansion of all legs that could act as stimuli towards males to request new palpal insertions.

Theraphosidae spiders are commonly known as tarantulas, comprising the largest spiders. It is the most diverse family among the Mygalomorphae; moreover, the subfamily Theraphosinae is the richest one and is endemic to the Neotropics. Usually, the pattern of sexual behavior follows the general descriptions presented above. Males detect contact pheromones that enhance their courtship, which includes displaying seismic signals through the substrate. Females of some species respond to the courtship by indicating their location and/or receptivity. The mating position is typical for mygalomorphs, with the exception of the species *Sickius longibulbi* Soares and Camargo, 1948, in which the male tries to knock the female down, pushing her entire body until she is lying on her dorsum (Bertani et al. 2008). This remarkable behavior and mating position is related to the absence of spermathecae on the females; thus, males need to reach the oviduct to increase the chance of fertilizing the female's eggs. The number of palpal insertions varies between one to sixteen, depending on the species, and most of them are characterized by a mating system with polyandry and polygyny (Costa and Pérez-Miles 2002; Ferretti et al. 2013). However, recently a case of monandry and polygyny was reported in the tarantula species *Eupalaestrus weijenberghi* (Thorell, 1894), in which males perform just one palpal insertion during each mating (Pérez-Miles et al. 2007). Finally, although earlier studies proposed sexual cannibalism as a rule for theraphosids, current studies indicate that post-mating attacks are only occasional (Costa and Pérez-Miles 2002; Ferretti et al. 2013) (See Chap. 2).

Conclusions

The chemical sexual communication in Mygalomorphae is a key feature of courtship and mating. In this way, when males detect chemical cues deposited in female silk threads they initiate courtship behavior. The behavioral patterns of courtship allow specific sexual recognition, and guide individuals into a proper mating position. The female's responses to male courtship indicate her sexual receptiveness, attract the male, and help the male in locating her. After they contact each other, the mating takes place in a face-to-face position, and usually males raise and clasp the females to allow them to make the palpal insertions attempts. Although it is a sporadic behavior, an active courtship during mating suggests that maybe females are testing male copulatory abilities. During mating, the females usually remain in a cataleptic state while males make the palpal insertions. After mating, males escape cautiously.

Usually, the sexual behavior of Mygalomorphae spiders has been considered as "simple", that is, males just walk searching for females and when they randomly find a female, the mating should occur in most cases. This point of view is changing as the number of studies has grown during the last 20 years, suggesting that the sexual behavior of Mygalomorphae is far from simple. Such complexity may be represented by specific courtship behavior of males and the active role of females leading to an intricate intersexual communication before mating, by the copulatory courtship performed by males, and also by the complex patterns of palpal insertions. Based on these statements, and the fact that information on sexual behavior is known for only five of the 14 families currently described, this may be viewed as a promising field of research where new questions about the mechanisms of sexual behavior operating on these spiders should be developed.

Araneomorphae: Oonopidae

Spiders of the family Oonopidae or goblin spiders are among the ten most diverse families of the world, with 1780 described species (World Spider Catalog 2017). They are small spiders, up to 3 millimeters long, inhabiting canopies and leaf litter of tropical and subtropical forests (Jocqué and Dippenaar-Schoemann 2006; Harvey and Edward 2007; Fannes et al. 2008; Izquierdo et al. 2012; Platnick et al. 2012). The taxonomy of the family has been studied in the last 10 years in the context of the PBI project (http://research.amnh.org/oonopidae/), resulting in an important increase in the number of known species and genera.

The family is diagnosed by the presence of six eyes (but there are even eyeless species in other regions of the world), the absence of a female palp claw, and by the fusion of the male testes into one single, unpaired structure (Burger and Michalik 2010). The family is also recognized by a dimorphic pattern in the number of tarsal organ receptors, with the first two legs having one more receptor than legs III and IV (Platnick et al. 2012) and paired tracheal spiracles anteriorly positioned (Platnick et al. 1991). Some groups of species have a strongly sclerotized body with dorsal and ventral abdominal scutes, although soft-bodied representatives are also very common (see Platnick and Dupérré 2010, 2011).

Oonopidae is part of the big clade Synspermiata, which includes all the haplogyne spiders without cribellum characterized by transferring sperm by synspermia (a group of sperm cells in the last stage of division surrounded by a protective capsule) (Michalik and Ramírez 2014). The family is at the same time part of the Dysderoidea with other three subfamilies (Segestriidae, Dysderidae, and Orsolobidae), which share the characteristic respiratory system composed of two book lungs and two tracheae (Platnick et al. 1991). Each trachea is tube-shaped and enters into the prosoma by ramifying in multiple small branches. Both book lungs and trachea are open to the exterior through spiracles, anteriorly and posteriorly to the epigastric fold respectively.

Three subfamilies are recognized: Orchestininae, which includes only the genus *Orchestina*; Sulsulinae, comprised of the genera *Sulsula, Dalmasula, Xiombarg, Unicorn, Cortestina,* and *Puan*; and Oonopinae, which includes all the remaining genera. This division has been proposed based on the study of the tarsal organ morphology and DNA sequences (Platnick et al. 2012; De Busschere et al. 2014). As a result of this work, the genus *Orchestina* and the subfamily Sulsulinae have been proposed as basal members of the family by sharing some similar characters with other families of Dysderoidea. Oonopinae is composed of members with and without sclerotized scutums, and a low degree of body sclerotization, which might be considered as a plesiomorphic trait according to the results of De Busschere et al. (2014).

The general biology of the group is incipient. Many aspects of the life history of these spiders are unknown, especially reproductive biology. The first studies in this field were carried out by Bristowe (1930) and Gerhardt (1933) in species of the genera *Oonops, Xestaspis,* and *Silhouettella.* Other observations were occasionally performed at the field, but with no additional data on detailed behaviors (see for example Harvey 1987). However, later studies provided detailed information on morphology and behaviors related to different aspects of reproduction and sexual selection, some of them summarized in the following sections.

Sexual Behavior in *Orchestina*: A Basal Oonopid Spider

As in many other oonopids, the sexual behavior of this group is barely known, and only one species from the Neotropics has been studied. The first detailed observations on this genus were carried out on a species from Northern Argentina, also distributed in Bolivia and Brazil, by Burger et al. (2010). In that study, general patterns of sexual behavior were described and detailed information of the genital morphology was provided. The courtship is very simple; the male initiates the contact with the female by detecting traces of silk threads that she has left on the substrate. In that experiment, the females had previously built a small cell with a few threads of silk and remained inside. Then, when the male contacted the threads of the female's web, he initiated searching behavior. No male was observed performing sperm induction prior to copulation. In that moment, the male advanced towards the female and both spiders started touching each other with front legs. Next, the male pushed the female back and adopted the copulatory position, and started to creep under her. This position has been observed in many theraphosids and other haplogyne spiders: the male is under the female with its clypeus facing the female's sternum. This position has been interpreted as "primitive", also occurring in the Oonopinae genus *Oonops* (Bristowe 1929; Gerhardt 1930, see also von Helversen 1976), which are presumably more basal oonopids. The "derived" copulatory position taken by other oonopids such as members of *Xestaspis, Grymeus,* and *Silhouettella* has evolved in convergence in different spider families (Gerhardt 1933; Bristowe 1930; von Helversen 1976; Harvey 1987; Burger 2007).

After this sequence, the male inserts both pedipalps at the same time, moving them alternately during the copulation (up and down, back and forth) one palp at a time, whereas the other palp remains motionless. Females are polyandrous (copulate with more than one male). First copulations last around 18 min, whereas second copulations last 15 min, and third copulations 23 min. Pedipalp movements stop around 2 min before the end of copulation. At the moment of separation the male moves back, and the two spiders touch each other with their first legs. Then the male remains close to the female's web showing self-grooming, but does not interact directly with the female. When there is some sort of contact, the female reaction is to go after the male, showing some agonistic predisposition. No sperm re-induction was observed. Although the male stays close to the female after copulation, it is hard to interpret this behavior as post-copulatory mate guarding, as has been reported for other spiders such as pholcids (Eberhard 1992a; Huber and Eberhard 1997).

Sexual Behavior in Other Species of Oonopidae

In this section, we summarize some important facts about sexual behavior of Oonopidae species from outside the Neotropics. *Silhouettella loricatula* (Roewer, 1942) belongs to the subfamily Oonopinae, characterized by the presence of abdominal scutums. It is known for Europe, Central Asia, and North Africa (World Spider Catalog 2017). The copulatory behaviors and morphological aspects of this species were studied in a series of papers by Matthias Burger and collaborators (see references below). As for *Orchestina,* the courtship in this species is quite simple, starting with a series of leg palpating and reaching the mating position very fast (Burger and Carrera 2011). Unlike *Orchestina,* the mating position in this species is derived: both spiders are facing in the same direction but the male is under the female with his sternum pointing to the venter of the female abdomen (Gerhardt 1933). Both pedipalps are inserted at the same time and rhythmic movements are observed during the entire copulation, although they do not show a clear pattern. During the copulation, the male vibrates his abdomen up and down, and soon after the insertion the pedipalps are rotated and only the tips of emboli are inserted. After some minutes during which both spiders remain motionless, the male moves away performing a somersault.

As in *Orchestina, S. loricatula* females accept to copulate with subsequent males, although in such cases females turn more aggressive, indicating a reduction in their sexual receptivity. This aggression has been reported for other oonopids and haplogyne families as well, and has been suggested as a mechanism for sexual female selection (Huber 1994, 1995; Huber and Eberhard 1997; Schäfer and Uhl 2002; Burger et al. 2006; Burger 2010). In addition, some females accept to remate with previously rejected males, indicating the existence of a selecting mechanism. In this species, a first copulation was the longest with almost 1.33 h, whereas second copulations lasted 1.26 h. However, historical records reported 5 h for first copulation in the same species (Bristowe 1930).

The most peculiar characteristic on the behavior of *S. loricatula* is that females are able to dump sperm from a previous male while they are copulating with a subsequent male (Burger 2007). The sperm is stored in the female receptacle and surrounded by a sac during the copulation (Burger et al. 2006; Burger 2007). It is not clear whether the male, female, or both sexes are implicated in producing the sac. In any case, it has been suggested that the expulsion of the sac may be favored by muscular contractions of the female genitalia, although a cooperation of rhythmical movements of the male pedipalp during the copulation is not discarded. The sperm dumping has been proposed as a mechanism of sexual selection, which would allow the females to evaluate and select some males against others. In *S. loricatula* it has been also argued that a male may extend the copulations as a strategy for guarding the female against other males, in order to protect its own sperm and paternity (Wynn and Vahed 2004; Linn et al. 2007). This mate-guarding strategy has been reported for other families such as Salticidae, Araneidae, and Pholcidae, among others (Jackson 1986; Merrigan 1995; Elgar 1998; Calbacho-Rosa et al. 2010, Elias et al. 2014).

Based on study of the female genital morphology of *S. loricatula* it has been proposed that other oonopids with a similar structure may dump sperm as well (Burger et al. 2003; Burger 2009, 2010). However, no direct observations have been made on those species, whereas no evidence of cryptic female choice was found for *Neoxyphinus termitophilus* (under *N. oglobini*), *Dysderina* sp., and *Heteroonops spinimanus* (Burger 2011). In addition to sperm dumping, other mechanisms such as sperm translocation into different sections of the female genitalia have been proposed, based on locking mechanisms of the uterus and muscular contractions (Burger 2011; Burger et al. 2003, 2006, 2010).

Araneomorphae: Pholcidae

Pholcids are sedentary, long-legged web spiders that are highly diverse in tropical and subtropical regions of the world (Huber 2014). To date, there are around 1600 described species (World Spider Catalog 2017), inhabiting a wide range of places, from leaf litter to caverns, under stones and logs, and in human buildings (Huber 2014). These spiders are recognized by the length of their legs in relation to their body, reaching up to 15 centimeters in some species (Huber and Astrin 2009), although short legs are also common in the subfamily Ninetinae. The morphology of the group is also unique. Male pedipalps are characterized by the presence of a modified cymbium, the procursus, which is used during copulation (see Huber 1998b, 2002). Modified setae on chelicerae and clypeus and apophyses on the chelicerae are also common in males. Legs are pseudo-segmented in both sexes, and females may have a sclerotized structure on the epigastric region, the epigynum, which is not common in haplogyne but characteristic of entelegyne spiders (Huber 2014).

Among other four families, pholcids are members of the "lost tracheae clade", a group that has lost the posterior trachea (Wheeler et al. 2016). In addition, these families are part of the Synspermiata, a big group in which sperm cells are transferred as fused spermatids sharing a common shield, although many other pholcids transfer their sperm cells as cleistosperm, a mechanism extended among entelegynes (Alberti and Weinmann 1985; Michalik and Ramírez 2014). The taxonomy of the family is well studied, especially for Africa (Huber and Warui 2012; Huber 2013; Huber and Kwapong 2013), Australia (Huber 2001), and for the Neotropical region (Huber 2000, 2014; World Spider Catalog 2017). At present, there are five subfamilies with unknown monophyletic support: Ninetinae, Modisiminae, Pholcinae, Smeringopinae, and Arteminae. Ninetinae has been proposed as the basal member of the family (Huber 2011, 2014), and together with Modisiminae and Pholcinae contains Neotropical representatives (Huber 2011). The only known exception for the Neotropical region is the species *Holocnemus pluchei* (Scopoli, 1763) (Smeringopinae), which is a common synanthropic species with a worldwide distribution.

One main feature of the family is its web, with the exception of Ninetinae for which the web is barely known (Huber 2014). The webs are dome-shaped, and the spiders are found hanging with the dorsal face of their bodies pointing to the floor (Eberhard and Briceño 1985; Eberhard 1992b). When disturbed, pholcids are able to escape and hide under logs, leaves, rocks, or crevices. In addition, long-legged pholcid species vibrate or whirl their entire body as a way of distracting or to blur their contours and confuse predators (Jackson et al. 1992, 1993).

Cryptic coloration is also a strategy to avoid predators, especially during the day (Huber 2014). The eggsac, formed by only a few threads of silk, is carried with the chelicerae until the hatching of the nymphs; brood-care is an extended behavior among pholcids.

Sexual Selection and Functional Morphology

Most studies about sexual selection and reproductive biology have been conducted using pholcid species from the Old World (Huber 1994, 1995, 1996, 1999; Kaster and Jakob 1997; Uhl 1998; Schäfer and Uhl 2002; Uhl et al. 2005; Calbacho-Rosa et al. 2010, 2012, 2013). Studies involving Neotropical species of pholcids for years have been restricted to their taxonomic descriptions and to brief behavioral descriptions (Huber and Eberhard 1997; Huber and Perez-Gonzales 2001; Huber 2005; Huber et al. 2005; Peretti et al. 2006; Huber et al. 2013).

Sexual dimorphism in body size and color is infrequent; however, in *Mecolaesthus longissimus* Simon, 1893 the opisthosoma of males are twice as longer than those of females (Huber 2005). There are two hypotheses related to sexual selection mechanisms for the emergence of this dimorphic characteristic. One of these considers abdomen length as an honest signaling trait of the male quality that can be used by females during mate choice. The other hypothesis predicts advantages provided by longer abdomen during male–male competition for access to the female.

On the other hand, Huber et al. (2005) showed that in two species of *Mesabolivar* (*M. samatiaguassu* Huber, Brescovit and Rheims, 2005 and *M. cuarassu* Huber, Brescovit and Rheims, 2005) males have long, frontal cheliceral apophyses and females have external genitalia strongly protruding with a median pocket and a pair of projections. The authors suggested that male cheliceral apophyses are inserted into the epigynal pockets, and that the length of the apophyses is directly linked to the length of the female epigynal projections. These female projections may have evolved by cryptic female choice, enabling those males with exaggerated apophyses to overcome such a barrier (Huber et al. 2005). Huber and Pérez-González (2001) reported the first case of female genital polymorphism in the Cuban pholcid *Ciboneya antria* Huber and Pérez, 2001. In this species, two forms of female genitalia were found: 'macrogynes', which have the frontal part of the external genitalia stretched forward, reaching under the prosoma, and slightly smaller 'microgynes', which have the frontal part of the epigynum slightly bent backwards. The authors suggested alternative mating strategies as the origin of such polymorphism, although details on the natural history of the species are needed to confirm this.

Sexual Behavior

With regard to pholcid sexual behavior, copulation only occurs if the female adopts a horizontal position, allowing the male to insert both pedipalps into the center of the female gonopore. Females are polyandrous, and they copulate with several males before oviposition. The different patterns of behaviors that males perform during copulation are associated with functions related to sperm competition and cryptic female choice, such as the removal of rival sperm, sperm transfer, and female stimulation, which are associated with the differential fertilization success of males.

There are different patterns of male courtship: bursts of quick dorsoventral vibrations of opisthosoma, leg-shaking (Huber 1994, 1995; Huber and Eberhard 1997; Calbacho-Rosa et al. 2013), chelicerae movements (with a possible stridulatory function) (Huber 1995; Huber and Eberhard 1997; Dutto et al. 2011; Stefani et al. 2012), and tug of web with all legs as in *Gertschiola neuquena* Huber, 2000 (Izquierdo, Cargnelutti, Calbacho-Rosa, Peretti, pers. obs.). Receptive females respond with tapping movements and move towards the approaching males. For example, in *Anopsicus zeteki* (Gertsch, 1939), male courtship is conspicuous and consists of leg tapping, brief vibrations of the entire body, and occasional abdomen-bobbing. In some cases, both sexes intensely repair the web during the courtship, resulting in longer courtships than when males do not perform this behavior (Huber 1998b). In *Coryssocnemis viridescens* Kraus, 1955, male courtship was almost nonexistent and males approached the female tapping with the anterior legs (Huber 1998b). In *Mesabolivar delclaroi* Machado and Brescovit, 2012, courtship was short, with interspersed beats of the male's first pair of legs on the web, alternated with chelicerae movements (Stefani et al. 2012). The movement of the male's che-

Table 12.2 Female and male courtship behavioral units performed by Neotropical pholcids, divided into non-genital (leg-shaking, abdomen vibration, and stridulation) and genital movements (male pedipalps)

	Non-genital courtship			Genital courtship			
Pholcid species	Leg-shaking	Abdominal vibration	Stridulation	Male pedipalp movements			References
Physocyclus globosus	Yes	Yes	Yes (female)	Rhythmic	Alternate	Non-symmetric	(a) (b)
Anopsicus zeteki	Yes	Yes	No	Rhythmic	Simultaneous	Symmetric	(c)
Coryssocnemis viridescens	Yes*	Yes	No	Rhythmic	Simultaneous	??	(c)
Modisimus culicinus	??	??	??	??	Simultaneous	Symmetric	(d)
Modisimus guatuso	Yes	Yes	No	Rhythmic	Alternate	Symmetric	(d)
Metagonia rica	??	??	??	First non-rhythmic, then rhythmic.	??	??	(e)
Gertschiola neuquena	No	No	No	Rhythmic	Alternate	??	(f)

Only species with enough information are included, for other species see text. For genital courtship, we used the following definitions: *symmetric* — both pedipalps inserted in the same way, *non-symmetric* — both palps inserted with a different orientation, *rhythmic* — pedipalps are moved in a regular pattern, *non-rhythmic* — the pattern of movements varies in each phase, *simultaneous* — each pedipalp is moved at the same time, and *alternate* — pedipalps are moved one at once

*Females: all femora raised from each side alternately (a) Huber and Eberhard (1997); (b) Peretti et al. (2006); (c) Huber (1998b); (d) Huber (1998c); (e) Huber (1997b); (f) Cargnelutti, Calbacho-Rosa, Izquierdo, Peretti, unpublished data.

licerae during courtship can be interpreted as an intra-specific signal to show his reproductive intentions to the female (Huber and Eberhard 1997; Schäfer and Uhl 2002). In *Metagonia rica* Gertsch, 1986, successful courtship consisted of rhythmic movements of the male opisthosoma, but sometimes the male tapped the female with his anterior legs (Huber 1997b).

Before genital insertion, males rotated their pedipalps 180°, and approached the females. Copulation in these animals consists of a single long insertion of male pedipalps at the same time, both symmetrically or asymmetrically. The pedipalps are kept inserted in the female genital cavity during copulation, and the lateral movements may be rhythmic twists [for example, *Physocyclus globosus* (Taczanowski, 1874) and *Anopsicus zeteki*] or non-rhythmic (as in *Metagonia rica*), either simultaneously or alternately (Table 12.2). During the entire copulation, the pattern of such movements changes gradually, increasing during the first minutes and then decreasing or even stopping (as in *Physocyclus globosus, Anopsicus zeteki,* and *Metagonia rica*). The duration of copulations may change according to the female reproductive status (virgin or mated), as in *Coryssocnemis viridescens* and *Metagonia rica* (Huber 1997b, 1998b). In some cases, other associated structures of male pedipalp may be implicated, such as bristles on the cymbium in *Mesabolivar delclaroi* that are used to rub the female's abdomen (Stefani et al. 2012).

As showed above, males' pedipalp movements during copulation are both common and diverse in pholcids, and they may be involved in several adaptive functions. The straightforward function for males' pedipalp movements is sperm transfer, though the same action may be used by males to stimulate females as a form of genitalic copulatory courtship (Eberhard 1991, 1996, 2009, 2011). Female stimulation, for example, may induce the female to use the current male's sperm for fertilization instead of previous mating males' sperm (Calbacho-Rosa and Peretti 2015). The mechanisms behind this process may be the ejection of rival sperm by male pedipalp movements during the copulation (sperm removal) or even during or after copulation by an active role of the female (sperm-dumping) (Eberhard 1991, 2009, 2011; Huber and Eberhard 1997; Schäfer and Uhl 2002). In pholcid Neotropical species, this behavior has been observed for *Mesabolivar globulosus* (Nicolet, 1849) (F. Cargnelutti, L. Calbacho-Rosa, and A. V. Peretti, *pers. obs.*), and in cosmopolitan species such as *Holocnemus pluchei* (Calbacho-Rosa et al. 2013), *Pholcus phalangioides* (Schäfer and Uhl 2002), and *Physocyclus globosus* (Huber and Eberhard 1997).

Additionally to genital courtship, males perform courtship-like non-genital behavior during copulation. Such behaviors include biting, tapping, rubbing, squeezing, leg shaking, vibrating the abdomen, and feeding the female (Eberhard 1991, 1994, 1996). During copulation of *Anopsicus zeteki*, the male usually contacts the female legs while performing abdominal movements that consist of short highly rhythmical bursts (Huber 1998b). In *Coryssocnemis viridescens*, males also perform irregular up-and-down movements with their abdomens (Huber 1998b). In *Modisimus guatuso* Huber, 1998, males perform rhythmic movements with their legs and abdomens during copulation, while both sexes tap on their partners with their anterior legs. The modified hairs on the male chelicerae contact the female epigynum at this stage, and may provide the male with information regarding his position towards the female, or can also function to stimulate the female (Huber 1998c). In *Mesabolivar delclaroi* and *Mesabolivar globulosus*, the bristles located in the genital male palpal cymbia rub the female's abdomen whenever she moves (A.V. Peretti and L. Calbacho-Rosa, pers. obs.; Stefani et al. 2012).

Another non-genitalic behavior is stridulation, which is used by *P. globosus* females as a communicatory signal to modulate the male copulatory movements. A strong positive correlation between female stridulation and male palpal squeezing during copulation has been found (Peretti et al. 2006). Finally, a special case of non-genitalic contact structure is the frontal lobe in the pholcid male *Modisimus culicinus* (Simon, 1893). Clypeal glands open at the lobe and during copulation the female mouth is in contact with the lobe, suggesting gustatory courtship (Huber 1997a).

Araneomorphae: Tetragnathidae

The orb-weaving family Tetragnathidae is a worldwide-distributed group of spiders that catch their prey by spinning horizontal orb-webs with an open hub near the water or on top of the vegetation (Levi 2005). They are known as 'long-jawed spiders', and they belong to a big group of spiders classified within the superfamily Araneoidea (Levi and Coddington 1983; Schütt 2000; Agnarsson 2004; Wheeler et al. 2016). This large clade is composed of other well-known families as Araneidae (orb-weavers), Theridiidae (cobweb weavers), and Linyphiidae, among others (Hormiga et al. 1995; Scharff and Coddington 1997; Álvarez-Padilla et al. 2009; Dimitrov and Hormiga 2009, 2011; Gregorič et al. 2015; Dimitrov et al. 2016; Wheeler et al. 2016). All these families share the presence of a triplet of one flagelliform gland and two aggregate gland spigots on the posterior lateral spinnerets. Each flagelliform spigot produces an axial line that is coated by viscid glue produced by the aggregate glands (Wheeler et al. 2016). Tetragnathids are characterized by the absence of cribellum, the presence of paracymbium, simple male pedipalps with a conductor and embolus coiling together, and the absence of aciniform gland spigots on the median posterior spinnerets (Álvarez-Padilla and Hormiga 2011).

They are represented by 49 genera and 994 species (World Spider Catalog 2017), although its diversity may be underestimated. The family is also well known from Baltic amber mainly from the Cretaceous around 135 million years ago, and four genera have extinct and extant representatives: *Azilia*, *Cyrtognatha*, *Homalometa*, and *Tetragnatha* (Dunlop et al. 2016).

According to two main hypotheses, the internal relationships of the family are variable. Álvarez-Padilla et al. (2009) included three subfamilies (Tetragnathinae, Leucauginae and Metainae) plus the "*Nanometa*" clade. Recently, Dimitrov and Hormiga (2011) proposed a slightly different classification, with the same three subfamilies but adding Diphiinae. In the same year, Álvarez-Padilla and Hormiga (2011) discussed those results in terms of the validity of Diphiinae as a group.

The genitalia are highly diverse through the family. Tetragnathinae has reverted to haplogyne condition (genitalia with one single duct working as copulatory and fertilization site) from entelegyne ancestors (two different ducts serving for copulation and fertilization). Females may have one or a few central, membranous sacs and two spermathecae at both sides, entirely functional as in *Tetragnatha* or vestigial as in *Cyrtognatha* (Álvarez-Padilla and Hormiga 2011). Male pedipalps are simple, with only one tegular apophysis (the conductor) (Hormiga et al. 1995; Griswold et al. 1998). Variations are found in the shape of embolus, which may be tubular as in many Tetragnathinae, filiform, or lamelliform as in other groups (Álvarez-Padilla and Hormiga 2011). A synapomorphy of the subfamily Tetragnathinae is the presence of a median constriction in the cymbium, which may also have one or more processes.

Although they are interesting and abundant species, their behavior has been poorly studied. The reproductive strategies have been explored from different aspects, from morphology (Michalik et al. 2006; Álvarez-Padilla and Hormiga 2011; Cabra-García et al. 2014) to sexual selection (Aisenberg 2009; Aisenberg and Eberhard 2009; Aisenberg and Barrantes 2011; Méndez and Eberhard 2014). One interesting particularity of some genera such as *Leucauge*, *Tetragnatha*, and *Pachygnatha* is the contralateral insertion of the male pedipalps during mating: the right structures of the copulatory bulb are inserted into the left spermathecae, whereas the left structures are inserted into the right spermatheca (Huber and Senglet 1997; Aisenberg and Barrantes 2011; Álvarez-Padilla and Hormiga 2011).

Finally, some aspects of the biology of these spiders are very common in many species of the family. As mentioned above, they are normally found near the water or in environments with rivers surrounded by vegetation and rocks (Levi 2005). The activity of the spiders is concentrated during the sunset and night, although some groups are diurnal. In some cases, they remain hidden, helped by their mimetic coloration, with their legs extended longitudinally close to the body.

Pre-copulatory Behavior

Although the family Tetragnathidae has received a lot of attention in taxonomic and phylogenetic studies, only a few studies have focused on their reproductive biology and in Neotropical species. One of such species is *Tetragnatha elongata* Walckenaer, 1841. In this spider, courtship seems to be very subtle. Males position themselves at the edge of the female's web and tap the silk for a few seconds and rest, then they repeat this behavior until the females respond by vibrating their webs. If these vibrations are slow rhythmic pulses, the male approaches the female and the couple clasp their cheliceral fangs assuming a ventral-to-ventral mating position. Finally, the males use their third leg pair to push the females into mating position (Danielson-François et al. 2002). Apparently the brief courtship is a widespread behavior of the genus (Danielson-François et al. 2002; Danielson-François and Bukowski 2005;

Cargnelutti et al. 2015). Notice that although *T. elongata* inhabits also the Neotropics (Central America, Cuba, Jamaica; World Spider Catalog 2017), this study was performed with a North-American population.

In *Leucauge mariana* (Taczanowski, 1881) and *L. argyra* (Walckenaer, 1841), the courtship is more complex than in other species of the family. In *L. mariana* the courtship includes jerking, rocking, abdomen bobbing, palpal rubbing, twanging, line tapping, and tapping on the female. As in *Tetragnatha*, the female responds by orienting to the male and adopting the mating posture. Finally, the couple engage their cheliceral fangs, with the female clasping male chelicerae by closing her fangs (Huber and Eberhard 1997; Aisenberg 2009, Aisenberg and Eberhard 2009). On the other hand, the courtship in *L. argyra* includes jerking, burst of palpal rubbing, burst of twanging, tapping the web, tapping on the female, and foreleg rubbing. The couple interlocks their chelicerae as has been described for *L. mariana* (Aisenberg and Barrantes 2011). Aisenberg (2009) explains that those males of *L. mariana* who performed leg tapping on the female in higher frequencies and for a longer time increase their chances that females will collaborate in producing a genital mating plug. These results show that females choose males taking into account their performance during courtship, but this may not occur in other species such as *T. elongata*. However, this may be difficult to test because female choice can occur after copulation in a cryptic way (Danielson-François et al. 2002).

On the other hand, there are only a few studies focused on other aspects of precopulatory selection such as male–male competition. Field observations in *T. elongata* show that males are able to fight near the female's web for the access to copulate. However, it is not clear whether the female is selecting or merely accepting winners (Danielson-François et al. 2002). This behavior has been also reported from field observations in *Tetragnatha argentinesis* Mello-Leitão, 1931 and *Tetragnatha longidens* Mello-Leitão, 1945 (A. Aisenberg, pers. com.). It has been proposed that longer, sexually dimorphic legs in males of *L. mariana* may be used during male–male competition, although other functions such as adaptation to wandering life is also possible (Aisenberg 2009). In any case, more studies are needed to determine if some behaviors of male–male competition are under the influence of sexual selection.

Post-copulatory Sexual Selection

Research focused on post-copulatory sexual selection in Tetragnathidae has been more numerous, both in processes of sperm competition (Eberhard and Huber 1998; Aisenberg 2009; Aisenberg and Barrantes 2011; Méndez and Eberhard 2014; Danielson-François and Drobot 2016), and cryptic female choice mechanisms (Eberhard 1994; Aisenberg and Eberhard 2009; Aisenberg 2009; Aisenberg and Barrantes 2011; Barrantes et al. 2013; Aisenberg et al. 2015).

Males of *Leucage mariana* (Mexico, Hispaniola to Peru; World Spider Catalog 2017) have the ability of developing genital plugs, as long as females collaborate with them by producing and supplying their own substances, complicating copulation success for an upcoming male (Eberhard and Huber 1998). In order to counteract these strategies, males of *L. mariana* have developed different tactics such as hooking the plug and pulling it, breaking and penetrating the plug, or dissolving the plug by injecting substances below it. All these possibilities involve the use by males of a hook-shaped structure of the conductor of the pedipalps (Méndez and Eberhard 2014).

Danielson-François and Drobot (2016) reported that males of *Tetragnatha elongata* (North, Central America, Cuba and Jamaica; World Spider Catalog 2017) during mating scrape their pedipalps across female genitalia, with hematodochae inflation, but without the corresponding embolus insertion. These types of behaviors, called 'flubs', have been described as "mistakes" when males try to insert their pedipalps, but according to the authors this behavior could have the function of removing masses of sperm from previous rival males. Finally, males of *L. mariana* and *L. argyra* (USA to Brazil; World Spider Catalog 2017) perform oversight of penultimate females, which means that they guard subadult females that are close to molt, increasing their chances of paternity (Aisenberg 2009; Aisenberg and Barrantes 2011).

Copulatory courtship has been reported in different species of tetragnathids. In particular, males of *L. argyra* and *Leucauge* sp. perform a gentle tapping with legs I, II, and III on the female abdomen while they remove one palp from the epigynum and before inserting the other palp (Eberhard 1994). In *L. mariana*, tapping is also common while the embolus is inserted (Eberhard 1994). In turn, females of *L. mariana* seems to bias their collaboration in the production of genital plug to those males that execute more rhythmic thrusts with their frontal legs on the legs of the female, and perform shorter palpal insertions (Aisenberg and Eberhard 2009). Moreover, if during the first copulation males perform a low number of short insertions and/or commit numerous flubs during palpal insertions, females will be more predisposed to accept second matings with other males (Aisenberg 2009).

In an experiment conducted in *L. mariana,* the setae present in the chelicerae of males and females were shaved, and in both cases the individuals were exposed to normal sexual partners, evaluating the responsiveness to rematings and genital plug formation (Aisenberg et al. 2015). The absence of stimulation by male setae, as well as the absence in the female of the ability to receive stimulation in the chelicerae, reduced the chances of male paternity. When cheliceral setae were modified, females accepted subsequent copulations, the formation of genital plugs decreased, and the interruptions of copulations increased (Aisenberg et al. 2015). According to Aisenberg and Barrantes (2011), females of *L. argyra* determine copulation duration by forming genital plugs, which function as adhesive traps for males. Males get stuck to the plug substance and can even end up cannibalized by females. Finally, Cargnelutti et al. (2015) observed the presence of a white substance on the epigynum of *T. longidens* (Argentina, Brazil) and *T. argentinensis* (Argentina) (World Spider Catalog 2017) after copulation, which disappeared after 24 h. These authors suggest that they could be new cases of female sperm-dumping; however, further studies are necessary to test this hypothesis.

Araneomorphae: Lycosidae

The family Lycosidae is composed by 2418 species (World Spider Catalog 2017), subdivided into 11 subfamilies: Allocosinae, Artoriinae, Evippinae, Lycosinae, Pardosinae, Piratinae, Sosippinae, Tricasinae, Venoniinae, Wadicosinae, and Zoicinae (Piacentini 2014). Our knowledge about the reproductive aspects of Neotropical representatives of the Lycosidae family suffers from the (common) paradox of knowing so much about only a few species. Three subfamilies deserve special attention: Lycosinae, Allocosinae, and Sosippinae. Most of the investigations concerning Lycosidae reproductive biology were performed using biological models from these groups, thanks to the efforts of researchers from Brazil and Uruguay, but mostly by researchers from the *Instituto de Investigaciones Biológicas Clemente Estable* (Montevideo, Uruguay).

Lycosinae

Lycosinae is the richest Lycosidae subfamily, with ca. 60% of the described species (World Spider Catalog 2017). For the Neotropics, the species *Schizocosa malitiosa* (Tullgren, 1905) has been used as a model system from back in the 1970s up to today, following the theoretical advances throughout the time.

An interesting characteristic of the reproductive biology of *S. malitiosa* is the complexity of the copulatory stage, which may be classified into two patterns (Costa 1979; Costa and Sotelo 1994; Costa and Toscano-Gadea 2003). The first pattern consists of a series of repeated insertions with a single palp, followed by a change of side, repeated insertions with the other palp, and so on. The second pattern is characterized by single insertions, alternating between both palps. In terms of insertion contribution, the first pattern represents 82% of total insertions, while the remaining 18% occurs during the second pattern (Costa 1979). Experimental manipulations demonstrated that sperm transfer does occur during both insertion patterns (Costa and Toscano-Gadea 2003), without differences in terms of progeny production.

Such discrepancies, associated with the long duration of the copulatory stage, led researchers to suspect the existence of different roles for each insertion pattern, roles extending beyond sperm transfer. A leading hypothesis was the stimulatory role of insertions for cryptic female choice (Eberhard 1996). Aisenberg and Costa (2005), in an elegant experimental approach, demonstrated that *S. malitiosa* females' sexual receptivity is strongly inhibited by the presence of sperm and associated fluids within their reproductive tracts. With regard to the associated fluids transferred together with the sperm, Aisenberg and Costa (2005) proposed that they acted as receptivity-inhibiting substances (RIS), based on the evidence provided by themselves but also on evidence of RIS occurrence in insects (Gillot 2003). Moreover, Aisenberg and Costa (2005) suggested that cryptic female choice would be triggered only after previous sperm transfer, i.e., that the stimulation obtained through insertions would be insufficient to entice cryptic female choice in this species. Further experiments demonstrated that the first insertion pattern has a stronger influence over the female's reluctance induced by RIS (Estramil and Costa 2007; Aisenberg et al. 2008), supporting the hypothesis of multiple roles for different insertion patterns.

Aisenberg et al. (2008) cleverly noticed a difference in reporting the inducing of a reproductive refractory state in the females mediated by RIS, instead of attributing an active control of the mating process to one of the sexes. From the perspective of males, the inhibition promoted by RIS could be seen as a result of reproductive strategy against polyandry, which theoretically would be a case of sexual conflict, as proposed by Arnqvist and Rowe (2005). Alternatively, RIS effects could be a consequence of a female's mate choice based on the quality or amount of RIS transferred, resulting in a cryptic female choice strategy as proposed by Eberhard (1996), favoring females in their mate-choice preferences.

Another important source of evidence provided by the studies using *S. malitiosa* takes into account the role of female draglines for male mate-search decisions (Baruffaldi and Costa 2010; Baruffaldi et al. 2010). Males of *S. malitiosa* can discriminate the sexual status of females based solely on the information available in their draglines (Baruffaldi and Costa 2010), which would be the source for selection of males able to better process such information. Such signaling was attributed to a pheromone deposited by females, in order to improve mate search by the males (Baruffaldi and Costa 2010). Moreover, this mate search-enticing pheromone becomes inactive over time in natural conditions, allowing males to discriminate recent from old draglines, which may result in optimal energy expenditure during mate search (Baruffaldi et al. 2010).

Allocosinae

In this section, we will focus on two cases from the Allocosinae subfamily, the Neotropical species *Allocosa senex* (Mello-Leitão, 1945) and *Allocosa marindia* Simó, Lise, Pompozzi and Laborda 2017. These are sympatric and synchronic species, distributed across Uruguay, Argentina, and Brazil (Capocasale 1990; Simó et al. 2017), which are adapted to live in sandy coasts of different water bodies (Jorge et al. 2015). They are whitish, a quality that allows them to be camouflaged within the environment they inhabit. Their phenology is adjusted for the summer season, when they are active during the night. When individuals of both species are inactive, they take refuge within their tube-like burrows. These are retreats with silk-lined walls dug in the sand (Costa 1995).

Although they share many behavioral characteristics common to spiders of the family, such as maternal care during the entire development of the nymphs and during the few days after their emergence (Dondale 1986; Piacentini 2014), both species show a reversal in sexual size dimorphism; females are smaller than males (Aisenberg et al. 2007; Aisenberg 2014). Another important pattern is the reversal of typical sex roles performed by males and females (Aisenberg et al. 2007;

Aisenberg 2014). Females of *A. senex* are the mobile sex which searches for males, probably detecting the presence of male volatile pheromones (Aisenberg et al. 2010), and which initiates courtship. Copulation of both species of *Allocosa* occur inside males' burrows, contrasting with most Neotropical lycosids studied, in which individuals generally copulate on more exposed areas above the ground (Costa 1975; Costa and Capocasale 1984; Sordi 1996; Stefani et al. 2011; Costa and González 2015), or rarely on silk (Capocasale 1982; González et al. 2013).

The copulation proceeds by a sequence of mounts performed by the male over the female, up to approximately nine mounts and dismounts (Aisenberg et al. 2007). During each mount, one or more palpal insertions occur alternately, and each insertion consists of numerous ejaculations, similar to reports on other species of the sub-family Lycosinae (Costa and Capocasale 1984; Costa and González 2015) and Sossipinae (Capocasale 1982; Aisenberg et al. 2011). After the last mounts, the male leaves its burrow and block its entrance by covering it with silk and sand, completely hiding the entrance for potential visual predators. One important aspect of this blocking process takes into account the necessary participation of the female, which helps by adding silk from inside. This collaborative action may be interpreted as a shared reproductive effort from both sexes, probably selected for increasing the success rate of oviposition and hatching of the nymphs. The females only leave the burrow when the offspring are ready to disperse (Postiglioni et al. 2008), though they remain attached over the female's abdomen during the first days after leaving the burrow.

Most of the sexual selection studies involving Allocosinae have focused on the role of pre-copulatory mate choice over the reproductive strategies of the species. *Allocosa* is probably the Neotropical model with the best description about pre-copulatory mutual mate choice (Aisenberg et al. 2007, 2009, 2011). From the perspective of females, mate choice is based on burrow quality, an extended male trait also interpreted as a fundamental resource transferred to the female in order to proceed with the further steps of the reproductive process of the species. In this sense, the burrow may be classified as a nuptial gift provided by the male to the female (see below other cases of nuptial gift offering in spiders). Thus, females prefer males that offer the largest burrows, which will serve as a refuge during mating, and will be the offspring nest.

From the perspective of the males, mate choice is based on both the mating status of the females (virgin or mated) and on their physiological condition, which are generally assessed by the evaluation of body condition (Aisenberg et al. 2009). It is possible to assume a heuristic approach for such male mate choice, where the reproductive status may be used as a first criterion for selection, followed by the body condition criteria, which could define the amount of investment exerted in the process. At least for *A. senex*, if the females do not fulfill the male requirements, an extreme sex role reversal occurs in the form of sexual cannibalism of the female by the male, i.e., the unsuccessful courting female may become a meal to the male (Aisenberg et al. 2011).

The mutual mate choice has a strong influence over those traits targeted for selection by the preferences of the opposite sex. For example, male reproductive effort during burrow construction seems to be associated to the availability of partners and to the degree of preference exerted by them, and male decisions related to reproductive investment may be adjusted in accordance with female availability. Carballo et al. (2017) found that *A. senex* males rejected during an encounter extend their burrows more often than males with no exposure to females, i.e., burrow extension is a function of adjustment to female availability and not a consequence of burrow maintenance. In addition, females tend to accept more promptly males that have enlarged their burrows in relation to their first encounter.

Another striking aspect of the reproductive behavior of these species is the occurrence of sex role reversal during copulatory courtship observed in *A. senex*, consisting in body movements performed by the females (body shakes) during copulation. The intensity of these shakes, or at least the frequency of them, has an inverse correlation to the latency to ejaculation by the males (Garcia-Diaz et al. 2015), i.e., an increase in the number of shakes anticipates the ejaculation process. This finding shed some light on the role of intersexual copulatory communication, where this copulatory courtship could be acting as a signal to motivate palpal insertion and ejaculation, and/or inhibit cannibalistic tendencies of males in this species.

Some other aspects concerning the reproductive biology of *Allocosa* seem to be very promising. First, the role of mutual mate choice within *Allocosa* may not be excluded as a driver for species isolation. For example, Bollatti et al. (2017) found that the sexual behavior of *A. senex* shows a pattern of geographical variation that could lead to divergences in the traits involved in mate choice. Given this, the *Allocosa* model may offer a unique opportunity to investigate the role of sexual selection over the diversification of species. Second, it was observed that in *A. marindia* the sex ratio is biased towards females, and such a pattern could be associated with infection by the endosymbiotic bacteria *Wolbachia* (Lerette et al. 2015). This bacterium is able to manipulate the reproductive development of its hosts towards the production of females, in order to increase its own transmission (Charlat et al. 2003; Duron et al. 2008; Goodacre and Martin 2013). Such bias towards females has unknown consequences over the emerging reproductive strategies in the infected populations, so this field seems very promising for further studies.

Sosippinae

Sexual selection studies involving the Neotropical Sosippinae species have mostly been based on *Aglaoctenus lagotis* (Holmberg, 1876). This species belongs to a group of lycosids that abandoned the cursorial habit to build and live in a funnel-web, from where it performs its life-history activities (González et al. 2014). Compared to other lycosid models, our knowledge about the reproductive biology of *A. lagotis* is relatively broad given the amount of available papers. For example, Stefani et al. (2011) provide a description of the mating behavior, male mate choice abilities, and the consequences of maternal care upon offspring survival. Moreover, González (2015) brought important evidence about the phenotypic variability of the sexual behavior of the species based on inter-populational comparisons (see also González et al. 2013, 2014, 2015a, b).

Stefani et al. (2011) described the main stages of the reproductive behavior of *A. lagotis* males based on samples from the Southeast region of Brazil. Their study is based on the premise that male reproductive success results from the interaction of two independent abilities: the capacity to process female cues during mate search and mate choice stages, and the ability to properly stimulate a female during courtship (Stefani et al. 2011). Mechanistically, males seem to perform mate choice based on chemical cues found within the female funnel-web, preferring webs of virgin females instead of recently mated females. Once a female has been found, the reproductive process consists of three stages. The first stage consists of a seismic courtship performed by the male over the female's funnel-web, eliciting a female approach in case of receptivity. The courtship is followed by the second stage, the pre-copulation, which consists in brief moments of direct contact between the male's forelegs and the female's hindlegs. The process ends with copulation, where the males perform a single insertion with each palp, but with several haematodochal expansions in each insertion. Stefani et al. (2011) observed a low frequency of sexual cannibalism towards the males, but without exploring the adaptive consequences of it.

Across the wide distributional range of *A. lagotis*, ranging from Colombia to Argentina (World Spider Catalog 2017), it is possible to observe a conspicuous morphological variation among populations (Santos and Brescovit 2001). Moreover, González et al. (2013) showed that such morphological variability also applies for the reproductive biology of *A. lagotis*. These authors observed differences in the reproductive behavior in terms of duration of specific behavioral units, as well as the occurrence of exclusive units for morphologically divergent populations. Further studies developed by the same research team have demonstrated a phenological isolation among those populations, which in turn results in a temporal reproductive isolation mechanism (González et al. 2014). However, phenology seems to be acting in concert with other isolation mechanisms, as described by González et al. (2015a). These authors observed the occurrence of behavioral isolation between two "forms" of *A. lagotis* (Central Argentina form vs Southern Uruguay form), independent from whether the samples of each form were from allopatric or sympatric populations. Curiously, González et al. (2015a) observed an asymmetry in terms of male courtship towards a heteromorphic female for the sympatric populations, i.e., while the males of one form do not court a heteromorphic female, 80% of the males of the opposite form court a heteromorphic female. Divergence between Argentinian and Uruguayan forms seems to be applied to their mating systems as well, since one of the forms tend to show higher polyandry levels than the other (González, Costa, Peretti, pers. com.).

As we mentioned before, we know several aspects about the reproductive biology of *A. lagotis*, but at the same time this evidence may become sparse if eventually the available data was obtained from a group of cryptic species. The observed patterns from the comparisons involving the Argentinean and the Uruguayan populations could be sufficient for considering them as taxonomic independent entities, waiting for a detailed taxonomic screening searching for morphological cryptic species. Moreover, despite the interpretative emphasis on the role of natural selection upon the observed differences, the *A. lagotis* model may provide important evidence helping to understand the role of sexual selection in the diversification of the group. It would be interesting if further studies could take into account whether the observed variation in male reproductive behavior has some correlation with female mate choice preferences.

Araneomorphae: Trechaleidae: Paratrechalea

Nuptial gift-giving behavior is a sexual strategy broadly defined as the transfer of materials or substances, other than gametes, from one sex to another during any step of the reproductive process in order to improve donor fitness [Vahed (1998) and Lewis et al. (2014) provide a definition that avoids any assumption about gift effect over recipient fitness]. Strong emphasis has been devoted to the analysis of nuptial gifts in insect models, a contingency strictly based on the amount of empirical evidence available. Most probably, because nuptial gifts are less frequent in other taxa, few vertebrates (Mougeot et al. 2006) and arachnids (reviewed in Nitzsche 2011, and Albo et al. 2014a) were considered during the emergence of studies about this remarkable sexually selected trait. Here we will focus on the adaptive interpretations of nuptial gifts within spiders, and how recent studies in new described models involving Neotropical spider species may help us to develop an integrative theoretical rationale about nuptial gifts as an evolutionary stable strategy. Our goal is to offer comments about this reproductive strategy beyond the usual comparison among the known cases. By doing this, we intend to present arguments connecting emerging patterns that become only clear after the description of Neotropical models.

Spiders' Nuptial Gifts, with Emphasis on Prey Gifts

Taking the broad definition of nuptial gifts provided above, the described cases of spiders' nuptial gifts in general follow two patterns. First, as in the great majority of all other animals, only males provide the donations [see Vahed (1998) for cases of female offering]. Second, and except for one case (spiders from the genus *Allocosa*, Aisenberg et al. 2007, 2009, 2011), spiders' nuptial gifts have been classified as nuptial feeding, since the donations are assimilated through the female's oral tract (reviewed in Albo et al. 2014a).

Among the existing nuptial gift classes, nuptial prey gifts are one of the most important contributions from studies in Neotropical models. This particular type of nuptial gift was initially described as a prey wrapped in silk that males offer to females during the courtship. Nowadays, it is known that males can also offer inedible items, such as prey leftovers or plant parts, so-called worthless gifts (Albo et al. 2011a).

Over a long period of time, prey gifts in spiders were recorded only for the Pisauridae family (Nitzsche 2011), especially focusing on the Palearctic species *Pisaura mirabilis* (Clerck, 1757). Hence, the Neotropical spider species of the *Paratrechalea* genus (Trechaleidae family) (Costa-Schmidt et al. 2008; Nitzsche 2011) appear to be good candidates for performing comparative studies. In this sense, this reproductive strategy acquired more attention, transforming what was considered an exception into a behavioral trait with phylogenetic information (Albo et al. 2017).

Most of the existing studies in *Paratrechalea* provide direct and/or indirect evidence for the association of prey gifts with pre-copulatory and copulatory steps of the reproductive process (Costa-Schmidt et al. 2008; Albo et al. 2009; Brum et al. 2012; Klein et al. 2012, 2014; Trillo et al. 2014), while a few studies indicate indirect evidence supporting spiders' prey gift participation during post-copulatory sexual selection processes (Albo and Peretti 2015; Costa-Schmidt 2015).

Prey Gift Importance During the Pre-copulatory Phase

As in all secondary sexual traits, there is a close connection of nuptial gifts with both inter- and intrasexual selection. This is because females will choose males, in part, based on this trait, while males will use it to compete among themselves for accessing available mates. Pre-copulatory intrasexual competition usually leads to the expression of different strategies, which helps males to increase their access to reproductive opportunities. In the case of the gift-giving behavior, males can engage in physical fights independently of the gift (Nitzsche 2011), but they can also compete indirectly to access females by being faster and more effective during courtship. In other words, males benefit if they are ready to offer a gift and start courtship as soon as they find a female. This interpretation is supported by empirical evidence showing that males from *Paratrechalea ornata* (Mello-Leitão, 1943) redirect their foraging effort towards reproduction once they encounter female cues. Fast gift construction favors males to be the first ones courting. Such pressure drives males to even start silk-wrapping before they contact the female, when they detect pheromones from females' silk in the substrate (Albo et al. 2009), a pattern also suggested for the Palaearctic model *Pi. mirabilis* (Lang 1996; Albo et al. 2011b).

In parallel, as gift-carrying males are preferred over males without gifts (Albo and Costa 2010), they will also benefit from attracting more females while they court. In this context, direct evidence taken from insects supports the importance of sensory exploitation as an evolutionary path for the emergence of such traits (Vahed 2007; Warwick et al. 2009). It has been commonly suggested for the Palearctic spider *P. mirabilis* that the nuptial gift exploits foraging motivation of females (Bilde et al. 2007). There is still lack of evidence for this hypothesis in the Neotropical species *Pa. ornata*, but it has been shown that silk-wrapping confers advantages to males by attracting females. This attraction would occur via chemical substances associated with the silk of the gift, such as pheromones or aphrodisiac substances (Brum et al. 2012), as well as through visual signals, since the silk turns the gift into a white package reinforcing this function (Trillo et al. 2014). Thus, the prey gift itself may be interpreted as a multimodal signal structure. Additionally, once the female accepts and grabs the gift, both sexes hold it with their chelicerae, resulting in a peculiar situation for mutual mate choice based on multimodal signaling. Female mate choice during the pre-copulatory phase results in a close-range interaction between the sexes, mediated mostly by the physical structure of the prey gift. In summary, at this phase the gift in this species is an important trait mediating attraction and evaluation of the partner.

Prey Gift Importance During the Copulatory Phase

The prey gift role during mating can also be visualized from two aspects: mating position and sperm transfer. Similarly to the Palearctic species *P. mirabilis*, in the Neotropical *Pa. ornata* the gift is held by the female within her chelicerae during the entire mating process, which includes a courtship phase when the male is also grabbing the gift, and a sperm-transfer phase when the male releases the gift from his chelicerae and mounts the female (Costa-Schmidt et al. 2008). During the copulatory courtship phase, males and females simultaneously hold the prey gift, which restrains male and female position. This allows sexes to be in close contact, exchanging tactile and chemical signals and probably evaluating each other. As in most courtships in *Pa. ornata*, the intensity and the amount of time invested in such phases seem to be highly variable. During the sperm-transfer phase, the female still holds the gift in her chelicerae, while the male holds the gift with the third pair of legs, releasing the prey gift from his chelicerae and mounting over the female. This subtle copulatory pattern raises several questions concerning the role of the third leg pair during sperm transfer. Since the end of copulation is commonly determined by the female, it may be possible for males that by grabbing the gift in this way, they can hold it avoiding female gift stealing during mating. In addition, males would potentially evaluate the female's behavior, allowing them to infer the female's level of aggressiveness. Even though sexual cannibalism is low in this species (Albo and Costa 2010), spiders are voracious predators and it is in the male's interests to perceive female behavior that can be risky for them. Finally, another possibility is that by holding the prey gift with the third leg pair, males properly maintain the mating position. None of the three ideas have been tested for the Neotropical *Pa. ornata*, but there is some evidence for the Palearctic *Pi. mirabilis*. In this latter species, it has been shown that gift size and shape influence the duration of sperm transfer (Lang 1996; Stålhandske 2001; Andersen et al. 2008).

Prey Gift Importance During the Post-copulatory Phase

Post-copulatory selection is directly related to polyandrous species, a situation where the males must cope with sperm competition and/or cryptic female choice. Polyandry is the rule among gift-giving spiders mainly because females can acquire direct nutritive benefits from multiple gifts (Toft and Albo 2015). Only recently, indirect evidences linking prey gifts with cryptic female choice became available (Albo et al. 2013; Albo and Peretti 2015). In the Neotropical *Pa. ornata* there is a positive correlation between mating duration and sperm stored by females (Albo and Peretti 2015). There is no evidence that the gift itself is favored via post-copulatory processes, but there are indications that eventually males offering non-nutritive or worthless gifts would be disfavored in the amount of sperm stored (Albo and Peretti 2015). On the other hand, there is a clear lack of information with regard to how prey gifts could be associated to sperm-competition processes in both the Palearctic and the Neotropical species, but offering a speculative view, it could probably be associated with decisions taken by the male during the copulatory phase. Even though there is no evidence of sperm depletion, the reproductive investment of males can be significant, as it starts during gift construction, and follows during sperm transfer. It is known that male investment is affected by body condition, poorly fed males being limited in the amount of silk used (Trillo et al. 2014), and this behavior can be restricted even after males reach a good physiological condition (Macedo-Rego et al. 2016). However, understanding whether prey gift quality has some correlation to the amount/quality of the sperm transferred by males is an open line of inquiry. One possible hypothesis could be that there is a negative correlation between these variables: high-quality prey gifts allow males to transfer less sperm and invest most of the copulation time in stimulatory movements in order to influence cryptic female choice.

Adaptive Functions of Prey Gifts

Great effort has traditionally been placed on the adaptive interpretation of prey gifts, probably due to their intuitive direct benefit appeal in relation to other sources of nuptial gifts. There are three main hypotheses concerning the adaptive role of prey gifts within arthropods (Vahed 1998): the parental-effort hypothesis, the mating-effort hypothesis, and the sexual cannibalism defense hypothesis. These ideas are not mutually exclusive; for instance, while mating effort could be acting during the early stages of reproduction (e.g., during mate choice), paternal investment may be acting afterwards, if the nuptial gift has a nutritive value for female and offspring fitness. Here, we present an up-to-date scenario of the available evidence for the Neotropical *Pa. ornata* in comparison with the Palaearctic spider *Pi. mirabilis*. In addition to not being mutually exclusive, a general pattern shows that most of the available information points towards the mating-effort hypothesis (Albo and Costa 2010; Albo et al. 2014b; Trillo et al. 2014).

Assuming that some substances donated by males are nutritive for females (Boggs 1995), the gift can supply important food resources, increasing female fecundity or offspring fitness, and representing a type of paternal investment (Thornhill 1976a; Gwynne 1984; Simmons and Parker 1989). Studies in *Pa. ornata* have failed to show any positive effect on female fecundity or longevity acquired from the gift, a situation also applying to studies in the Palearctic spider. However, recent studies in this last species have verified that females can access fitness benefits when they get multiple food gifts (Prokop and Maxwell 2009; Toft and Albo 2015). In fact, it is predicted that the positive effects of direct benefits may be much more expected in situations when females are under a stressful physiological regime (food-deprived, for example), where the prey gift may improve female's condition and consequently her investment in reproduction.

However, the paternal-investment hypothesis becomes weak in species in which empirical evidence showed that gifts are not nutritive for females (Warwick et al. 2009; Gershman et al. 2012). In this scenario, males can manipulate mating time, increasing the amount of sperm transferred, inducing female remating refractory period, and accelerating latency of oviposition to maximize their mating effort (Thornhill 1976b; Simmons and Gwynne 1991; Eberhard 1996; Wolfner 1997; Heifetz et al. 2001; Arnqvist and Rowe 2005; Sakaluk et al. 2006). These aspects seem to be present in the spider *Pa. ornata,* giving support to the hypothesis in this species.

Finally, in predator species it has been suggested that the gift can act as a shield protecting males from sexual cannibalism (Kessel 1955, Bristowe 1958). This hypothesis generally lacks empirical support. In *Pa. ornata*, authors rule out the notion that prey gifts have any connection to sexual cannibalism avoidance (Albo and Costa 2010), but there is no clear evidence to discard it. In fact, we cannot exclude the possible influence of cannibalism avoidance during the early stages of gift-giving behavior evolution, as suggested by Bilde et al. (2007). Moreover, the importance of prey gifts as a source of defense against cannibalism has recently been demonstrated by Toft and Albo (2016) for the Palaearctic *Pi. mirabilis*.

Prey Gifts Modulating Alternative Reproductive Tactics

The sexual selection framework provides an outstanding potential for the evolution and expression of alternative polymorphic traits and the subsequent co-evolution of preferences. This is because the variance in reproductive success among individuals promotes wide and diverse evolutionary strategies to gain fitness advantages (Andersson and Simmons 2006). Alternative reproductive tactics (ARTs) are known in many taxa (Taborsky 1994; Gross 1996; Alonzo and Warner 2000; Oliveira et al. 2008; Fitzpatrick et al. 2016), being characterized by the emergence of different behavioral strategies among males to attract and mate with females. Such behavioral variation can appear as a genetic polymorphism with alternative tactics, or as a conditional strategy used by all individuals in the population (Neff and Svensson 2013).

Alternative reproductive tactics have been described for the Neotropical *Pa. ornata*, in which nuptial gift content can vary from nutritive to worthless items wrapped in silk (Albo et al. 2014b). In this species, deception by worthless gifts occurs at a very high proportion in the field (70%), contrasting with the Palearctic spider *Pi. mirabilis* (38% of worthless gifts, Albo et al. 2011a) and the known examples described in the literature (Neff and Svensson 2013). Studies on this topic are relatively new, and there is not much information with regard to the evolution of these alternative tactics. However, it seems that *Pa. ornata* behavior fits as a conditional strategy, which involves males switching from one tactic to another at certain thresholds. The male decision could be determined by features such as size, age, or aggressiveness, which are condition-dependent, and modeled by individual ability to get resources from the environment (Gross 1996; Neff and Svensson 2013).

Concluding Remarks

Sexual selection theory has solid foundations, but also has its own theoretical frontiers, challenging those who are deeply compromised in studying it. Not surprisingly, spiders as biological models are helping us to take some steps further into a better understanding of questions with regard to how the reproductive behavior of the individuals of a population can be influenced by the surrounding conditions. Here, we selected some representative Neotropical spider models to understand the emergence and adaptive maintenance of the observed reproductive patterns and/or strategies.

However, the same information provided throughout the sections of this chapter permits us to suggest the occurrence of an implicit pattern, one related to how scientific evidence is validated, especially when evidence refutes the established theory. This implicit pattern is related to the fundamental process of establishing a new biological model when basic information regarding the life history of the model provides the background for further investigations. This process is clearly present in all of the models described in this chapter, but with a slight variation among the taxonomic groups, which may be helpful if the involved strategies can be directly compared to other well-known models, usually outside the Neotropics. For example, the recording of nuptial gift-giving spiders from the Trechaleidae family in the late 2000s (Costa-Schmidt et al. 2008) was intensively investigated with regard to its basic assumptions, before shedding some light on the evolutionary emergence and adaptive maintenance of a reproductive strategy that was only known for the Pisauridae family (see Maria Jose Albo's papers on the subject). There were, however, several models where the researchers needed to start without any reference, becoming somehow the reference themselves (e.g., Anita Aisenberg's papers regarding the sexual strategy of *Allocosa* and *Leucauge*).

In terms of future prospects, many of the projections by Huber (2005) continue as interesting areas of investigation, which can be followed by the valuable synthesis of the entire process provided by Schneider and Andrade (2011). Nowadays, a lot of effort is being applied to understand how each stage of the reproductive process of the involved models influence each other, e.g., how courtship strategies may be connected to fertilization success. Great efforts have been made in this direction, though several important aspects remain as a "black box", one with huge consequences if the real behavior inside such black boxes diverges from our premises. A representative example is the assumption of minimal variation of the ejaculation processes by the males in experimental studies. Even though males may control the rate of sperm transfer by controlling the internal pressure over their palps, a constant rate of sperm transfer is often assumed. The consequences of such control would be tremendous, which could be linked to processes such as male cryptic choice.

Huber (2005) also points out our "myopic" efforts in understanding intersexual communication, by referring to the modest investments in investigation of the role of chemical and vibratory communication. Animal communication often relies on a multimodal structure with potential synergetic interactions among the involved sensory modalities. In addition to this intrinsic complexity of animal communication, it would be welcomed if researchers would invest their efforts equally on both signal producer and signal receptor. In terms of signal production, we assume that inter-individual variation is at the core of several reproductive processes, mainly mate-choice strategies. However, little is known about the inter-individual variation in terms of cognitive ability to access the available information, consisting in an almost unexplored venue that may help us to understand the emergence of mate-choice preferences and/or alternative mating strategies.

In addition to recognizing the existence of a natural variation among the specimens of a population, researchers have also found consistent correlations involving the decisions made by the specimens along the components of their life history. Such patterns are now being investigated within the domain of behavioral syndromes, which may provide us with additional evidence in order to understand the evolution and adaptive maintenance of sexual strategies.

As we mentioned before, this chapter is just an approximation of an extensive and inspiring subject, completely biased by our personal academic background. We just hope that our point of view may be useful in describing this specific moment of sexual selection studies involving Neotropical spider species, which somehow converge into the same descriptions made in other macroecological contexts (e.g., Schneider and Andrade 2011; Macedo and Machado 2014, and chapters therein).

Acknowledgements This compilation was only possible thanks to the work and help of several colleagues. It would be impossible to name all of them, and we extend our appreciation to all those who somehow shared their information with ourselves (professors, grad and undergrad students, independent researchers, etc.). We, as authors, take all responsibility for any inconsistency presented here. LECS received a post-doctoral grant from Capes/Brazil (PNPD 20132737); FB, FC, LCR, MI, SC, NF, and AVP received financial support from FONCYT, SECYT-UNC, and CONICET, Argentina.

References

Agnarsson I (2004) Morphological phylogeny of cobweb spiders and their relatives (Araneae, Araneoidea, Theridiidae). Zool J Linn Soc Lond 141:447–626

Aisenberg A (2009) Male performance and body size affect female re-mating occurrence in the orb-web spider *Leucauge mariana* (Araneae, Tetragnathidae). Ethology 115:1127–1136

Aisenberg A (2014) Adventurous females and demanding males: sex role reversal in a Neotropical spider. In: Macedo R, Machado G (eds) Sexual selection and breeding biology using Neotropical animals systems and models. Elsevier, Amsterdam, pp 163–182

Aisenberg A, Costa FG (2005) Females mated without sperm transfer maintain high sexual receptivity in the wolf spider *Schizocosa malitiosa*. Ethology 111:545–558

Aisenberg A, Eberhard WG (2009) Female cooperation in plug formation in a spider: effects of male copulatory courtship. Behav Ecol 20:1236–1241

Aisenberg A, Barrantes G (2011) Sexual behavior, cannibalism, and mating plugs as sticky traps in the orb weaver spider *Leucauge argyra* (Tetragnathidae). Naturwissenschaften 98:605–613

Aisenberg A, Viera C, Costa FG (2007) Daring females, devoted males, and reversed sexual size dimorphism in the sand-dwelling spider *Allocosa brasiliensis* (Araneae, Lycosidae). Behav Ecol Sociobiol 62:29–35

Aisenberg A, Estramil N, Toscano-Gadea C, González M (2008) Timing of female sexual unreceptivity and male adjustment of copulatory behaviour under competition risk in the wolf spider *Schizocosa malitiosa*. J Ethol 27:43–50

Aisenberg A, González M, Laborda A, Postiglioni R, Simó M (2009) Reversed cannibalism, foraging, and surface activities of *Allocosa alticeps* and *Allocosa brasiliensis*: two wolf spiders from coastal sand dunes. J Arachnol 37:135–138

Aisenberg A, Baruffaldi L, González M (2010) Behavioural evidence of male volatile pheromones in the sex-role reversed wolf spiders *Allocosa brasiliensis* and *Allocosa alticeps*. Naturwissenschaften 97:63–70

Aisenberg A, Costa FG, González M (2011) Male sexual cannibalism in a sand-dwelling wolf spider with sex role reversal. Biol J Linn Soc 103:68–75

Aisenberg A, Barrantes G, Eberhard WG (2015) Post-copulatory sexual selection in two tropical orb-weaving *Leucauge* spiders. In: Peretti AV, Aisenberg A (eds) Cryptic female choice in arthropods. Springer, Berlin, pp 79–108

Alberti G, Weinmann C (1985) Fine structure of spermatozoa of some labidognath spiders (Filistatidae, Segestriidae, Dysderidae, Oonopidae, Scytodidae, Pholcidae, Araneae, Arachnida) with remarks on spermiogenesis. J Morphol 185:1–35

Albo MJ, Bidegaray-Batista L, Bechsgaard J, Silva ELC, Bilde T, Pérez-Miles F (2017) Molecular phylogenetic analyses show that Trechaleidae and Lycosidae are sister groups. Arachnology 17:169–176

Albo MJ, Costa-Schmidt LE, Costa FG (2009) To feed or to wrap? Female silk cues elicit male nuptial gift construction in a semiaquatic trechaleid spider. J Zool 277:284–290

Albo MJ, Costa FG (2010) Nuptial gift-giving behaviour and male mating effort in the Neotropical spider *Paratrechalea ornata* (Trechaleidae). Anim Behav 79:1031–1036

Albo MJ, Winther G, Tuni C, Toft S, Bilde T (2011a) Worthless donations: male deception and female counter play in a nuptial gift-giving spider. BMC Evol Biol 11:329

Albo MJ, Toft S, Bilde T (2011b) Condition dependence of male nuptial gift construction in the spider *Pisaura mirabilis* (Pisauridae). J Ethol 29:473–479

Albo MJ, Bilde T, Uhl G (2013) Sperm storage mediated by cryptic female choice for nuptial gifts. Proc R Soc B 280:20131735

Albo MJ, Toft S, Bilde T (2014a) Sexual selection, ecology, and evolution of nuptial gifts in spiders. In: Macedo R, Machado G (eds), Sexual selection: perspectives and models from the Neotropics. Academic, London, pp 183–200

Albo MJ, Melo-González V, Carballo M, Baldenegro F, Trillo MC, Costa FG (2014b) Evolution of worthless gifts is favoured by male condition and prey access in spiders. Anim Behav 92:25–31

Albo MJ, Peretti AV (2015) Worthless and nutritive nuptial gifts: mating duration, sperm stored and potential female decisions in spiders. PLoS One 10(6):e0129453. https://doi.org/10.1371/journal. pone.0129453

Alonzo SH, Warner RR (2000) Female choice, conflict between the sexes and the evolution of male alternative reproductive behaviours. Evol Ecol Res 2:149–170

Álvarez-Padilla F, Dimitrov D, Giribet G, Hormiga G (2009) Phylogenetic relationships of the spider family Tetragnathidae (Araneae, Araneoidea) based on morphological and DNA sequence data. Cladistics 25:109–146

Álvarez-Padilla F, Hormiga G (2011) Morphological and phylogenetic atlas of the orb-weaving spider family Tetragnathidae (Araneae: Araneoidea). Zool J Linn Soc Lond 162:713–879

Andersen T, Bollerup K, Toft S, Bilde T (2008) Why do males of the spider *Pisaura mirabilis* wrap their nuptial gifts in silk: female preference or male control? Ethology 114:775–781
Andersson M (1994) Sexual selection. Princeton University Press, Princeton
Andersson M, Simmons LW (2006) Sexual selection and mate choice. Trends Ecol Evol 21:296–302
Arnqvist G, Rowe L (2005) Sexual Conflict. Princeton University Press, Princeton
Austad SN (1984) Evolution of sperm priority patterns in spiders. In: Smith RL (ed) Sperm competition and the evolution of mating systems. Harvard University Press, Cambridge, pp 223–249
Baerg WJ (1928) The life cycle and mating habits of the male tarantula. Q Rev Biol 3:109–116
Barrantes G, Aisenberg A, Eberhard WG (2013) Functional aspects of genital differences in *Leucauge argyra* and *L. mariana* (Araneae, Tetragnathidae). J Arachnol 41:59–69
Baruffaldi L, Costa FG (2010) Changes in male sexual responses from silk cues of females at different reproductive states in the wolf spider *Schizocosa malitiosa*. J Ethol 28:75–85
Baruffaldi L, Costa FG, Rodríguez A, González A (2010) Chemical communication in *Schizocosa malitiosa*: evidence of a female contact sex pheromone and persistence in the field. J Chem Ecol 36:759–767
Bertani R, Fukushima CS, Silva Júnior PI (2008) Mating behaviour of *Sickius longibulbi* (Araneae, Theraphosidae, Ischnocolinae), a spider that lacks spermathecae. J Arachnol 36:331–335
Bilde T, Tuni C, Elsayed R, Pekar S, Toft S (2007) Nuptial gifts of male spiders: sensory exploitation of the female's maternal care instinct or foraging motivation? Anim Behav 73:267–273
Birkhead TR (2010) How stupid not to have thought of that: post-copulatory sexual selection. J Zool 281:78–93
Boggs CL (1995) Male nuptial gifts: behaviour consequences and evolutionary implications. In: Leather SR, Haerdie J (eds) Insect reproduction. CRC Press, Boca Raton, pp 215–242
Bolatti F, Garcia Diaz V, Peretti AV, Aisenberg A (2017) Geographical variation in sexual behavior and body traits in a sex role reversed wolf spider. Sci Nat. https://doi.org/10.1007/s00114-017-1460-x
Bristowe WS (1929) The mating habits of spiders, with special reference to the problems surrounding sex dimorphism. Proc Zool Soc Lond 1929:309–358
Bristowe WS (1930) A supplementary note on the mating habits of spiders. Proc Zool Soc Lond 1930:395–413
Bristowe WS (1958) The world of spiders. Collins, London
Brum PED, Costa-Schmidt LE, Araújo AM (2012) It is a matter of taste: chemical signals mediate nuptial gift acceptance in a Neotropical spider. Behav Ecol 23:442–447
Bücherl W (1952) Aranhas do Rio Grande do Sul. Mems Inst Butantan 24:127–155
Burger M (2007) Sperm dumping in a haplogyne spider. J Zool 273:74–81
Burger M (2009) Female genitalia of goblin spiders (Arachnida: Araneae: Oonopidae): a morphological study with functional implications. Invertebr Biol 128:340–358
Burger M (2010) Complex female genitalia indicate sperm dumping in armored goblin spiders (Arachnida, Araneae, Oonopidae). Zoology 113:19–32
Burger M (2011) Functional morphology of female goblin spider genitalia (Arachnida: Araneae: Oonopidae) with notes on fertilization on spiders. Zool Anz 250:123–133
Burger M, Nentwig W, Kropf C (2003) Complex genital structures indicate cryptic female choice in a haplogyne spider (Arachnida, Araneae, Oonopidae, Gamasomorphinae). J Morphol 255:80–93
Burger M, Graber W, Michalik P, Kropf C (2006) *Silhouettella loricatula* (Arachnida, Araneae, Oonopidae): a haplogyne spider with complex female genitalia. J Morphol 267:663–677
Burger M, Michalik P (2010) The male genital system of goblin spiders: evidence for the monophyly of Oonopidae (Arachnida: Araneae). Am Mus Novit 3675:1–13
Burger M, Izquierdo M, Carrera P (2010) Female genital morphology and mating behavior of *Orchestina* (Arachnida: Araneae: Oonopidae). Zoology 113:100–109
Burger M, Carrera PC (2011) Copulatory behavior of the goblin spider *Silhouettella loricuatula* (Arachnida: Araneae: Oonopidae). Bull Br Arachnol Soc 15:173–178

Cabra-García J, Hormiga G, Brescovit AD (2014) Female genital morphology in the secondarily haplogyne spider genus *Glenognatha* Simon, 1887 (Araneae, Tetragnathidae), with comments on its phylogenetic significance. J Morphol 275:1027–1040

Calbacho-Rosa L, Córdoba-Aguilar A, Peretti AV (2010) Occurrence and duration of post-copulatory mate guarding in a spider with last sperm precedence. Behaviour 147:1267–1283

Calbacho-Rosa L, Moreno-García MA, Lanz-Mendoza H, Peretti AV, Córdoba-Aguilar A (2012) Reproductive activities impair immunocompetence in *Physocyclus dugesi* (Araneae: Pholcidae). J Arachnol 40:18–22

Calbacho-Rosa L, Galicia-Mendoza I, Dutto MS, Córdoba-Aguilar A, Peretti AV (2013) Copulatory behavior in a pholcid spider: males use specialized genitalic movements for sperm removal and copulatory courtship. Naturwissenschaften 100:407–416

Calbacho-Rosa L, Peretti AV (2015) Copulatory and post-copulatory sexual selection in haplogyne spiders, with emphasis on Pholcidae and Oonopidae. In: Peretti AV, Aisenberg A (eds) Cryptic female choice in arthropods. Springer, Berlin, pp 109–144

Capocasale RM (1982) Las especies del género *Porrimosa* Roewer, 1959 (Araneae, Hippasinae). J Arachnol 10:145–156

Capocasale RM (1990) Las especies de la subfamilia Hippasinae de América del Sur (Araneae, Lycosidae). J Arachnol 18:131–134

Carballo M, Baldenegro F, Bollatti F, Peretti AV, Aisenberg A (2017) No pain, no gain: male plasticity in burrow digging according to female rejection in a sand-dwelling wolf spider. Behav Processes 140:174–180

Cargnelutti F, Bollatti F, Peretti AV, Aisenberg A (2015) Comportamiento sexual en *Tetragnatha* sp.: ¿Un caso de expulsión femenina de esperma? I Reunión Argentina de Biología Evolutiva. Libro de resúmenes, p 78

Carranza J (2009) Defining sexual selection as sex-dependent selection. Anim Behav 77:749–751

Charlat S, Hurst GDD, Mercot H (2003) Evolutionary consequences of *Wolbachia* infections. Trends Genet 19:217–223

Copperi S, Ferretti N, Pompozzi G, Pérez-Miles F (2012) Can't you find me? Female sexual response in an Argentinean tarantula (Araneae, Theraphosidae). Rev Colomb Entomol 38:164–166

Costa FG (1975) El comportamiento precopulatorio de *Lycosa malitiosa* Tullgren (Araneae, Lycosidae). Rev Bras Biol 35:359–368

Costa FG (1979) Análisis de la cópula y de la actividad postcopulatoria de *Lycosa malitiosa* Tullgren (Araneae, Lycosidae). Rev Bras Biol 39:361–376

Costa FG (1995) Ecología y actividad diaria de las arañas de la arena *Allocosa* spp (Araneae, Lycosidae) en Marindia, localidad costera del Sur del Uruguay. Rev Bras Biol 55:457–466

Costa FG (1998) El comportamiento sexual de las arañas (con particular énfasis en los licósidos sudamericanos). Rev Etol n. especial:11–24

Costa-Schmidt LE (2015) Cryptic female choice and nuptial prey gifts in a spider model. In: Peretti AV, Aisenberg A (eds) Cryptic female choice in arthropods. Springer, Berlin, pp 145–167

Costa FG, Capocasale RM (1984) *Lycosa carbonelli* sp. nov., una etoespecie gemela, simpátrida de *Lycosa thorelli* (Keyserling) (Araneae, Lycosidae). J Arachnol 11:423–431

Costa FG, Pérez-Miles F (1992) Notes on mating and reproductive success of *Ceropelma longisternalis* (Araneae, Theraphosidae) in captivity. J Arachnol 20:129–133

Costa FG, Sotelo JR (1994) Stereotypy and versatility of the copulatory pattern of *Lycosa malitiosa* (Araneae, Lycosidae) at cool versus warm temperatures. J Arachnol 22:200–204

Costa FG, Pérez-Miles F (1998) Behavior, life cycle and webs of *Mecicobothrium thorelli* (Araneae, Mygalomorphae, Mecicobothriidae). J Arachnol 26:317–329

Costa FG, Pérez-Miles F (2002) Reproductive biology of Uruguayan theraphosids (Araneae, Theraphosidae). J Arachnol 30:571–587

Costa FG, Toscano-Gadea CA (2003) Experimental interruption and re-initiation of mating in a wolf spider: an analysis of behavioural patterns and resultant progeny. Ethol Ecol Evol 15:173–181

Costa FG, Montes de Oca L, Perdomo C, Ortiz-Villatoro D, Baruffaldi L, Pérez-Miles F (2013) Is more better? Sexual confusion during courtship between two sympatric and synchronic tarantulas: *Acanthoscurria suina* and *Eupalaestrus weijenberghi*. Fortschr Zool 30:577–580

Costa FG, Curbelo B, Pérez-Miles F (2015) Long-term persistence and water resistance of female sex cues in the tarantula *Eupalaestrus weijenberghi* (Araneae: Theraphosidae). Arachnology 16:311–313

Costa FG, González M (2015) El comportamiento sexual de *Hogna bivittata* y *Hogna* sp. (Araneae, Lycosidae), dos especies simpátridas del sur de Uruguay. Boletín de la Sociedad Zoológica del Uruguay 24:91–102

Costa-Schmidt LE, Carico JE, Araújo AM (2008) Nuptial gifts and sexual behavior in two species of spider (Araneae, Trechaleidae, *Paratrechalea*). Naturwissenschaften 95:731–739

Coyle FA (1986) Courtship, mating, and the function of male-specific structures in the mygalomorph spider genus *Euagrus* (Araneae, Dipluridae). In: Eberhard WE, Lubin YD, Robinson BC (eds) Proceedings of the Ninth International Congress of Arachnology, Panamá. Smithsonian Institution Press, Washington DC, pp 33–38

Danielson-François A, Drobot Y (2016) Dual function of male genitalia in *Tetragnatha* sperm removal and sperm transfer. XX International Congress of Arachnology. p 56

Danielson-François A, Fetterer CA, Smallwood PD (2002) Body condition and mate choice in *Tetragnatha elongata* (Araneae, Tetragnathidae). J Arachnol 30:20–30

Danielson-François AM, Bukowski TC (2005) Female mating history influences copulation behavior but not sperm release in the orb-weaving spider *Tetragnatha versicolor* (Araneae, Tetragnathidae). J Insect Behav 18:131–148

Darwin C (1871) The descent of man and selection in relation to sex. John Murray, London

Davies NB, Krebs JR, West SA (2012) An introduction to behavioural ecology, 4th edn. Wiley-Blackwell, Oxford

De Busschere C, Fannes W, Henrard A, Gaublomme E, Jocqué R, Baert L (2014) Unraveling the goblin spiders puzzle: rDNA phylogeny of the family Oonopidae (Araneae). Arthropod Syst Phylo 72:177–192

Dimitrov D, Hormiga G (2009) Revision and cladistic analysis of the orbweaving spider genus *Cyrtognatha* Keyserling, 1881 (Araneae, Tetragnathidae). Bull Am Mus Nat Hist 317:1–140

Dimitrov D, Hormiga G (2011) An extraordinary new genus of spiders from Western Australia with an expanded hypothesis on the phylogeny of Tetragnathidae (Araneae). Zool J Linnean Soc 161:735–768

Dimitrov D, Benavides LR, Arnedo MA, Giribet G, Griswold CE, Scharff N, Hormiga G (2016) Rounding up the usual suspects: a standard target-gene approach for resolving the interfamilial phylogenetic relationships of ecribellate orb-weaving spiders with a new family-rank classification (Araneae, Araneoidea). Cladistics 33:221. https://doi.org/10.1111/cla.12165

Dippenaar-Schoeman AS (2002) Baboon and trapdoor spiders of Southern Africa: an identification manual, Plant Protection Research Institute Handbook N° 13. Agricultural Research Council, Pretoria

Dondale CD (1986) The subfamilies of wolf spiders (Araneae: Lycosidae). Actas X Congreso Internacional de Aracnología Barcelona 1:327–332

Dugatkin LA (2014) Principles of animal behavior, 3rd edn. W. W. Norton and Company, New York

Dunlop JA, Penney D, Jekel D (2016) A summary list of fossil spiders and their relatives. In: World spider catalog. Natural History Museum Bern, online at http://wsc.nmbe.ch, version 16.5, Accessed on January 2017

Duron O, Bouchon D, Boutin S, Bellamy L, Zhou L, Engelstädter J, Hurst GD (2008) The diversity of reproductive parasites among arthropods: *Wolbachia* do not walk alone. BMC Biol 6:27. https://doi.org/10.1186/1741-7007-6-27

Dutto MS, Calbacho-Rosa L, Peretti AV (2011) Signalling and sexual conflict: female spiders use stridulation to inform males of sexual receptivity. Ethology 117:1040–1049

Eberhard WG (1991) Copulatory courtship and cryptic female choice in insects. Biol Rev 66:1–31

Eberhard WG (1992a) Notes on the ecology and behavior of *Physocyclus globosus* (Araneae, Pholcidae). Bull. Br. Arachnol Soc 9:38–42

Eberhard WG (1992b) Web construction by *Modisimus* sp. (Araneae, Pholcidae). J Arachnol 20:25–34
Eberhard WG (1994) Evidence for widespread courtship during copulation in 131 species of insects and spiders, and implications for cryptic female choice. Evolution 48:711–733
Eberhard WG (1996) Female control: sexual selection by cryptic female choice. Princeton University Press, Princeton
Eberhard WG (2004) Why study spider sex: special traits of spiders facilitate studies of sperm competition and cryptic female choice. J Arachnol 32:545–556
Eberhard WG (2009) Post-copulatory sexual selection: Darwin's omission and its consequences. Proc Natl Acad Sci U S A 106:10025–10032
Eberhard WG (2011) Experiments with genitalia: a commentary. Trends Ecol Evol 26:17–21
Eberhard WG (2015) Cryptic female choice and other types of post-copulatory sexual selection. In: Peretti AV, Aisenberg A (eds) Cryptic female choice in arthropods: patterns, mechanisms and prospects. Springer, Cham, pp 1–26
Eberhard WG, Briceño RD (1985) Behavior and ecology of four species of *Modisimus* and *Blechroscelis* (Araneae, Pholcidae). Rev Arachnol 6:29–36
Eberhard WG, Huber B (1998) Courtship, copulation, and sperm transfer in *Leucauge mariana* (Araneae, Tetragnathidae) with implications for higher classification. J Arachnol 26:342–368
Elgar MA (1998) Sperm competition and sexual selection in spiders and other arachnids. In: Birkhead TR, Møller AP (eds) Sperm competition and sexual selection. Academic, London, pp 307–337
Elias DO, Andrade MCB, Kasumovic MM (2011) Dynamic population structure and the evolution of spider mating systems. Adv Insect Physiol 41:65–114
Elias DO, Lee N, Hebets EA, Mason AC (2006) Seismic signal production in a wolf spider: parallel versus serial multi-component signals. J Exp Biol 209:1074–1084
Elias DO, Sivalinghem S, Mason AC, Andrade MCB, Kasumovic MM (2014) Mate-guarding courtship behaviour: tactics in a changing world. Anim Behav 97:25–33
Estramil N, Costa FG (2007) Female sexual receptivity after partial copulations in the wolf spider (*Schizocosa malitiosa*). J Zool 271:148–153
Fannes W, De Bakker D, Loosveldt K, Jocqué R (2008) Estimating the diversity of arboreal oonopid spider assemblages (Araneae, Oonopidae) at Afrotropical sites. J Arachnol 36:322–330
Ferretti N, Ferrero A (2008) Courtship and mating behavior of *Grammostola schulzei* (Schmidt 1994) a burrowing tarantula from Argentina. J Arachnol 36:480–483
Ferretti N, Pompozzi G, Pérez-Miles F (2011) Sexual behavior of *Acanthogonatus centralis* (Araneae: Mygalomorphae: Nemesiidae) from Argentina, with some notes on their burrows. J Arachnol 39:533–536
Ferretti N, Pompozzi G, Copperi S, Pérez-Miles F, González A (2012) Copulatory behavior of Microstigmatidae (Araneae: Mygalomorphae): a study with *Xenonemesia platensis* from Argentina. J Arachnol 40:252–255
Ferretti N, Pompozzi G, Copperi S, González A, Pérez-Miles F (2013) Sexual behaviour of mygalomorph spiders: when simplicity becomes complex, an update of the last 21 years. Arachnology 16:85–93
Ferretti N, Copperi S, Schwerdt L, Pompozzi G (2015) First record of *Vitalius roseus* (Mello-Leitão, 1923) (Araneae: Theraphosidae: Theraphosinae) in Argentina: distribution map, natural history, and sexual behavior. Arachnology 16:241–243
Fitzpatrick JL, Earn DJD, Bucking C, Craig PM, Nadella S, Wood CM, Balshine S (2016) Postcopulatory consequences of female mate choice in a fish with alternative reproductive tactics. Behav Ecol 27:312–320
Foelix RF (2011) Biology of spiders, 3rd edn. Oxford University Press, New York
Garcia-Diaz V, Aisenberg A, Peretti AV (2015) Communication during copulation in the sex-role reversed wolf spider *Allocosa brasiliensis*: female shakes for soliciting new ejaculations? Behav Process 116:62–68
Gerhardt U (1929) Zur vergleichenden sexualbiologie primitives Spinnen, insbesondere der Tetrapneumonen. Z Morph Ökol Tiere 14:699–764

Gerhardt U (1930) Biologische Untersuchungen an südfranzösischen Spinnen. Z Morph Ökol Tiere 19:184–227

Gerhardt U (1933) Neue Untersuchungen zur Sexualbiologie der Spinnen, insbesondere an Arten der Mittelmeerländer und der Tropen. Z Morph Ökol Tiere 27:1–75

Gershman SN, Mitchell C, Sakaluk SK, Hunt J (2012) Biting off more than you can chew: sexual selection on the free amino acid composition of the spermatophylax in decorated crickets. Proc R Soc B 279:2531–2538

Gillot C (2003) Male accessory gland secretions: modulators of female reproductive physiology and behavior. Annu Rev Entomol 48:163–184

González M (2015) Aspectos reproductivos de *Aglaoctenus lagotis*: estudio interpoblacional de una araña lobo sedentaria de gran variabilidad fenotípica. PhD Thesis, Universidad Nacional de Córdoba, Argentina

González M, Peretti AV, Viera C, Costa FG (2013) Differences in sexual behavior of two distant populations of the funnel-web wolf spider *Aglaoctenus lagotis*. J Ethol 31:175–184

González M, Costa FG, Peretti AV (2014) Strong phenological differences between two populations of a Neotropical funnel-web wolf spider. J Nat Hist 48:2183. https://doi.org/10.1080/00222933.2014.908974

González M, Peretti AV, Costa FG (2015a) Reproductive isolation between two populations of *Aglaoctenus lagotis*, a funnel-web wolf spider. Biol J Linn Soc 114:646–658

González M, Costa FG, Peretti AV (2015b) Funnel-web construction and estimated immune costs in *Aglaoctenus lagotis* (Araneae: Lycosidae). J Arachnol 43:158–167

Goodacre S, Martin O (2013) Endosymbiont infections in spiders. In: Nentwig W (ed) Spider Ecophysiology. Springer, Heidelberg, pp 93–105

Gowaty PA (2015) Standing on Darwin's shoulders: the nature of selection hypotheses. In: T. Hoquet (ed) Current perspectives on sexual selection, history, philosophy and theory of the life sciences 9, doi: https://doi.org/10.1007/978-94-017-9585-2_6

Gregorič M, Agnarsson I, Blackledge TA, Kuntner M (2015) Phylogenetic position and composition of Zygiellinae and *Caerostris*, with new insight into orb-web evolution and gigantism. Zool J Linnean Soc 175:225–243

Griswold CE, Coddington JA, Hormiga G, Scharff N (1998) Phylogeny of the orb-web building spiders (Araneae, Orbiculariae: Deinopoidea, Araneoidea). Zool J Linnean Soc 123:1–99

Gross MR (1996) Alternative reproductive strategies and tactics: diversity within sexes. Trends Ecol Evol 11:92–98

Gwynne DT (1984) Male mating effort, confidence of paternity, and insect sperm competition. In: Smith RL (ed) Sperm competition and the evolution of animal mating systems. Academic, London, pp 117–149

Harvey MS (1987) *Grymeus*, a new genus of pouched oonopid spider from Australia (Chelicerata: Araneae). Mem Natl Mus Victoria 48:123–130

Harvey MS, Edward KL (2007) Three new species of cavernicolous goblin spiders (Araneae, Oonopidae) from Australia. Rec Aust Mus 24:9–17

Hebets EA, Barron AB, Balakrishnan CN, Hauber ME, Mason PH, Hoke KL (2016) A systems approach to animal communication. Proc R Soc B 283:20152889. https://doi.org/10.1098/rspb.2015.2889

Heifetz Y, Tram U, Wolfner MF (2001) Male contributions to egg production: the role of behaviour gland products and sperm in *Drosophila melanogaster*. Proc R Soc B 268:905–908

Hormiga G, Eberhard WG, Coddington JA (1995) Web-construction behaviour in Australian *Phonognatha* and the phylogeny of nephiline and tetragnathid spiders (Araneae, Tetragnathidae). Aust J Zool 43:313–364

Huber BA (1997a) Evidence for gustatorial courtship in a haplogyne spider (*Hedypsilus culicinus:* Pholcidae: Araneae). Neth J Zool 47:95–98

Huber B (1997b) On American '*Micromerys*' and *Metagonia* (Araneae, Pholcidae), with notes on natural history and genital mechanics. Zool Scr 25:341–363

Huber BA (1994) Genital morphology, copulatory mechanism and reproductive biology in *Psilochorus simoni* (Berland, 1911) (Pholcidae, Araneae). Neth J Zool 44:85–99

Huber BA (1995) Copulatory mechanism in *Holocnemus pluchei* and *Pholcus opilionoides*, with notes on male cheliceral apophyses and stridulatory organs in Pholcidae (Araneae). Acta Zool (Stockh) 76:291–300

Huber BA (1996) Genitalia, fluctuating asymmetry, and patterns of sexual selection in *Physocyclus globosus* (Araneae, Pholcidae). Rev Suisse Zool (vol. hors serie):289–284

Huber BA (1998a) Spider reproductive behaviour: a review of Gerhardt's work from 1911–1933, with implications for sexual selection. Bull Br Arachnol Soc 11:81–91

Huber BA (1998b) Genital mechanics in some Neotropical pholcid spiders (Araneae: Pholcidae) with implications for systematics. J Zool 244:587–599

Huber B (1998c) Notes on the Neotropical spider genus *Modisimus* (Pholcidae, Araneae), with descriptions of thirteen new species from Costa Rica and neighboring countries. J Arachnol 26:19–60

Huber BA (1999) Sexual selection in pholcid spiders (Araneae, Pholcidae): artful chelicerae and forceful genitalia. J Arachnol 27:135–141

Huber BA (2000) New world pholcid spiders (Araneae: Pholcidae): a revision at generic level. Bull Amer Mus Nat Hist 254:1–348

Huber BA (2001) The pholcids of Australia (Araneae, Pholcidae): taxonomy, biogeography, and relationships. Bull Amer Mus Nat Hist 260:1–144

Huber BA (2002) Functional morphology of the genitalia in the spider *Spermophora senoculata* (Pholcidae, Araneae). Zool Anz 241:105–116

Huber BA (2005) Sexual selection research on spiders: progress and biases. Biol Rev 80:363–385

Huber BA (2011) Phylogeny and classification of Pholcidae (Araneae): an update. J Arachnol 39:211–222

Huber BA (2013) West African pholcid spiders: an overview, with descriptions of five new species (Araneae, Pholcidae). Eur J Taxon 59:1–44

Huber BA (2014) Pholcidae. In: Roig-Junent S, Claps LE, Morrone JJ (eds) Biodiversidad de Artrópodos Argentinos, vol 3. Sociedad Entomológica Argentina, San Miguel de Tucumán, pp 131–140

Huber BA, Astrin JJ (2009) Increased sampling blurs morphological and molecular species limits: revision of the Hispaniolan endemic spider genus *Tainonia* (Araneae: Pholcidae). Invertebr Syst 23:281–300

Huber BA, Brescovitt AD, Rheims CA (2005) Exaggerated female genitalia in two new spider species (Araneae: Pholcidae), with comments on genital evolution by female choice versus antagonistic coevolution. Insect Syst Evol 36:286–292

Huber BA, Eberhard WG (1997) Courtship, copulation, and genital mechanics in *Physocyclus globosus* (Araneae, Pholcidae). Can J Zool 74:905–918

Huber BA, Kwapong P (2013) West African pholcid spiders: an overview, with descriptions of five new species (Araneae, Pholcidae). Eur J Taxon 59:1–44. https://doi.org/10.5852/ejt.2013.59

Huber BA, Pérez-González A, Astrin JJ, Blumea C, Baptista R (2013) *Litoporus iguassuensis* Mello-Leitão, 1918 (Araneae, Pholcidae): camouflaged retreat, sexual dimorphism, female color polymorphism, intra-specific genital variation, and description of the male. Zool Anz 252:511–521

Huber BA, Pérez-González A (2001) Female genital dimorphism in a spider (Araneae: Pholcidae). J Zool 255:301–304

Huber BA, Senglet A (1997) Copulation with contralateral insertion in entelegyne spiders (Araneae: Entelegynae: Tetragnathidae). Neth J Zool 47:99–102

Huber BA, Warui CM (2012) East African pholcid spiders: an overview, with descriptions of eight new species (Araneae, Pholcidae). Eur J Taxon 29:1–44

Izquierdo MA, Ferretti N, Pompozzi G (2012) On *Puan*, a new genus of goblin spiders from Argentina (Araneae, Dysderoidea, Oonopidae). Am Mus Novit 3757:1–22

Jackson RR (1986) Cohabitation of male and juvenile females: a prevalent mating tactic of spiders. J Nat Hist 20:1193–1210

Jackson RR, Pollard SD (1990) Intraspecific interactions and the function of courtship in mygalomorph spiders: a study of *Porrhothele antipodiana* (Araneae, Hexathelidae) and a literature review. New Zeal J Zool 17:499–526

Jackson RR, Rowe RJ, Campbell GE (1992) Antipredator defences of *Psilochorus sphaeroides* and *Smeringopus pallidus* (Araneae, Pholcidae), tropical web-building spiders. J Zool 228:227–232

Jackson RR, Jakob EM, Willey MB, Campbell GE (1993) Anti-predator defences of a web-building spider, *Holocnemus pluchei* (Araneae, Pholcidae). J Zool 229:347–352

Jocqué R, Dippenaar-Schoeman AS (2006) Spiders families of the world. Royal Museum for Central Africa, Tervuren

Jorge C, Laborda Á, Días MA, Aisenberg A, Simó M (2015) Habitat preference and effects of coastal fragmentation in the sand-dwelling spider *Allocosa brasiliensis* (Lycosidae, Allocosinae). Open J Anim Sci 5:309

Kaster JL, Jakob EM (1997) Last-male sperm priority in a haplogyne spider (Araneae: Pholcidae): correlations between female morphology and patterns of sperm usage. Ann Entomol Soc Am 90:254–259

Kessel EL (1955) The mating activities of balloon flies. Syst Zool 4:97–104

Klein AL, Trillo MC, Albo MJ (2012) Sexual receptivity varies according to female age in a Neotropical nuptial gift-giving spider. J Arachnol 40:138–140

Klein AL, Trillo MC, Costa FG, Albo MJ (2014) Nuptial gift size, mating duration and remating success in a Neotropical spider. Ethol Ecol Evol 26:29. https://doi.org/10.1080/03949370.2013.850452

Kokko H, Klug H, Jennions MD (2014) Mating systems. In: Shuker DM, Simmons LW (eds) The evolution of insect mating systems, The Royal Entomological Society. Oxford University Press, Oxford, pp 42–58

Lang A (1996) Silk investment in gifts by males of the nuptial feeding spider *Pisaura mirabilis* (Araneae: Pisauridae). Behaviour 133:697–716

Lerette S, Aisenberg A, Costa FG, Sotelo-Silveira JR, Bidegaray-Batista L (2015) Detección de la bacteria endosimbionte *Wolbachia* en la araña lobo *Allocosa alticeps* y posibles efectos sobre la proporción sexual. Bol Soc Zool Uruguay 24:171–176

Levi HW (2005) Tetragnathidae. In: Ubick D, Paquin P, Cushing PE, Roth V (eds) Spiders of North America: an identification manual. American Arachnological Society, USA, p 232

Levi HW, Coddington J (1983) Progress report on the phylogeny of the orb-weaving family Araneidae and the superfamily Araneoidea (Arachnida: Araneae). Verh Naturwiss Ver Hamburg 26:151–154

Lewis S, Vahed K, Koene JM, Engqvist L, Bussière LF, Perry JC, Gwynne D, Lehmann GUC (2014) Emerging issues in the evolution of animal nuptial gifts. Biol Lett 10:20140336

Linn CD, Molina Y, Difatta J, Christenson TE (2007) The adaptive advantage of prolonged mating: a test of alternative hypotheses. Anim Behav 74:481–485

Macedo R, Machado G (2014) Sexual selection: perspectives and models from the Neotropics. Elsevier, Amsterdam

Macedo-Rego RC, Costa-Schmidt LE, Santos E, Machado G (2016) Negative effects of prolonged dietary restriction on male mating effort: nuptial gifts as honest indicators of long-term male condition. Sci Rep 6:21846. https://doi.org/10.1038/srep21846

Méndez V, Eberhard WG (2014) Removal of genital plugs and insemination by males with normal and experimentally modified palps in *Leucauge mariana* (Araneae: Tetragnathidae). J Arachnol 42:284–292

Merrigan M (1995) The behavioral reproductive strategies of a pholcid spider, *Crossopriza lyoni* (Araneae: Pholcidae). MSc thesis, Tulane University, New Orleans

Michalik P, Ramírez MJ (2014) Evolutionary morphology of the male reproductive system, spermatozoa and seminal fluid of spiders (Araneae, Arachnida) - current knowledge and future directions. Arthropod Struct Dev 43:291–322

Michalik P, Sacher P, Alberti G (2006) Ultrastructural observations of spermatozoa of several Tetragnathid spiders with phylogenetic implications (Araneae, Tetragnathidae). J Morphol 267:129–151

Mougeot F, Arroyo BE, Bretagnolle V (2006) Paternity assurance responses to first-year and adult male territorial intrusions in a courtship-feeding raptor. Anim Behav 71:101–108

Neff BD, Svensson EI (2013) Polyandry and alternative mating tactics. Philos Trans R Soc B 368:20120045. https://doi.org/10.1098/rstb.2012.0045
Nitzsche ROM (2011) Courtship, mating and agonistic behaviour in *Pisaura mirabilis* (Clerck, 1757). Bull Br Arachnol Soc 15:93–120
Oliveira RF, Taborsky M, Brockman HJ (2008) Alternative reproductive tactics: an integrative approach. Cambridge University Press, Cambridge
Parker GA (1970) Sperm competition and its evolutionary consequences in the insects. Biol Rev 45:525–567
Partan SR, Marler P (2005) Issues in the classification of multimodal communication signals. Am Nat 166:231–245
Peretti AV (2014) Sexual selection in Neotropical species: rules and exceptions. In: Macedo R, Machado G (eds) Sexual selection and breeding biology using Neotropical animals systems and models. Elsevier, Amsterdam, pp 33–52
Peretti AV, Aisenberg A (2015) Cryptic female choice in arthropods: patterns, mechanisms and prospects. Springer, Berlin
Peretti AV, Eberhard WG (2010) Cryptic female choice via sperm dumping favours male copulatory courtship in a spider. J Evol Biol 23:271–281
Peretti AV, Eberhard WG, Briceño D (2006) Copulatory dialogue: female spiders sing during copulation to influence male genitalic movements. Anim Behav 72:413–421
Pérez-Miles F, Capocasale R (1982) Hallazgo de una tercera especie del género *Pycnothelopsis*: *Pycnothelopsis tacuariensis* sp. nov. (Araneae, Pycnothelidae). Comun Zool Mus Hist Nat Montevideo 9:1–7
Pérez-Miles F, Costa FG (1992) Interacciones intra e intersexuales en *Grammostola mollicoma* (Araneae, Theraphosidae) en condiciones experimentales. Bol Soc Zool Uruguay 7:71–72
Pérez-Miles F, Costa FG, Toscano-Gadea C, Mignone A (2005) Ecology and behavior of the "road tarantulas" *Eupalaestrus weijenberghi* and *Acanthoscurria suina* (Araneae, Theraphosidae) from Uruguay. J Nat Hist 39:483–498
Pérez-Miles F, Postiglioni R, Montes de Oca L, Baruffaldi L, Costa F (2007) Mating system in the tarantula spider *Eupalaestrus weijenberghi* (Thorell, 1894): evidences of monandry and polygyny. Zoology 110:253–260
Piacentini LN (2014) Análisis filogenético de las arañas lobo (Araneae: Lycosidae) utilizando caracteres morfológicos y moleculares. Tesis Doctoral de la Universidad de Buenos Aires, Argentina
Platnick NI, Coddington JA, Forster RR, Griswold CE (1991) Spinneret morphology and the phylogeny of Haplogynae spiders (Araneae, Araneomorphae). Am Mus Novit 3016:1–73
Platnick NI, Dupérré N (2010) The Andean goblin spiders of the new genera *Niarchos* and *Scaphios* (Araneae, Oonopidae). Bull Am Mus Nat Hist 345:1–120
Platnick NI, Dupérré N (2011) The goblin spider genus *Pescennina* (Araneae, Oonopidae). Am Mus Novit 3716:1–64
Platnick NI, Abrahim N, Álvarez-Padilla F, Andriamalala D, Baher BC, Baert L, Bonaldo AB, Brescovit AD, Chousou-Polydouri N, DupérréN EB, Fannes W, Gaublomme E, Gillespie RG, Grismado CJ, Griswold CE, Harvey MS, Henrard A, Hormiga G, Izquierdo MA, Jocqué R, Kranz-Baltensperger Y, Kropf C, Ott R, Ramírez MJ, Raven RJ, Rheims CA, Ruiz GRS, Santos AJ, Saucedo A, Sierwald P, Szüts T, Ubick D, Wang XP (2012) Tarsal organ morphology and the phylogeny of goblin spiders (Araneae, Oonopidae), with notes on basal genera. Am Mus Novit 3736:1–52
Postiglioni R, Costa FG (2006) Reproductive isolation among three populations of the genus *Grammostola* from Uruguay (Araneae, Theraphosidae). Iheringia 96:71–77
Postiglioni R, González M, Aisenberg A (2008) Permanencia en la cueva masculina y producción de ootecas en dos arañas lobo de los arenales costeros. Proceedings IX Jornadas de Zoología del Uruguay, 145
Prokop P, Maxwell MR (2009) Female feeding regime and polyandry in the nuptially feeding nursery web spider, *Pisaura mirabilis*. Naturwissenschaften 96:259–265

Quirici V, Costa FG (2005) Seismic communication during courtship in two burrowing tarantula spiders: an experimental study on *Eupalaestrus weijenberghi* and *Acanthoscurria suina*. J Arachnol 33:199–208

Quirici V, Costa FG (2007) Seismic sexual signal design of two sympatric burrowing tarantula spiders from meadows of Uruguay: *Eupalaestrus weijenberghi* and *Acanthoscurria suina* (Araneae, Theraphosidae). J Arachnol 35:38–45

Roughgarden J (2015) Sexual Selection: is anything left? In: Hoquet T (ed) Current perspectives on sexual selection, history, philosophy and theory of the life sciences 9, doi:https://doi.org/10.1007/978-94-017-9585-2_5

Ryan MJ, Rand AS (1993) Species recognition and sexual selection as a unitary problem in animal communication. Evolution 47:647–657

Sakaluk SK, Avery RL, Weddle CB (2006) Cryptic sexual conflict in gift-giving insects: chasing the chase-away. Am Nat 167:94–104

Santos A, Brescovit A (2001) A revision of the south American spider genus *Aglaoctenus* Tullgren, 1905 (Araneae, Lycosidae, Sosippinae). Andrias 15:75–90

Schäfer MA, Uhl G (2002) Determinants of paternity success in the spider *Pholcus phalangioides* (Araneae, Pholcidae): the role of male and female mating behaviour. Behav Ecol Sociobiol 51:368–377

Scharff N, Coddington JA (1997) A phylogenetic analysis of the orb-weaving spider family Araneidae (Arachnida, Araneae). Zool J Linn Soc Lond 120:355–434

Schmidt G, Krause RH (1994) Eine neue Vogelspinnen-Spezies aus Mexico, *Brachypelmides klaasi* sp. nov. (Araneida, Theraphosidae, Theraphosinae). Stud Neotrop Fauna Environ 29:7–10

Schneider J, Andrade M (2011) Mating behavior and sexual selection. In: Herberstein ME (ed) Spider behaviour: flexibility and versatility. Cambridge University Press, Cambridge, pp 215–274

Schütt K (2000) The limits of the Araneoidea (Arachnida: Araneae). Aust J Zool 48:135–153

Schwerdt L, Copperi S (2014) A contribution to the knowledge of burrows and reproductive biology of *Stenoterommata platensis* Holmberg (Mygalomorphae: Nemesiidae). Munis Entomol Zool 9:84–88

Selden PA, Gall J (1992) A Triassic mygalomorph spider from the northern Vosges, France. Palaeontology 35:211–235

Shuker DM (2014) Sexual selection theory. In: Shuker DM, Simmons LW (eds) The evolution of insect mating systems, The Royal Entomological Society. Oxford University Press, Oxford, pp 20–41

Simmons LW (2001) Sperm competition and its evolutionary consequences in the insects. Princeton University Press, Princeton/Oxford

Simmons LW, Gwynne DT (1991) The refractory period of female katydids (Orthoptera: Tettigoniidae): sexual conflict over the re-mating interval? Behav Ecol 2:276–282

Simmons LW, Parker GA (1989) Nuptial feeding in insects: mating effort versus paternal investment. Ethology 81:332–343

Simó M, Lise AA, Pompozzi G, Laborda Á (2017) On the taxonomy of southern south American species of the wolf spider genus *Allocosa* (Araneae: Lycosidae: Allocosinae). Zootaxa 4216:261–278

Sordi S (1996) Ecologia de populações da aranha *Porrimosa lagotis* (Lycosidae) nas reservas Mata de Santa Genebra, Campinas (SP) e Serra do Japi, Jundiaí (SP). PhD Thesis, Universidade Estadual de Campinas, São Paulo, Brasil

Stålhandske P (2001) Nuptial gift in the spider *Pisaura mirabilis* maintained by sexual selection. Behav Ecol 12:691–697

Stefani V, Del-Claro K, Silva LA, Guimaraes B, Tizo-Pedroso E (2011) Mating behavior and maternal care in the tropical savanna funnel-web spider *Aglaoctenus lagotis* Holmberg (Araneae: Lycosidae). J Nat Hist 45:1119–1129

Stefani V, Garcia K, Vecchia C, Silva LA, Guimarães B, Tizo-Pedroso E, Machado EO, Brescovit AD, Del-Claro K (2012) Mating behaviour, nympho-imaginal development and description

of a new *Mesabolivar* species (Araneae: Pholcidae) from the Brazilian dry forest. J Nat Hist 46:2117–2129
Taborsky M (1994) Sneakers, satellites, and helpers: parasitic and cooperative behavior in fish reproduction. Adv Study Behav 23:1–100
Thornhill R (1976a) Sexual selection and paternal investment in insects. Am Nat 110:153–163
Thornhill R (1976b) Sexual selection and nuptial feeding behaviour in *Bittacus apicalis* (Insecta:Mecoptera). Am Nat 110:529–548
Thornhill R (1983) Cryptic female choice and its implications in the scorpionfly *Harpobittacus nigriceps*. Am Nat 122:765–788
Toft S, Albo MJ (2015) Optimal numbers of matings: the conditional balance between benefits and costs of mating for females of a nuptial gift-giving spider. J Evol Biol 28:457–467
Toft S, Albo MJ (2016) The shield effect: nuptial gifts protect males against pre-copulatory sexual cannibalism. Biol Lett 12:20151082
Trillo MC, Melo-González V, Albo MJ (2014) Silk wrapping of nuptial gifts as visual signal for female attraction in a crepuscular spider. Naturwissenschaften 101:123–130
Uhl G (1998) Mating behaviour in the cellar spider, *Pholcus phalangioides*, indicates sperm mixing. Anim Behav 56:1155–1159
Uhl G (2002) Female genital morphology and sperm priority patterns in spiders (Araneae). In: Toft S, Scharff N (eds) European arachnology 2000. Aarhus University Press, Aarhus, pp 145–156
Uhl G, Schmitt S, Schäfer MA (2005) Fitness benefits of multiple mating versus female mate choice in the cellar spider (*Pholcus phalangioides*). Behav Ecol Sociobiol 59:69–76
Vahed K (1998) The function of nuptial feeding in insects: a review of empirical studies. Biol Rev 73:43–78
Vahed K (2007) All that glisters is not gold: sensory bias, sexual conflict and nuptial feeding in insects and spiders. Ethology 113:105–127
von Helversen O (1976) Gedanken zur Evolution der Paarungsstellung bei den Spinnen (Arachnida, Araneae). Entomol Germ 3:13–28
Warwick S, Vahed K, Raubenheimer D, Simpson SJ (2009) Free amino acids as phagostimulants in cricket nuptial gifts: support for the 'Candymaker' hypothesis. Biol Lett 5:194–196
Wheeler W, Coddington JA, Crowley LM, Dimitrov D, Goloboff PA, Griswold CE, Hormiga G, Prendini L, Ramírez MJ, Sierwald P, Almeida-Silva L, Álvarez-Padilla F, Arnedo MA, Benavides-Silva LR, Benjamin SP, Bond JE, Grismado CJ, Hasan E, Hedin M, Izquierdo MA, Labarque FM, Ledford J, Lopardo L, Maddison WP, Miller JA, Piacentini LN, Platnick NI, Polotow D, Silva-Dávila D, Scharff N, Szuts T, Ubick D, Vink CJ, Wood HM, Zhang J (2016) The spider tree of life: phylogeny of Araneae based on target-gene analyses from an extensive taxon sampling. Cladistics. https://doi.org/10.1111/cla.12182
Wolfner MF (1997) Tokens of love: functions and regulation of *Drosophila* male accessory gland products. Insect Biochem Mol Biol 27:179–192
World Spider Catalog (2017) World spider catalog. Nat Hist Mus Bern, Online http://wsc.nmbe.ch, version 17.5. Accessed on Sept 2017
Wynn H, Vahed K (2004) Male *Gryllus bimaculatus* guard females to delay them from mating with rival males and to obtain repeated copulations. J Insect Behav 17:53–66
Yañez M, Locht A, Macías-Ordoñez R (1999) Courtship and mating behaviour of *Brachypelma klassi* (Araneae, Theraphosidae). J Arachnol 27:165–170
Yoward PJ (1996) Spider sperm competition: the conduit/cul-de-sac hypothesis—a route to understanding or a dead end? D. Phil. thesis, University of York

Chapter 13
Parental Care and Sociality

Carmen Viera and Ingi Agnarsson

Abstract Spiders are famously aggressive and cannibalistic, and nearly all are solitary. Only about 20–25 out of over 46,000 known species display highly social behavior. Nevertheless, sociality has arisen in multiple families independently in spiders, probably via the 'maternal care route', with an apparent concentration of social species in the Neotropics. We review aspects of reproduction and maternal care and how these may interplay with the evolution and maintenance of social cooperative behavior, focusing on Neotropical spiders. We also discuss the behavioral, ecological, and evolutionary contexts in which these behaviors have evolved in spiders, and highlight the unique opportunities that exist for research due to the multiple independent evolutionary experiments that replicated origins of sociality offer. We ponder why social species appear concentrated in the Neotropics, with the outstanding example found in the genus *Anelosimus*. Curiously, highly social *Anelosimus* are restricted to the Neotropics, while the genus is distributed globally and ubiquitously displays extended maternal care. We discuss traits that are shared among these independently derived social species and thus form a part of a social 'syndrome'. Such traits include absence of dispersal, inbreeding, biased sex ratios, and even shared patterns of colony composition of individuals differing in personality type. Ecologically, social Neotropical spiders are mostly restricted to tropical lowland and mid-elevation forests where prey size tends to be greater than in areas where sub-social species are found. They are especially common in areas of high rainfall, where their very dense 3-dimensional webs may not only allow capture of large prey, but also serve as a predator defense, for examples where ants are particularly common. Neotropical social spiders receive benefits from collaboration in web construction, care of young, nest defense, and prey capture, where they can handle much larger prey than other similarly sized spiders, and more effectively fend off predators. Colonies seem to benefit from a mix of personality types within colonies with both bold and shy individuals being crucial to colony success, but with

C. Viera (✉)
Entomología, Biología Animal, Facultad de Ciencias, Universidad de la República, Montevideo, Uruguay

Laboratorio Ecología del Comportamiento, Instituto de Investigaciones Biológicas Clemente Estable, Montevideo, Uruguay
e-mail: anelosimus@gmail.com

I. Agnarsson
Department of Biology, University of Vermont, Burlington, VT, USA

C. Viera, M.O. Gonzaga (eds.), *Behaviour and Ecology of Spiders*,
https://doi.org/10.1007/978-3-319-65717-2_13

larger colonies having more shy individuals and thus characterized by lower overall aggression. While sociality seems to offer short-term benefits in certain environments, a switch to an inbred breeding system that is tightly linked to sociality in spiders seems responsible for a loss of genetic variability that may restrict diversification due to vulnerability to climate change, disease, and parasitism.

Social behavior, characterized by long-term cooperation among individuals, is relatively rare, yet is found in many animal groups; the best known examples occur in eusocial insects such as ants, bees, and termites (Wilson 1971, 1975). In such societies, individuals are often morphologically specialized (castes), and most contribute various tasks to the colony as non-reproductive workers, with a single or a few queens bearing all the young of the colony. Many eusocial lineages are characterized by haplodiploidy (males come from unfertilized eggs and are thus haploid), a trait that may facilitate social evolution through increasing relatedness among sisters (Hamilton 1964a, b) and/or by enabling the queen to control sex ratios and produce more of the sex contributing more to the colony: female workers (Gardner and Ross 2013). Other animals can be strictly asocial, where aggression is more characteristic than cooperation among members of the same species. Such is the case for many spiders that are solitary and aggressive, often cannibalistic, and cooperation is absent (Avilés 1997; Bilde and Lubin 2011; Foelix 1982; Yip and Rayor 2014). Perhaps most animal species show behaviors somewhere in between these extremes, ranging from elementary care of young by the mother (simple maternal care, or 'transient subsociality' as defined by Yip and Rayor 2014) to highly cooperative behaviors that involve multiple reproductive individuals, rather than a single queen (cooperative sociality or quasisociality, hereafter 'sociality') (Avilés 1997; Avilés and Purcell 2012; Bilde and Lubin 2011; Kullmann 1972; Lubin and Bilde 2007; Yip and Rayor 2014). Spiders are an interesting group in the study of sociality, in part because they show this broad range of behaviors, and in part because they lack the apparent (though poorly understood) common correlate of animal sociality, haplodiploidy (Avilés 1997). It has long been apparent that maternal care, where the mother cares for her newly hatched offspring, is a trait shared by species ranging from mostly solitary to those that are highly cooperative (e.g., Burgess 1978; Vollrath 1982; Yip and Rayor 2014). During this stage, juveniles are typically not aggressive towards one another. This observation underlies the prominent hypothesis on the origin of sociality through the extension of juvenile tolerance and web sharing from early instars to adulthood, dubbed the 'maternal care hypothesis' (Agnarsson 2002, 2004; Avilés 1986, 1997; Avilés and Gelsey 1998; Avilés and Tufino 1998; Burgess 1978; Grinsted et al. 2014; Smith 1986, 1987; Uetz 1983; Vollrath 1982). Indeed, "subsociality is maternal care that spans several, rather than few, juvenile instars" (Agnarsson 2004: 471), and sociality is then characterized by the absence of dispersal from the natal nest of these cohabiting, tolerant, and cooperative individuals.

Comparative studies across spider species that display the range of social behaviors could cast light on social evolution (e.g., Settepani et al. 2016), and on the role of early maternal care and tolerance among juveniles in social origins (Avilés 1997; Kullmann 1972). Indeed, the maternal care hypothesis makes an explicit phylogenetic prediction: that social species should phylogenetically nest within clades characterized by shorter-term maternal care and subsociality.

Fortunately, the phylogeny of many of the social spiders is fairly well understood (Agnarsson 2006, 2012b; Agnarsson et al. 2006a, 2007, 2013a, 2015, 2016; Agnarsson and Rayor 2013; Johannesen et al. 2007, 2009b; Liu et al. 2016; Ruch et al. 2015), facilitating such tests (Fig. 13.1). The results have supported the predictions of the maternal care hypothesis in the groups containing the highest number of social species: Theridiidae (Agnarsson 2006; Agnarsson et al. 2006a, 2007) and *Stegodyphus* (Johannesen et al. 2007; Johannesen et al. 2009a, 2009b). These phylogenies imply multiple independent origins of sociality, even among close relatives, and in all cases, highly social lineages are nested within clades that show ancestral maternal care, typically extended maternal care that is characterized as subsociality (see Agnarsson et al. 2006a). Therefore, there is little doubt that the origin of web-sharing sociality in spiders has its root in early maternal care. The most prominent alternative hypothesis, that sociality may have arisen from foraging groups of non-relatives, see for example Avilés (1997), in contrast, has not been supported by any explicit tests, and does not explain this strong phylogenetic concordance between adult cooperation and maternal care of juveniles.

Another interesting taxonomic and phylogenetic pattern is that there seems to be a particular concentration of both species and social origins in the Neotropics (Figs. 13.1 and 13.5) (Agnarsson 2012a, Agnarsson et al. 2006a, Avilés 1997, Avilés et al. 2001). Furthermore, the environment of social spiders, what abiotic factors may facilitate cooperation, and how social level varies with such factors has been particularly well studied in the Neotropics (Avilés et al. 2007, Guevara and Avilés 2007, 2009; Purcell 2011, Purcell and Avilés 2007, 2008). Thus, a focus on available knowledge with regard to parental care and sociality in Neotropical spiders (see also Avilés et al. 2001) seems useful in elucidating the ecological and evolutionary correlates of social behavior in the context of presumably 'preadaptive' maternal care. We do not include in our discussion territorial social spiders (e.g., Fig. 13.2), as these have no evolutionary connection to maternal care of young; for a summary of the biology of some Neotropical territorial social spiders see Avilés (1997) and Avilés et al. (2001).

While cooperative behavior has evolved repeatedly in spiders, it is found in only a tiny fraction, less than 0.05%, of spider species (Agnarsson et al. 2006a; Avilés 1997; Bilde and Lubin 2011; World Spider Catalog 2017). The degree of maternal care and cooperation varies across species, but may include collaborating in web construction, prey capture and feeding, nest defense, and cooperative care of egg-sacs and brood (Fig. 13.4). In social species showing high levels of cooperative brood care, some females may lack any reproductive output. However, the characterization of these species as eusocial (Buskirk 1981, Rypstra 1993, Vollrath 1986) is more misleading than useful. First, there is no caste system—though there may be some division of labor (Holbrook et al. 2014, Settepani et al. 2013)—in social spiders (Avilés 1997, Avilés et al. 2001, Avilés and Tufino 1998), and second there is

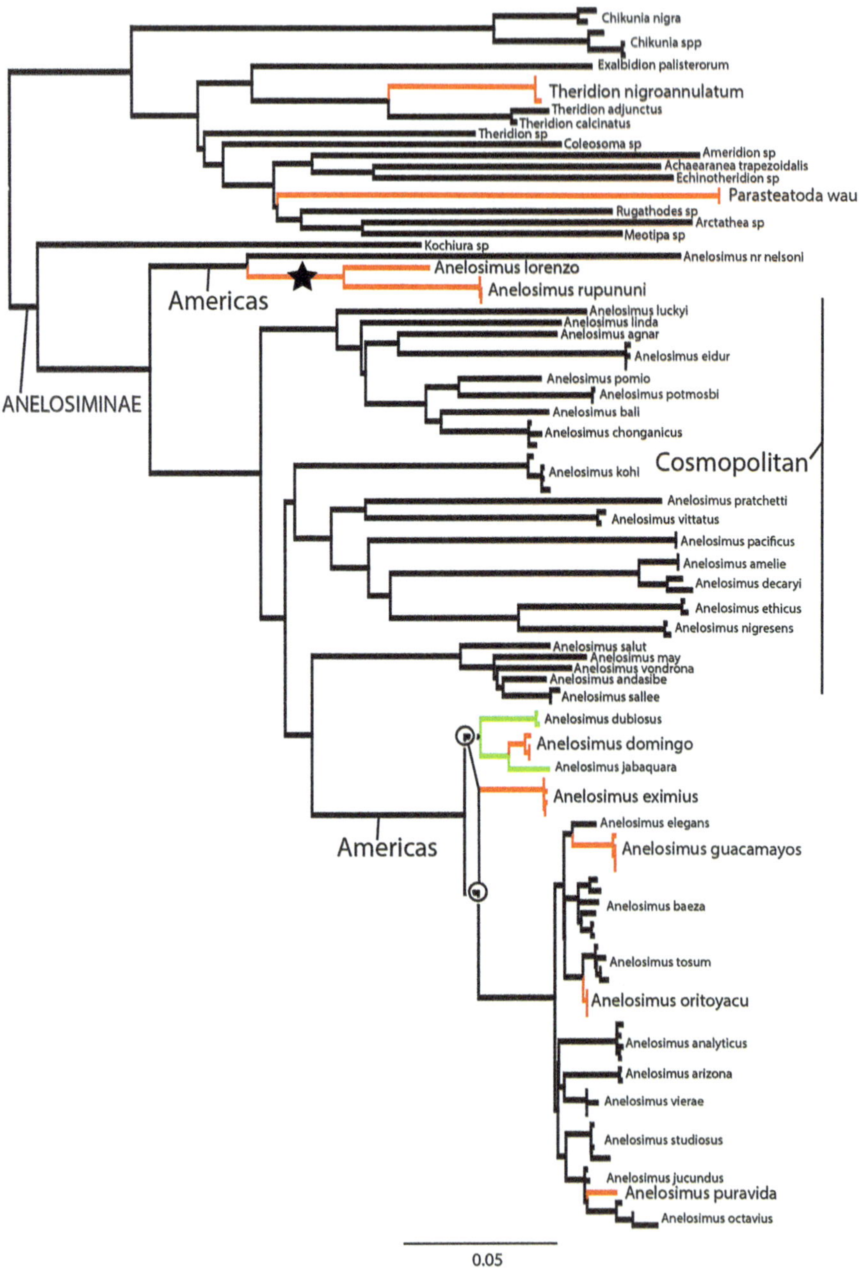

Fig. 13.1 Phylogeny of Anelosiminae and Theridiinae species. This phylogeny represents a tiny branch of the spider tree of life, yet is one that contains a clustering of about half the social spider species, each on its own 'spindly' branch. The maximum likelihood analysis (for details see Agnarsson (2014) and information available from the authors) contained nine of ten social theridiid species (*red branches*) but only a fraction of sub-social and solitary species of these subfamilies.

Fig. 13.2 *Parawixia bistriata* (Araneidae) from Southeastern Brazil. (**a**) Group of immature spiders resting during the day. (**b**, **c**) Prey capture. (**d**) Spiders leaving the retreat to build their orb webs at night (Photos: M.O. Gonzaga)

Fig. 13.1 (continued) Nevertheless, social species are scattered, and only a single social pair is recovered (*star*), presumably indicating speciation within a social lineage (*Anelosimus rupununi* and *A. lorenzo*). Sociality in nine species is thus best explained by eight independent social origins. In *Anelosimus*, sociality has only evolved (or been discovered) in the Americas. The social *T. nigroannulatum* is also American, whereas *P. wau* is from Papua New Guinea. Two species that show intermediate social levels are indicated with *green branches* (*A. dubiosus* and *A. jabaquara*). The social *A. eximius*, remains difficult to place, with two alternative topologies (indicated with *circles*) having the highest likelihoods (see also Agnarsson et al. 2007). The resolution of its placement will ultimately affect the number of reconstructed origins of sociality as yet no evidence available suggesting that the proportion of non-reproductive females in social colonies is higher than the proportion of females of solitary species that fail to reproduce (Agnarsson 2006). More than half of the social spiders belong to the family Theridiidae, and the majority of these to the genus *Anelosimus* (Agnarsson 2005, 2006; Agnarsson et al. 2015, 2016). *Anelosimus* species generally range from social to subsocial (e.g., Avilés 1997; Lubin and Bilde 2007; Yip and Rayor 2014), to species with a very brief period of maternal care of young (e.g., Agnarsson et al. 2006b; Ito and Shinkai 1993). The biology of social spiders in the Neotropics has been the subject of many studies (e.g., Agnarsson 2006; Agnarsson et al. 2006b, 2013a; Albo et al. 2007; Avilés 1986, 1993a, 1994, 1997, 2000; Avilés et al. 2000, 2001, 2006, 2007; Avilés and Maddison 1991; Avilés and Purcell 2011; Brach 1975, 1976, 1977; Buskirk 1981; Cangialosi 1990a, 1990b; Coddington and Agnarsson 2006; Gonzaga and Vasconcellos-Neto 2001, 2002a, b; Grinsted et al. 2014; Guevara and Avilés 2015; Harwood and Avilés 2013; Krafft 1985; Krafft and Pasquet 1991; Kullmann 1972; Levi and Smith 1982; Lichtenstein and Pruitt 2015; Majer et al. 2013; Marques et al. 1998; Nentwig 1985; Nentwig and Christenson 1986; Overal and Silva 1982; Pasquet and Krafft 1989, 1992; Pasquet et al. 1997; Pruitt et al. 2011, 2012; Rypstra and Tirey 1989; Saffre and Deneubourg 2002; Saffre et al. 1999, 2000; Samuk and Avilés 2013; Smith and Hagen 1996; Uetz 1983; Vakanas and Krafft 2001, 2004; Vasconcelos-Netto and Mello 1998; Venticinque and Fowler 1998, 2001; Venticinque et al. 1993; Viera and Albo 2008; Viera et al. 2006, 2007a, b, c; Viera and Garcia 2009; Vollrath and Parker 1992; Vollrath and Rohde-Arndt 1983; Vollrath and Windsor 1986). These studies reveal many shared characteristics of social spiders, despite each social lineage representing an independent evolutionary 'experiment'. Typically, social spiders have nests with multiple egg-laying females where offspring stay to breed in the natal nest, while subsocial nests consist of a mother and her offspring who disperse at or before adulthood. A dramatic shift in mating system, from outbred to strongly inbred with subdivided population structure, therefore characterizes social spiders (Agnarsson et al. 2013a; Avilés and Bukowski 2006; Avilés and Purcell 2012b; Bilde et al. 2005; Johannesen et al. 2009a; Lubin et al. 2009; Ruch et al. 2009). Differential survival of colony lineages as well as individuals within colonies (multilevel selection), and inbreeding, may have many consequences, including sex ratio bias and loss of genetic variability (Agnarsson et al. 2013a; Avilés 1997;Avilés and Bukowski 2006; Avilés and Purcell 2012b; Bilde et al. 2005; Johannesen et al. 2009a; Lubin et al. 2009; Riechert and Roeloffs 1993; Ruch et al. 2009). Remarkably, nearly all social spiders indeed have highly female-biased sex ratios (Avilés 1986, 1987, 1997; Avilés and Maddison 1991; Avilés et al. 2000; Elgar and Godfray 1987; Lubin 1991; Smith 1986, 1987; Vollrath 1986). There is little to no evidence for dispersal of females between colonies, or for mixing among colony lineages (e.g., Agnarsson et al. 2010b; Avilés 2000; Avilés and Gelsey 1998; Leborgne et al. 1994; Pasquet and Krafft 1989; Vollrath 1982); however, limited male dispersal has been detected (Lubin et al. 2009; Smith et al. 2016). Rather, colony formation is typically by swarm dispersal by multiple individuals, or colony 'budding' into two or more daughter colonies (Avilés 1997, 2000; Lubin and Robinson 1982; Saffre and Deneubourg 2002). Individual females may also form new colonies, and individual males may rarely disperse among colonies. Limited levels of dispersal, for example, have been clearly demonstrated in the African *Stegodyphus* (Berger-Tal et al. 2016; Schneider et al. 2001; Smith et al. 2016). Other traits shared by many social spiders include colony composition of individuals differing in 'boldness', and an interplay between boldness, aggression towards prey, and colony size (Pruitt et al. 2011, 2012).

Spiders are diverse on all continents except Antarctica, but this diversity is relatively poorly known outside Europe and North America. Global biodiversity inventories—point estimates based on 1-hectare plots—indicate that species richness is concentrated in the tropics, where the Neotropics are especially diverse (Agnarsson et al. 2013b; Coddington et al. 1991, 1996, 2009; Colwell and Coddington 1994). The number of spider species displaying sociality is also especially high in the Neotropics (Fig. 13.5). Whether that is simply a function of higher diversity in the region, or some other factors, is unclear (see below). One obvious bias is the intensity of study: the Neotropics are by far the best studied tropical region with respect to cooperative spiders, in large part thanks to the efforts of Leticia Avilés and colleagues (e.g., Avilés 1993b, 1994, 1997; Avilés et al. 2001, 2006, 2007; Avilés and Purcell 2011). However, such bias does not easily explain some apparent patterns such as the exclusive occurrence of sociality in Neotropical *Anelosimus*, despite global distribution of the genus (Agnarsson 2012b, Agnarsson et al. 2016) (Figs. 13.1 and 13.5). Hence, the Neotropics for some reason have an inordinate number of spiders that differ from the typical spider in terms of aggression and cooperation. Here we summarize some recent research on parental care and cooperative behavior in Neotropical spiders, to highlight the wealth of recent research in the area and the contribution of this region to understanding of the broader issues of origin of sociality.

Phylogenetics: The Taxonomic Distribution and Origin of Sociality

Analyzing behaviors such as maternal care and sociality using phylogenetic tools is essential to address some basic sociobiology questions. These include whether sociality has evolved multiple times, and if so, in what taxonomic and behavioral contexts. Furthermore, phylogenetics can help answer questions with regard to what the evolutionary causes and consequences of the switch to cooperative behavior are—and in the case of spiders, an associated immediate shift to inbreeding (e.g., Agnarsson et al. 2006a; Avilés 1997; Johannesen et al. 2007). Phylogenetic work to date has yielded two striking patterns regarding the taxonomic distribution of social species, they are simultaneously phylogenetically clustered (non-randomly distributed within a small portion of spider families) and 'spindly' (occurring only on isolated phylogenetic branches within these clusters) (Fig. 13.1) (Agnarsson et al. 2006a; Johannesen et al. 2007). Below, we consider the special case of Neotropical social spiders.

Yip and Rayor (2014) offer an excellent review of subsocial spiders and provide a useful framework to discuss the range of behaviors from basic maternal care such as construction of an eggsac, to cooperative behavior where some females forgo reproduction. They refer to as 'transient subsocial' those species whose maternal care is limited to care of egg and recently emerged instars, prior to the stage at which they begin to feed (Yip and Rayor 2014). Typically, this is limited to protection of the eggsac and spiderlings as they emerge from it. A large, but unknown, number of spiders offer some protection of the eggsac; this may well be an ancestral behavior for most spiders and will thus not be a focus of this review. Instead, we focus on those species that Yip and Rayor (2014) label 'subsocial', showing maternal care beyond the stage at which the juveniles start to feed, up to species showing high levels of sociality like *Anelosimus eximius* (Figs. 13.3 and 13.4) and *Theridion nigroannulatum* (Fig. 13.4).

It is interesting to look at the number and distribution of both subsocial (maternal care) and social spider species worldwide and across the spider tree of life. Yip and Rayor (2014) reviewed the literature and found information on 70 species that they consider subsocial and list in their Table 1. They furthermore estimate an additional 14 subsocial species of Eresidae, bringing the total to 84, plus an unknown number among the genus *Anelosimus*. They included in their list 17/74 *Anelosimus* species (World Spider Catalog 2017); a further eight are social (Agnarsson 2006), but most likely the remainder are all subsocial (under the Yip and Rayor definition). Therefore, there is reasonably good evidence for subsociality in at least 130 spider species. Of course, the actual number is probably vastly greater; we simply lack field observations

Fig. 13.3 *Anelosimus eximius* from Pará, Brazil (Photo: M.O. Gonzaga)

Fig. 13.4 Colonies of *Anelosimus eximius* (*above*) spanning less than a meter (*left*) and over 2 m diameter (*right*) and containing hundreds to over a thousand individuals respectively. Below are females of *Theridion nigroannulatum* collectively defending eggsacs (Photos: Ingi Agnarsson, Matjaž Kuntner)

on most spider species. Nevertheless, if we consider these known species several intriguing patterns emerge. First, these are spread across many spider families and genera, and Yip and Rayor estimated at least 18 independent origins of subsociality in spiders. This is remarkably close to the number of estimated independent origins of quasisociality (Agnarsson et al. 2006a; Avilés 1997), even though not all subsocial lineages have yielded social species. Second, subsocial spiders are found throughout the world; and while they are more common in subtropical or tropical areas, they range to higher latitudes, e.g., in Northern USA and Europe. In contrast, social species are almost entirely constricted to the tropics, apart from populations of the otherwise subsocial *A. studiosus* that is socially polymorphic in certain areas of the USA (Jones

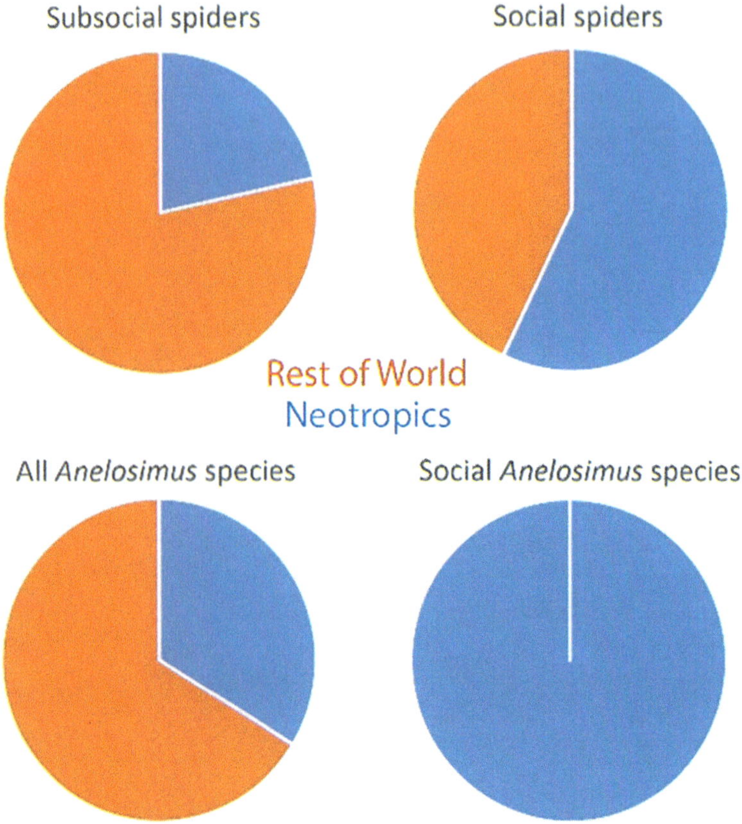

Fig. 13.5 The proportion of social and subsocial spiders in the Neotropics versus the rest of the world. Above, the distribution of subsocial species (*left*), with less than a quarter occurring in the Neotropics, versus social species (*right*) well over half of which are Neotropical. Below, the distribution of all *Anelosimus* species, of which some 30% are found in the Neotropics (*left*), three of all eight social species (*right*)

and Parker 2000, 2002). Subsocial species are not particularly common in the Neotropics, with about 28/130 (21%) of the known species occurring there (Fig. 13.5). Again, in contrast, social species are disproportionally Neotropical where more than half of the social species are found (12/21) (Fig. 13.5).

What could be the reason that the Neotropics have generated more social species, even though the 'preadaptive' subsocial trait is not particularly common there? One obvious possibility is knowledge bias. Of the 12 social Neotropical species, five (*Aebutina binotata*, *Tapinillus* sp., *Theridion nigroannulatum*, *Anelosimus guacamayos* and *A. oritoyacu*) were discovered—or for the first time characterized as social—relatively recently by Leticia Avilés (e.g., Avilés 1993b; Avilés et al. 2001, 2006a; Avilés and Purcell 2011). Her efforts researching social spiders in the Neotropics are probably not matched on any other continent, at least not in lowland rainforest areas where sociality is concentrated (social *Stegodyphus* species are well studied in drier areas of Africa and India (Kraus and Kraus 1988). Without the 'Leticia factor', the diversity of Neotropical social spiders would not stand out as clearly. However, there is other evidence that sociality may have evolved more frequently in the Neotropics than expected, based on the distribution of subsocial lineages. The best evidence comes from ecological and phylogenetic research on the genus *Anelosimus*. *Anelosimus* contains 74 species that range in behavior from 'solitary' as defined by Agnarsson et al. (2006b) to social (Avilés et al. 2001, 2007), or in other words, in dispersal from natal nest from 2nd to 7th instar to absence of dispersal altogether. Probably none of the species fit into the least social category of Yip and Rayor named 'transient subsocial' (dispersal of young prior to commencement of feeding), as in all studied species at least 2nd instar (out of eggsac) spiderlings have been observed in the natal nest. Out of these 74 species, only 25 are found in the Neotropics (~34%), yet all eight social *Anelosimus* are Neotropical (Fig. 13.4), representing no less than 6–7 independent origins of sociality (Fig. 13.1) (Agnarsson 2006; Agnarsson et al. 2007). What could explain the inordinate number of social origins in Neotropical *Anelosimus*? Neotropical *Anelosimus* are certainly better studied than those from other tropical regions; however, some detailed studies on *Anelosimus* elsewhere (Agnarsson 2012b; Agnarsson et al. 2010a, 2015, 2016; Agnarsson and Kuntner 2005) leave no doubt that the vast majority of *Anelosimus* species currently known outside the Neotropics are 'only' subsocial. Another possibility could be phylogenetic constraints. For example, if all the social species belonged to a single Neotropical clade, that clade might be characterized by an (as yet unknown) trait that facilitated sociality. However, this is not the case, as social Neotropical *Anelosimus* belong to at least two, distantly related clades (Fig. 13.1) (Agnarsson et al. 2007). The least inclusive clade that contained both of these Neotropical lineages would contain all remaining globally distributed *Anelosimus* lineages (Fig. 13.1). This observation is curious, and merits further scrutiny. However, we can conclude here that, for whatever reason, the Neotropics are rich in social spider species, and are an exciting area where much work has been done and where opportunities exist for a broad range of future studies into the origin and evolution of maternal care and sociality. In the following sections we further explore some of these topics.

Ecology and Sociogeography of Cooperative Spiders

The geographical distribution of species across habitats and landmasses is a central theme of major biological disciplines such as ecology and biogeography (Levin 2009, Losos and Ricklefs 2010). The factors at play are many and diverse, including dispersal ability, geographical history, phylogenetic constraints, and major stochastic events, as well as the ecology and behavior of species. The extent to which behavior of individuals, other than dispersal behavior per se, affects species distribution and diversity has received relatively little attention. For example, their means of dispersal may afford, or limit, opportunities to cross barriers and colonize distant landmasses. However, other types of behavior such as degree of cooperative behavior and breeding system may also shape species distributions; what might be referred to as 'sociogeography'. For example, eusociality has allowed insects to dominate many ecosystems, and it has been argued that their extraordinary ecological success in the tropics in fact constrains their diversity (Roubik 1989). The breeding system also relates to biogeographical patterns, e.g., selfing species are more likely to successfully colonize islands. In Neotropical social spiders, Avilés et al. (2007: 783) suggested some sociogeographic patterns. One observation was that "Interestingly, all social spider species appear to occupy a set of habitats more restricted than those available to the phylogenetic lineages in which they occur" (Avilés et al. 2007: 784). As for the particular distribution, they suggested that in the Neotropics "…the absence of subsocial *Anelosimus* species in the lowland rain forest may be due to an increased probability of maternal death in this habitat due to greater predation and/or precipitation, while absence of a sufficient supply of large insects at high elevations or latitudes may restrict social species to low- to mid-elevation tropical moist forests. We refer to these as the 'maternal survival' and 'prey size' hypotheses, respectively, and suggest that both in combination may explain the geographical distribution of sociality in the genus." Uetz and Hodge 1990) found that spiders in prey-poor environments were less social than spiders in prey-rich environments. Similarly, populations of *Anelosimus eximius* have both a greater proportion of solitary females and smaller average nest sizes toward the upper end of its elevational range (1000–1300 m) than do populations in the lowlands where prey are larger and more abundant (Purcell and Avilés 2007).

The environmental and biogeographical parallels between altitude and latitude are well known (Stevens 1992; Jimenez-Castillo et al. 2007; Swenson and Enquist 2007). However, altitude and latitude only 'explain' the distribution of sociality to the extent that they correlate with environmental and ecological factors that favor or disfavor cooperative behaviors (Avilés et al. 2007). One may suppose that multiple factors play a role, and that these may not always vary in the same manner with altitude and latitude. For example, the widespread species *Anelosimus studiosus* ranges from southern S. America up to northern USA. The species shows clear altitudinal trends across this range. It is restricted to low altitudes at high latitudes, reaches over 3000 m near the equator, and is found at a broad range of altitudes at mid-latitudes (Fig. 13.6). Notably, this species is mostly absent in lowland rainforests where social species are prominent. Even more curiously, in North America, social phenotypes occur in certain populations of this normally subsocial species. Social phenotypes are apparently absent in Florida at 26°N, but appear at about 30°N and increase in occurrence towards 36°N in Tennessee (Riechert and Jones 2008). In either case, it is hypothesized that cooperative behavior is favored in environments where a single female has a high probability of dying before her offspring are able to care for themselves (Jones et al. 2007; Bilde et al. 2007)—the 'maternal survival' hypothesis (Avilés et al. 2007). The maternal survival hypothesis may help explain the rarity of subsocial *Anelosimus* species in 'social habitats', such as lowland rainforests, where they may experience higher predation risk and frequent web

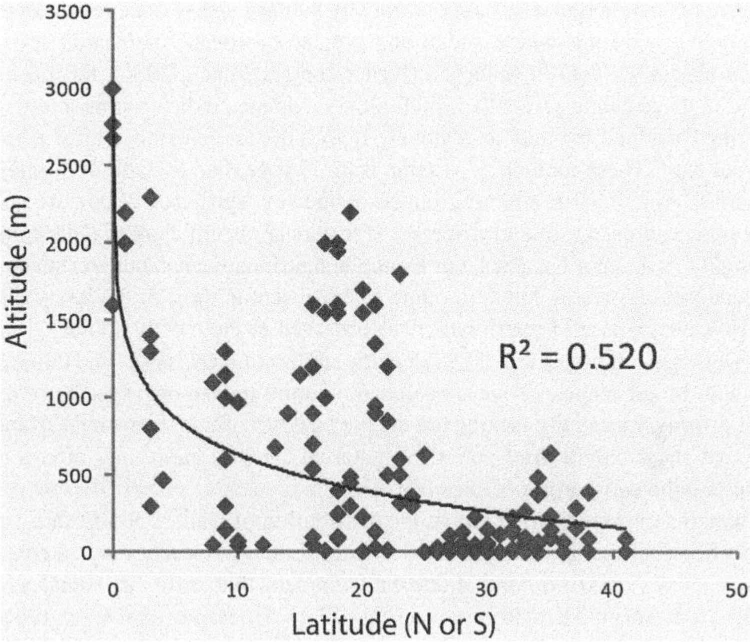

Fig. 13.6 The distribution of the socially polymorphic *A. studiosus* in relation to altitude and latitude. Note the near absence of records from the lowland tropics

damage due to frequent precipitation (see Avilés et al. 2007). In turn, one of the key ecological variables that has been hypothesized to explain the absence of social species from high latitudes and altitudes is prey size (Powers and Avilés 2007, Purcell and Avilés 2008, Yip et al. 2008, Guevara and Avilés 2007, 2009). As colonies grow in both number of spiders and volume of the web, web surface per spider decreases, as does the number of intercepted prey per capita (Yip et al. 2008). Larger colonies therefore increasingly need to capture larger prey items that solitary individuals cannot handle. Accordingly, social species may be absent where sufficiently large prey items are rare. Recently, the hypothesis has received direct support based on an array of empirical data (Powers and Avilés 2007, Purcell and Avilés 2008, Yip et al. 2008, Guevara and Avilés 2007, 2009). Not only do social spider habitats have larger prey than habitats of subsocial species, but also the prey actually caught by social spiders is larger than that which subsocial species catch.

Much work has focused on the 'typical' subsocial and social *Anelosimus* species. However, a few species that show unusually short duration of maternal care, and are thus solitary for most of their life cycle, have received less attention. Species such as *A. crassipes* in Japan, (Ito and Shinkai 1993), and *A. pacificus* in Central America (Agnarsson et al. 2006b) have only a brief period of maternal care post emergence of juveniles from eggsac (still fitting Yip and Rayor's broad definition of subsociality). Similarly, *A. decaryi* and *A. amelie* from Madagascar and Mayotte (Agnarsson et al. 2010), and *A. potmosbi* and *A. pomio* from Papua New Guinea all share this brief existence of maternal care. These six, mostly solitary species occupy coastal habitats, often beachfront, where social and typical subsocial *Anelosimus* are nearly entirely absent [*A. kohi* in Malaysia (Agnarsson and Zhang 2006). has been documented in these habitats]. Further, preliminary evidence indicates that a few species occurring in inland habitats at relatively high latitudes have only brief periods of maternal care. These include *A. vittatus* from Europe, and a clade of species from southern S. America (the *ethicus* group as defined by Agnarsson 2005). We note that initial observations on time of dispersal of instars are insufficient to determine level of sociality in observed species. For example, matriphagy characterizes many of the subsocial–social species, but is (thought to be) absent in most of the less social species. However, a case of matriphagy was observed in the mostly solitary *A. nigrescens* (Dias et al., *in prep.*). In fact, systematic studies of *Anelosimus* and other spiders that show broad ranges of social behavior within and among species remain an urgent priority. Especially lacking are detailed studies of transient social species and those that show only a brief period of maternal care, as these may offer a unique insight into the early origin of sociality.

It remains unclear to what extent the distribution of solitary *Anelosimus* species follows a latitudinal and/or altitudinal pattern. Latitude clearly plays a role in the example of *A. studiosus,* and *Anelosimus* spiders that differ in social structure clearly differ in global distributions (Fig. 13.7). However, and more broadly, it remains an open question whether the pattern of distribution of solitary, subsocial, and social *Anelosimus* observed in the Americas holds when examining species worldwide. In particular, why no permanently social *Anelosimus* species occurs outside the Americas remains perplexing (Figs. 13.1 and 13.5).

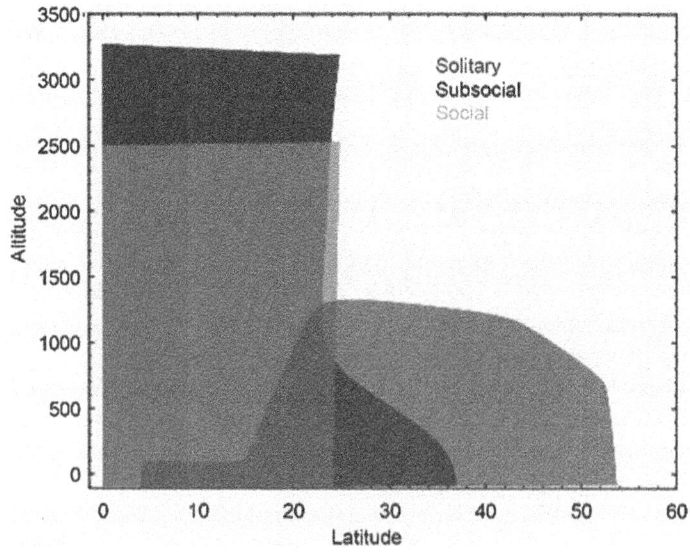

Fig.13.7 A course schematic showing the distribution of solitary, subsocial, and social *Anelosimus* worldwide, in relation to altitude and latitude

Reproductive Aspects of Subsocial Spiders

According to Trivers (1974), parental effort is all the investment that affects the survival success of the offspring. In spiders, parental care is almost synonymous with maternal care, because male contribution to the survival of offspring is very rare, with the exception of *Manogea porracea* (Araneidae), which present amphisexual care (Moura et al. 2017) (Fig. 13.8). In this spider, males protect their brood and keep web integrity in the absence of the mothers. Maternal care is common in many species of spiders, including solitary ones, but the prolongation in time of this behavior is what underlies the formation of more permanent groups.

In a broad sense, nearly all spiders show maternal care, since they build protective sacs for their eggs (Foelix 2011). Another level of investment is added when the spiders take care of the eggsac and the emerged spiderlings during the first instars. These are considered as "solitary" by Wilson (1971) and "transient subsocial behavior" by Yip and Rayor (2014). Agnarsson et al. (2006a) considers the subsocial spiders to be those in which the offspring cooperate in prey capture and web building. We follow here the definition of subsociality of Yip and Rayor (2014)

Fig. 13.8 *Manogea porracea* male close to eggsacs and spiderlings (Photo: M.O. Gonzaga)

in their excellent and exhaustive review about maternal care and subsocial spiders: "... offspring stay together with the parent beyond the age at which they begin to feed, but disperse prior to their own egg-laying and display no alloparental care among adults". Below, we address the importance of extended maternal care in the evolution of sociality from a behavioral context. Furthermore, we consider all reproductive behavior including courtship, mating, and post-mating behaviors which affects parental care and the evolution of sociality.

In social spiders, females cooperate in care of eggsacs and all the spiderlings since their emergence from the eggsac. In many cases (e.g., *Anelosimus eximius*, *A. domingo, A. lorenzo, Achaearanea wau* and *Stegodyphus dumicola*), a group of females lay the eggsacs more or less synchronously, and then cooperate in taking care of them (Lubin and Robinson 1982, Avilés and Salazar 1999). The high cost of this behavior is shared among all the adult females, not only by the mothers. All females collaborate in colony tasks to assure brood survival and reproductive success. In subsocial spiders, such cooperation among adult females is absent, presumably due to innate intolerance among adult females, as typical of most spiders.

Detailed studies of reproductive behavior can provide us with information about how it relates to parental care and other strategies of survival. Hence, it is necessary to understand the role of all the reproductive cycle in the evolution of sociality. We present a case study focusing on the parental care of the subsocial *Anelosimus vierae* from Uruguay. For a better understanding of maternal behavior in this spider—as a model subsocial spider—we describe here all the stages prior to motherhood, beginning with dispersion and male's sexual tactics. We analyze the fights of males for access to females, spermatic induction, courtship, and mating. Finally, we summarize other maternal behaviors.

Dispersal from the Maternal Nest

Avilés and Gelsey (1998) highlight the lack of the dispersal phase characterizing the transition from subsocial to social spiders. In subsocial spiders, the colony is a family group lasting until the dispersal of the new generation, usually at the subadult stage. Avilés and Gelsey (1998) found that subadults of both sexes dispersed during the mating season. The same process has been studied in other subsocial species, such as *A. studiosus* and *A. vierae*. The sex ratio prior to dispersal in *A. jucundus* and *A. studiosus* is 1:1, but in *A. vierae* it is approximately 2:1 female-biased (Viera et al. 2007a). Not all individuals necessarily disperse; rather, one or more females can remain in the natal nest. This process opens the possibility of 'new' colonies—natal nests of non-dispersing females—starting the next cycle as a multi-female colony. Thus, polymorphism in colony type and size is observed in the field. Gonzaga and Vasconcellos-Neto (2001) found that in the *A. jabaquara*, a species showing levels of sociality somewhat intermediate between 'typical' subsocial and social species, large adult females are more likely to leave the natal nest, while those females remaining tend to be relatively small. They conclude that the high costs of dispersal and colony foundation may favor dispersal of large females. Jones and Parker (2000, 2002) analyzed the cost and benefits associated with delayed dispersal in *A. Studiosus*, and found more benefits than costs to both mother and offspring. Most females disperse at the subadult stage, while one or a few females remain in the maternal nest. Males all abandon the natal nest at or near adulthood, a common strategy in most organisms to avoid inbreeding but mostly absent in highly social spider species (Viera et al. 2007a). As seen in many other subsocial species such as *Anelosimus arizona* (Powers and Avilés 2003), Ferreira et al. (*in prep.*) observed in *A. viera* that the amount of available food resources is an important factor determining the timing of dispersal—females dispersed later from well-fed colonies.

Cooperative Behavior Prior to Dispersal

Trophalaxy is a complex and frequent behavior between mothers and offspring in social arthropods. It seems widespread in social spiders but rarer in subsocial species (Avilés and Gelsey 1998). One extraordinary case of regurgitation among subadult *A. vierae* has been observed (Viera et al. 2005). Subadult females provided supplementary food to their brothers previous to their dispersal. Males, on average, reach adulthood in six or seven molts, earlier than females, which need seven or eight molts (Viera et al. 2007a). Subadult sisters potentially regurgitate food to their brothers to accelerate the male's maturity. Alternatively, altruistic food sharing among all colony individuals may yield greater overall survival. To test these alternatives, we designed an experiment grouping subadult males and females of different body condition, mixing starved individuals with individuals fed ad libitum. The results demonstrated that regurgitation was always from overfed to starved individuals, and strongly biased from females towards males. Males fed by females reached greater size and had a relatively longer first pair of legs than males which did not receive food via regurgitation. The allometric growth of front legs make sense; since the first legs are used in male–male ritualized fights, males with longer first legs have a higher probability of winning contests (Gómez et al. 2015). Thus, food donations from females to their brothers may increase inclusive fitness by assuring better access of brothers to females.

Intolerance among Adult Females

Like other subsocial spiders, individuals of *A. vierae* show relatively high inter-individual tolerance, which breaks down at adulthood, when adult females become aggressive among themselves. This aggression limits the degree to which nests are founded by more than one adult female, and thus the level of sociality the species displays. Within subsocial colonies, collaboration and cooperation is extensive among juveniles and with their mother. But unlike social spiders, adult subsocial females do not collaborate in key tasks such as prey capture and cooperative care of young. Furthermore, the presence of another female inhibits a female from attempting capture of prey. Female intolerance and territoriality were described by Furey (1998) in *A. studiosus,* and have also been observed by Viera et al. (2007a) in *A. vierae*. Adult females which remain in the natal nest show intolerance among themselves, and agonistic behavior similar to inter-male contests (Viera et al., in prep.) Experiments in *A. viera* using the methodology proposed by Susan Riechert (*com. pers.*) demonstrated that within multi-female colonies, each female is territorial and avoids any contact with other females, especially when guarding eggsacs (Tambasco et al., *in prep.*). In these experiments, females were forced to encounter other and fight females, and were able to steal eggsacs from other females after winning a contest. This intolerance and aggressive behavior lead to avoidance, and lack of any cooperative behavior related to social benefits such as prey capture and caring for brood. Lack of cooperation among subadult females results in a higher per-capita cost of brood care than in social species.

Preparing for Mating

The sexual behavior of social species is poorly known. We assume that sexual behavior will be relatively simple due to inter-individual tolerance, such as absence of sexual cannibalism and other aggressive interactions toward the males. In many solitary spiders, males avoid female aggression by behaviors such as prolonged and complex courtship, and providing nuptial gifts. However, aggression is not absent between males. They fight for access to females, and may use waiting strategies to get eventual mating, as was also observed by Lubin and Bilde (2007) in the African social spider *Stegodyphus dumicola*. Male fights are ritualized and, at least in experimental conditions, can escalate, to result in serious injuries and death (Albo et al. 2007). Ritualized fights begin with the contenders using the first legs to push into a position facing each other, probably to compare strength. Generally, the larger male wins the contest, but the smaller animal (loser) can flee the arena before a fight escalates. However, frequently the loser male remains completely still near the couple (winning male and female) as a 'satellite male', and awaits an opportunity for a later mating once the winning male has left (Fig. 13.9). Females readily accept these satellite males. The potentially dangerous waiting is rewarded by reproductive success, because both first and second males apparently fertilize eggs in equal proportions, with no evidence of sperm priority (Lorieto et al. 2010).

Behavioral patterns involved in male–male conflicts include silk thread tension, vibration, persecution, ritualized fighting, and grappling. The two last behaviors occur in few cases, and when both males have similar body condition. In the ritualized fighting, the males confront venter-to-venter and facing upwards, contacting leg tarsi and vibrating their bodies intensively, but causing no damage. In the grapple, males face each other, crossing legs and biting each other chelicerae to chelicerae, usually finishing with injuries or death to one of the individuals (Albo et al. 2007; Rojas and Viera 2016).

Fig. 13.9 Couple of *Anelosimus vierae* and a satellite male (Photo: C. Rojas)

Both the first and second mating male show high percentages of paternity. These results make it difficult to explain the escalated fights for access to female, suggesting that other factors may have greater importance. For example, that there may be cryptic female choice in this species (Lorieto et al. 2010), or that the waiting strategy is costly in other ways, such as in risk of predation.

Sexual Strategies

Males can encounter two kinds of scenarios when they are searching for females; individual nests (uni-female nest) with one subadult or adult female, and communal nests (multi-female nest) containing two or more females of different instar subadults (pre-penultimate, penultimate, and adult female) (Albo et al. 2007). The scenario determines the sexual strategies and preferences of males (Viera and Albo 2008; Rojas and Viera 2015). Viera and Albo (2008) made an experimental design simulating a multi-female nest of *A. vierae,* allowing males to choose between females of different age and reproductive status. These multi-female nests consisted of one mature female, one recently molted adult female, and one subadult (penultimate) female. Females of different reproductive status were attractive to males, as they courted at least one female per colony irrespective of reproductive state. When two males were exposed to nests containing only one subadult female, they fought for access to the females described above.

After winning a contest and before mating with the female, the males—like all spider males—must perform sperm induction. Although sperm induction occurs in every sexual encounter, it has not been well described except some cases in big spiders (Mygalomorphae) and in six species of the *Theridion varians* group (Theridiidae) by Knoflach (1998). Rojas and Viera (2016) made a detailed description of the sperm induction behavior in *A. vierae*. This behavior can occur prior to or after courtship, before re-mating with the same female, or before mating with another female. The sperm-induction description was done observing males after first mating. The duration of the entire process of sperm induction (sperm web construction + emergence of sperm drop + filling the palps) takes approximately 5 min. The sperm induction can happen before, in the middle, or at the end of interactions with a female.

Adult females assume a characteristic mating posture upon accepting a male (Fig. 13.10). Non-mature females are receptive and accept courtship and mating attempts, assuming the adult female mating posture, resulting in a peculiar behavior called pseudocopulation common in cobweb spiders (Knoflach 1998, Albo et al. 2007) or non-conception behavior, observed also in *Anelosimus studiosus* (Pruitt et al. 2011). The possible function or advantage of this behavior has been explored for both sexes involved (Rojas and Viera, *in prep.*). This behavior was found to be very common in subadult females, perhaps to retain in the nest males to mate with. Since the sex ratio is biased toward females, the probabilities of encounters decline through the reproductive season. On the other hand, males that remain in the nest with subadult females can copulate with virgin females without fights with other males. We observed in experimental conditions that females can easily accept males to mate with if they were pseudocopulated before, without male individual recognition (Viera and Rojas, *in prep.*). Furthermore, we did not find differences in the time of maturity between pseudocopulated and non-pseudocopulated subadult females, indicating that earlier maturation is not elicited by pseudocopulating behavior. Males also courted more frequently pseudocopulated females than the naïve females. This situation was observed by Pruitt and Riechert (2011) in *A. studiosus,* where prior sexual experience facilitated mating. According to Burghardt (2005), the pseudocopulation is probably sexual proof, and it is expected to reduce the latency to mating. For males that fight to access females, it is advantageous to be accepted more quickly, in order to avoid other males (Foellmer and Fairbairn 2005; Albo et al. 2007; Rojas and Viera 2015). Pruitt and Riechert (2011) found in *A. studiosus* that females invest more energy in the brood from the males with which they pseudocopulated than from those from males with which they did not.

Fig. 13.10 Mating of *Anelosimus vierae*, showing the female receptive posture (Photo: C. Rojas)

Maternal Behavior

Although the existence of brood recognition by the mothers is not known, only mature and copulated females care for foreign eggsacs. (Viera et al. 2007c). Mothers can adopt foreign eggsacs, implied by cleaning maneuvers and permanent contact with them. This behavior is energetically expensive, as eggsacs are large and females stop eating during this process (21 days) in *A. vierae* (Viera et al. 2007b) The next stages of brood care are very demanding and need a great energy and time investment, also in other subsocial species like *A. studiosus* (Fig. 13.11).

The spiderlings are not able to open the eggsac to hatch. Their natural or 'adopted' mothers must make a hole in the eggsac to allow the spiderlings to exit (Fig. 13.12).

Although mothers cannot recognize their own brood, they have an internal clock that is activated externally by movements of spiderlings inside of the eggsacs. This precision mechanism is adjusted at the oviposition event. We have in laboratory condition experiments exchanged eggsacs of different age, and the mothers opened

Fig.13.11 (**a**) *Anelosimus studiosus* female with eggsac. (**b**) *A. vierae* with eggsac (Photo A: M.O. Gonzaga; B: M. Trillo)

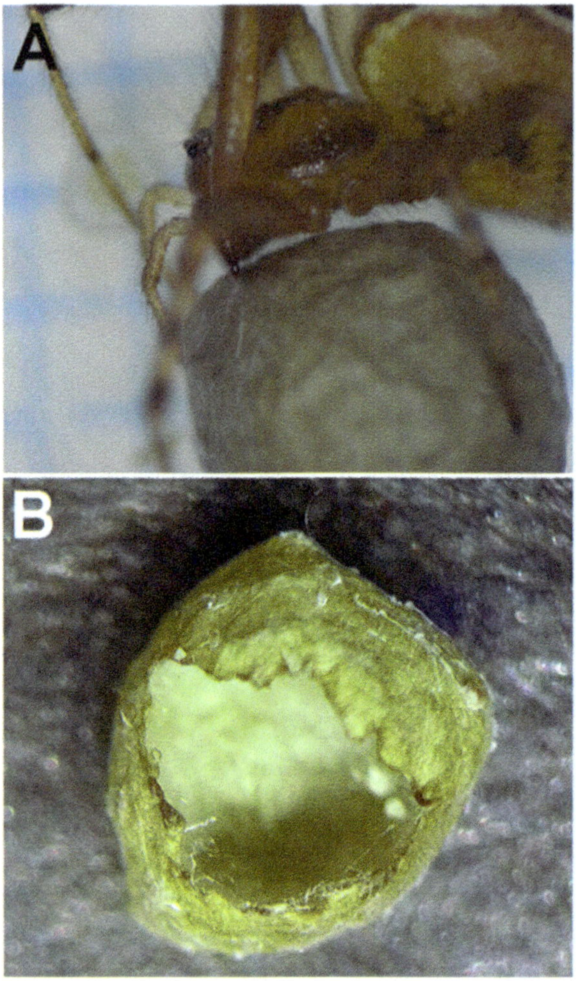

Fig. 13.12 (**a**) Mother opening the eggsac (*Anelosimus vierae*). (**b**) Eggsac showing a large hole (Photo A: M. Santana; B: C.Rojas)

the eggsacs when the time coincided with the expected time of hatching based on the date of their own eggsac-laying (Viera et al. 2007c). The mother cares for her brood continuously after the exit of spiderlings, and in this period we can point to the differences with the behavior of solitary spiders. Most solitary spiders show maternal behavior in protecting the eggsac, but maternal care ceases when the spiderlings emerge from the eggsacs. In some solitary spiders there is further maternal care; for example, in *Lycosa* the mother carries the spiderlings on her body for a while. But the effort and the complexity of parental investment by the subsocial spiders are remarkable, and individual investment is greater than in the social spiders, which can share the labors with other females of the same nest.

When the brood comes out of the eggsacs, the spiderlings are unable to feed by themselves and need their mother's help. Providing nutrients to the young during maternal care is one of the mechanisms that can affect growth and survival of the young, and their reproductive success. Salomon et al. (2011) found that maternal nutrition affects offspring performance via maternal care in a subsocial spider. The subsocial spider mother's investment in future reproduction by feeding her young was clearly illustrated by Reinhold (2002) for many taxa, highlighting the important role of extended maternal care.

Initially, the spiderlings are fed by regurgitation from the mother; and later the mother kills prey, tears it apart, or predigests it, and offers it to the spiderlings. During this process, the mother does not feed herself, waiting to eat after satiating her brood. Furthermore, mothers regurgitate to the spiderlings in a 'frozen' posture with open chelicerae, avoiding possible injuries (Fig. 13.13).

This process does not occur in adult virgin or mated females without brood (Viera et al. 2007c). Such females are tolerant and can care for the eggsacs, but do not open them nor feed the spiderlings by regurgitation. This intolerance among adult females in subsocial species, according to the maternal survival hypothesis, may help to explain the rarity of subsocial *Anelosimus* species in 'social habitats', such as lowland rainforests where they may experience higher predation risk and frequent web damage due to frequent precipitation (see Avilés et al. 2007).

The ability of brood to feed without the mothers´ help is very important in uni-female nests where there are not other females that can help. Spiderlings have been found to depend on their mother until the 4th instar in the subsocial *A. studiosus* (Brach, 1977). In *A. vierae*, Ghione et al. (2004) found that the spiderlings are capable of feeding by themselves already at the first instar, while they improve their prey-capture efficiency in later instars. This early independence from mothers makes these subsocial spiders similar to solitary spiders, as the premature death of the mothers would not rule out the survival of the young. In multi-females nests, subadult females can collaborate in cooperative capture and donate it to the younger ones. However, the presence of the mother in subsocial spiders is absolutely necessary for the care and opening of the eggsac. A final effort of maternal investment is to offer the body as food for its young, a common behavior in social spiders and others showing maternal care beyond the first feeding of spiderlings, such as in *Amaurobius ferox* (Kim et al. 2000)

Fig. 13.13 Mother of *Anelosimus vierae* feeding spiderlings

To summarize, subsocial spiders are a good model to examine the traits that underlie the evolution of permanent cooperative sociality. Multipronged approaches focusing on (1) phylogeny, (2) reproductive behavior, especially maternal care, and (3) ecological and abiotic factors relating to sociality are necessary to gain a holistic understanding of the evolution of social life in arthropods. A synthesis combining these approaches may also help to explain why no permanently social *Anelosimus* species occur outside the Americas. However, we do not attempt such a synthesis here, as more behavioral studies on reproductive aspects are needed to develop a species-level database for comparative purposes, as are further studies of biotic and abiotic factors in shaping the origin and distribution of sociality. Thorough comparative studies uniting these approaches with an explicit focus on evolutionary replica—independent origins of sociality—will probably offer the next major insights into social evolution in spiders and other animals.

Acknowledgements We are grateful to two anonymous reviewers who improved the final version of the chapter, and to Marcelo O. Gonzaga, Carolina Rojas, Mariana Trillo, and Martín Santana for photos. Viera also thanks the support of Sistema Nacional de Investigación (ANII), Comisión Sectorial de Investigación Cientifica (CSIC) and Programa de Desarrollo de las Ciencias Básicas, UdelaR, Uruguay.

References

Agnarsson I (2002) Sharing a web — on the relation of sociality and kleptoparasitism in Theridiid spiders (Theridiidae, Araneae). J Arachnol 30:181–188

Agnarsson I (2004) Morphological phylogeny of cobweb spiders and their relatives (Araneae, Araneoidea, Theridiidae). Zool J Linnean Soc 141:447–626

Agnarsson I (2005) Revision and phylogenetic analysis of American *ethicus* and *rupununi* groups of *Anelosimus* (Araneae, Theridiidae). Zool Scr 34:389–413

Agnarsson I (2006) A revision of the New World *eximius* lineage of *Anelosimus* (Araneae, Theridiidae) and a phylogenetic analysis using worldwide exemplars. Zool J Linnean Soc 146:453–593

Agnarsson I (2012a) A new phylogeny of Anelosimus and the placement and behavior of *Anelosimus vierae* n. sp from Uruguay (Araneae: Theridiidae). J Arachnol 40:78–84

Agnarsson I (2012b) Systematics of new subsocial and solitary Australasian *Anelosimus* species (Araneae: Theridiidae). Invertebr Syst 26:1–16

Agnarsson I and Zhang JX (2006) New species of Anelosimus (Araneae: Theridiidae) from Africa and Southest Asia, with notes on sociality and color polymorphism. Zootaxa 1147:1–34

Agnarsson I (2014) Félagslyndar köngulær á þróunarfræðilegum blindgötum. Náttúrufræðingurinn 84: 132-140 [in Icelandic with English abstract: 'social spiders on evolutionary blind alleys']

Agnarsson I, Kuntner M (2005) Madagascar: an unexpected hotspot of social *Anelosimus* spider diversity (Araneae : Theridiidae). Syst Entomol 30:575–592

Agnarsson I, Rayor LS (2013) A molecular phylogeny of the Australian huntsman spiders (Sparassidae, Deleninae): implications for taxonomy and social behaviour. Mol Phylogenet Evol 69:895–905

Agnarsson I, Avilés L, Coddington JA, Maddison WP (2006a) Sociality in theridiid spiders: repeated origins of an evolutionary dead end. Evolution 60:2342–2351

Agnarsson I, Barrantes G, May-Collado LJ (2006b) Notes on the biology of *Anelosimus pacificus* Levi, 1963 (Theridiidae, Araneae) — evidence for an evolutionary reversal to a less social state. J Nat Hist 40:2681–2687
Agnarsson I, Maddison WP, Avilés L (2007) The phylogeny of the social *Anelosimus* spiders (Araneae : Theridiidae) inferred from six molecular loci and morphology. Mol Phylogenet Evol 43:833–851
Agnarsson I, Kuntner M, Coddington JA, Blackledge TA (2010a) Shifting continents, not behaviours: independent colonization of solitary and subsocial spider lineages on Madagascar (Araneae, Theridiidae). Zool Scr 39(1):75–87
Agnarsson I, Maddison WP, Avilés L (2010b) Complete separation along matrilines in a social spider metapopulation inferred from hypervariable mitochondrial DNA region. Mol Ecol 19:3052–3063
Agnarsson I, Avilés L, Maddison WP (2013a) Loss of genetic variability in social spiders: genetic and phylogenetic consequences of population subdivision and inbreeding. J Evol Biol 26:27–37
Agnarsson I, Coddington J, Kuntner M (2013b) Spider systematics — state of the art and prospects. In: Penney D (ed) Spider research in the 21st century. Siri Scientific Press, Manchester
Agnarsson I, Jencik BB, Veve GM, Hanitriniaina S, Agostini D, Goh SP, Pruitt J, Kuntner M (2015) Systematics of the Madagascar Anelosimus spiders: remarkable local richness and endemism, and dual colonization from the Americas. Zookeys:13–52. https://doi.org/10.3897/zookeys.509.8897
Agnarsson I, Gotelli NJ, Agostini D, Kuntner M (2016) Limited role of character displacement in the coexistence of congeneric *Anelosimus* spiders in a Madagascan montane forest. Ecography 39:743–753
Albo MJ, Viera C, Costa FG (2007) Pseudocopulation and male–male conflict elicited by subadult females of the subsocial spider *Anelosimus cf. studiosus* (Theridiidae). Behaviour 144:1217–1234
Avilés L (1986) Sex-ratio bias and possible group selection in the social spider *Anelosimus eximius*. Am Nat 128:1–12
Avilés L (1987) Female biased sex ratios in social spiders: a result of group selection? In: Eder J, Rembold H (eds) chemistry and biology of social insects. Verlag J Peperny, Munich, p 349
Avilés L (1993a) Interdemic selection and the sex-ratio — a social spider perspective. Am Nat 142:320–345
Avilés L (1993b) Newly-discovered sociality in the Neotropical spider *Aebutina binotata* Simon (Dictynidae). J Arachnol 21:184–193
Avilés L (1994) Social behaviour in a web-building lynx spider, *Tapinillus* sp. (Araneae: Oxyopidae). Biol J Linnean Soc 52:163–176
Avilés L (1997) Causes and consequences of cooperation and permanent-sociality in spiders. In: Choe JC, Crespi BJ (eds) The evolution of social insects and arachnids. Cambridge University Press, Cambridge, pp 476–498
Avilés L (2000) Nomadic behaviour and colony fission in a cooperative spider: life history evolution at the level of the colony? Biol J Linnean Soc 70:325–339
Avilés L, Bukowski T-C (2006) Group living and inbreeding depression in a subsocial spider. Proc R Soc Lond B 273:157–163
Avilés L, Gelsey G (1998) Natal dispersal and demography of a subsocial *Anelosimus* species and its implications for the evolution of sociality in spiders. Can J Zool 76:2137–2147
Avilés L, Maddison W (1991) When is the sex-ratio biased in social spiders — chromosome-studies of embryos and male meiosis in *Anelosimus* species (Araneae, Theridiidae). J Arachnol 19:126–135
Avilés L, Purcell J (2011) Anelosimus Oritoyacu, a cloud forest social spider with only slightly female-biased primary sex ratios. J Arachnol 39:178–182
Avilés L, Purcell J (2012) The evolution of inbred social systems in spiders and other organisms: from short-term gains to long-term evolutionary dead ends? In: Adv Study Behav, vol 44, pp 99–133

Avilés L, Salazar P (1999) Notes on the social structure, life cycle, and behaviour of *Anelosimus rupununi*. J Arachnol 27:497–502
Avilés L, Tufino P (1998) Colony size and individual fitness in the social spider *Anelosimus eximius*. Am Nat 152:403–418
Avilés L, McCormack J, Cutter A, Bukowski T (2000) Precise, highly female-biased sex ratios in a social spider. Proc R Soc Lond B 267:1445–1449
Avilés L, Maddison WP, Salazar PA, Estevez G, Tufino P, Canas G (2001) Arañas sociales de la Amazonia ecuatoriana, con notas sobre seis especies sociales no descritas previamente. Rev Chil Hist Nat 74:619–638
Avilés L, Maddison W-P, Agnarsson I (2006) A new independently derived social spider with explosive colony proliferation and a female size dimorphism. Biotropica 38:743–753
Avilés L, Agnarsson I, Salazar PA, Purcell J, Iturralde G, Yip EC, Powers KS, Bukowski TC (2007) Natural history miscellany — altitudinal patterns of spider sociality and the biology of a new midelevation social Anelosimus species in Ecuador. Am Nat 70:783–792
Berger-Tal R, Berner-Aharon N, Aharon S, Tuni C, Lubin Y (2016) Good reasons to leave home: proximate dispersal cues in a social spider. J Anim Ecol 85:1035–1042
Bilde T, Lubin Y (2011) Group living in spiders: cooperative breeding and coloniality. In: Herberstein ME (ed) Spider behavior flexibility and versatility. Cambridge University Press, Cambridge, pp 275–306
Bilde T, Lubin Y, Smith D, Schneider JM, Maklakov AA (2005) The transition to social inbred mating systems in spiders: role of inbreeding tolerance in a subsocial predecessor. Evolution 59:160–174
Bilde T, Coates KS, Birkhofer K, Bird T, Maklakov AA, Lubin Y, Avilés L (2007) Survival benefits select for group living in a social spider despite reproductive costs. J Evol Biol 20(6):2412–2426
Brach V (1975) The biology of the social spider *Anelosimus eximius* (Araneae Theridiidae). Bull South Calif Acad Sci 74:37–41
Brach V (1976) Subsocial behavior in the funnel-web wolf spider *Sosippus floridanus* (Araneae: Lycosidae). Florida Entomol 59:225–229
Brach V (1977) *Anelosimus studiosus* (Araneae: Theridiidae) and the evolution of quasisociality in theridiid spiders. Evolution 31:154–161
Burgess JW (1978) Social behavior in group-living spider species. Symp Zool Soc London 42:69–78
Burghard GM (2005) The genesis of animal play:testing the limits. Bradford (Ed). MIT, Cambridge, Massachusetts, London, p 491
Buskirk RE (1981) Sociality in the Arachnida. In: Hermann HR (ed) Social insects, vol 2. Academic Press, New York, pp 281–367
Cangialosi KR (1990a) Kleptoparasitism in colonies of the social spider *Anelosimus eximius* (Araneae, Theridiidae). Acta Zool Fennica 190:51–54
Cangialosi KR (1990b) Social spider defense against kleptoparasitism. Behav Ecol Sociobiol 27:49–54
Coddington JA, Agnarsson I (2006) Subsociality in Helvibis Thorelli keyserling 1884 (Araneae, Theridiidae, Theridiinae) from French Guiana. J Arachnol 34:642–645
Coddington JA, Griswold CE, Davila DS, Penaranda E, Larcher SF (1991) Designing and testing sampling protocols to estimate biodiversity in tropical ecosystems. In: The unity of evolutionary biology: Proceedings of the Fourth International Congress of Systematic and Evolutionary Biology, 2 vols. pp 44–60
Coddington JA, Young LH, Coyle FA (1996) Estimating spider species richness in a southern Appalachian cove hardwood forest. J Arachnol 24:111–128
Coddington JA, Agnarsson I, Miller JA, Kuntner M, Hormiga G (2009) Undersampling bias: the null hypothesis for singleton species in tropical arthropod surveys. J Anim Ecol 78:573–584
Colwell RK, Coddington JA (1994) Estimating terrestrial biodiversity through extrapolation. Philos Trans R Soc Lond B 345:101–118
Elgar MA, Godfray HCJ (1987) Sociality and sex ratios in spiders. Trends Ecol Evol 2:6–7

Foelix RF (1982) Biology of spiders. Harvard University Press, Cambridge, p 306
Foelix RF (2011) Biology of spiders, 3rd edn. Oxford University Press, New York, p 412
Foellmer MW, Fairbairn DJ (2005) Competing dwarf males: sexual selection in a orb-weaving spider. J Evol Biol 18:629–641
Furey RE (1998) Two cooperatively social populations of the theridiid spider *Anelosimus studiosus* in a temperate region. Anim Behav 55:727–735
Gardner A, Ross L (2013) Haplodiploidy, sex-ratio adjustment, and eusociality. Am Nat 181:60–67
Ghione S, Costa FG, Viera C (2004) Ability to capture prey in early instars of the subsocial spider *Anelosimus studiosus* (Henz, 1850) from Uruguay (Araneae, Theridiidae). Bull Br Arachnol Soc 13(2):60–62
Gómez D, Rojas C, Viera C (2015) The influence of regurgitation by sister on male's growth in the subsocial spider *Anelosimus vierae* (Araneae, Theridiidae). Arachnol 16:252–254
Gonzaga MO, Vasconcellos-Neto J (2001) Female body size, fecundity parameters and foundation of new colonies in *Anelosimus jabaquara* (Araneae, Theridiidae). Insect Soc 48:94–100
Gonzaga MD, Vasconcellos-Neto J (2002a) Collective prey capture and feeding behaviours of *Anelosimus jabaquara* Levi 1956 (Araneae : Theridiidae). Behaviour 139:573–584
Gonzaga MD, Vasconcellos-Neto J (2002b) Influence of collective feeding on weight gain and size variability of *Anelosimus jabaquara* Levi 1956 (Araneae : Theridiidae). Behaviour 139:1431–1442
Grinsted L, Breuker CJ, Bilde T (2014) Cooperative breeding favors maternal investment in size over number of eggs in spiders. Evolution 68:1961–1973
Guevara J, Avilés L (2007) Multiple techniques confirm elevational differences in insect size that may influence spider sociality. Ecol 88:2015–2023
Guevara J, Avilés L (2009) Elevational changes in the composition of insects and other terrestrial arthropods at tropical latitudes: a comparison of multiple sampling methods and social spider diets. Insect Conserv Divers 2:142–152
Guevara J, Avilés L (2015) Ecological predictors of spider sociality in the Americas. Glob Ecol Biogeogr 24:1181–1191
Hamilton W (1964a) The genetical evolution of social behaviour. II J Theor Biol 7:17–52
Hamilton W (1964b) The genetical evolution of social behaviour. I J Theor Biol 7:1–16
Harwood G, Avilés L (2013) Differences in group size and the extent of individual participation in group hunting may contribute to differential prey-size use among social spiders. Biol Lett 9(6):20130621. https://doi.org/10.1098/rsbl.2013.0621
Holbrook CT, Wright CM, Pruitt JN (2014) Individual differences in personality and behavioural plasticity facilitate division of labour in social spider colonies. Anim Behav 97:177–183
Ito C, Shinkai A (1993) Mother-young interactions during the brood-care period in *Anelosimus crassipes* (Araneae: Theridiidae). Acta Arachnol 42:73–81
Jiménez-Castillo M, Wiser SK, Lusk CH (2007) ORIGINAL ARTICLE: Elevational parallels of latitudinal variation in the proportion of lianas in woody floras. J Biogeogr 34(1):163–168
Johannesen J, Lubin Y, Smith DR, Bilde T, Schneider JM (2007) The age and evolution of sociality in Stegodyphus spiders: a molecular phylogenetic perspective. Proc R Soc Lond B 274:231–237
Johannesen J, Moritz RFA, Simunek H, Seibt U, Wickler W (2009a) Species cohesion despite extreme inbreeding in a social spider. J Evol Biol 22:1137–1142
Johannesen J, Wickler W, Seibt U, Moritz RFA (2009b) Population history in social spiders repeated: colony structure and lineage evolution in *Stegodyphus mimosarum* (Eresidae). Mol Ecol 18:2812–2818
Jones TC, Parker PG (2000) Costs and benefits of foraging associated with delayed dispersal in the spider *Anelosimus studiosus* (Araneae, Theridiidae). J Arachnol 28:61–69
Jones TC, Parker PG (2002) Delayed juvenile dispersal benefits both mother and offspring in the cooperative spider *Anelosimus studiosus* (Araneae : Theridiidae). Behav Ecol 13:142–148
Jones TC, Riechert SE, Dalrymple SE, Parker PG (2007) Fostering model explains variation in levels of sociality in a spider system. Anim Behav 73(1):195–204

Kim KW, Roland C, Horel A (2000) Functional value of matriphagy in the spider *Amaurobius ferox*. Ethology 106:729–742
Knoflach B (1998) Mating in *Theridion varians* Hahn and related species (Araneae, Theridiidae). J Nat Hist 32:545–604
Krafft B (1985) Les araignées sociales. Recherche 16:884–892
Krafft B, Pasquet A (1991) Synchronized and rhythmical activity during the prey capture in the social spider *Anelosimus eximius* (Araneae, Theridiidae). Insect Soc 38:83–90
Kraus O, Kraus M (1988) The genus Stegodyphus (Arachnida, Araneae). Sibling species, species groups, and parallel origin of social living. Verh NatVereins in Hamburg 30:151–254
Kullmann E (1972) Evolution of social behavior in spiders (Araneae, Eresidae and Theridiidae). Amer Zool 12:419–426
Leborgne R, Krafft B, Pasquet A (1994) Experimental study of foundation and development of *Anelosimus eximius* colonies in the tropical forest of French Guiana. Insect Soc 41:179–189
Levi HW, Smith DRR (1982) A new colonial {\iAnelosimus} spider from Suriname (Araneae: Theridiidae). Psyche 89:275–278
Levin SA (2009) The Princeton guide to ecology. Princeton University Press, Princeton
Lichtenstein JLL, Pruitt JN (2015) Similar patterns of frequency-dependent selection on animal personalities emerge in three species of social spiders. J Evol Biol 28:1248–1256
Liu J, May-Collado LJ, Pekar S, Agnarsson I (2016) A revised and dated phylogeny of cobweb spiders (Araneae, Araneoidea, Theridiidae): a predatory cretaceous lineage diversifying in the era of the ants (hymenoptera, Formicidae). Mol Phylogenet Evol 94:658–675
Lorieto V, Ghione S, Viera C (2010) An indirect aproach to study sperm precedence in a subsocial spider. Ethol Ecol Evol 22:1–10
Losos J, Ricklefs R (2010) The theory of island biogeography revisited. Princeton Univ Press, Princeton
Lubin YD (1991) Patterns of variation in female-biased colony sex ratios in a social spider. Biol J Linn Soc 43:297–311
Lubin Y, Bilde T (2007) The evolution of sociality in spiders. Advances in the study of behavior, vol 37. Elsevier academic press Inc, San Diego, pp 83–145
Lubin YD, Robinson MH (1982) Dispersal by swarming in a social spider. Science 216:319–321
Lubin Y, Birkhofer K, Berger-Tal R, Bilde T (2009) Limited male dispersal in a social spider with extreme inbreeding. Biol J Linn Soc 97:227–234
Majer M, Agnarsson I, Svenning JC, Bilde T (2013) Social spiders of the genus *Anelosimus* occur in wetter, more productive environments than non-social species. Naturwissenschaften 100:1031–1040
Marques ESA, Vasconcelos-Netto J, Britto-de-Mello M (1998) Life history and social behavior of *Anelosimus jabaquara* and *Anelosimus dubiosus* (Araneae, Theridiidae). J Arachnol 26:227–237
Moura RR, Vasconcellos-Neto J, Gonzaga M d O (2017) Extended male care in Manogea porracea (Araneae: Araneidae): the exceptional case of a spider with amphisexual care. Anim Behav 123:1–9
Nentwig W (1985) Social spiders catch larger prey: a study of *Anelosimus eximius* (Araneae: Theridiidae). Biol Bull 17:79–85
Nentwig W, Christenson TE (1986) Natural history of the non-solitary sheetweaving spider *Anelosimus jucundus* (Araneae: Theridiidae). Zool J Linnean Soc 87:27–36
Overal WL, Silva RF (1982) Population dynamics of the quasisocial spider *Anelosimus eximius* (Araneae: Theridiidae). Breed, MD, Michener, CD & Evans, HE (Eds) the biology of social insects proceedings of the ninth congress of the International Union for the Study of social insects, boulder, Colorado, August: 181–182
Pasquet A, Krafft B (1989) Colony distribution of the social spider *Anelosimus eximius* (Araneae, Theridiidae) in French Guiana. Insect Soc 36:173–182
Pasquet A, Krafft B (1992) Cooperation and prey capture efficiency in a social spider, *Anelosimus eximius* (Araneae, Theridiidae). Ethology 90:121–133

Pasquet A, Trabalon M, Bagneres AG, Leborgne R (1997) Does group closure exist in the social spider *Anelosimus eximius*? Behavioural and chemical approach. Insect Soc 44:159–169
Powers KS, Avilés L (2003) Natal dispersal patterns of a subsocial spider *Anelosimus* cf. *jucundus* (Theridiidae). Ethology 109:725–737
Powers KS, Avilés L (2007) The role of prey size and abundance in the geographical distribution of spider sociality. J Anim Ecol 76(5):995–1003
Pruitt JN, Iturralde G, Avilés L, Riechert SE (2011) Amazonian social spiders share similar within-colony behavioural variation and behavioural syndromes. Anim Behav 82:1449–1455
Pruitt JN, Oufiero CE, Avilés L, Riechert SE (2012) Iterative evolution of increased behavioral variation characterizes the transition to sociality in spiders and proves advantageous. Am Nat 180:496–510
Purcell J (2011) Geographic patterns in the distribution of social systems in terrestrial arthropods. Biol Rev 86:475–491
Purcell J, Avilés L (2007) Smaller colonies and more solitary living mark higher elevation populations of a social spider. J Anim Ecol 76:590–597
Purcell J, Avilés L (2008) Gradients of precipitation and ant abundance may contribute to the altitudinal range limit of subsocial spiders: insights from a transplant experiment. Proc R Soc Lond B 275:2617–2625
Reinhold K (2002) Maternal effects and the evolution of behavioral and morphological characters: a literature review indicates the importance of extended maternal care. J Hered 93:400–405
Riechert SE, Roeloffs RM (1993) Evidence for and consequences of inbreeding in the cooperative spiders. In: Thornhill N (ed) The natural history of inbreeding and outbreeding. University of Chicago Press, Chicago, pp 283–303
Riechert SE, Jones TC (2008) Phenotypic variation in the social behaviour of the spider Anelosimus studiosus along a latitudinal gradient. Anim Behav 75(6):1893–1902
Rojas-Buffet C, Viera C (2015) Reproductive behaviour in the uruguayan subsocial spider *Anelosimus vierae*: a review. Bol Soc Zool Uruguay 24:206–216
Rojas-Buffet C, Viera C (2016) Loading the male pedipalps: sperm induction in a subsocial spider. J Arachnol 44:96–98
Roubik DW (1989) Ecology and natural history of tropical bees. Cambridge University Press, New York
Ruch J, Heinrich L, Bilde T, Schneider JM (2009) The evolution of social inbreeding mating systems in spiders: limited male mating dispersal and lack of pre-copulatory inbreeding avoidance in a subsocial predecessor. Biol J Linn Soc 98:851–859
Ruch J, Riehl T, May-Collado LJ, Agnarsson I (2015) Multiple origins of subsociality in crab spiders (Thomisidae). Mol Phylogenet Evol 82:330–340
Rypstra AL (1993) Prey size, social competition, and the development of reproductive division-of-labor in social spider groups. Am Nat 142:868–880
Rypstra AL, Tirey RS (1989) Observations on the social spider *Anelosimus domingo* (Araneae Theridiidae) in southwestern Peru. J Arachnol 17:368–371
Saffre F, Deneubourg JL (2002) Swarming strategies for cooperative species. J Theor Biol 214:441–451
Saffre F, Mailleux AC, Deneubourg JL (1999) Dragline attachment pattern in the neotropical social spider *Anelosimus eximius* (Araneae: Theridiidae). J Insect Behav 12:277–282
Saffre F, Mailleux AC, Deneubourg JL (2000) Exploratory recruitment plasticity in a social spider (*Anelosimus eximius*). J Theor Biol 205:37–46
Salomon M, Mayntz D, Toft S, Lubin C (2011) Maternal nutrition affects offspring performance via maternal care in a subsocial spider. Behav Ecol Sociobiol 65:1191–1202
Samuk K, Avilés L (2013) Indiscriminate care of offspring predates the evolution of sociality in alloparenting social spiders. Behav Ecolo Sociobiol 67:1275–1284
Schneider JM, Roos J, Lubin Y, Henschel JR (2001) Dispersal of Stegodyphus Dumicola (Araneae, Eresidae): they do balloon after all! J Arachnol 29:114–116

Settepani V, Grinsted L, Granfeldt J, Jensen JL, Bilde T (2013) Task specialization in two social spiders, Stegodyphus Sarasinorum (Eresidae) and *Anelosimus eximius* (Theridiidae). J Evol Biol 26:51–62
Settepani V, Bechsgaard J, Bilde T (2016) Phylogenetic analysis suggests that sociality is associated with reduced effectiveness of selection. Ecol Evol 6:469–477
Smith DR (1986) Population genetics of a cooperative spider, *Anelosimus eximius* (Theridiidae). Barrientos. JA [Ed] Actas 1:1–428
Smith DR (1987) Genetic variation in solitary and cooperative spiders of the genus Anelosimus (Araneae: Theridiidae). In: Eder J, Rembold H (eds) Chemistry and biology of social insects. Verlag J Peperny, Munich, pp 347–348
Smith DR, Hagen RH (1996) Population structure and interdemic selection in the cooperative spider *Anelosimus eximius*. J Evol Biol 9:589–608
Smith DR, Su YC, Berger-Tal R, Lubin Y (2016) Population genetic evidence for sex-specific dispersal in an inbred social spider. Ecol Evol 6:5479–5490
Stevens GC (1992) The elevational gradient in altitudinal range: an extension of Rapoport's latitudinal rule to altitude. Am Nat 140(6):893–911
Swenson NG, Enquist BJ (2007) Ecological and evolutionary determinants of a key plant functional trait: wood density and its community-wide variation across latitude and elevation. Am J Bot 94(3):451–459
Trivers RL (1974) Parent-offspring conflict. Int Comp Biol 14:249–264
Uetz GW (1983) Sociable spiders. Nat Hist 92:62–68
Uetz G W, Hodge MA (1990) Influence of habitat and prey availability on spatial organization and behavior of colonial web building spider. Nat Geogr 6:22–40
Vakanas G, Krafft B (2001) Coordination of behavioral sequences between individuals during prey capture in a social spider, *Anelosimus eximius*. J Insect Behav 14:777–798
Vakanas G, Krafft B (2004) Regulation of the number of spiders p articipating in collective prey transport in the social spider *Anelosimus eximius* (Araneae, Theridiidae). C R Biol 327:763–772
Vasconcelos-Netto J, Mello M-B-D (1998) Life history and social behavior of Anelosimus Jabaquara and *Anelosimus dubiousus* (Araneae, Theridiidae). J Arachnol 26:227–237
Venticinque EM, Fowler HG (1998) Sheet web regularity: fixed isometric relationships in the social spider *Anelosimus eximius* (Araneae: Theriididae). Cienc Cult 50:371–373
Venticinque EM, Fowler HG (2001) Local extinction risks and asynchronies: the evidence for a metapopulation dynamics of a social spider, *Anelosimus eximius* (Araneae, Theridiidae). In: ROJ B, Gascon C, Lovejoy TE, RCG M (eds) Lessons from Amazonia: the ecology and conservation of a fragmented Forest. Yale University Press, New Haven, p 187
Venticinque EM, Fowler HG, Silva CA (1993) Modes and frequencies of colonization and its relation to extinctions, habitat and seasonality in the social spider *Anelosimus eximius* in the Amazon (Araneidae: Theridiidae). Psyche 100:35–41
Viera C, Albo MJ (2008) Males of a subsocial spider choose among females of different ages and the same reproductive status. Ethol Ecol Evol 20:35–41
Viera C, Garcia LF (2009) A new record of a social spider, *Anelosimus lorenzo* fowler and Levi, 1979 (Araneae, Theridiidae), from a temperate zone (Uruguay). Gayana 73:237–240
Viera C, Ghione S and Costa FG (2005) Regurgitation among penultimate juveniles in the subsocial spider Anelosimus cf. studiosus (Theridiidae): are males favored? J Arachnol 34:258–260.
Viera C, Ghione S, Costa FG (2006) Regurgitation among penultimate juveniles in the subsocial spider *Anelosimus* cf. *studiosus* (Theridiidae): are males favored? J Arachnol 34:258–260
Viera C, Costa FG, Ghione S, Benamu-Pino MA (2007a) Progeny, development and phenology of the sub-social spider *Anelosimus cf. studiosus* (Araneae, Theridiidae) from Uruguay. Stud Neotrop Fauna Environ 42:145–153
Viera C, Ghione S, Costa F-G (2007b) Post-embryonic development of the sub-social spider *Anelosimus* cf. *studiosus* (Araneae, Theridiidae). Bull Brit Arachnol Soc 14:30–32
Viera C, Ghione S, Costa FG (2007c) Mechanisms underlying egg-sac opening in the subsocial spider *Anelosimus* cf. *studiosus* (Araneae Theridiidae). Ethol Ecol Evol 19:61–67

Vollrath F (1982) Colony Foundation in a social spider Zeitschrift Fur Tierpsychologie. J Comp Ethol 60:313–324
Vollrath F (1986) Eusociality and extraordinary sex-ratios in the spider *Anelosimus eximius* (Araneae, Theridiidae). Behav Ecol Sociobiol 18:283–287
Vollrath F, Parker GA (1992) Sexual dimorphism and distorted sex ratios in spiders. Nature 360:156–159
Vollrath F, Rohde-Arndt D (1983) Prey capture and feeding in the social spider *Anelosimus eximius*. Z Tierpsychol 61:334–340
Vollrath F, Windsor DM (1986) Subsocial and social *Anelosimus*: a comparison, especially of nest defense. In: Eberhard WG, Lubin YD, Robinson BC (eds) Proceedings of the Ninth International Congress of Arachnology, Panamá,1983. Smithsonian Inst. Press, Washington, D.C., pp 295–298
Wilson EO (1971) The insect societies. Harvard University Press, Cambridge, p 548
Wilson EO (1975) Sociobiology: the new synthesis. Harvard University Press, Princeton, p 697
World Spider Catalog (2017). World spider catalog. Natural History Museum Bern, online at http://wsc.nmbe.ch, version 17.5, Accessed on Jan 2017
Yip EC, Rayor LS (2014) Maternal care and subsocial behaviour in spiders. Biol Rev 89:427–449
Yip EC, Powers KS, Aviles L (2008) Cooperative capture of large prey solves scaling challenge faced by spider societies. Proc Natl Acad Sci 105(33):11818–11822

Chapter 14
Plasticity and Cognition in Spiders

Hilton F. Japyassú

Abstract Spiders can be a particularly important model for the study of cognition. Their close interaction with niche-constructed environmental features, such as webs, cocoons, draglines or retreats, allows for the experimental manipulation of these silken structures, and thus for a controlled study of the cognitive machinery that underlie the use and construction of these structures. There are contrasting theories about cognition, and we explore particularly the opposition between the traditional approach, the one that requires information to be processed solely within the central nervous system (CNS), and the extended cognition approach, which is less restrictive. Here we review the literature on spider cognition with an eye to the experimental data that allows the contrast between these theories of cognition, and conclude that spiders evolved to process information prior to reaching the nervous system: they use their webs to decide whether to attack or not a prey item, and we can experimentally alter their decision by manipulating web properties, such as radii tension. The experimental manipulation of web threads also alters the attentional state of the web building spider so that she predictably ignores important cues for decisions taken during the building process. Together, the experimental evidence shows that spiders extend their cognitive machinery outside the bounds of their CNS, making use of the external silken structures to offload cognitive processing. This insight may help to explain graded changes in brain/body allometry, because smaller animals could rely more on extended cognition so as not to be behaviourally limited by a relatively small brain. Extended cognition could also help explain the emergence of new levels of organisation, particularly the transition from solitary to social life. In general, extended cognition emerges as a natural bridge between two traditionally separate research agendas: the area of cognitive development (learning mechanisms) and that of evolution through natural selection.

H.F. Japyassú (✉)
Instituto de Biologia, Universidade Federal da Bahia, Salvador, BA, Brazil
e-mail: japyassu@ufba.br

C. Viera, M.O. Gonzaga (eds.), *Behaviour and Ecology of Spiders*,
https://doi.org/10.1007/978-3-319-65717-2_14

Cognition, Plasticity, and Evolution

Although cognition clearly leads to plasticity, the reverse is not necessarily the case. Early ethologists provide not only examples of so-called instinctive behavior, most of them rigid performances elicited only in the presence of specific innate releasing stimuli, but also plenty of cases of alternative instinctive responses to alternative stimuli. Together, these early examples demonstrate that classic ethology framed instinct and plasticity in a continuum from rigid or stereotyped to ever more labile responses (Japyassú and Malange 2014).

The more a behavioral system allows for alternative responses, the more there is room for what we call cognitive performances, that is, performances that require flexible information[1] processing abilities, with concomitant behavioral adjustments to environmental changes (Japyassú 2008). These adjustments could be adaptive, or alternatively could be unforeseen responses to new ecological challenges (Penna-Gonçalves et al. 2008), particularly when animals are exposed to environments outside of the normal range for the species (Ghalambor et al. 2007). Finally, these new responses could be stable, either through self-organization and/or learning processes; and as stable new behavioral characteristics, these responses could drive evolution to new local optima (Japyassú 2010) through evolutionary processes such as phenotypic accommodation or assimilation (West-Eberhard 2003).

Models of Cognition

We do not discuss any general definition of cognition; instead we embrace a broad and prevalent definition of cognition as the acquisition, processing, storage, and use of information (Shettleworth 2010). Because we stick to a semantic conception of information[2], the acquisition of information is the act of making sense about some aspect of the world. Within this general framework, there is plenty of room for disagreement, and many distinct models of cognition are available in the literature.

[1] Information has at least three broad meanings: the statistical, the semantic, and the physical (Harms 2006). We use the semantic sense to characterize "knowledge of" or "meaning" in both the referential properties of symbols and instructional aspects of knowledge in natural biological systems.

[2] The actual meaning of a piece of information depends not only on the referent (the external object), but also on the internal state of the system. In the first case, meaning involves a denotative relation between a sign and its counterpart in the external world (the referent). In the second case, meaning involves a connotative relation between the sign and the internal elements of the system, a relationship that ensures an interpretation, that leads to a procedure or a path of action within the system (Harms 2004, 2006). This second, interpretive side of information requires a characterization of the connectivity between the internal elements of the system, and is thus by definition a relational conception of information. This idea of a system of mutual relations is also relevant to naturalize important properties of any cognitive system, such as agency and normativity (Moreno and Mossio 2015).

One well-known controversy revolves around information processing. Some consider that cognition operates through a general-purpose learning mechanism, one that can solve very different kinds of problems, such as finding a rewarding food patch or finding the appropriate sexual partner. Others conceive cognition as a specialized learning device, one that is evolved to solve a single problem or a single class of problems. In this last conception, the mind comprises a myriad of neural modules, each processing adaptively one single kind of information. The first conception is prevalent among behaviorists, whereas the second is common among evolutionary biologists and psychologists, as well as among behavioral biologists (Laland and Brown 2011; Sanderson 2014).

These conceptions have opposing opinions about the quantity of previous information a cognitive mechanism should have to learn to solve problems. General-purpose learning mechanisms have no a priori information about any particular problem; instead, these mechanisms should extract information from the problem at hand, should find regularities and patterns while experiencing successive instances of that same kind of problem, thus learning the problem structure, and creating memories that help to solve similar tasks in the future. Specialized learning mechanisms, on the other hand, have a priori information that help the system to find the solution to the task. For example, although wandering jumping spiders from the genus *Portia* do learn how to enter safely into the orb-web of a prey-spider, they optimize the trial and error learning procedure by trying, from scratch, reasonable frequencies and intensities of the aggressive mimicry signals (Jackson and Nelson 2011). In this way, these spiders should have previous information on the most effective range of signal frequencies before starting to learn.

Notwithstanding the differences between these opposing conceptions of cognition, both the general and the specialized conceptions share a common assumption. Both agree that cognition is something that happens within the brain, or the central nervous system (CNS). This is the traditional approach to cognition, whereby central cognitive processing is postulated in abstraction from bodily mechanisms, be it sensory processing or motor control. In contrast, there is a constellation of theories about cognition that downplay the importance of the brain or CNS, arguing instead that cognition extends in various ways to the non-CNS body, or even to the nearby environment. Thus, distinct perspectives oppose themselves to the above conceptions, postulating complementary or alternative theories, such as embodied, situated, embedded, extended, or enacted cognition (Wilson and Foglia 2017).

Embodied cognition theory postulates that the physical structure of the body is part of the solution to ecological problems that animals face, and thus that bodily information helps to reduce the requirements for CNS information processing (Shapiro 2010). As an example, robot bipedal locomotion is much less computationally intensive if the robot physical body has a design that narrows the range of possible movements (Matsushita et al. 2005; Pfeifer et al. 2006). Thus, one can trade information processing for adaptive morphology, and reduce the necessity for central cognition by building upon the natural properties of the materials that constitute a robot, or an animal, as researchers on morphological computation and soft robotics have been keen to demonstrate (Pfeifer et al. 2014). Embedded cognition is one strand of the situated cognition theories. It extends cognition to encompass the natural and social environment. The focus is on the strategies that organisms use to off-load cognitive processing onto the environment. This could be as simple as flagging food-caching sites to facilitate later food retrieval, or as complex as using cloud-computation to process big-data. This specialized use of external devices opens room for the idea of a cognitive performance that is distributed across the physical, social, or cultural environment (Hutchins 1995).

The thesis of extended cognition is the claim that cognitive systems themselves extend beyond the boundary of the individual organism (Clark and Chalmers 1998). The practical difference between embedded and extended approaches is that, in the embedded or situated approaches, you use the environment (for example, a computer) to process part of the information so that your brain and body can solve the problem with less effort; now, in the extended approach, the claim is that, in some cases, the environment not only helps, but instead becomes so intimately connected to the agent that it becomes part of the bodily information processing system (as if the computer was an integral part of your body, like a chip implanted on the brain).

Finally, in the enacted approach, what constitutes cognition is the dynamic coupling between environment and the biological system (Thompson 2007). This approach is similar to, but does not go so far as the dynamic systems approach (Thelen and Smith 1994) which almost eliminates the distinctiveness of cognition, or informational processes, from other biological processes.

This is not the place to go into the details of any of these theories about cognition. Instead, following Japyassú and Laland (2017), we will contrast the centralized and extended approaches to cognition. By centralized, we mean all the traditional cognitive science theories that consider cognition as something that happens within the CNS. This is in contrast to embodied, embedded, enacted, extended, and dynamic system theories, that posit cognitive relevance to the extra-CNS elements that help to solve problems. We shall call these theories collectively the extended cognition approach, because of their emphasis on extending cognition from the encapsulated brain to its external world.

We will discuss findings on spider cognition that allow the distinction between these contrasting approaches, so that we can have an empirical grasp of the theoretically rich debate on cognition. To distinguish the boundaries of the cognitive system, to decide if it is restricted to brain functioning or extends to the body or nearby environment, we will take advantage of the mutual manipulability criterion (MM, Kaplan 2012). In a nutshell, the MM specifies that two components are part of one same system if they reciprocally alter each other. For example, if manipulating something (external to the CNS) results in cognition changes and, in the opposite direction, altering the CNS internal cognitive processes result in changes in this external part of the system, then we shall conclude that this external part is indeed a constituent of the cognitive system and, accordingly, cognition extends.

Spider Extended Cognition

The idea that there is either innate or learned information within the CNS for the performance of a behavior is a trivial one. Phylogenetic or developmental memories are considered relevant for animal adaptive responses. Nevertheless, there is evidence that spiders solve problems building upon information that is not within their brains (Japyassú 2008). The interaction between unusual external features of the environment (such as an experimentally modified web) and the normal spider brain not only results in novel, unforeseen, and adaptive foraging behavior in the very first performance, but this new behavior is also a stable output of the behavioral system (Penna-Gonçalves et al. 2008). Stable behavioral outputs can result from the self-organization (see below) of the system's components (brain/body/environment) in the very moment of the interaction of the animal with the environment, in this case, the altered web. The take-home lesson here is that the information for adaptive responses is not necessarily available in the brain beforehand: that information can emerge at the exact moment of the performance.

Self-organization is a process well known for producing emergent properties, that is, properties that only appear after the interaction between the components of the system. For example, social behavior in ants always results in the correct, collective choice of the shortest trail to alternative food sources, but no individual ant knows which is the best alternative; the best choice emerges from systematic interactions among the individuals of the colony (Sumpter 2010). The point we are exploring in this section is a kind of downgrading of this same logic, from the social to the individual level. Emergent properties could appear in the interaction between one individual and its surrounding environment, and these properties would be particularly important when the relevant features of this environment occur regularly in his ecological niche.

Webs and silk threads are a regular feature of the ecological niche of spiders. Spiders use silk for a variety of functions, from egg-sac protection, to courtship, defense, territorial marking, or foraging. Silk is an old spider adaptation, one that defines the whole taxonomic group, and thus a niche-constructed environmental feature that is particularly well suited to co-evolve with spider behavior and cognition.

Foraging is one of the main functions of spider webs and silk, and cognition is clearly involved in foraging. For example, web spiders memorize the characteristics of a single captured prey, such as the prey type, size, and location (Ades 1988; Rodríguez and Gamboa 2000; Rodríguez and Gloudeman 2011; Rodríguez et al. 2013). Also, web-builders benefit from specialized, rapid one-trial learning during the formation of search images for avoiding dangerous prey, such as ants (Henaut et al. 2014). They can even learn to change web properties, such as web asymmetry (Nakata 2012) or the size of a region of the web (Heiling and Herberstein 1999), so as to optimize future prey capture on the basis of previous foraging experience. Finally, web-building spiders even show a sense of numerosity connected to foraging. Spiders wrap successively captured prey in one single silk package, and store this package for eating later. Packages imperceptibly removed from the web are actively searched for by the spider, in a clear indication of memory. More to the point, packages with a larger number of prey items (irrespective of the prey size) are more intensively searched for than packages with a smaller number of prey items (Rodríguez et al. 2015). A sense of numerosity requires the ability of abstraction, which is an impressive cognitive feat for tiny spiders.

As shown above, there is clear evidence for the conjunction between cognition and silk thread manipulation. Our point is to evaluate if this cognitive machinery extends to web threads, through the application of the mutual manipulability criterion (MM). The available information in the literature allows the application of the MM in exemplary cases: prey capture and web building. As we see below, these exemplary cases provide positive evidence for extended cognition.

Prey capture Animals always have to choose between alternative paths of action, and these decisions frequently rely on memories of the consequences of previous and similar decisions. The decision about proceeding to the capture of an ensnared prey item depends on the evaluation of its profitability and of the costs involved. Many spiders ignore small prey items, particularly when they are sated, but hungry spiders hunt these less profitable prey. This decision involves adaptive processing of information by the web threads. Web-building spiders can actively focus attention on a particular web portion. They do that by pulling more strongly on the web threads from the more profitable areas of the trap, a behavior that has been shown to lead to enhanced capture success in these web regions (Nakata 2010). Hungry spiders pull the radii more intensely than sated spiders, and thus respond more promptly to less profitable prey, such as fruit flies (Watanabe 2000). Also, enhanced attention to specific web areas can be artificially induced by experimentally augmenting radii tension, and spiders respond more quickly to stimuli coming from the tensed region of the web (Watanabe 2000, Nakata 2010). These attentional changes can have lasting results, as spiders can learn to focus on particularly profitable web areas, tensioning these areas more strongly as soon as the web has been built anew (Nakata 2013).

If we use the MM criteria (Kaplan 2012), it becomes clear that spider cognition extends to the web threads, because they change, and are changed by, CNS cognition. CNS cognition can alter thread tension (for example, hungry spiders tense web radii), but the reverse also is true; radii tension changes result in attentional changes, making the spider systematically ignore some prey. Radii tension modulates the decision to either attack or ignore a prey item, and the spider uses thread tension to process prey information adaptively.[3] Spiders are able to tune their webs to become more sensitive to distinct kinds of stimuli. In this sense, web threads cannot be understood as passive transmitters, or even passive filters of vibratory information. Thread properties are adjustable, and can process the same information in distinct

[3]The radial threads modulate the resonance and the attenuation of prey vibrations, as well as the velocity of their propagation, and thereby promote signal transformation through the web (Landolfa and Barth 1996). Tense threads increase the amplitude of some, and reduce the amplitude of other prey vibration frequencies (Mortimer et al. 2015).

ways, thus conveying different outputs to one same input. Spiders change thread properties in a way that is functionally similar to the way they change the properties of actual neural networks (for example, facilitating synapses), so the web must be considered parcel and part of their cognitive system. This intimate connection between the nervous system and web threads is further reinforced if one considers the informational liaisons between web-building algorithms and web structure.

Web building Not only while sensing prey through the web, but also while building the web, the spider uses the structure and spatial distribution of threads as cognitive devices that reduce the difficulty of the very web-building process. Cognition is ubiquitous in the building process: spiders adjust mesh size to cope with distinct prey (Murakami 1983; Sandoval 1994; Schneider and Vollrath 1998; Heiling and Herberstein 2000), learning from experience with previous webs the structural changes to be accomplished in subsequent buildings (Heiling and Herberstein 1999; Venner et al. 2000). If forced to build its vertical web in a horizontal cage, *Argiope argentata* will first build a very irregular horizontal and planar structure, but subsequent buildings become progressively similar to regular orbs (Nogueira and Ades 2012), in a compelling example of long-term web-building learning.

Attention is necessary while building webs, because the spider has to evaluate multiple cues to decide the position of the next threads in the emerging trap. External cues, such as prey-induced vibratory stimuli and prey nutrients (Pasquet et al. 1994; Blamires et al. 2011), wind intensity (Wu et al. 2013), gravity (Witt et al. 1976; Eberhard 1987; Vollrath 1988a, b), and humidity (Baba et al. 2014), all inform distinct aspects of the final web. Spiders also use internal cues to guide web building, such as the amount of silk supply, spider size, weight (Eberhard 1988a), and leg length (Witt et al. 1968; Vollrath 1987). Finally, the spider relies on cues put in place in the building process (position, angle, distances along threads), and on memory of these cues in previous assessments, to decide the next building steps (Eberhard 1972, 1982, 1988b, 2012a, 2012b). As a minor example of the complexities involved in web-building, we will detail the decision about the distance between successive sticky spiral segments on one specific radius of the orb. This decision involves the assessment of many distinct cues, such as reference points (the position of the inner loop of sticky spiral; the position of the outer loop of temporary spiral), the distance from the hub, the angle of the radius with gravity, the distance between radii, the measurement of distances (such as the actual temporary spiral to inner loop distance, Eberhard and Wcislo 2011). The spider also has to compare actual distances with either short-term memories of similar distances in the previous sticky spiral segment attachment, or less recent memories concerning the attachment of the previous sticky spiral loop, on the same radius (see review at Eberhard and Wcislo 2011; Eberhard, *in prep.*). Sometimes the spider ignores some cues in favor of others, for example when faced with conflict between distinct cues, either in natural or in experimental webs (Eberhard 2011; Eberhard and Hesselberg 2012; see review at Japyassú and Laland 2017). For example, when fixing one sticky spiral segment, the spider senses some cues on the radii (the position of the inner sticky spiral loop, and the position of the outer temporary spiral loop), to calculate the distance between

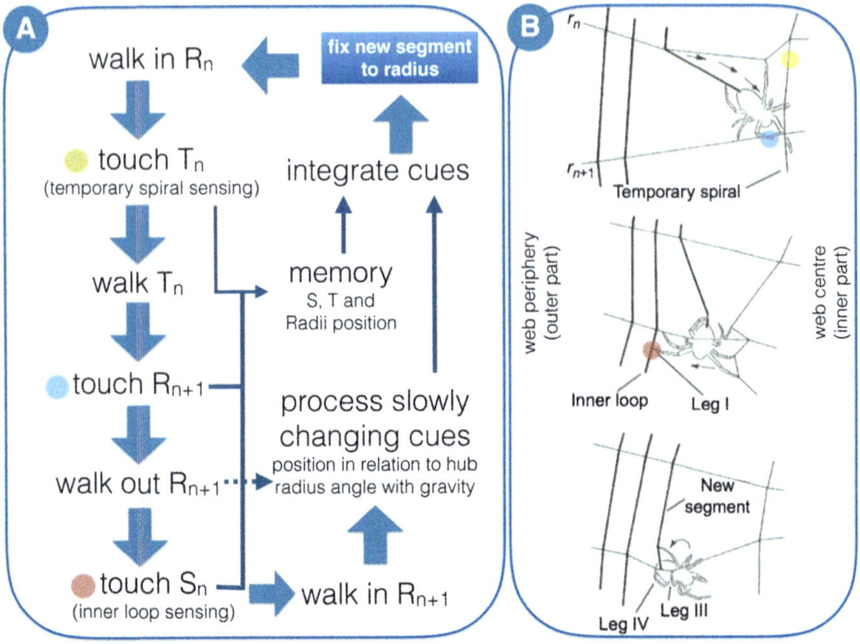

Fig. 14.1 Cycle of actions necessary to build the current segment of the adhesive spiral. Steps and processes within the cycle (**a**), with the illustration of some of the behaviors involved (**b**, adapted from Eberhard and Wcislo 2011). The cycle (*blue arrows* in **a**) begins and ends with the fixation of the current adhesive spiral segment (*blue box*). The spider fixes (the current segment) over the current radius (R_n, **a**; spider behavior displayed at **b**, *top figure*), and then in the next radius (R_{n+1}, **a**; spider behavior displayed at **b**, *bottom figure*). The spider performs successive actions (*large blue arrows*, **a**), while assessing the position of some rapidly changing cues (*coloured balls*). Slowly and rapidly changing cues are stored, compared to each other (to obtain distances and rates of change) and then integrated (*continuous thin blue lines*, **a**) to determine the position of the next adhesive segment fixation (in R_{n+1}). When confronted with conflicting cues, the spider may ignore some cues (inner loop sensing, *dotted blue line*, **a**). These cycles are repeated until the completion of the capture area (Modified from Fig.1 in Japyassú and Laland 2017)

the actual sticky segment and the previous, inner sticky spiral loop, thus producing a regular spacing between successive sticky spiral loops (Fig. 14.1). Nevertheless, this strategy would fail to produce regular spacing in experimental (or natural) webs with a partially destroyed spiral loop. In this case, the distances walked through the previous radius would be much smaller than the distances walked in the next radius (the one with the destroyed spiral segment), introducing a conflict between the actual and the previous cues. Facing this conflict, the spider ignores some cues (inner loop sensing) and favors others (temporary spiral sensing), avoiding the production of an irregular mesh of sticky spiral loops. Thus, due to the complexity of the task, involving the assessment of multiple cues, the experimental manipulation of the actual configuration of threads during the web-building process can actively change the spider's attention, leading the spider more prone to ignore some cues.

The application of the MM criteria (Kaplan 2012) shows that web structure is part and parcel of the web-building cognitive machinery. Changing the spider's cognitive state results in changes in the web. For example, well-fed spiders build orbs less frequently (Vollrath and Samu 1997), with smaller capture areas (Mayntz et al. 2009; Baba and Miyashita 2006; *but see* Vollrath and Samu 1997), and also webs with an added structure, the barrier web (Baba and Miyashita 2006). Also, the injection of drugs or the ingestion of natural substances causes changes in the web geometry (Hesselberg and Vollrath 2004; Albín et al. 2014), and it is thus clear that CNS cognitive machinery is causally connected to web structure. The reverse is also true, as experiments showing that the removal of threads (experimental webs) during the web-building process alters the spider attentional state, reviewed above, clearly demonstrate. As a result, web-building cognition extends to the web itself, as the very structure of the web can change the internal cognitive workings in the spider central nervous system.

Implications for the Evolution of Cognition

If cognition extends to external features of the environment, then the cognitive capacity of any animal should be measured not only by the volume or number of its neurons and/or synapses, but should instead also include the environmental information processing capacity. This could help explain findings that have been difficult to accommodate within the traditional view of cognition as CNS information processing. For example, it has long been shown that there are graded changes in brain–body allometry: larger animals such as mammals or reptiles cannot possibly be as small as ants or spiders, because they would have prohibitively large brains (Eberhard and Wcislo 2011). Extended cognition fits smoothly as a reasonable explanation for these otherwise incomprehensible taxa-specific brain–body allometry rules.

If smaller animals have proportionally smaller brains, they should show a relatively impaired behavioral performance, when compared to larger-brained animals, according to the traditional view of cognition. Nevertheless, this is not what the experimental evidence dictates. Tiny youngsters build as regular and functional webs as adult spiders (Hesselberg 2010; Eberhard and Wcislo 2011), small bees are capable of cognitive feats such as the formation of concepts (Giurfa et al. 2001), and the same is true about spiders that show a sense of numerosity (Rodríguez et al. 2015). From the perspective of the traditional, central cognition approach, such cognitive feats are not expected from such small-brained animals. Nevertheless, from the extended cognition approach, tiny animals are not expected to be behaviorally limited, as long as they are able to export cognitive processing to the environment, as spiders do.

For cognition to extend, the environment around the animal has to be predictable, so that the animal can be confident that some processes will regularly occur outside of its body. In this situation, CNS cognition can evolve so as to complement environmental processing, or even better, it can evolve to control the environment so as to modulate this environmental processing. This is most probably the case when the animal actively changes the environment through ecological engineering, and thus niche construction emerges as a major evolutionary process facilitating brain/environment cognitive coupling, and driving an evolutionary path that could in the end result in the incorporation of the external environmental features to the animal itself, so that the environmental feature becomes almost indissociable from the animal. This is actually the case in many social animals, like ants, termites, bees, and wasps, which are almost defined by the regular structures (nests, mounds, hives, etc.) they create.

The first step in this body/environment coupling process is the self-organization of brain neural networks and environmental features. The environmental effect of the action of one brain network can eventually feed back into the workings of another, previously unrelated brain network, creating new brain–environment feedback loops. These loops could stabilize the newly emerged (brain/environment) system in a new state, that is, could stabilize a new behavioral performance.[4] If the new state is adaptive, the brain/environment coupling is selected for, and the emergent performance becomes predictable in the long run. The important point here is that the coupling enables a reduction in CNS information processing through environmental processing, and thus small animals can evolve in the direction of complex performance even when anatomically restricted to having a small brain.

This process that begins with the self-organization of a new behavior, and continues through niche-construction to co-adapt the distributed, environmental

[4]Although it is notoriously difficult to detect novelty in a lifelong, complete repertoire of actions (because some performances could be simply rare in place of nonexistent), sometimes novelty is the only possibility, for example when the behaviour is impossible without a particular experimental manipulation. This is the case of the reeling attack tactic, whereby the spider reels a dry thread so that an entangled prey comes close enough to be wrapped. Reeling attack is the default foraging strategy for a whole family of cobweavers, but orbweavers cannot possibly attack through reeling under natural conditions, because their orbweb's radii are firmly attached to the frame (and thus cannot be reeled). Surprisingly, orbweavers on experimental orbwebs (with a radii artificially cut free from the frame) do promptly reel-attack their prey in the very first trial; this new behaviour is stable, occurring predictably in the experimental orbwebs, and in all the species studied (Penna-Gonçalves et al. 2008). Since orbweavers never attack naturally through reeling, and considering this behaviour is impossible in normal orbwebs, this experimental result requires explanation, because these spiders cannot possibly have an adapted neural network for controlling a reeling attack. The explanation is rather simple: orbweavers do reel threads in natural circumstances, but only when building their webs, and never in a foraging context (prey attack). Thus, the cut-free radius of the experimentally modified orbweb provides the opportunity for the spider to perform a known behavior within a novel, prey-attack context. This is precisely the case of self-organization discussed above. A novelty (predatory reeling in orbweavers) emerges and stabilizes through an environmental (cut-free radius) modification that allows a feedback between two existing neural networks. The cut-free radius allows the co-occurrence of a (natural) web-building behaviour (reeling) with an attack behaviour (prey-wrapping), with the consequent emergence of a new foraging tactic: the reel-attack. This exemplifies how self-organization can produce new and stable behaviors; in this case, the evolutionary appearance of the reel-attack requires only the evolution of a specific environmental feature (a detachable radius), and this is precisely what occurred in the transition from ancestral orbwebs to derived cobwebs.

information processing to CNS information processing, can proceed even further in social animals to the formatting of social interactions. Thus, the evolution of cognition through self-organization and niche construction helps the evolution of adaptive and distributed social information processing, and thus prompts the emergence of new levels of biological organization.

This cognitive route to the emergence of the social level mimics the emergence of the organismic level (with the onset of bodily integration through neural systems) from the simpler perceptual mechanisms of single-celled animals (Pezzulo & Levin 2015; Baluška and Levin 2016). Also, it is the basis for a nascent theoretical integration between traditionally separate research agendas: the area of cognitive development (learning mechanisms) and that of evolution through natural selection (Power *et al.* 2015; Watson and Szathmáry 2016). Extended cognition may after all be the proximate process instructing the evolution of new levels of biological organization.

References

Ades C (1988) Memória e aprendizagem em aranhas. Biotemas 1:2–27

Albin A, Lacava M, Viera C (2014) Effects of gonyleptidine on orb web spider Araneus Lathyrinus (Holmberg 1875). Arachnol 16:154–156

Baba Y, Miyashita T (2006) Does individual internal state affect the presence of a barrier web in Argiope bruennichii (Araneae: Araneidae)? J Ethol 24(1):75–78

Baba YG, Kusahara M, Maezono Y, Miyashita T (2014) Adjustment of web-building initiation to high humidity: a constraint by humidity-dependent thread stickiness in the spider Cyrtarachne. Naturwissenschaften 101(7):587–593

Baluška F, Levin M (2016) On having no head: cognition throughout biological systems. Front Psychol 7:902

Blamires SJ, Chao YC, Liao CP, Tso IM (2011) Multiple prey cues induce foraging flexibility in a trapbuilding predator. Anim Behav 81(5):955–961

Clark A, Chalmers D (1998) The extended mind. Analysis 58:10–23

Eberhard WG (1972) The web of Uloborus diversus (Araneae: Uloboridae). J Zool 166(4):417–465

Eberhard WG (1982) Behavioral characters for the higher classification of orb-weaving spiders. Evolution 36:1067–1095

Eberhard WG (1987) Effects of gravity on temporary spiral construction by Leucauge mariana (Araneae: Araneidae). J Ethol 5(1):29–36

Eberhard WG (1988a) Behavioral flexibility in orb-web construction: effects of supplies in different silk glands and spider size and weight. J Arachnol 295–302

Eberhard WG (1988b) Memory of distances and directions moved as cues during temporary spiral construction in the spider Leucauge mariana (Araneae: Araneidae). J Insect Behav 1(1):51–66

Eberhard WG (2011) Are smaller animals behaviorally limited? Lack of clear constraints in miniature spiders. Anim Behav 81:813–823

Eberhard WG, Wcislo WT (2011) Grade changes in brain-body allometry: morphological and behavioural correlates of brain size in miniature spiders, insects and other invertebrates. Adv Insect Physiol 40:155

Eberhard WG (2012a) Cues guiding placement of the first loop of the sticky spiral in orbs of Micrathena duodecimspinosa (Araneidae) and Leucauge mariana (Tetragnathidae). Arachnology 16(7):224–227

Eberhard WG (2012b) Correlations between Leg positions and spaces between sticky lines in the orbs of Micrathena duodecimspinosa (Araneae: Araneidae). Arachnology 15(7):235–240
Eberhard WG, Hesselberg T (2012) Cues that spiders (Araneae: Araneidae, Tetragnathidae) use to build orbs: lapses in attention to one set of cues because of dissonance with others? Ethology 118(7):610–620
Ghalambor CK, McKay JK, Carroll SP, Reznick DN (2007) Adaptive versus non-adaptive phenotypic plasticity and the potential for contemporary adaptation in new environments. Funct Ecol 21:394–407
Giurfa M, Zhang S, Jenett A, Menzel R, Srinivasan MV (2001) The concepts of 'sameness' and 'difference' in an insect. Nature 410(6831):930–933
Harms WF (2004) Information and meaning in evolutionary processes. Cambridge University Press, Cambridge
Harms WF (2006) What is information? Three concepts. Biol Theory 1:230–242
Heiling AM, Herberstein ME (1999) The role of experience in web-building spiders (Araneidae). Anim Cogn 2:171–177
Heiling AM, Herberstein ME (2000) Interpretations of orb-web variability: a review of past and current ideas. Ekologia(Bratislava)/Ecology(Bratislava) 19:97–106
Hesselberg T, Vollrath F (2004) The effects of neurotoxins on web-geometry and web-building behaviour in Araneus diadematus Cl. Physiol Behav 82(2):519–529
Hesselberg T (2010) Ontogenetic Changes in Web Design in Two Orb-Web Spiders. Ethology 116(6):535–545
Hénaut Y, Machkour-M'Rabet S, Lachaud JP (2014) The role of learning in risk-avoidance strategies during spider–ant interactions. Anim Cogn 17:185–195
Hutchins E (1995) Cognition in the wild. MIT Press, Cambridge
Jackson RR, Nelson XJ (2011) Reliance on trial and error signal derivation by Portia Africana, an araneophagic jumping spider from East Africa. J Ethol 29:301–307
Japyassú HF (2008) Cognições mínimas. In: Vianna B (ed) Biologia da libertação. Mazza Edições, Belo Horizonte, pp 97–113
Japyassú HF (2010) Fenótipos amplificáveis em pequenas cognições. Rev Etol 9:63–71
Japyassú HF, Malange J (2014) Plasticity, stereotypy, intra-individual variability and personality: handle with care. Behav Processes 109:40–47
Japyassú, H. F., & Laland, K. N. (2017). Extended spider cognition. Animal Cognition, 1-21
Kaplan DM (2012) How to demarcate the boundaries of cognition. Biol Philos 27:545–570
Laland KN, Brown GR (2011) Sense and nonsense: evolutionary perspectives on human behavior. Oxford University Press, Oxford
Landolfa MA, Barth FG (1996) Vibrations in the orb web of the spider Nephila clavipes: cues for discrimination and orientation. J Comp Physiol A Neuroethol Sens Neural Behav Physiol 179(4):493–508
Matsushita K, Lungarella M, Paul C, Yokoi H (2005) Locomoting with less computation but more morphology. In Robotics and Automation, 2005. ICRA 2005. Proceedings of the 2005 IEEE International Conference on (pp. 2008–2013). IEEE
Mayntz, D., Toft, S., & Vollrath, F. (2009). Nutrient balance affects foraging behaviour of a trap-building predator. Biology letters, rsbl20090431
Moreno A, Mossio M (2015) Biological autonomy: a philosophical and theoretical enquiry, vol 12. Springer, Berlin
Mortimer B, Holland C, Windmill JF, Vollrath F (2015) Unpicking the signal thread of the sector web spider Zygiella x-notata. J R Soc Interface 12(113), 20150633
Murakami Y (1983) Factors determining the prey size of the orb-web spider, Argiope amoena (L. Koch) (Argiopidae). Oecologia 57(1–2):72–77
Nakata K (2010) Attention focusing in a sit-and-wait forager: a spider controls its prey-detection ability in different web sectors by adjusting thread tension. Proc R Soc B 277:29–33

Nakata K (2012) Plasticity in an extended phenotype and reversed up–down asymmetry of spider orb-webs. Anim Behav 83:821–826

Nakata K (2013) Spatial learning affects thread tension control in orb-web spiders. Biol Lett 9(4):20130052

Nogueira SDC, Ades C (2012) Evidence of learning in the web construction of the spider Argiope argentata (Araneae: Araneidae). Rev Etol 11(1):23–36

Pasquet A, Ridwan A, Leborgne R (1994) Presence of potential prey affects web-building in an orb-weaving spider Zygiella x-notata. Anim Behav 47(2):477–480

Penna-Gonçalves V, Garcia CRM, Japyassú HF (2008) Homology in a context dependent predatory behavior in spiders (Araneae). J Arachnol 36:352–359

Pezzulo G, Levin M (2015) Remembering the body: applications of computational neuroscience to the top-down control of regeneration of limbs and other complex organs. Integr Biol 7:1487–1517

Pfeifer R, Iida F, Gómez G (2006) Morphological computation for adaptive behavior and cognition. In: International congress series, vol 1291. Elsevier, Kitakyushu (Japan), pp 22–29

Pfeifer R, Iida F, Lungarella M (2014) Cognition from the bottom up: on biological inspiration, body morphology, and soft materials. Trends Cogn Sci 18:404–413

Power DA, Watson RA, Szathmáry E, Mills R, Powers ST, Doncaster CP, Czapp B (2015) What can ecosystems learn? Expanding evolutionary ecology with learning theory. Biol Direct 10:69

Rodríguez RL, Gamboa E (2000) Memory of captured prey in three web spiders (Araneae, Araneidae, Linyphiidae, Tetragnathidae). Anim Cogn 3:91–97

Rodríguez RL, Gloudeman MD (2011) Estimating the repeatability of memories of captured prey formed by Frontinella Communis spiders (Araneae: Linyphiidae). Anim Cogn 14:675–682

Rodríguez RL, Kolodziej RC, Höbel G (2013) Memory of prey larders in golden orb-web spiders, *Nephila clavipes* (Araneae: Nephilidae). Behaviour 150:1345–1356

Rodríguez RL, Briceño RD, Briceño-Aguilar E, Höbel G (2015) Nephila Clavipes spiders (Araneae: Nephilidae) keep track of captured prey counts: testing for a sense of numerosity in an orb-weaver. Anim Cogn 18:307–314

Sanderson SK (2014) Human nature and the evolution of society. Westview Press, Boulder

Sandoval, C. P. (1994). Plasticity in web design in the spider Parawixia bistriata: a response to variable prey type. Functional Ecology, 701-707

Schneider JM, Vollrath F (1998) The effect of prey type on the geometry of the capture web of Araneus diadematus. Naturwissenschaften 85(8):391–394

Shapiro LA (2010) Embodied cognition. Routledge, London

Shettleworth SJ (2010) Cognition, evolution, and behavior. Oxford University Press

Sumpter DJ (2010) Collective animal behavior. Princeton University Press, Princeton

Thelen E, Smith LB (1994) A dynamic systems approach to the development of cognition and action. MIT Press, Cambridge

Thompson E (2007) Mind in life: biology, phenomenology, and the sciences of mind. Harvard University Press, Cambridge

Venner S, Pasquet A, Leborgne R (2000) Web-building behaviour in the orb-weaving spider Zygiella xnotata: influence of experience. Anim Behav 59(3):603–611

Vollrath F (1987) Altered geometry of webs in spiders with regenerated leg. Nature 328:247–248

Vollrath F (1988a) Untangling the spider's web. Trends Ecol Evol 3(12):331–335

Vollrath F (1988b) Spiral orientation of Araneus diadematus orb-webs built during vertical rotation. J Comp Physiol A 162(3):413–419

Vollrath F, Samu F (1997) The effect of starvation on web geometry in an orb-weaving spider. Bull Br Arachnol Soc 10:295–297

Watanabe T (2000) Web tuning of an orb-web spider, Octonoba Sybotides, regulates prey-catching behavior. Proc R Soc B 267:565–569

Watson RA, Szathmáry E (2016) How can evolution learn? Trends Ecol Evol 31:147–157

West-Eberhard MJ (2003) Developmental plasticity and evolution. Oxford University Press, New York

Wilson RA, Foglia L (2017) Embodied Cognition, in The Stanford Encyclopedia of Philosophy (Spring 2017 Edition), Edward N. Zalta (ed.), forthcoming URL = https://plato.stanford.edu/archives/spr2017/entries/embodied-cognition/

Witt PN, Reed C, Peakall DB (1968) A spider's web: problems in regulatory biology. Springer, New York

Witt, P. N., Scarboro, M. B., Daniels, R., Peakall, D. B., & Gause, R. L. (1976). Spider web-building in outer space: evaluation of records from the Skylab spider experiment. Journal of Arachnology, 115-124

Wu CC, Blamires SJ, Wu CL, Tso IM (2013) Wind induces variations in spider web geometry and sticky spiral droplet volume. J Exp Biol 216(17):3342–3349

Chapter 15
Anti-Predator Strategies

Felipe M. Gawryszewski

Abstract Despite being widely known as a diverse group of predators, spiders are also a regular prey item of several vertebrate and invertebrate predators. Some of these organisms (e.g., wasp species and araneophagic spiders) are spider-hunting specialists. A number of morphological structures and behaviours in spiders have been proposed to be anti-predator adaptations. They comprise strategies such as background matching, disruptive patterns, web decorations, mimicry, masquerading, aposematism, urticating bristles, spines, retreats, barrier webs, group living, and dropping from webs. In this chapter, spider anti-predator strategies are presented, and the correlational and causal evidence of anti-predator adaptations are critically discussed in light of potential costs and benefits they may entail. Studies involving Neotropical species are presented to illustrate most strategies.

Spiders are distributed worldwide and can be found in almost all terrestrial ecosystems (Foelix 2010). Their diversity, abundance, and spatial distribution may explain the diversified range of their predators. Predators consist of other spiders (Gonzaga et al. 1998; Moura et al. 2017); wasps, especially Pompilidae (Gonzaga and Vasconcellos-Neto 2006; Restrepo-Giraldo et al. 2012), Sphecidae (Coville and Coville 1980; Blackledge et al. 2003), and Crabronidae (Gonzaga and Vasconcellos-Neto 2005a; Araújo and Gonzaga 2007; Buschini et al. 2008, 2010a, b; Pitilin et al. 2012); birds (Stiles 1995; Gunnarsson 2007); and lizards (Schoener and Toft 1983; Spiller and Schoener 1998), among others.

Considering the diversity of spiders, variety of spider predators, diversity of ecosystems occupied by spiders, and the long evolutionary history of this group, it is not a surprise that the anti-predator devices and behaviours of spiders are also extremely diverse. These include morphological structures, construction of shelters, construction of complex three-dimensional silk structures also used to capture prey, colour patterns similar to those of the substrate used for rest, aggressive displays, and a number of other strategies. The suite of strategies used by each spider group

F.M. Gawryszewski (✉)
Universidade Federal de Goiás, Goiás, Brazil
e-mail: f.gawry@gmail.com

C. Viera, M.O. Gonzaga (eds.), *Behaviour and Ecology of Spiders*,
https://doi.org/10.1007/978-3-319-65717-2_15

Fig. 15.1 (**a**) Shelter constructed by *Araneus omnicolor* (Araneidae) using a dry leaf. The spider is being attacked by the parasitoid wasp *Hymenoepimecis veranii* (see Chap. 16 for more details on this interaction). (**b**) Similar shelter used by *Tidarren haemorrhoidale* (Theridiidae). Prey items are usually transported to the interior or close to the shelter to be consumed (Photos: M.O. Gonzaga)

may be related to its susceptibility to distinct predators. Thus, the most frequent predator of diurnal orb-weavers that remain visually exposed in the centre of their bidimensional webs, for example, seems to be hunting wasps (Rayor 1996; Blackledge and Pickett 2000; Blackledge et al. 2003; Gonzaga and Vasconcellos-Neto 2005a). Birds, on the other hand, have greater effects (at least in temperate regions) on webless and three-dimensional web-structure builders than on orb-weavers (Gunnarsson and Wiklander 2015).

Broadly, anti-predator defences can be divided into strategies to avoid detection and recognition, and defence strategies used after detection and recognition (Stevens and Merilaita 2009). The former encompass mechanisms such as camouflage, masquerade, and construction of shelters (Fig. 15.1), whereas the latter include strategies such as flight, aggressive displays, and leg autotomy. In this chapter, I first briefly describe some anti-predator strategies and then discuss the correlational and causal evidence supporting them as defences against predators. When available, specific reviews are cited. Moreover, I recommend other general reviews on the topic of anti-predatory strategies in spiders, including Cloudsley-Thompson (1995), Nelson and Jackson (2011), and Pekár (2014). Last, I present ecological and evolutionary factors related to the evolution and maintenance of anti-predator strategies in spiders. Studies involving Neotropical species are presented to illustrate most of these strategies.

Crypsis, Colour Change, and Masking Behaviour

Crypsis encompasses a range of strategies that reduce the probability of predators detecting prey (sensu Stevens and Merilaita 2009). Two of these strategies are more frequently mentioned in the literature: background matching (frequently referred to as crypsis) and disruptive colouration. Background matching is the possession of patterns that resemble the typical background where the individual encounters its predators, whereas disruptive colouration is the possession of contrasting colours that break up the animal silhouette, making it more difficult to be detected (Stevens and Merilaita 2009). Background matching (including anachoresis) is the most frequently proposed defensive strategy in spiders (Pekár 2014). Nonetheless, experiments testing the defensive benefit of crypsis are scant, especially in the Neotropics. In an experiment conducted in Taiwan using the orb-weaver *Nephila pilipes* (Araneidae), for instance, a conspicuous, manipulated yellow model attracted more wasp predators than models resembling the spider's colour patterns and black models (Fan et al. 2009). Nephilids from the Neotropics also frequently present brightly coloured patterns, but it remains to be tested whether their colouration has a cost or benefit in terms of predation risk. A two-species comparison also offered some evidence for the selection of crypsis. A light-coloured species of *Lycosa* (Lycosidae) from southern Florida (USA) was more abundant in light-coloured sandy areas, whereas a dark-coloured species was typically found in the adjacent grassy areas (Richman et al. 1995). Evidence from the Neotropical region includes studies showing that some spiders are less prone to abandon substrates that, to a human observer, matched their colouration (Peixoto et al. 2012), and/or actively select matching substrate (Messas et al. 2014; Souza et al. 2015).

Colour change and masking behaviour are special cases of crypsis. Several Thomisidae species are able to change colour over several days. *Misumena vatia* (Thomisidae; Holarctic distribution), for instance, is a flower-dwelling species that can change colour from white to yellow depending on the colour of the flower it is sitting on (Packard 1905; Gabritschevsky 1927). However, to date, there is no direct evidence of the defensive benefit of colour change in thomisids. Other species are able to quickly darken their abdomen by contracting specialised cells containing white guanocytes (Edmunds and Edmunds 1986). *Argiope* species (Araneidae; a genus with Neotropical representatives), for instance, exhibit this behaviour and drop their web after a predator attack. At least for a human observer, the darkening of the abdomen makes the spider more difficult to be detected (Edmunds and Edmunds 1986). Other spiders, such as *Homalonychus* (Homalonychidae; found in desert areas in the EUA and Mexico), *Sicarius* (Sicariidae; found mainly in deserts and dry areas in South America and Africa) and *Stephanopis* (Thomisidae; includes several Neotropical species), have specialised hairs in their cuticle that trap material from the environment (sand, bark debris), a phenomenon usually referred to as masking behaviour (Duncan et al. 2007; Gawryszewski 2014). There has been no experimental study on the effect of masking behaviour on predator avoidance, but for

Stephanopis spp. the presence of bark debris appears to improve the background matching of the spiders, especially against dark-coloured bark (Gawryszewski 2014).

Mimicry and Masquerade

Although the distinction between mimicry and masquerade is sometimes blurred, mimicry as defined by Skelhorn et al. (2009) refers to strategies that indirectly affect the fitness of the model through the action of a third organism, usually a predator. Masquerade, on the other hand, refers to a phenomenon that either does not have a fitness consequence for the model, or in which its fitness is directly affected by the mimicry (Skelhorn et al. 2009). Moreover, mimicry and masquerade differ from crypsis because they are strategies that are efficient even out of context. A bird-dropping spider, for instance, can be misclassified by the predator in any background, whereas a background matching pattern is effective only in that particular context.

In a literature survey, Pekár (2014) found that spiders are mimics of several animal taxa, such as Coleoptera, Lepidoptera, Diptera, Heteroptera, and even Anura. Nonetheless, the majority of species (82%) are mimics of ants (Oliveira and Sazima 1984, 1985; Oliveira 1986, 1988; Pekár 2014; Fig. 15.2). Ant-mimicry has evolved in nine spider families [Aphantochilidae, Araneidae, Corinnidae, Eresidae, Gnaphosidae, Salticidae, Theridiidae, Thomisidae, and Zodariidae (Mclver and Stonedahl 1993)]. Most mimic species of spiders are believed to be Batesian, but there are some cases of aggressive mimicry (Mclver and Stonedahl 1993). To a human observer, many ant-mimic spiders resemble the colour, shape, and behaviour of ants (Mclver and Stonedahl 1993; Cushing 2012). In addition, chemical mimicry of ant cuticular hydrocarbons has also been documented for some spider species (Lenoir et al. 2001). Experimental data support the defensive advantage of ant-mimicry in many species, either because many predators avoid preying upon ants, or because of a dilution effect of living in a group (see Mclver and Stonedahl 1993, Cushing 1997, 2012 for more details).

Fig. 15.2 *Myrmecium* sp. (Corinnidae) from Amazon Forest in Jacareacanga, PA, Brazil (Photo: M.O. Gonzaga)

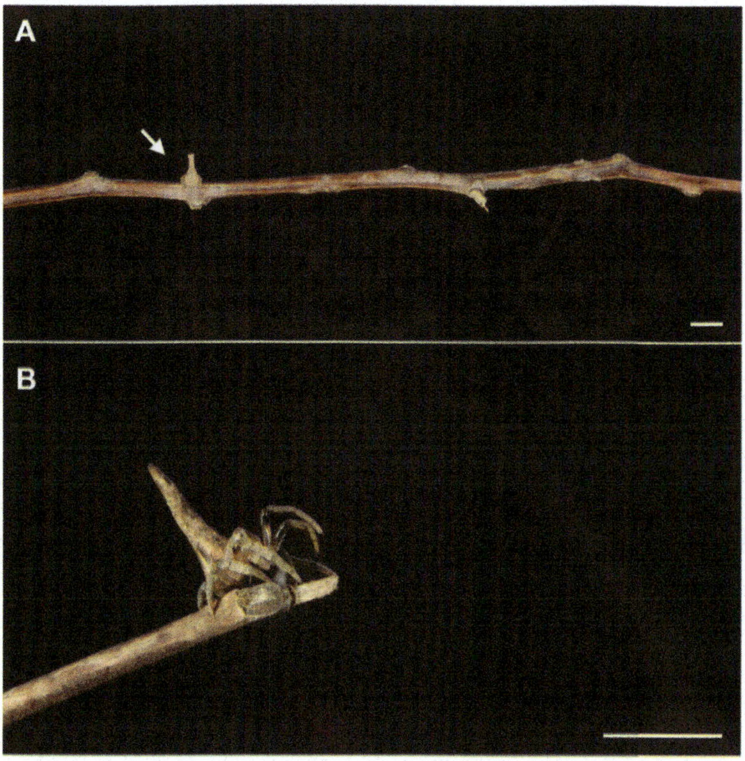

Fig. 15.3 *Wixia abdominalis* (Araneidae). (**a**) Resting position on a *Eucalyptus* branch. (**b**) Details of a subadult female. *Scales*: *A*: 0.5 cm, *B*: 1 cm (Photos: M.O. Gonzaga)

Masquerade has been proposed for several spider species (Cloudsley-Thompson 1995; Pekár 2014), but its benefit as a defensive strategy has been seldom tested. Many *Mastophora* species (Araneidae; several Neotropical species) are believed to resemble uninteresting objects, such as a bird dropping (Cloudsley-Thompson 1995). Similarly, the araneid *Wixia abdominalis* closely resembles a *Eucalyptus* twig and is usually found resting on branches of these cultivated trees in southeastern Brazil (Fig. 15.3) (Xavier et al. 2017). The Neotropical crab spider *Epicadus heterogaster* (Thomisidae) resembles a flower and is commonly found resting on leaves (Fig. 15.4). This could be a case of aggressive mimicry similar to that of the flower mantis (O'Hanlon et al. 2014), but a defensive benefit is also plausible. *Cyclosa ginnata* (Araneidae), when sitting against the discoid-shaped web decoration in their orb-webs, have a colour and shape similar to a bird dropping to a human observer, and similar colours from a hymenopteran perspective (Liu et al. 2014). The presence of the spider plus the web decoration has been shown to reduce wasp attacks when compared to attacks on spiders without web decorations (Liu et al. 2014). Several other species of *Cyclosa* and *Allocyclosa* from the Neotropics also present colour patterns very similar (at least to the human eye) to those of their

Fig. 15.4 *Epicadus heterogaster* (Thomisidae) resting on the abaxial surface of a leaf in Serra do Japi, Jundiaí, SP, Brazil (Photo: M.O. Gonzaga)

stabilimenta (e.g., Gonzaga and Vasconcellos-Neto 2012; see section "Silk-Mediated Defences" for more on defensive functions of web decorations).

Leg Autotomy

Leg autotomy has been described in at least ten spider families (Pholcidae, Therediidae, Lyniphiidae, Araneidae, Agelenidae, Lycosidae, Thomisidae, Salticidae, and Filistatidae), representing a large diversity of spider guilds (Fleming et al. 2007). Typically, from 5% to 20% of individuals in a population have at least one leg absent (Fleming et al. 2007). The autotomy typically occurs either between the coxa and throcanther (Wood 1926; Amaya and Klawinski 2001). The leg does not have to be forcefully pulled to be autotomised, and autotomy may be induced by simply pinching the leg (Wood 1926) or by the injection of venom (Eisner and Camazine 1983). Leg autotomy frequently happens in encounters with predators (Formanowicz 1990; Punzo 1997), but it may also occurs in intraspecific interactions, especially in male–male encounters (Dodson and Beck 1993), and when a dangerous prey counter-attacks (Eisner and Camazine 1983).

The advantage of leg autotomy is clear: better limp alive than die eight-legged. The efficacy of leg autotomy as an anti-predator mechanism, however, probably depends on the type predator performing the attack (Formanowicz 1990, Punzo 1997). Moreover, such an extreme defensive mechanism is expected to come at a great cost. Indeed, several experiments have demonstrated the potential cost of leg autotomy in terms of running speed (Amaya and Klawinski 2001; Apontes and

Brown 2005), foraging efficiency (Steffenson et al. 2014), and mating (Taylor et al. 2008). Surprisingly, however, some studies did not find a strong negative effect of leg autotomy (Johnson and Jakob 1999; Brueseke et al. 2001; Taylor et al. 2008; Wrinn and Uetz 2008). In the orb-weaver *Zygiella x-notata* (Araneidae), for instance, there was no difference in longevity, egg sac production, web capture area, and prey capture rate between intact individuals and individuals with at least one leg lost (Pasquet et al. 2011). Nonetheless, eight-legged individuals produced webs with a longer total stick thread and with lower interspiral distances than six or seven-legged individuals. In addition, five-legged individuals were not able to build a proper web (Pasquet et al. 2011).

Group Living

One of the possible advantages involved in the evolution and maintenance of group living is the reduced risk of predation. The reduced risk of predation may arise from three factors: (1) the probability that animals in groups may repel predators more efficiently, (2) a dilution effect, in which the individual risk of predation is reduced as the size of the group increases, and (3) earlier predator detection in animals living in groups, allowing more time for flight (Krebs and Davies 2004).

A few spider species exhibit some form of grouping behaviour (see Chap. 13), which varies from simple aggregation of individual webs to complex social interactions (Whitehouse and Lubin 2005). Spiders that live in large groups may attract more parasitoids and predators than smaller groups or solitary individuals (Uetz and Hieber 1994). However, at least after a certain group size in the Mexican orb-weaver *Metepeira incrassata* (Araneidae), the dilution effect comes into place, thereby reducing the individual risk of predation (Uetz and Hieber 1994). Interestingly, silk threads can be an efficient mode of warning signal transmission of approaching enemies because webs are interconnected, which allows spiders to evade predators early (Uetz et al. 2002). Moreover, the position in the colony also matters: the risk of predation in *M. incrassata* is reduced in the centre of the colony, and individuals actively try to position themselves at the centre (Rayor and Uetz 1990). Similarly, Alves-Costa and Gonzaga (2001) observed that larger individuals of *Philoponella vittata* (Uloboridae), from the Brazilian Amazon Forest, are usually found in central positions of aggregations constructed using webs of several host spiders to support their own. They evaluated the frequency of prey capture in peripheral and central positions, finding no nutritional advantage to spiders located in the core of these aggregations. The authors suggested that other advantages, possibly the reduced risk of attacks by predators and parasitoids, may explain why large individuals fight for the central positions.

Maintenance of groups in some neotropical spider species, such as *Parawixia bistriata* (Araneidae), may also result in protection against predators. Spiders of this species remain in aggregations during the day, protected within a dense mesh of threads (Fig. 15.5), and leave this structure at night to construct their orb webs

Fig. 15.5 Aggregation of *Parawixia bistriata* (Araneidae). The spiders remain grouped during the day within a dense structure of threads, and build their individual webs at night, attached to a communal frame. This species occurs in areas covered by Cerrado vegetation in Brazil and also in Bolivia, Paraguay, and Argentina (Photo: M.O. Gonzaga)

attached to a communal frame. There is no information available on predators attacking these groups, but the collective refuge seems to be strong enough to restrict the access of birds and other potential predators. Only one unidentified species of a parasitoid wasp (Ichneunonidae) is known to enter the aggregations to attack these spiders (M.O. Gonzaga, *personal communication*).

Spines and Urticating Bristles

Spines — pointy abdominal cuticular projections — vary from short and pointy to extremely long, reaching up to three times the abdominal length, as in the Amazonian araneid *Micrathena cyanospina* (Levi 1985). Evidence of spines as a defensive mechanism is available only from indirect results. In the Caribbean Islands, the abundance of orb-web spiders is negatively affected by the presence of lizards; however, the spiny *Gasteracantha cancriformis* (Araneidae) (Fig. 15.6) is not affected by these — potential — predators (Schoener and Toft 1983). Moreover, anecdotal evidence suggests that spines are particularly suited to providing protection against avian predators, but not wasps (Peckham 1889). Interestingly, the long spines found in the New World orb-weavers *Micrathena* and *Chaetacis* (Araneidae) appear to have evolved independently eight times (Magalhães and Santos 2012). Authors propose that these long spines are antipredator devices against wasps and hummingbirds, as it would be physically impossible or energetically unprofitable to capture these spiders (Magalhães and Santos 2012).

Fig. 15.6 *Gastheracantha cancriformes* (Araneidae) from Uberlândia, MG, Brazil (Photo: M.O. Gonzaga)

Urticating bristles — or urticating hairs — are widespread and exclusive to the Theraphosidae of the Americas; approximately 90% of species in this region have urticating bristles (Marshall and Uetz 1990). Urticating bristles are typically found on the dorsal opisthosoma, and cause mechanical irritation of the victim (Cooke et al. 1972). Tarantulas have three types of behaviours involving urticating hairs: brushing hairs off towards the predator, direct contact of abdominal hairs with the predator body, and adding urticating hairs to silk. In Aviculariinae, the spiders usually direct their abdomens to the potential aggressor and transfer the urticating setae when the target is touched (Bertani and Marques 1996). In Theraphosinae, on the other hand, the light weight and aerodynamics of the setae usually allow them to float through the air and reach the respiratory tract of the aggressor (Cooke et al. 1972; Bertani and Marques 1996; Perafán et al. 2016). Moreover, different types of hairs have been proposed to be more efficient against vertebrate or invertebrate predators (Bertani and Guadanucci 2013). At least in *Theraphosa lebondi* and *Megaphobema* spp., egg sacs with urticating hairs are effective only against invertebrate predators (Marshall and Uetz 1990).

Aposematism

Spiders are potentially dangerous to predators, especially species with large fangs, spines, urticating hairs, and/or powerful venom. Despite that, aposematism has been proposed for very few species (Pekár 2014), such as the orb-weavers with spines, hardened cuticle, and, at least to human eyes, conspicuous colours (Edmunds and Edmunds 1986; Cloudsley-Thompson 1995; Vanderhoff et al. 2008; Gawryszewski and Motta 2012); the Theraphosids with large fangs, urticating hairs, and apparently conspicuous patterns (Pocock 1899; Cott 1966); and Theridiidae species with potent venoms and bright marks in the abdomen (Vetter 1980). Experimental data for aposematism exist only for the black widow *Latrodectus mactans* (Theridiidae). Birds were less likely to attack *L. mactans* spider models with the typical red marks than spider models without red marks (Brandley et al. 2016). The red and black colour patterns of the South American *Parawixia bistriata* (Fig. 15.5) are also suggestive of aposematism. After maturation, individuals became solitary, and their colouration changes from the red and black conspicuous pattern to an apparently cryptic colouration (Fernández Campón 2013).

Silk-Mediated Defences

Spiders have at their disposal a building material: silk. Other than capturing prey, silk-mediated structures are also employed to protect spiders against predation. The evolution of 3-dimensional webs from 2-dimensional orb-weaver ancestors could have been driven by wasp predators. A between-species analysis found correlational evidence that wasps are less likely to capture 3-dimensional web-builder species than 2-dimensional web-builder species (Blackledge et al. 2003). Interestingly, the evolution of these 3-dimensional builders coincides with the appearance of wasp predators in the fossil record (Blackledge et al. 2003).

In addition, some orb-weavers add a 3-dimensional barrier web in front of the orb web. These barrier webs are believed to have a defensive function by mechanically protecting spiders against approaching wasps and birds. In a study of orb-weavers from Ghana, barrier webs were more frequently found in smaller species, and in smaller individuals within species (Edmunds and Edmunds 1986). Experimental data support the defensive hypothesis of barrier webs for the horizontal orb-web of *Cyrtophora moluccensis* (Araneidae; Eastern Hemisphere distribution) against wasp predators (Blamires et al. 2013). However, the hypothesis was not supported for the vertical orb-web of *Thelacantha brevispina* (Araneidae; found in India, Southeast Asia, and Australia; Tseng et al. 2011). Moreover, the barrier web of *C. moluccensis* reduces the prey capture rate of individuals (Blamires et al. 2013), but the opposite occurs in *T. brevispina* (Tseng et al. 2011).

Several spiders build retreats (e.g., funnel-webs, trapdoors, and leaf retreats). Retreats probably reduce the risk of predation, either by concealing the spider, or by providing a refuge in case of approaching predators, but experimental evidence for the anti-predatory function of retreats is still scarce. In an experimental set-up excluding lizards in Australia, the abundance of orb-weaver species with retreats was not affected by the treatment, whereas the abundance of species without retreats with web heights lower than 20 cm was negatively affected by the presence of lizards (Manicom et al. 2008).

Orb-weavers frequently add decorations to their webs. Web decorations can be cruciate, discoid, tufts made of silk, or some environmental detritus added to the web (prey carcass, moults; Herberstein and Tso 2011). There is a lively debate on the function of these decorations. The evidence for cruciate, circular, or tufted decorations as a defensive mechanism is inconclusive. Some experimental evidence supports the defensive mechanism of these decorations, whereas other researchers found that the presence of these decorations increase the risk of predation (*see* Bruce 2006 for more details). Silk web decorations could potentially protect spiders by improving spider camouflage, by making the spider look bigger and deterring certain gape-limited predators, or by acting as a barrier when a predator attacks (see Bruce 2006 for more detail). On the other hand, the literature in general supports a defensive function of detritus decorations. *Cyclosa* species (Araneidae) typically attach detritus to the centre of the web, and the spider rests between these pellets of debris. To the human observer, the presence of debris makes the spider quite camouflaged. The presence of detritus in webs of the neotropical *Cyclosa fililineata* and *C. morretes* (Araneidae) reduced the attack rate on spider models (Gonzaga and Vasconcellos-Neto 2005b; Fig. 15.7). In addition, the presence of debris in the east Asian *Thelacantha brevispina* (Araneidae) web may also deflect wasp attacks, giving the spider more time to evade the danger (Tseng and Tso 2009).

Other Responses to Predators

Spiders exhibit a variety of other defensive behaviours to avoid predation. Many orb-weavers when faced with cues of a potential predator approaching may vibrate the web, change sides on the web, drop the web and remain motionless on the ground, or run to the web edge (Edmunds and Edmunds 1986; Blackledge and Pickett 2000). In the presence of predator olfactory cues, juveniles of *Argiope versicolor* decreased the frequency of web decorations, and built webs with lower total thread length and lower capture area when compared with control spiders (Li and Lee 2004). The wolf spider *Pardosa milvina* (Lycosidae) diminished its activity when cues from a potential predator were present. This response was only present with a predator bigger than itself or when a small predator was in great numbers (Persons and Rypstra 2001). Moreover, males of *P. milvina* modulate their courtship display depending on the risk of predation (Rypstra et al. 2016). Furthermore, deimatic displays, either visual or vibratory, have been proposed for some spider species (Cloudsley-Thompson 1995); and Mastophora dizzydeani, found in Colombia and Peru, regurgitate an unpleasant odour when hand-manipulated (Eberhard 1980).

The nocturnal habits of many species may also have, at least in part, a defensive function by avoiding diurnal visually oriented predators. In Taiwan, *Cyclosa moluccensis* (Araneidae), for instance, attracted several wasp predators during the day, but at night no predator approached the spider (Blamires et al. 2013).

Fig. 15.7 (**a**) *Cyclosa morretes* in a position adjacent to its linear detritus stabilimentum in the Brazilian Atlantic Forest (Parque Nacional Itatiaia, RJ). (**b**) *Cyclosa* sp. from Manaus, AM, Brazil and its complex detritus stabilimentum resembling a large spider (Photos: M.O. Gonzaga)

Furthermore, the Panamanian *Eustala oblonga* (Araneidae) gains protection against predators by inhabiting acacia plants with mutualistic ants (Styrsky 2014). Finally, non-sex-limited polymorphism of several spider species can be driven by a negative frequency-dependent selection by predators (apostatic selection), where the probability of predation of different colour morphs in a population is negatively correlated to its frequency (Oxford and Gillespie 1998).

Ecological and Evolutionary Context

The sensory systems that predators use to track and capture their prey will influence the evolution of anti-predator adaptation in spiders. Several of these nocturnal predators, including spiders, use non-visual clues to track and capture their prey (Barth 2000). A nocturnal spider of which the main predator is another poor-vision spider may have a series of non-visual anti-predator defences that are not obvious to humans, such as olfactory or vibratory camouflage (Ruxton 2009). Some evidence shows this occurs on the other side of this story; spider predator specialists, such as *Portia,* use movement patterns that reduce the probability of being detected by its prey (Jackson and Hallas 1986).

In addition, diurnal visually orientated predators may have significant differences in their sensory systems. Wasps and birds, for instance, have different colour vision systems. Sphecidae wasps, like most insects, have three different photoreceptors, with peak absorbance curves in the UV, blue, and green regions of the light spectrum (Peitsch et al. 1992). Birds, on the other hand, have four different photoreceptors for colour vision, with absorbance curves in the UV/violet, blue, green, and red regions of the light spectrum (Hart 2001). This means that a red patch could potentially be extremely conspicuous to an avian predator but relatively inconspicuous to a wasp predator. Thus, the red patch in *Ladrotectus* is most probably a warning signal to birds, and perhaps other vertebrate predators, but not to wasps (Brandley et al. 2016). Nonetheless, most spiders are predated by a suite of predators, which may have important consequences for the evolution of defences in this group. The imperfect mimicry of certain spider species, for instance, could have been driven by selective forces from ant-eating and ant-averse predators (Pekár et al. 2011).

Visual anti-predation strategies (e.g., mimicry, masquerade, aposematism) are expected to be found more frequently in diurnal spiders, because these spiders are more likely predated by visually orientated predators such as wasps and birds. Mimicry and aposematism are indeed more frequently proposed for diurnal than nocturnal spiders (Pekár 2014). Surprisingly, however, the same study found that masquerade is more frequently proposed to occur in nocturnal than diurnal spiders. This probably occurs because nocturnal spiders use masquerade during the daytime to avoid visually orientated predators (Pekár 2014). In addition, diurnal predators could be important selective forces on the maintenance and evolution of the nocturnal habit of spider species. Wasp predators and many birds, for instance, are not active at night. However, most spiders are nocturnal, and one of the main spider predators are spiders themselves. Therefore, the nocturnal habit certainly changes the type of spider predators, but does not necessarily reduce the risk of predation.

Spiders that inhabit visually homogeneous habitats are more likely to be detected by visually oriented predators than spiders than inhabit heterogeneous habitats. Moreover, theoretical and experimental data predict that disruptive colour patterns are more efficient in heterogeneous habitats, whereas background matching is efficient only on the specific matching background (Cuthill et al. 2005; Schaefer and Stobbe 2006; Stevens et al. 2006). Therefore, a colour pattern that resembles the background has an implicit cost: outside that particular background, this colour pattern may make the individual more conspicuous to a predator, which could limit the species distribution to places where crypsis is effective (Ruxton et al. 2004). This is particularly true when the background is homogeneous. Diurnal spiders that use leaves or flowers to ambush their prey are particularly susceptible to this problem. Flower-dwelling crab spiders that ambush flower visitors are faced with the problem of living in a homogeneous habitat that varies sharply in colour from ambush site to ambush site. Perhaps in response to this cost, many crab spiders are able to change their body colour, often to match the flower background colouration (Packard 1905; Gabritschevsky 1927; Llandres et al. 2013).

Therefore, habitat generalist spiders are, in general, faced with the problem of being in a variety of different backgrounds. In response to this cost, habitat generalists may present disruptive colouration, which would reduce predation risk more efficiently in a greater variety of backgrounds (Cuthill et al. 2005, Schaefer and Stobbe 2006, Stevens et al. 2006); or, alternatively, these species may exhibit masquerade (crypsis independent of context), colour polymorphism, or colour change. This same rationale can be used for non-visual forms of camouflage. Cursorial spiders that go across different terrains may have difficulty in masking vibratory and olfactory cues used by predators. In addition, because of the movements of cursorial spiders, they may be under stronger selection by birds than sit-and-wait predators (Gunnarsson 2007). Therefore, the nocturnal habit of some cursorial spiders may have been driven, at least in part, by the selection of diurnal bird predators.

Concluding Remarks

The diversity and widespread nature of anti-predator defences in spiders are, perhaps, a testimony of the relevance of predators on spider evolution. The challenge of covering this immense variety of defensive strategies in spiders has already been alluded to elsewhere (Cloudsley-Thompson 1995; Nelson and Jackson 2011). Similarly, this chapter by no means covers in detail all defensive strategies in neotropical spiders. Nonetheless, some general patterns of the study of defence in spiders emerged. Firstly, many of the proposed defensive strategies are still in need of experimental tests (e.g., crypsis, masquerade, aposematism). Secondly, the literature would benefit from a more detailed appraisal of the costs and benefits of defensive strategies. Thirdly, the sensory system and attack behaviour of predators should be taken into consideration when studying defensive strategies in spiders. Finally, non-visual forms of predator avoidance could be interesting avenues for future research projects.

References

Alves-Costa CP, Gonzaga MO (2001) Prey capture and spatial distribution of *Philoponella vittata* (Araneae Uloboridae) in host webs. Ethol Ecol Evol 13:239–246

Amaya CC, Klawinski PD (2001) The effects of leg autotomy on running speed and foraging ability in two species of wolf spider (Lycosidae). Am Midl Nat 145:201–205

Apontes P, Brown CA (2005) Between-sex variation in running speed and a potential cost of leg autotomy in the wolf spider *Pirata sedentarius*. Am Midl Nat 154:115–125

Araújo MS, Gonzaga MO (2007) Individual specialization in the hunting-wasp *Trypoxylon* (*Trypargilum*) *albonigrum* (Hymenoptera, Crabronidae). Behav Ecol Sociobiol 61:1855–1863

Barth F (2000) How to catch the wind: spider hairs specialized for sensing the movement of air. Naturwissenschaften 87:51–58

Bertani RER, Guadanucci JPL (2013) Morphology, evolution and usage of urticating setae by tarantulas (Araneae: Theraphosidae). Zoologia 30:403–418

Bertani RER, Marques OAV (1996) Defensive behaviors in Mygalomorph spiders: release of urticating hairs by some Aviculariinae (Araneae, Theraphosidae). Zool Anz 234:161–165

Blackledge TA, Pickett K (2000) Predatory interactions between mud-dauber wasps (Hymenoptera, Sphecidae) and *Argiope* (Araneae, Araneidae) in captivity. J Arachnol 28:211–216

Blackledge TA, Coddington JA, Gillespie RG (2003) Are three-dimensional spider webs defensive adaptations? Ecol Lett 6:13–18

Blamires SJ, Hou C, Chen LF, Liao CP (2013) Three-dimensional barricading of a predatory trap reduces predation and enhances prey capture. Behav Ecol Sociobiol 67:709–714

Brandley N, Johnson M, Johnsen S (2016) Aposematic signals in North American black widows are more conspicuous to predators than to prey. Behav Ecol 27:1104–1112

Bruce MJ (2006) Silk decorations: controversy and consensus. J Zool 269:89–97

Brueseke MA, Rypstra AL, Walker SE (2001) Leg autotomy in the wolf spider *Pardosa milvina*: a common phenomenon with few apparent costs. Am Midl Nat 146:153–160

Buschini MLT, Borba NA, Brescovit AD (2008) Patterns of prey selection of *Trypoxylon* (*Trypargilum*) *lactitarse* Saussure (Hymenoptera: Crabronidae) in southern Brazil. Braz J Biol 68:519–528

Buschini MLT, Borba NA, Brescovit AD (2010a) Prey selection in the trap-nesting wasp *Trypoxylon* (*Trypargilum*) *opacum* Brèthes (Hymenoptera: Crabronidae). Braz J Biol 70:529–536

Buschini MLT, Caldas TR, Borba NA, Brescovit AD (2010b) Spiders used as prey by the hunting wasp *Trypoxylon* (*Trypargilum*) *agamemnon* Richards (Hymenoptera: Crabronidae). Zool Stud 49:169–175

Cloudsley-Thompson JL (1995) A review of the anti-predator devices of spiders. Bull Br Arachnol Soc 10:81–96

Cooke J, Roth VD, Miller F (1972) The urticating hairs of theraphosid spiders. Am Mus Novit 2498:1–43

Cott H (1966) Adaptive coloration in animals. Methuen, London. 1957 (1966 printing). First edition reprinted with minor corrections

Coville RE, Coville PL (1980) Nesting biology and male behavior of *Trypoxylon* (*Trypargilum*) *tenoctitlan* in Costa Rica (Hymenoptera, Sphecidae). Ann Entomol Soc Am 73:110–119

Cushing PE (1997) Myrmecomorphy and myrmecophily in spiders: a review. Fla Entomol 80:165–193

Cushing PE (2012) Spider–ant associations: an updated review of myrmecomorphy, myrmecophily, and myrmecophagy in spiders. Psyche 2012:1–23

Cuthill I, Stevens M, Sheppard J, Maddocks T, Párraga C, Troscianko T (2005) Disruptive coloration and background pattern matching. Nature 434:72–74

Dodson GN, Beck MW (1993) Pre-copulatory guarding of penultimate females by male crab spiders, *Misumenoides formosipes*. Anim Behav 46:951–959

Duncan RP, Autumn K, Binford GJ (2007) Convergent setal morphology in sand-covering spiders suggests a design principle for particle capture. Proc R Soc B 274:3049–3057

Eberhard WG (1980) The natural history and behavior of the bolas spider, *Mastophora dizzydeani* sp. n.(Araneidae). Psyche 87:3–4
Edmunds J, Edmunds M (1986) The defensive mechanisms of orb weavers (Araneae: Araneidae) in Ghana, West Africa. In: Eberhard WG, Lubin Y, Robinson BC (eds) Proceedings of the 9th International Conference of Arachnology. Smithsonian Institution Press, Washington, DC, pp 73–89
Eisner T, Camazine S (1983) Spider leg autotomy induced by prey venom injection: an adaptive response to "pain"? Proc Natl Acad Sci USA 80:3382–3385
Fan C-M, Yang E-C, Tso I-M (2009) Hunting efficiency and predation risk shapes the color-associated foraging traits of a predator. Behav Ecol 20:808–816
Fernández Campón F (2013) Substrate preference in a colonial spider: is substrate choice affected by color morph? Entomol Sci 17:130–133
Fleming PA, Muller D, Bateman PW (2007) Leave it all behind: a taxonomic perspective of autotomy in invertebrates. Biol Rev 82:481–510
Foelix R (2010) Biology of spiders. Oxford University Press, Oxford
Formanowicz DR (1990) The antipredator efficacy of spider leg autotomy. Anim Behav 40:400–409
Gabritschevsky E (1927) Experiments on color changes and regeneration in the crab-spider, *Misumena vatia*. J Exp Zool 42:117–267
Gawryszewski FM (2014) Evidence suggests that modified setae of the crab spiders *Stephanopis* spp. fasten debris from the background. Zoomorphology 133:205–215
Gawryszewski FM, Motta PC (2012) Colouration of the orb-web spider *Gasteracantha cancriformis* does not increase its foraging success. Ethol Ecol Evol 24:23–38
Gonzaga MO, Vasconcellos-Neto J (2005a) Orb-web spiders (Araneae: Araneomorphae; Orbiculariae) captured by hunting-wasps (Hymenoptera: Sphecidae) in an area of Atlantic Forest in south-eastern Brazil. J Nat Hist 39:2913–2933
Gonzaga MO, Vasconcellos-Neto J (2005b) Testing the functions of detritus stabilimenta in webs of *Cyclosa fililineata* and *Cyclosa morretes* (Araneae : Araneidae): do they attract prey or reduce the risk of predation? Ethology 111:479–491
Gonzaga MO, Vasconcellos-Neto J (2006) Nesting characteristics and spiders (Arachnida: Araneae) captured by *Auplopus argutus* (Hymenoptera, Pompilidae) in an area of Atlantic Forest in southeastern Brazil. Entomol News 117:281–287
Gonzaga MO, Vasconcellos-Neto J (2012) Variation in the stabilimenta of Cyclosa fililineata Hingston, 1932, and Cyclosa morretes Levi, 1999 (Araneae: Araneidae), in southeastern Brazil. Psyche:396594
Gonzaga MO, Santos AJ, Dutra GF (1998) Web invasion and araneophagy in *Peucetia tranquillini* Mello-Leitão, 1922 (Araneae: Oxyopidae). J Arachnol 26:249–250
Gunnarsson B (2007) Bird predation on spiders: ecological mechanisms and evolutionary consequences. J Arachnol 35:509–529
Gunnarsson B, Wiklander K (2015) Foraging mode of spiders affects risk of predation by birds. Biol J Linn Soc 115:58–68
Hart NS (2001) The visual ecology of avian photoreceptors. Prog Retin Eye Res 20:675–703
Herberstein ME, Tso I-M (2011) Spider webs: evolution, diversity and plasticity. In: Herberstein ME (ed) Spider behaviour: flexibility and versatility. Cambridge University Press, Cambridge, pp 57–98
Jackson RR, Hallas S (1986) Comparative biology of *Portia africana*, *P. albimana*, *P. fimbriata*, *P. labiata*, and *P. shultzi*, araneophagic, web-building jumping spiders (Araneae: Salticidae): utilisation of webs, predatory versatility, and intraspecific interactions. N Z J Zool 13:423–489
Johnson S, Jakob E (1999) Leg autotomy in a spider has minimal costs in competitive ability and development. Anim Behav 57:957–965
Krebs JR, Davies NB (2004) An introduction to Behavioural ecology, 3rd edn. Wiley-Blackwell, Malden
Lenoir A, D'Ettorre P, Errard C, Hefetz A (2001) Chemical ecology and social parasitism in ants. Annu Rev Entomol 46:573–599

Levi HW (1985) The spiny orb-weaver genera *Micrathena* and *Chaetacis* (Araneae: Araneidae). Bull Mus Comp Zool 150:429–618

Li D, Lee WS (2004) Predator-induced plasticity in web-building behaviour. Anim Behav 67:309–318

Liu M-H, Blamires SJ, Liao C-P, Tso I-M (2014) Evidence of bird dropping masquerading by a spider to avoid predators. Sci Rep 4:5058

Llandres AL, Figon F, Christidès J-P, Mandon N, Casas J (2013) Environmental and hormonal factors controlling reversible colour change in crab spiders. J Exp Biol 216:3886–3895

Magalhaes I, Santos AJ (2012) Phylogenetic analysis of *Micrathena* and *Chaetacis* spiders (Araneae: Araneidae) reveals multiple origins of extreme sexual size dimorphism and long abdominal spines. Zool J Linn Soc Lond 166:14–53

Manicom C, Schwarzkopf L, Alford RA, Schoener TW (2008) Self-made shelters protect spiders from predation: a field experiment. PNAS 105:14903–14907

Marshall SD, Uetz GW (1990) Incorporation of urticating hairs into silk: a novel defense mechanism in two Neotropical tarantulas (Araneae, Theraphosidae). J Arachnol 18:143–149

McIver JD, Stonedahl G (1993) Myrmecomorphy: morphological and behavioral mimicry of ants. Annu Rev Entomol 38:351–379

Messas YF, Souza HS, Gonzaga MO, Vasconcellos-Neto J (2014) Spatial distribution and substrate selection by the orb-weaver spider Eustala perfida Mello-Leitão, 1947 (Araneae, Araneidae). J Nat Hist 48:2645–2660

Moura RR, Vasconcellos-Neto J, Gonzaga MO (2017) Extended male care in *Manogea porracea* (Araneae: Araneidae): the exceptional case of a spider with amphisexual care. Anim Behav 123:1–9

Nelson JX, Jackson RR (2011) Flexible use of anti-predator defences. In: Herberstein ME (ed) Spider behaviour: flexibility and versatility. Cambridge University Press, Cambridge, pp 99–126

O'Hanlon JC, Holwell GI, Herberstein ME (2014) Pollinator deception in the orchid mantis. Am Nat 183:126–132

Oliveira PS (1986) Ant-mimicry in some spiders from Brazil. Bulletin de la Société Zoologique de. France 111:297–311

Oliveira PS (1988) Ant-mimicry in some Brazilian salticid and clubionid spiders (Araneae: Salticidae, Clubionidae). Biol J Linn Soc 33:1–15

Oliveira PS, Sazima I (1984) The adaptive bases of ant-mimicry in a neotropical aphantochilid spider (Araneae: Aphantochilidae). Biol J Linn Soc 22:145–155

Oliveira OS, Sazima I (1985) Ant-hunting behaviour in spiders with emphasis on *Strophius nigricans* (Thomisidae). Bull Br Arachnol Soc 6:309–312

Oxford G, Gillespie RG (1998) Evolution and ecology of spider coloration. Annu Rev Entomol 43:619–643

Packard A (1905) Change of color and protective coloration in a flower-spider (*Misumena vatia* Thorell). J N Y Entomol Soc 13:85–96

Pasquet A, Anotaux M, Leborgne R (2011) Loss of legs: is it or not a handicap for an orb-weaving spider? Naturwissenschaften 98:557–564

Peckham EG (1889) Protective resemblances in spiders. Occas Pap Nat Hist Soc Wis 1:61–113

Peitsch D, Fietz A, Hertel H, de Souza J, Ventura DF, Menzel R (1992) The spectral input systems of hymenopteran insects and their receptor-based color-vision. J Comp Physiol A 170:23–40

Peixoto PEC, Souza JC, Schramm JE Jr (2012) To be or not to be…a flower? A test of possible cues influencing hunting site selection in subadult females of the crab spider *Epicadus heterogaster* (Guerin 1812) (Araneae: Thomisidae). Stud Neotropical Fauna Envirom 47:73–79

Pekár S (2014) Comparative analysis of passive defences in spiders (Araneae). J Anim Ecol 83:779–790

Pekár S, Jarab M, Fromhage L, Herberstein ME (2011) Is the evolution of inaccurate mimicry a result of selection by a suite of predators? A case study using myrmecomorphic spiders. Am Nat 178:124–134

Perafán C, Galvis W, Gutiérrez M, Perez-Milles F (2016) *Kankuamo*, a new theraphosid genus from Colombia (Araneae, Mygalomorphae), with a new type of urticating setae and divergent male genitalia. Zookeys 601:89–109

Persons MH, Rypstra AL (2001) Wolf spiders show graded antipredator behavior in the presence of chemical cues from different sized predators. J Chem Ecol 27:2493–2504

Pitilin RB, Araújo MS, Buschini MLT (2012) Individual specialization in the hunting-wasp *Trypoxylon* (*Trypargilum*) *agamenon* Richards (Hymenoptera: Crabronidae). Zool Stud 51:655–662

Pocock RI (1899) XII. — The genus *Pœcilotheria*: its habits, history, and species. J Nat Hist 3:82–96

Punzo F (1997) Leg autotomy and avoidance behavior in response to a predator in the wolf spider, *Schizocosa avida* (Araneae, Lycosidae). J Arachnol 25:202–205

Rayor L (1996) Attack strategies of predatory wasps (Hymenoptera: Pompilidae; Sphecidae) on colonial orb web-building spiders (Araneidae: *Metepeira incrassata*). J Kansas Entomol Soc 69:67–75

Rayor L, Uetz GW (1990) Trade-offs in foraging success and predation risk with spatial position in colonial spiders. Behav Ecol Sociobiol 27:77–85

Restrepo-Giraldo C, Rodriguez J, Pitts JP (2012) Temporal activity patterns of the spider wasp Pepsis montezuma Smith (Hymenoptera: Pompilidae) in a disturbed lower montane rainforest (Manizales, Colombia). Psyche:516419

Richman DB, Meister JS, Whitcomb WH, Murray L (1995) A comparison of populations of wolf spiders (Araneae, Lycosidae) on two different substrates in southern Florida. J Arachnol 23:151–156

Ruxton GD (2009) Non-visual crypsis: a review of the empirical evidence for camouflage to senses other than vision. Philos Trans R Soc B 364:549–557

Ruxton GD, Sherratt TN, Speed MP (2004) Avoiding attack. First. Oxford University Press, Oxford

Rypstra AL, Walker SE, Persons MH (2016) Cautious versus desperado males: predation risk affects courtship intensity but not female choice in a wolf spider. Behav Ecol 27:876–885

Schaefer H, Stobbe N (2006) Disruptive coloration provides camouflage independent of background matching. Proc R Soc Lond B 273:2427–2432

Schoener TW, Toft CA (1983) Spider populations: extraordinarily high densities on islands without top predators. Science 219:1353–1355

Skelhorn J, Rowland HM, Ruxton GD (2009) The evolution and ecology of masquerade. Biol J Linn Soc 99:1–8

Souza HS, Messas YF, Gonzaga MO, Vasconcellos-Neto J (2015) Substrate selection and spatial segregation by two congeneric species of *Eustala* (Araneae: Araneidae) in southeastern Brazil. J Arachnol 43:59–66

Spiller D, Schoener TW (1998) Lizards reduce spider species richness by excluding rare species. Ecology 79:503–516

Steffenson MM, Formanowicz DR, Brown CA (2014) Autotomy and its effects on wolf spider foraging success. Ethology 120:1128–1136

Stevens M, Merilaita S (2009) Animal camouflage: current issues and new perspectives. Philos Trans R Soc B 364:423–427

Stevens M, Cuthill I, Windsor AMM, Walker HJ (2006) Disruptive contrast in animal camouflage. Proc R Soc Lond B 273:2433–2438

Stiles FG (1995) Behavioral, ecological and morphological correlates of foraging for arthropods by the hummingbirds of a tropical wet forest. Condor 97:853–878

Styrsky JD (2014) An orb-weaver spider exploits an ant–acacia mutualism for enemy-free space. Ecol Evol 4:276–283

Taylor PW, Roberts JA, Wignall AE, Uetz GW (2008) Foreleg autotomy reduces mating success of male *Schizocosa ocreata* wolf spiders. J Insect Behav 21:193–202

Tseng L, Tso I-M (2009) A risky defence by a spider using conspicuous decoys resembling itself in appearance. Anim Behav 78:425–431

Tseng HJ, Cheng R-C, Wu SH, Blamires SJ, Tso I-M (2011) Trap barricading and decorating by a well-armored sit-and-wait predator: extra protection or prey attraction? Behav Ecol Sociobiol 65:2351–2359

Uetz GW, Hieber CS (1994) Group size and predation risk in colonial web-building spiders: analysis of attack abatement mechanisms. Behav Ecol 5:326–333

Uetz GW, Boyle J, Hieber CS, Wilcox RS (2002) Antipredator benefits of group living in colonial web-building spiders: the "early warning" effect. Anim Behav 63:445–452

Vanderhoff EN, Byers CJ, Hanna CJ (2008) Do the color and pattern of *Micrathena gracilis* (Araneae: Araneidae) attract prey? Examination of the prey attraction hypothesis and crypsis. J Insect Behav 21:469–475

Vetter RS (1980) Defensive behavior of the black widow spider *Latrodectus hesperus* (Araneae: Theridiidae). Behav Ecol Sociobiol 7:187–193

Whitehouse MEA, Lubin Y (2005) The functions of societies and the evolution of group living: spider societies as a test case. Biol Rev 80:347–361

Wood FD (1926) Autotomy in arachnida. J Morphol Physiol 42:143–195

Wrinn KM, Uetz GW (2008) Effects of autotomy and regeneration on detection and capture of prey in a generalist predator. Behav Ecol 19:1282–1288

Xavier GM, Moura RR, Gonzaga MO (2017) Orb web architecture of Wixia abdominalis O. Pickard-Cambridge, 1882 (Araneae: Araneidae): intraorb variation of web components. J Arachnol 45:160–165

Chapter 16
Host Behavioural Manipulation of Spiders by Ichneumonid Wasps

Marcelo O. Gonzaga, Thiago G. Kloss, and Jober F. Sobczak

Abstract Modified webs constructed by spiders parasitized by Ichneumonid wasps were first mentioned in literature in 1771. These initial observations were restricted to the description of the cocoon web spun by an unidentified spider species presenting a cocoon attached. Only in the year 2000 was the subject intensively studied in another host/parasitoid system. The interaction between *Leucauge argyra* (Tetragnathidae) and *Hymenoepimecis argyraphaga* (Ichneumonidae) was carefully described by W. G. Eberhard in Costa Rica. Web modifications, in this case, are even more extensive than those previously recorded. Cocoon webs spun by *L. argyra* are composed of just a few strong lines. Spirals are absent, and the cocoon remains suspended attached to the hub of the structure. From these studies up to now, several other cases have been described in the Neotropics. The initial doubt about the generality of host manipulation involving the *Polysphincta* genus-group was solved, but many questions arise from the subsequent studies. We still know almost nothing about the mechanisms involved in manipulation, for example. Recent studies suggest that it involves the injection of some substance by the parasitoid onto its host because the removal of the attached larva leads to a restoration of the original web patterns. Another interesting aspect is the investigation of how the structure of normal webs affects the design of cocoon webs. Finally, it is important to demonstrate that specific alterations in normal webs result in benefits to the manipulative wasp. The objectives of this chapter are to present an overview of recent discoveries involving these interactions, a brief historical summary of the researching efforts in the Neotropical region, and perspectives for future studies.

M.O. Gonzaga (✉)
Instituto de Biologia, Universidade Federal de Uberlândia, Uberlândia, Minas Gerais, Brazil
e-mail: mogonzaga@yahoo.com.br

T.G. Kloss
Universidade Federal do Espírito Santo, Centro de Ciências Agrárias, Departamento de Biologia, Alegre, Espírito Santo, Brazil

J.F. Sobczak
Universidade da Integração Internacional da Lusofonia Afro-Brasileira, Instituto de Ciências Exatas e da Natureza, Acarape, Ceará, Brazil

C. Viera, M.O. Gonzaga (eds.), *Behaviour and Ecology of Spiders*,
https://doi.org/10.1007/978-3-319-65717-2_16

The parasite is clearly visible on Broussard's face. In X-ray, the creature is a maze of complicated biology. But the shocking thing is that, in X-ray, we can see that Broussard's jaws are forced wide open, and THE PARASITE HAS EXTRUDED SOME KIND OF LONG TUBE, WHICH IS STUFFED INTO HIS MOUTH AND DOWN HIS THROAT, ending near his stomach…

- HUNTER: It doesn't make any sense. It paralyzes him… puts him into a coma… then keeps him alive.
- MELKONIS: We can't expect to understand a life form like this. We're out of our back yard. Things are different here…

The X-ray reveals a spreading dark blot in the vicinity of Broussard's chest. In the centre, the stain is completely opaque…

- ROBY: That tube must be depositing it in him…
- MELKONIS: Could be some kind of venom, or poison…
- HUNTER: This is horrible.

The excerpts above were transcribed from the original story 'Alien', written by Dan O'Bannon and Ronald Shusett, which was later turned into a script by Walter Hill and David Giler, replacing the characters Hunter and Melkonis with Lieutenant Ripley (Sigourney Weaver) and Ash (Ian Holm), respectively. Hunter and Melkonis were wrong in two very important ways: (1) it is quite understandable that the creature would keep its host alive while her egg (the dark stain that appears later in the X-ray) developed within it, and (2) things in space (at least in this specific situation) are not that different from what happens in our backyard. In fact, the author, Dan O'Bannon, said that the habits of parasitoid wasps that lay their eggs in caterpillars were the inspiration for the story. If Melkonis had better knowledge of the biology of these wasps then, he would already know exactly what was happening to his friend Broussard. And here, on our planet, that is what we are trying to understand in detail. The interactions between parasitoid insects (Fig. 16.1) and their hosts involve complex behaviours and physiological mechanisms, including the production of anaesthetics and other substances that alter the normal behaviour of hosts and put them under the control of their enemies. This is the subject of this chapter.

Host Behaviour Manipulation by Parasitoids

Host behaviour manipulation by parasites is a widespread phenomenon that has long aroused the curiosity of the scientific community (Cram 1931; van Dobben 1952; Holmes and Bethel 1972; Moore 1984; Barnard and Behnke 1990; Godfray 1994; Poulin 2000; Moore 2002; Thomas et al. 2005; Lafferty and Shaw 2013; Hafer 2016; Soghigian et al. 2017). For example, numerous publications report parasites of fish that alter the activity patterns and foraging locations of their intermediate hosts, making them especially susceptible to avian predators (Barber et al. 2000; Shaw et al. 2009). Thus, the parasites gain access to new avian hosts, enabling their life cycle to continue. Mice infected by *Toxoplasma gondii* lose their fear of

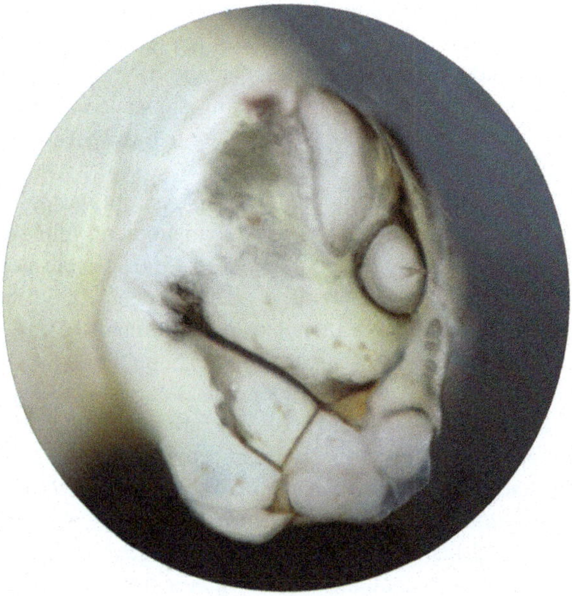

Fig. 16.1 Head of a *Hymenoepimecis bicolor* (Ichneumonidae) larva, a real version of the monster that terrorised the crew of Nostromo in the film 'Alien' (1979). This species, however, is an ectoparasitoid (it remains attached to the body of its host, not its interior) and attacks only the spider *Nephila clavipes* (Araneidae). The larva remains attached to the body of its host, making small perforations in its abdomen and feeding on its haemolymph. When it is ready for pupation, it induces a behavioural modification in its host leading to the construction of a web with specific architectural characteristics that ensure cocoon survival for a long period. After the spider constructs this web, the larva kills the spider, consumes its entire body, and then builds a cocoon, where it will complete its development and emerge as an adult

cats. Berdoy et al. (2000) demonstrated in a laboratory experiment that infected mice do not avoid locations marked with cat urine. When a cat ingests infected prey, the parasite is released into its digestive tract. The parasite then multiplies in the intestinal wall and produces oocysts. Other remarkable examples have been reported in ants (Yanoviak et al. 2008), crickets (Biron et al. 2006), isopods (Hansen and Poulin 2005), copepods (Hafer and Milinski 2016), and snails (Wesolowska and Wesolowski 2014), among other taxa (see Hughes et al. 2012). The behavioural manipulation caused by insect parasitoids, however, is comparatively less studied, although new and interesting examples have been reported more frequently in recent years.

Studies of host behavioural manipulation induced by parasitoid insects (especially hymenopterans) include cases of changes in habitat preferences and an increased food consumption rate of the host, care of parasitoid offspring by parasitized hosts, and construction of structures by hosts to support and protect the cocoon spun by parasitoid larvae (Wickler 1976; Godfray 1994; McLachlan 1999; Eberhard 2000a; Grosman et al. 2008; Matsumoto 2009; Sobczak et al. 2009; Gonzaga et al.

2010; Korenko and Pekár 2011). In addition, development, growth, and survival of the parasitized hosts have direct consequences for the development of immature stages of the parasitoids (Fritz 1982). Thus, parasitoids are frequently able to regulate the development of their hosts by injecting substances such as venoms and hormones, which alter host physiology and ensure that the parasitoid life cycle will not be interrupted (Vinson 1975).

Most cases of behavioural manipulation by parasitoid wasps described so far involve species of the subfamily Pimplinae (Hymenoptera, Ichneumonidae). This subfamily comprises approximately 95 genera and 1500 species, most of them ecto- or endoparasitoids of holometabolous insects (Dubois et al. 2002; Gauld and Dubois 2006). Among the several genera included in Pimplinae, the *Polysphincta* clade (sensu Gauld and Dubois 2006) (hereafter 'polysphinctine wasps') is impressive because of its spectacular way of life. This group is currently composed of 24 genera (Gauld and Dubois 2006; Pallacio et al. 2007; Matsumoto 2016) with a cosmopolitan distribution. All species are koinobiont ectoparasitoids of spiders (although there is no information on natural history for six genera — *Inbioia*, *Zabrachypus*, *Lamnatibia*, *Aravenator*, *Pterinopus*, and *Ticapimpla* — and only preliminary information on natural history of another genus, *Piogaster*). This means that the hosts, after being attacked, continue their normal activities while they are slowly being consumed by the parasitoid larvae (Dubois et al. 2002, Gauld and Dubois 2006). Immature stages of most species studied so far change some behaviour of their hosts, possibly through inoculation of substances that induce the construction of modified webs (Eberhard 2000a, b; Gonzaga and Sobczak 2007; Sobczak et al. 2009; Gonzaga et al. 2010; Takasuka et al. 2015).

The first study on parasitism of spiders by polysphinctine wasps to include the suggestion of behaviour manipulation was published by De Geer (1771). He described an orb web of an unidentified spider species, with a cocoon attached to it (Fig. 16.2). The description and illustration of this web indicate some characteristics later observed in other cases of host behaviour manipulation, such as a reduction in the number of radii and the absence of sticky spirals. After this initial description, some other authors included illustrations of modified webs in papers describing interactions between polysphinctines and spiders (e.g. Nielsen 1923); however, the first well-documented case of behavioural manipulation was described only in 2000: the construction of a simple and strong structure by *Leucauge argyra* (Tetragnathidae) when parasitized by *Hymenoepimecis argyraphaga* (Eberhard 2000a, b).

Eberhard (2000b) observed that third instar larvae of *H. argyraphaga* were able to induce their hosts to construct a modified web on the night they would be killed and consumed. In such cases, the larvae chemically induced the expression of the early steps of one specific subroutine of orb-web construction, suppressing all the subsequent behaviours that result in a normal circular orb (Eberhard 2001). On their last night alive, parasitized individuals presented bursts of activity. They added one to several radial lines in quick succession and then spent some time (up to 30 min) immobile before another burst. These spiders presented two behavioural patterns of adding radial lines in webs. In both patterns, they basically attached a dragline at the

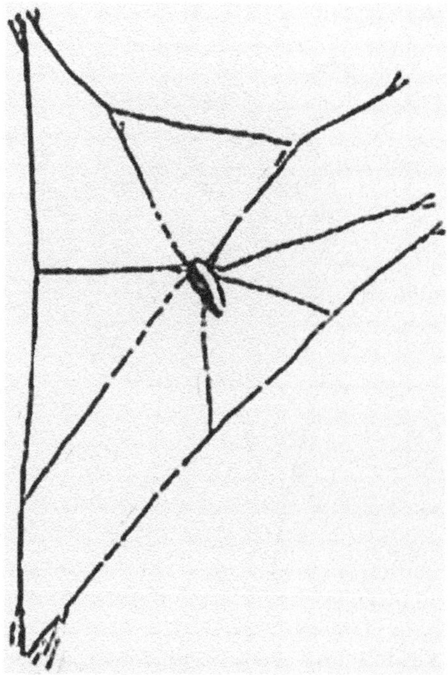

Fig. 16.2 Modified web described by De Geer (1771)

hub, walked toward the substrate along a radial line, walked along the substrate a short distance, and then attached the line that had been laid from the hub. Then they returned to the hub, walking along the same line or along another radial line that had been laid before, laying a second dragline. In the first behavioural pattern, the spider added lines without attachments to previous radial lines, and in the second, more common pattern, their draglines were attached to the radial lines on the way out and on the way back to the hub. As a result, the modified webs (cocoon webs) presented only a few strong lines, composed of approximately the same number of radial threads usually spun during the construction of normal webs (Fig. 16.3).

In the early 2000s, this was the only detailed information on a case of host manipulation by a species within the genus *Hymenoepimecis*. However, Gauld (2000) argued that *H. argyraphaga* possesses a number of plesiomorphic features, and that it possibly occupies a basal phylogenetic position within the group. He suggested that this position had important implications for understanding the evolution of biological traits within the group, making additional studies on host/parasitoid interactions involving other species of the genus critical to determining whether behavioural manipulation is a particularity of *H. argyraphaga* or a widespread trait within the genus, and perhaps, within polysphinctines. Other *Hymenoepimecis* species thus became excellent candidates for subsequent research on this subject.

Hymenoepimecis is currently composed of 20 valid species. The genus is exclusively Neotropical, and its distribution ranges from Mexico to southern Brazil, with

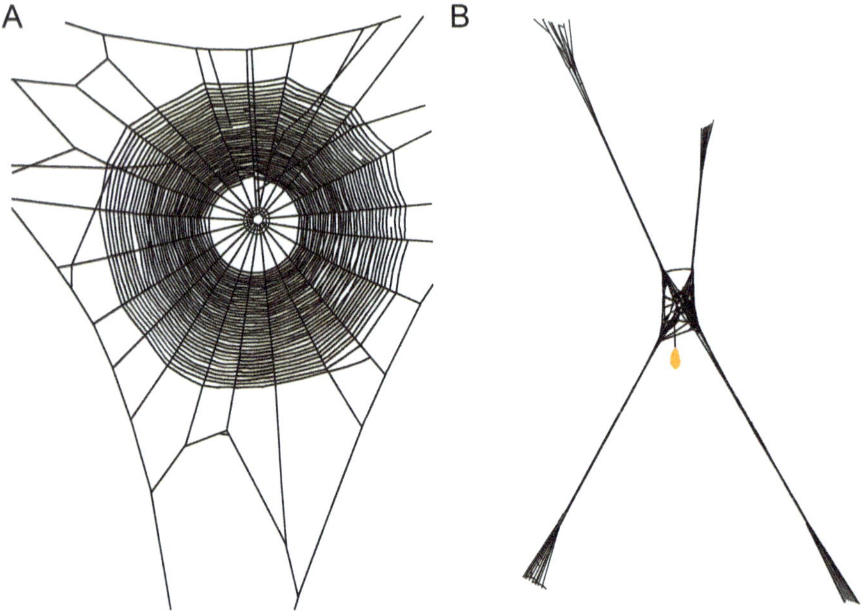

Fig. 16.3 (**a**) Web of an unparasitized adult individual of *Leucauge argyra*. (**b**) Dorsal view of the cocoon web (Modified from Eberhard (2001))

one species recorded in Cuba (*H. atriceps*). All species for which some information is available in the literature attack orb-weaver spiders of the families Araneidae and Tetragnathidae (Fincke et al. 1990; Eberhard 2000a, b; Eberhard 2001; Gauld and Dubois 2006; Gonzaga and Sobczak 2007; Sobczak et al. 2009; Gonzaga et al. 2010; Sobczak et al. 2012a, b). In the early 2000s, however, only nine species were known, and little information was available on their natural history.

The next case of behavioural manipulation involving one of these species was published by Gonzaga and Sobczak (2007). In this case, the interaction was between the spider *Araneus omnicolor* and the parasitoid *H. veranii* in southeastern Brazil. The authors described attacking and egg-laying behaviours, which included an event of infanticide, and structural differences between normal and cocoon webs. They also observed a reduction in the orb components (number of radii and spirals) of cocoon webs. Normal webs of *A. omnicolor* present an irregular three-dimensional structure attached to the orb, which is used to support a dead curled leaf that constitutes a shelter for spiders in resting positions. In cocoon webs, this three-dimensional structure remains intact; however, orbs are absent or very reduced. The authors argued that this reduction might decrease the probability of web rupture due to the interception of insects, until the emergence of the adult parasitoid.

Since these initial reports from Costa Rica and Brazil, nine other species of *Hymenoepimecis* have been observed attacking orb-weavers and inducing web modifications (Table 16.1). Sobczak et al. (2009) described two new species of *Hymenoepimecis* from southeastern Brazil attacking orb-weaver spiders. In *H. japi*,

Table 16.1 Spider hosts from the Neotropical region and their ichneumonid parasitoids

Spider host	Parasitoid wasp	Location	Source
Linyphiidae			
Dubiaranea sp.	*Eruga* sp.	Mulungu, CE, Brazil	Sobczak et al. (Unpublished data)
Tetragnathidae			
Leucauge argyra	*Hymenoepimecis argyraphaga*	Costa Rica	Eberhard (2000a, b, 2001)
Leucauge mariana	*Eruga* ca. *gutfreundi* *Hymenoepimecis tedfordi*	Costa Rica Costa Rica	Eberhard (2013)
Leucauge roseosignata	*Hymenoepimecis japi*	Jundiaí, SP, Brazil	Sobczak et al. (2009)
Leucauge volupis	*Hymenoepimecis jordanensis*	Estrela do Sul, MG, Brazil	Gonzaga et al. (2015b)
Leucauge henryi	*Hymenoepimecis manauara*	Manaus, AM, Brazil	Pádua et al. (2016)
Araneidae			
Nephila clavipes	*Hymenoepimecis bicolor* *Hymenoepimecis robertsae*	Jundiaí, SP, Brazil Santa Ana, San José, Costa Rica	Gonzaga et al. (2010)
Araneus omnicolor	*Hymenoepimecis veranii* *Hymenoepimecis neotropica*	Jundiaí, SP, Brazil Jundiaí, SP, Brazil	Gonzaga and Sobczak (2007) Sobczak et al. (2012a)
Araneus orgaos	*Hymenoepimecis veranii*	Jundiaí, SP, Brazil	Sobczak et al. (2014)
Araneus venatrix	*Hymenoepimecis silvanae*	Jundiaí, SP, Brazil	Sobczak et al. (2012b)
Manogea porracea	*Hymenoepimecis sooretama*	Linhares, ES, Brazil	Sobczak et al. (2009)
Mecynogea bigiba	*Hymenoepimecis japi*	Jundiaí, SP, Brazil	Unpublished data
Argiope argentata	*Acrotaphus chedalae*	Santa Teresa, ES, Brazil Jundiaí, SP, Brazil	Gonzaga and Sobczak (2011)
Argiope trifasciata	*Acrotaphus tibialis*		Eberhard (2013)
Eustala perfida	*Acrotaphus tibialis*	Jundiaí, SP, Brazil	Messas et al. (Unpublished data)
Cyclosa monteverdi	*Polysphincta gutfreundi*	Costa Rica	W.G. Eberhard (Unpublished data)
Cyclosa morretes	*Polysphincta janzeni*	Viçosa, MG, Brazil Ribeirão Grande, SP, Brazil Santa Teresa, ES, Brazil	Unpublished data Gonzaga et al. (2015b) Kloss et al. (2016a, b)
Cyclosa fililineata	*Polysphincta janzeni* *Polysphincta* nr. *purcelli*	Ribeirão Grande, SP, Brazil Santa Teresa, ES, Brazil	Gonzaga et al. (2015b) Kloss et al. (2016a, b)

(continued)

Table 16.1 (continued)

Spider host	Parasitoid wasp	Location	Source
Cyclosa sp.	*Polysphincta sp. nov.*	Rio Preto do Eva, AM, Brazil	Unpublished data
Allocyclosa bifurca	*Polysphincta gutfreundi*	San Jose Province, Costa Rica	Barrantes et al. (2008), Eberhard (2010a)
Parawixia bistriata	*Hymenoepimecis sp.*	Uberlândia, MG, Brazil	Unpublished data
Theridiidae			
Achaearanea tingo	*Zatypota alborhombarta*	Ribeirão Grande, SP, Brazil	Gonzaga et al. (2016)
Achaearanea cinnabarina	*Zatypota* nr. *riverai*	Jundiaí, SP, Brazil	Unpublished data
Anelosimus baeza	*Zatypota solanoi* *Zatypota* sp. nov	Jundiaí, SP, Brazil Mulungu, CE, Brazil	Unpublished data
Anelosimus nigrescens	*Zatypota solanoi*	Jundiaí, SP, Brazil	Unpublished data
Anelosimus jabaquara	*Zatypota solanoi*	Jundiaí, SP, Brazil	Unpublished data
Anelosimus octavius	*Zatypota sp. nr. solanoi*	Bebedero, Costa Rica	Eberhard (2010b)
Anelosimus nr. *studiosus*	*Zatypota sp. nr. solanoi*	San Pedro de Montes de Oca, Costa Rica	Eberhard (2010b)
Cryptachaea migrans	*Zatypota alborhombarta*	Santa Teresa, ES, Brazil Cariacica, ES, Brazil	Unpublished data Unpublished data
Cryptachaea rioensis	*Zatypota alborhombarta*	Santa Teresa, ES, Brazil Domingos Martins, ES, Brazil Conceição da Barra, ES, Brazil	Unpublished data Unpublished data Unpublished data
Cryptachaea migrans	*Zatypota morsei*	Santa Teresa, ES, Brazil Domingos Martins, ES, Brazil	Unpublished data Unpublished data
Theridion evexum	*Zatypota petronae*	San Jose Province, Costa Rica	Barrantes et al. (2008)

a parasitoid of *L. roseosignata*, when the parasitoid larva reaches the last instar before pupation, there is a significant modification in the host web design, similar to that observed by Eberhard (2001) in parasitized individuals of *L. argyra*. The modified web is composed of only three strong threads converging to a platform, located in the hub, which holds the cocoon; viscid spirals are completely absent. In the second species, *H. sooretama*, a parasitoid of *Manogea porracea* (Araneidae), there is no apparent modification of web architecture. However, the cocoon constructed by the parasitoid larva is attached to the web in a position far from the usual resting position of the spider. In this case, it seems that host behavioural changes are restricted to transporting the larva to a more protected area, a location covered by a denser mesh of threads (Sobczak et al. 2009).

All other studied species of *Hymenoepimecis* also change the behaviour of their hosts in some way (e.g., Eberhard 2013; Pádua et al. 2016). Gonzaga et al. (2015a), for example, observed that cocoon webs constructed by the *L. volupis* attacked by *H. jordanensis* are similar to those constructed by unparasitized, immature individuals, presenting a lower tangle that is absent in webs spun by adults. This structure may increase web stability, reducing the probability that the cocoon will fall to the ground. In this case, sticky spirals are also absent from the cocoon webs, but the reduction in spirals is not as subtle as that observed for parasitized *L. argyra*, which starts before the construction of the cocoon web. The same gradual effect on webs over the interval of several days was also observed in *Nephila clavipes* attacked by *H. bicolor* and *H. robertsae* (Gonzaga et al. 2010). Differences between the cocoon webs of *L. volupis* and the cocoon webs spun by other congeneric species suggest that the substance used for host manipulation may vary in concentration or composition (see the section "Mechanism of Manipulation" in this chapter).

Despite the great diversity within polysphinctines and the long time since the first behavioural record, and the existence of complex parasitoid–host interactions such as host behavioural manipulation by larvae, several aspects of such interactions remain poorly known. In the Neotropical region, most studies were carried out in Costa Rica and Brazil, and they involved species of *Hymenoepimecis*; however, information on certain species of *Zatypota*, *Polysphincta*, *Eruga*, and *Acrotaphus* is also available, which indicates that the phenomenon of host manipulation is widespread within polysphinctines (Table 16.1, Figs. 16.4, 16.5, and 16.6). Contributions from other geographic regions are currently mostly restricted to studies developed in Japan (Matsumoto and Konishi 2007; Matsumoto 2009; Takasuka et al. 2009, 2015; Takasuka and Matsumoto 2011a, b), the Czech Republic (Korenko and Pekár 2011; Korenko et al. 2011, 2014), Italy (Korenko and Pekár 2011; Korenko et al. 2014, 2015a, b), the Netherlands (Korenko et al. 2015b), and Canada (Bovee and Leech 2014). Most of these studies have focused on interactions involving the genus *Zatypota*, but there is also detailed information available on certain species of *Reclinervellus* (Matsumoto and Konishi 2007; Takasuka et al. 2015), *Brachyzapus* (Matsumoto 2009), and *Polysphincta* (Bovee and Leech 2014).

Mechanism of Manipulation

The exact mechanism(s) of behavioural manipulation of spider hosts remains to be described in detail, but some recent studies have indicated that it may involve the direct injection of hormones by the parasitoid larva when it reaches the last instar (Takasuka et al. 2015; Kloss et al. 2017). The idea that psychotropic substances promote alterations in web building behaviours has been around for a long time.

Fig. 16.4 Spider hosts carrying larvae or eggs of polysphinctines. (**a**) *Achaearanea cinnabarina*, (**b**) *Anelosimus baeza*, (**c**) *Steatoda* sp., (**d**) *Dubiaranea* sp., (**e**) *Anelosimus jabaquara*, (**f**) *Cryptachaea* sp., (**g, k**) *Achaearanea tingo*, (**h**) *Araneus omnicolor*, (**i**) *Araneus orgaos*, (**j**) *Araneus venatrix*, (**l**) *Araneus workmani*, (**m**) *Argiope argentata*, (**n**) *Eustala perfida*, (**o**) *Leucauge henryi*, (**p**) *Leucauge roseosignata*, (**q**) *Leucauge volupis*

Fig. 16.5 Normal webs of some spider hosts. (**a**) *Leucauge volupis*, (**b**) *Leucauge roseosignatha*, (**c**) *Nephila clavipes*, (**d**) *Anelosimus nigrescens*, (**e**) *Achaearanea tingo*, (**f**) *Araneus omnicolor*, (**g**) *Argiope argentata*

Fig. 16.6 Cocoon webs of (**a**) *Leucauge volupis*, (**b**) *Leucauge roseosignatha*, (**c**) *Nephila clavipes*, (**d**) *Anelosimus nigrescens*, (**e**) *Achaearanea tingo*, (**f**) *Araneus omnicolor*, (**g**) *Argiope argentata*

Witt (1971) and Eberhard (2000b, 2001) suggested that substances produced by the larva were responsible for the unusual structure of cocoon webs constructed by parasitised individuals of *L. argyra*. The similarities between cocoon webs spun by some host species (but not all of them — see Korenko and Pekár 2011 for exceptions) and the resting (or 'moulting') webs constructed before moulting were clues indicating that the manipulative compounds might be ecdysteroids or some precursor of moulting hormones.

The first experimental evidence suggesting that some chemical substance produced by the parasitoid larva is responsible for host behavioural alteration during web building was obtained by Eberhard (2010a). He observed that behavioural modification of *Allocyclosa bifurca* (Araneidae) by the ichneumonid wasp *Polysphincta gutfreundi* is gradual, and that the spider is able to recover its normal behaviour when the parasitoid larva is experimentally removed. These results suggest that the effects on spiders may depend on a cumulative or dose-dependent process rather than on injection of several distinct substances that are each responsible for influencing specific behaviours. Gonzaga et al. (2010) and Kloss et al. (2016a) tested an alternative hypothesis, considering that reduction in web investment might result from nutritional restrictions imposed by the parasitoid. They found no evidence supporting the idea that modification in web design arises as a by-product of nutritional deficiencies, and agreed that alterations in spider behaviour are probably caused by the injection of some substance by the larva.

Recently, Takasuka et al. (2015), studying the behavioural modifications of *Cyclosa argenteoalba* (Araneidae) induced by *Reclinervellus nielseni* (Ichneumonidae), analysed, in detail, the similarities between cocoon webs spun by parasitised individuals and the moulting webs. They hypothesised that the parasitoid larva evokes the innate moulting web construction behaviour of *C. argenteoalba*, suggesting that injection of chemical components corresponding to moulting hormones (ecdysteroids) into the spider body may be responsible for behavioural changes. They found a number of similar characteristics between the two types of webs (e.g., presence of fibrous thread decorations on the radii, absence of stabilimenta, and reduction in radii number), confirming that presence of the same substance(s) leading to moulting web and cocoon web construction is a valid possibility. However, they also found some important differences, such as repeated thread weavings that occur during cocoon web construction but not during moulting web construction. The authors attributed these differences to the distinct hormone concentrations in each situation.

The latest evidence in this direction was reported by Kloss et al. (2017). They also observed that moulting web structures of *C. morretes* and *C. fililineata* are similar to cocoon webs spun by these spiders when parasitized, respectively, by *P. janzeni* and *P.* sp. nr. *purcelli* (probably a new species). The authors compared the levels of 20-OH-ecdysone (20E) in unparasitized spiders, second-stage larvae, parasitized spiders carrying second-stage larvae, third-stage larvae, and parasitized spiders carrying third-stage larvae. The results indicated that the levels of the hormone in parasitized spiders carrying third-instar larvae (those with cocoon webs) were much higher than those observed in the other groups. They suggested that parasitoid larvae may directly inject the hormone into the spider body, or produce and inject a precursor chemical that is responsible for 20E synthesis in the spider host. However, the generality of this mechanism must be investigated, considering other cases of host behavioural alterations that result in webs that are very different from moulting webs (e.g., Eberhard 2000a, 2010a, b, 2013; Sobczak et al. 2009).

Host Selection

Little information is available on host selection and cues used for host location in most species of polysphinctines. Most reported cases are restricted to few observations, and a large data set is required to determine whether there is a preference for a restricted range of host body sizes and/or spider species used for egg laying. Thus, an analysis of records currently available in the literature indicates that most Neotropical polysphinctines are specialists, using only one or two host species (Table 16.1); however, this result is probably an artefact of limited investigation of habits of the great majority of these species. Some parasitoid species, such as *Zatypota solanoi*, are able to attack congeneric host species that occur in sympatry or in different locations across their distribution range. In the same way, some host species are attacked by distinct parasitoid species; however, additional studies are still needed to confirm the patterns that can be identified from the dataset presented in Table 16.1.

Lack of information is also a problem with regard to host size preferences in most cases. For a few, however, it is possible to recognise that wasps actively selected specific host ranges. Gonzaga and Sobczak (2007), for example, reported that *H. veranii* attacked relatively small individuals at a frequency higher than that expected based on the abundance of those individuals in the field, whereas large spiders are ignored. Similarly, Sobczak (2013) observed that large females of *N. clavipes* are rarely attacked by *H. bicolor*. Eventual attacks on these spiders often result in predation by the host. Finally, Fincke et al. (1990) showed that *H. robertsae* also selects intermediate-sized individuals of *N. clavipes* in Panama. Large individuals may provide more resources for the larvae, but immobilisation of a large spider may be associated with a high risk of failure. On the other hand, *H. jordanensis* prefers relatively large hosts (Gonzaga et al. 2015b). The authors argued that the range of host sizes used by polysphinctines is probably determined by their own body size (specifically, by the relationship between the size of the wasp and the spider), attacking behaviours, and venom characteristics and nutritional requirements of their developing larvae. In addition, distinct host species (and distinct instars within species) may have particular nutritional compositions, and host size selection may be influenced by the balance of lipid and proteins available. This last hypothesis, however, remains to be tested in further studies.

Host Immobilisation and Egg-Laying Processes

In the Neotropical region, attacks on host spiders and oviposition behaviours of polysphinctines have been described for eight species, representatives of the genera *Hymenoepimecis*, *Polysphincta*, and *Zatypota*. Observations suggest that attack strategies depend on the specific reactions of hosts to threats and the particular architectures of host webs, such as the presence of stabilimenta and barrier threads.

Within the genus *Hymenoepimecis*, the attacking behaviours of four species have been witnessed by researchers. *H. bicolor* attacking *N. clavipes* tend to hover around a target spider before executing a direct attack and then dart rapidly at the spider, grasping it with their legs (Eberhard 2000b; Sobczak 2013) (Fig. 16.7a). *Hymenoepimecis argyraphaga* attacks *Leucauge argyra* in a similar way; however, it has also been observed hanging immobile in a web radius in the free zone near the web hub, waiting until a spider that had left the web returned to its resting position (Eberhard 2000b). This is probably an alternative behaviour to gain access to spiders that had escaped the initial attack. *H. veranii* also uses web threads spun by the host to perform the attack. In this case, a female wasp stands immobile on barrier threads and waits until the target spider leaves its retreat to capture prey (Fig. 16.7b). At that moment, the wasp performs a direct attack (Gonzaga and Sobczak 2007). It inserts the tip of its ovipositor into the host spider's mouth, probably reaching the suboesophageal ganglion, and then inspects the host abdomen, searching for the presence of eggs previously deposited by other wasps. The venom has an immediate effect, rendering the attacked spider motionless for at least 18 min. After removing any egg of another female, the wasp lays its own egg on the dorsal surface of host's abdomen. Finally, there has been an observation of *H. sooterama* invading a web of *M. porracea* and walking on the web threads to the position occupied by the spider (Sobczak et al. 2009), but there is no record of an attack.

Attacks by *P. janzeni* on *C. morretes* and *P.* sp. nr. *purcelli* on *C. fililineata* are quite different from those of *Hymenoepimecis* spp. Females start the attacks by landing on the web hub, near the position occupied by the spider, but never directly on the spider. The spider jumps off the web immediately or moves quickly towards the web edge. After the initial attack, the wasp remains motionless at the web hub until the spider returns and touches the body of the wasp. Wasps wait for the hosts to return for periods ranging from 30 min to 14 h. The attack starts immediately after the spider touches the body of the wasp. After a struggle lasting a few seconds, the wasp remains positioned with its head facing the posterior part of the spider's abdomen and inserts its ovipositor into the spider's mouth, leading to immediate paralysis of the host. The wasp then inserts and withdraws its ovipositor from the spider's mouth repeatedly for approximately 5 min. Subsequently, the wasp appears to inspect the spider's body, repeatedly rubbing and jabbing the base of its ovipositor all over the host's abdomen (near the location where the egg will be attached) for approximately 4 min. After this behaviour, the wasp again proceeds to insert and withdraw its ovipositor into and out of the spider's mouth repeatedly for another 3 min. Finally, the wasp deposits a single egg on the anterodorsal surface of the host's abdomen and returns to the hub of the web, where it remains for at least 1 h. The function of this behaviour is not known; however, the wasp may remain on the web after oviposition as a strategy to detect possible egg removal by the spider after its recovery from paralysis (Kloss et al. 2016b).

Records of *Zatypota* in the Neotropics are restricted to a study published by Weng and Barrantes (2007) on *Z. petronae* behaviour, and some unpublished observations conducted by J. Sobczak and M. Gonzaga on *Z. solanoi* in Brazil. *Z. petronae* invades the retreat of the host spider *Theridion evexum* and attacks the spider

Fig. 16.7 (**a**) *Hymenoepimecis bicolor* immobilising its host, a female of *Nephila clavipes*. (**b**) *Hymenoepimecis veranii* waiting for the host to leave its shelter

inside its retreat. Details regarding *Z. petronae* behaviour during the host immobilization and egg-laying processes are not available in the literature (Weng and Barrantes 2007). *Z. solanoi* attacks *Anelosimus* species by invading their tridimensional webs and walking through the dense mesh of threads until it reaches a spider. A similar process of immobilisation, insertion of the ovipositor into the spider mouth, and inspection of the abdominal region was observed prior to egg laying. In one attack, we observed *Z. solanoi* killing a second instar larva attached to an *Anelosimus baeza* adult female.

Consequences for the Parasitoid

Modification in web architecture yields an increase in pupal survival (Kloss et al. 2016a; Sobczak 2013), because the modified structure is more stable and less efficient at intercepting insects than the normal web structure. Few studies, however, have compared the efficiency of modified and normal webs at preserving the integrity of the cocoons from their construction to the emergence of the adult wasp. Evidence obtained from the interaction between *H. bicolor* and *N. clavipes* and between *C. fililineata* and *P.* nr. *purcelli* indicate that host behavioural manipulation is important to ensure parasitoid survival during this period.

Sobczak (2013), for example, established three spider groups in individual enclosures in the field. The first group was composed of parasitized spiders that built modified webs, the second group was composed of parasitized spiders used for manipulation, and the third group was composed of unparasitized individuals. Just after their modified web construction, spiders in group 2 were transferred to the normal webs constructed by spiders in group 3. The original owners of those webs had been previously removed. Thus, the larvae had no option but to construct their cocoons in normal webs. After that, the enclosures were opened and the webs were exposed to normal field conditions, including strong winds and frequent heavy rains, which occur from January to May in the study area. Most cocoons (18 of 20) from group 1 (cocoons in modified webs) remained intact after 16 days of observations, whereas only one of 20 cocoons attached to normal webs were intact after the same period. The author observed that the collapse of normal webs and the cocoons falling to the ground were the main causes of pupal mortality in the latter group.

Kloss et al. (2016a) conducted a similar experiment with parasitized and unparasitized *C. fililineata* and *C. morretes* females in another area of Atlantic forest, also in southeastern Brazil. Survival of parasitoid individuals reintroduced to the same cocoon webs was higher than that of parasitoid larvae transferred from the original cocoon webs to unmodified webs, for both species. The author observed several mortality factors during the experiment, including web rupture due to rain resulting in cocoons falling to the ground and subsequent predation by indeterminate predators, web rupture due to falling branches, predation by araneophagic spider species (Mimetids), and predation by ants in damaged webs in which the cocoons came in contact with vegetation. Of these factors, the highest difference between groups was in mortality resulting from web rupture by the rain, indicating that web modification improves the stability of the structure holding the cocoon.

Conclusions

Systematic investigation on the subject of host behavioural manipulation involving spider hosts and polysphinctine wasps is a relatively new area of investigation, and many aspects such as host selection and the mechanism(s) involved in the phenomenon of manipulation are currently poorly understood. However, during the few years since the description of the cocoon webs spun by parasitized individuals of *L. argyra*, significant discoveries have been made. We know now, with reasonable confidence, that some substance injected by the larva into the host induces the construction of modified webs, which present structural characteristics that confer to them an increased stability and a reduced probability of rupture due to insect interception. Wasp survival during the pupal stage is certainly higher in cocoon webs as a direct result of this new architecture. We also know that cocoon webs, at least in some cases, are very similar to moulting webs, and the first evidence that moulting hormones may be involved in the process of manipulation has arisen in recent studies.

These and other recent discoveries are part of the initial efforts to clarify the patterns, origins, and ecological consequences of interactions between spiders and parasitoids. Certainly, there are many other cases (including genera without any available information to date) to be discovered and described. We hope to discover, in the next years, how these further findings will influence the patterns presented in this chapter. Being more optimistic than the character Melkonis mentioned at the beginning of this chapter, we believe that there is no reason not to expect to understand (at least partially) these life forms. Contribution from other research groups, especially within the highly biodiverse Neotropical region, is very important in this endeavour.

References

Barber I, Hoare D, Krause J (2000) Effects of parasites on fish behaviour: a review and evolutionary perspective. Rev Fish Biol Fish 10:131–165

Barnard CJ, Behnke JM (1990) Parasitism and host behaviour. Taylor and Francis, London

Barrantes G, Eberhard WG, Weng J (2008) Seasonal patterns of parasitism of the tropical spiders *Theridion evexum* (Araneae, Theridiidae) and *Allocyclosa bifurca* (Araneae, Araneidae) by the wasps *Zatypota petronae* and *Polysphincta gutfreundi* (Hymenoptera, Ichneumonidae). Rev Biol Trop 56:749–754

Berdoy M, Webster JP, Macdonald DW (2000) Fatal attraction in rats infected with *Toxoplasma gondii*. Proc R Soc Lond B 267:1591–1594

Biron DG, Ponton F, Marché L, Galeotti N, Renault L, Demey-Thomas E, Poncet J, Brown SP, Jouin P, Thomas F (2006) 'Suicide' of crickets harbouring hairworms: a proteomics investigation. Insect Mol Biol 15:731–742

Bovee J, Leech R (2014) *Araneus gemmoides* (Araneae, Araneidae) death by a parasitoid (Hymenoptera Ichneumonidae). Bull Entomol Soc Can 46:86–90

Cram ER (1931) Developmental stages of some nematodes of the Spiruroidea parasitic in poultry and game birds. USDA Technical Bulletin, Beltsville

De Geer C (1771) Memoires pour servir à l'histoire des insects. French Grefing and Hesselberg, Stockholm, Sweden.

Dubois J, Rollard C, Villemant C, Gauld ID (2002) The phylogenetic position of parasitoids of spiders within Pimplinae (Hymenoptera, Ichneumonidae). In: Samu F, Szinetár CS (eds) Proceedings of the 20th European Colloquium of Arachnology, Budapest

Eberhard WG (2000a) The natural history and behavior of *Hymenoepimecis argyraphaga* (Hymenoptera: Ichneumonidae) a parasitoid of *Plesiometa argyra* (Araneae, Tetragnathidae). J Hymenopt Res 9:220–240

Eberhard WG (2000b) Spider manipulation by a wasp larva. Nature 406:255–256

Eberhard WG (2001) Under the influence: webs and building behavior of *Plesiometa argyra* (Araneae, Tetragnathidae) when parasitized by *Hymenoepimecis argyraphaga* (Hymenoptera, Ichneumonidae). J Arachnol 29:354–366

Eberhard WG (2010a) Recovery of spiders from the effects of parasitic wasps: implications for fine-tuned mechanisms of manipulation. Anim Behav 79:375–383

Eberhard WG (2010b) New types of behavioral manipulation of host spiders by a parasitoid wasp. Psyche. https://doi.org/10.1155/2010/950614

Eberhard WG (2013) The polysphinctine wasps *Acrotaphus tibialis, Euga* ca. *gutfreund,* and *Hymenoepimecis tedfordi* (Hymenoptera, Ichneumonidae, Pimplinae) induce their host spiders to build modified webs. Ann Entomol Soc Am 106:652–660

Fincke OM, Higgins L, Rojas E (1990) Parasitism of *Nephila clavipes* (Araneae, Tetragnathidae) by an ichneumonid (Hymenoptera, Polysphinctini) in Panama. J Arachnol 18:321–329

Fritz RS (1982) Selection for host behavior modification by parasitoids. Evolution 36:283–288

Gauld ID (2000) The re-definition of pimpline genus *Hymenoepimecis* (Hymenoptera: Ichneumonidae) with a description of a plesiomorphic new Costa Rican species. J Hymenopt Res 9:213–219

Gauld ID, Dubois J (2006) Phylogeny of the *Polysphincta* group of genera (Hymenoptera: Ichneumonidae; Pimplinae): a taxonomic revision of spider ectoparasitoids. Syst Entomol 31:529–564

Godfray HCJ (1994) Parasitoids: behavioural and evolutionary ecology. Princeton University Press, Princeton

Gonzaga MO, Sobczak JF (2007) Parasitoid-induced mortality of *Araneus omnicolor* (Araneae, Araneidae) by *Hymenoepimecis* sp. (Hymenoptera, Ichneumonidae) in southeastern Brazil. Naturwissenschaften 93:223–227

Gonzaga MO, Sobczak JF (2011) Behavioral manipulation of the orb-weaver spider *Argiope argentata* (Araneae: Araneidae) by *Acrotaphus chedelae* (Hymenoptera: Ichneumonidae). Entomol Sci 14:220–223

Gonzaga MO, Sobczak JF, Penteado-Dias AM, Eberhard WG (2010) Modification of *Nephila clavipes* (Araneae, Nephilidae) webs induced by the parasitoids *Hymenoepimecis bicolor* and *H. robertsae* (Hymenoptera Ichneumonidae). Ethol Ecol Evol 22:151–165

Gonzaga MO, Cardoso JFC, Vasconcellos-Neto J (2015a) Do parasitoids explain differential abundance of two syntopic orb-weaver spiders (Araneae: Araneidae)? Acta Oecol 69:113–120

Gonzaga MO, Moura RR, Pêgo PT, Bang DL, Meira FA (2015b) Changes to web architecture of *Leucauge volupis* (Araneae: Tetragnathidae) induced by the parasitoid *Hymenoepimecis jordanensis* (Hymenoptera: Ichneumonidae). Behaviour 152:181–193

Gonzaga MO, Loffredo AP, Penteado-Dias AM, Cardoso JCF (2016) Host behavior modification of *Achaearanea tingo* (Araneae: Theridiidae) induced by the parasitoid wasp *Zatypota alborhombarta* (Hymenoptera: Ichneumonidae). Entomol Sci 19:133–137

Grosman AH, Janssen A, Brito EF, Cordeiro EG, Colares F, Fonseca JO, Lima ER, Pallini A, Sabelis MW (2008) Parasitoid increases survival of its pupae by inducing hosts to fight predators. PLoS One 3(6):e2276

Hafer N (2016) Conflicts over host manipulation between different parasites and pathogens: investigating the ecological and medical consequences. BioEssays 38:1027–1037

Hafer N, Milinski M (2016) Inter- and intraspecific conflicts between parasites over host manipulation. Proc R Soc Lond B 283:20152870

Hansen EK, Poulin R (2005) Impact of a microphallid trematode on the behaviour and survival of its isopod intermediate host: phylogenetic inheritance? Parasitol Res 97:242–246

Holmes JC, Bethel WM (1972) Modification of intermediate host behaviour by parasites. In: Canning EU, Wright CA (eds) Behavioural aspects of parasite transmission. Academic Press, London, pp 123–149

Hughes DP, Brodeur J, Thomas F (2012) Host manipulation by parasites. Oxford University Press, Oxford

Kloss TG, Gonzaga MO, Rouxinol JAM, Sperber CF (2016a) Host behavioural manipulation of two orb-weaver spiders by parasitoid wasps. Anim Behav 111:289–296

Kloss TG, Gonzaga MO, Rouxinol JAM, Sperber CF (2016b) Attack behavior of two wasp species of the *Polysphincta* genus group (Hymenoptera, Ichneumonidae) on their orb-weaver spider hosts (Araneae, Araneidae). J Insect Behav 29:315–324

Kloss TG, Gonzaga MO, Oliveira LL, Sperber CF (2017) Proximate mechanism of behavioral manipulation of an orb-weaver spider host by a parasitoid wasp. PLoS One 12(2):e0171336

Korenko S, Pekár S (2011) A parasitoid wasp induces overwintering behaviour in its spider host. PLoS One 6(9):e24628

Korenko S, Michalkova V, Zwakhals K et al (2011) Host specificity and temporal and seasonal shifts in host preference of a web-spider parasitoid *Zatypota percontatoria*. J Insect Sci 11:1–12

Korenko S, Isaia M, Satrapová J, Pekár S (2014) Parasitoid genus-specific manipulation of orb-web host spiders (Araneae, Araneidae). Ecol Entomol 39:30–38

Korenko S, Korenková B, Satrapová J et al (2015a) Modification of *Tetragnatha montana* (Araneae, Tetragnathidae) web architecture induced by larva of the parasitoid *Acrodactyla quadrisculpta* (Hymenoptera, Ichneumonidae, Polysphincta genus-group). Zool Stud 54:40

Korenko S, Satrapová J, Zwakhals K (2015b) Manipulation of araneid spider web architecture by the polysphinctine parasitoid *Zatypota picticollis* (Hymenoptera: Ichneumonidae: Pimplinae). Entomol Sci 18:383–388

Lafferty KD, Shaw JC (2013) Comparing mechanisms of host manipulation across host and parasite taxa. J Exp Biol 216:56–66

Matsumoto R (2009) "Veils" against predators: modified web structure of a host spider induced by an ichneumonid parasitoid, *Brachyzapus nikkoensis* (Uchida) (Hymenoptera). J Insect Behav 22:39–48

Matsumoto R (2016) Molecular phylogeny and systematics of the *Polysphincta* group of genera (Hymenoptera, Ichneumonidae, Pimplinae). Syst Entomol 41:854–864

Matsumoto R, Konishi K (2007) Life histories of two ichneumonid parasitoids of *Cyclosa octotuberculata* (Araneae): *Reclinervellus tuberculatus* (Uchida) and its new sympatric congener (Hymenoptera: Ichneumonidae: Pimplinae). Entomol Sci 10:67–278

McLachlan A (1999) Parasites promote mating success: the case of a midge and a mite. Anim Behav 57:1199–1205

Moore J (1984) Altered behavioral responses in intermediate hosts — an Acanthoceptalan parasite strategy. Am Nat 123:572–577

Moore J (2002) Parasites and the behavior of animals. Oxford University Press, Oxford

Nielsen E (1923) Contributions to the life history of the Pimpline spider parasites (*Polysphincta, Zaglyptus, Tromatobia*) (Hym. Ichneum.) Entomol Medd 14:137–205

Pádua DG, Salvatierra L, Sobczak JF, Oliveira ML (2016) Parasitism of *Hymenoepimecis manauara* Pádua and Oliveira (Hymenoptera: Ichneumonidae: Pimplinae) on *Leucauge henryi* Mello-Leitão (Araneae: Tetragnathidae) in Brazilian Amazonian. Biodivers Data J. https://doi.org/10.3897/BDJ.4.e11219

Pallacio E, Sääksjärvi IE, Vahtera V (2007) *Lamnatibia*, a new genus of the Polysphincta group of genera from Colombia (Hymenoptera: Ichneumonidae; Pimplinae). Zootaxa 1431:55–63

Poulin R (2000) Manipulation of host behaviour by parasites: a weakening paradigm? Proc R Soc Lond B 267:787–792

Shaw JC, Korzan WJ, Carpenter RE, Kuris AM, Lafferty KD, Summers CH, Øverli Ø (2009) Parasite manipulation of brain monoamines in California killifish (*Fundulus parvipinnis*) by the trematode *Euhaplorchis californiensis*. Proc R Soc Lond B 276:1137–1146

Sobczak JF (2013) Estudos biológicos e ecológicos da interação entre *Nephila clavipes* (Araneae, Nephilidae) e o parasitoide *Hymenoepimecis bicolor* (Hymenoptera, Ichneumonidae, Pimplinae). PhD Thesis, Universidade Federal de São Carlos

Sobczak JF, Loffredo APS, Penteado-Dias AM, Gonzaga MO (2009) Two new species of *Hymenoepimecis* (Hymenoptera: Ichneumonidae: Pimplinae) with notes on their spider hosts and behaviour manipulation. J Nat Hist 43:2691–2699

Sobczak JF, Loffredo APS, Carmago LF, Penteado-Dias AM (2012a) *Hymenoepimecis neotropica* (Brues and Richardson) (Hymenoptera, Ichneumonidae, Pimplinae) parasitoid of *Araneus omnicolor* (Keyserling) (Araneae, Araneidae): first host record and new occurrence to Brazil. Rev Bras Entomol 56:390–392

Sobczak JF, Loffredo APS, Penteado-Dias AM (2012b) Parasitism on *Araneus venatrix* (Koch, 1838) (Araneae: Araneidae) by *Hymenoepimecis silvanae* Loffredo and Penteado-Dias, 2009 (Ichneumonidae, Pimplinae) with description of male of the wasp. Braz J Biol 72:221–223

Sobczak JF, Sobczak JCMSM, Messas YF, Souza HS, Vasconcellos-Neto J (2014) A new record of a host-parasitoid interaction: *Hymenoepimecis veranii* Lofredo and Penteado-Dias, 2009 (Hymenoptera: Ichneumonidae) parasitizing *Araneus orgaos* Levi, 1991 (Araneae: Araneidae). J Insect Behav 27:753–758

Soghigian J, Valsdottir LR, Livdahl TP (2017) A parasite's modification of host behavior reduces predation on its host. Ecol Evol 7:1453–1461
Takasuka K, Matsumoto R (2011a) Lying on the dorsum: unique host-attacking behaviour of *Zatypota albicoxa* (Hymenoptera, Ichneumonidae). J Ethol 29:203–207
Takasuka K, Matsumoto R (2011b) Infanticide by a solitary koinobiont ichneumonid ectoparasitoid of spiders. Naturwissenschaften 98:529–536
Takasuka K, Matsumoto R, Ohbayashi N (2009) Oviposition behavior of *Zatypota albicoxa* (Hymenoptera, Ichenumonidae), an ectoparasitoid of *Achaearanea tepidariorum* (Araneae, Theridiidae). Entomol Sci 12:232–237
Takasuka K, Yasui T, Ishigami T, Nakata K, Matsumoto R, Ikeda K, Maeto K (2015) Host manipulation by an ichneumonid spider-ectoparasitoid that takes advantage of preprogrammed web-building behaviours for its cocoon protection. J Exp Biol 218:2326–2332
Thomas F, Adamo S, Moore J (2005) Parasitic manipulation: where are we and where should we go? Behav Process 68:185–199
Van Dobben WH (1952) The food of the cormorant in the Netherlands. Ardea 40:1–63
Vinson SB (1975) Biochemical coevolution between parasitoids and their host. In: Price PW (ed) Evolutionary strategies of parasitic insects and mites. Plenum Press, New York, pp 14–48
Weng JL, Barrantes G (2007) Natural history and larval behavior of the parasitoid *Zatypota petronae* (Hymenoptera: Ichneumonidae). J Hymenopt Res 16:326–335
Wesolowska W, Wesolowski T (2014) Do Leucochloridium sporocysts manipulate the behaviour of their snail hosts? J Zool 292:151–155
Wickler W (1976) Evolution-oriented ethology, kin selection, and altruistic parasites. Z Tierpsychol 42:206–214
Witt PN (1971) Drugs alter web-building of spiders: a review and evaluation. Behav Sci 16:98–113
Yanoviak SP, Kaspary M, Dudley R, Poinar G Jr (2008) Parasite-induced fruit mimicry in a tropical canopy ant. Am Nat 171:536–544

The manufacturer's authorised representative in the EU is Springer Nature Customer Service Centre GmbH, Europaplatz 3, 69115 Heidelberg, Germany. If you have any concerns regarding our products, please contact ProductSafety@springernature.com

Printed and bound by CPI Group (UK) Ltd, Croydon, CR0 4YY
30/09/2026
02225435-0011